MONEY AND FINANCIAL INSTITUTIONS
A GAME THEORETIC APPROACH

ECONOMISTS OF THE TWENTIETH CENTURY

General Editors: David Colander, *Christian A. Johnson Distinguished Professor of Economics, Middlebury College, Vermont, USA* and Mark Blaug, *Professor Emeritus, University of London, UK, Professor Emeritus, University of Buckingham, UK and Visiting Professor, University of Amsterdam, The Netherlands*

This innovative series comprises specially invited collections of articles and papers by economists whose work has made an important contribution to economics in the late twentieth century.

The proliferation of new journals and the ever-increasing number of new articles make it difficult for even the most assiduous economist to keep track of all the important recent advances. By focusing on those economists whose work is generally recognized to be at the forefront of the discipline, the series will be an essential reference point for the different specialisms included.

A list of published and future titles in this series is printed at the end of this volume.

Money and Financial Institutions

A Game Theoretic Approach

The Selected Essays of Martin Shubik Volume Two

Martin Shubik

Seymour H. Knox Professor of Mathematical Institutional Economics, Cowles Foundation for Research in Economics, Yale University, USA

ECONOMISTS OF THE TWENTIETH CENTURY

Edward Elgar

Cheltenham, UK • Northampton, MA, USA

Published by
Edward Elgar Publishing Limited
Glensanda House
Montpellier Parade
Cheltenham
Glos GL50 1UA
UK

Edward Elgar Publishing, Inc.
136 West Street
Suite 202
Northampton
Massachusetts 01060
USA

A catalogue record for this book
is available from the British Library

Library of Congress Cataloguing in Publication Data

Shubik, Martin.
Money and financial institutions : a game theoretic approach /
Martin Shubik.
p. cm. – (Economists of the twentieth century series)
(Selected essays of Martin Shubik ; v. 2)
Includes index.
1. Money. 2. Financial institutions. 3. Prices. 4. Game theory.
5. Shubik, Martin. I. Title. II. Series. III. Series: Shubik, Martin. Essays.
Selections ; v. 2.
HG220.A2S548 1999
332.4–dc21 99–22253
CIP

ISBN 1 84064 190 8

Printed and bound in Great Britain by MPG Books Ltd, Bodmin, Cornwall

Contents

Acknowledgements

The publishers wish to thank the following who have kindly given permission for the use of copyright material.

Academic Press for articles: 'On Market Games' (with Lloyd Shapley), *Journal of Economic Theory*, **1**, June 1969, pp. 1–17; 'The Noncooperative Equilibria of a Closed Trading Economy with Market Supply and Bidding Strategies' (with Pradeep Dubey), *Journal of Economic Theory*, **17**(1), February 1978, pp. 1–20; 'Efficiency Properties of Strategic Market Games: An Axiomatic Approach' (with Pradeep Dubey and Andreu Mas-Colell), *Journal of Economic Theory*, **22**(2), April 1980, pp. 339–62; 'A Strategic Market Game with Complete Markets' (with Rabah Amir, Siddharta Sahi and Shuntian Yao), *Journal of Economic Theory*, **51**(1), June 1990, pp. 126–43.

American Economic Association for articles: 'On the Core of an Economic System with Externalities' (with Lloyd S. Shapley), *The American Economic Review*, **LIX**(4) part 1, September 1969, pp. 678–84; 'Pecuniary Externalities: A Game Theoretic Analysis', *The American Economic Review*, **LXI**(4), September 1971, pp. 713–18; 'Commodity Money, Oligopoly, Credit and Bankruptcy in a General Equilibrium Model', *Western Economic Journal*, **X**(4), December 1972, pp. 24–38.

Blackwell Publishers for article: 'Extended Edgeworth Bargaining Games and Competitive Equilibrium', *Estratto da Metroeconomica*, **XX**(III), September–December 1968, pp. 299–312.

The Econometric Society for article: 'Quasi-cores in a Monetary Economy with Nonconvex Preferences' (with L.S. Shapley), *Econometrica*, **34**(4), October 1966, pp. 805–27.

Elsevier Science for articles: 'Perfect Competition in Strategic Market Games with Interlinked Preferences' (with Pradeep Dubey), *Economics Letters*, **17**, 1985, pp. 3–4; 'The Unique Minimal Cash Flow Competitive Equilibrium', *Economics Letters*, **25**, 1987, pp. 303–6; 'A Note on an Optimal Garnishing Rule' (with P. Dubey), *Economics Letters*, **27**, 1988, pp. 5–6.

The Geneva Association for article: 'The Capital Asset Pricing Model as a General Equilibrium With Incomplete Markets' (with John Geanakoplos), *The Geneva Papers on Risk and Insurance Theory*, **15**(1), March 1990, pp. 55–71.

The Institute of Management Sciences for article: 'On Matching Book: A Problem in Banking and Corporate Finance' (with M.J. Sobel), *Management Science*, **38**(6), June 1992, pp. 827–39.

Institute for Operations Research for article: 'Pure Competition, Coalitional Power, and Fair Division' (with Lloyd S. Shapley), *International Economic Review*, **10**(3), October 1969, pp. 337–62.

International Association for Research in Income and Wealth for article: 'On the Eight Basic Units of a Dynamic Economy Controlled by Financial Institutions', *Review of Income and Wealth*, **21**(2), 1975, pp. 183–201.

Journal of Economic Behavior and Organization for article: 'Society, Land, Love or Money: A Strategic Model of How to Glue the Generations Together', *Journal of Economic Behavior and Organization*, **2**, 1981, pp. 359–85.

North-Holland Publishing Company for article: 'Bankruptcy and Optimality in a Closed Trading Mass Economy Modelled as a Non-cooperative Game' (with Pradeep Dubey), *Journal of Mathematical Economics*, **6**, 1979, pp. 115–34.

Physica-Verlag for article: 'Competitive Outcomes in the Cores of Market Games' (with L.S. Shapley), *International Journal of Game Theory*, **4**(4), 1976, pp. 229–37.

Princeton University Press for articles: 'Edgeworth Market Games', in *Contributions to the Theory of Games, Volume IV*, A.W. Tucker and R.D. Luce (eds), Princeton: Princeton University Press, 1959, pp. 267–78; 'Concepts and Theories of Pure Competition' (with L.S. Shapley), in *Essays in Mathematical Economics in Honor of Oskar Morgenstern*, Martin Shubik (ed.), Princeton: Princeton University Press, 1967, pp. 63–79.

Springer-Verlag for articles: 'A Strategic Market Game with Price and Quantity Strategies' (with Pradeep Dubey), *Zeitschrift für Nationalökonomie*, **40**(1–2), 1980, pp. 25–34; 'A Strategic Market Game with a Mutual Bank with Fractional Reserves and Redemption in Gold (A Continuum of Traders)' (with Dimitrios Tsomocos), *Journal of Economics*, **55**(2), 1992, pp. 123–50; 'A Noncooperative Model of a Closed Trading Economy with Many Traders and Two Bankers', *Zeitschrift für Nationalökonomie*, **36**, 1976, pp. 49–60; 'The Optimal Bankruptcy Rule in a Trading Economy Using Fiat Money' (with Charles Wilson), *Zeitschrift für Nationalökonomie*, **37**(3–4), 1977, pp. 337–54.

The University of Chicago for article: 'Trade Using One Commodity as a Means of Payment' (with Lloyd Shapley), *Journal of Political Economy*, **85**(5), 1977, pp. 937–68.

Introduction

The occasion of publishing a large set of one's articles provides an opportunity for reflection and a consideration of the author's *apologia pro sua vita*. The test of one's own work lies with time and with others. Nevertheless the exercise of introspection provides an attempt to explain to oneself and to others what one was trying to do, is doing and still hopes to do. This might offer some insight on the gap between the wish and the reality as well as the self-assessment and the assessment of others.

In trying to classify my publications I find that they fall into several categories: general economics and political economy; oligopoly theory; the theory of money and financial institutions; pure game theory; game theory applied to economics, law, political science, management, experimental gaming, simulation, defense studies and operations research. I have concentrated my selection primarily on my economics publications, thus I have left out most of my work in defense studies, in operations research and in other behavioral sciences.

The period covered is from 1952 to 1992. Although I have worked on many topics, my basic long-term concerns have been with the understanding of several fundamental features of the economy, in particular the functioning of the price system and the purpose and evolution of money and financial institutions.

My early interests by the end of high school were in history, politics, society and the economy. Although by then I suspected that I had at best a mediocre talent for mathematics, I chose to go into the Mathematics and Physics program at the University of Toronto, not because I thought that I might become a mathematician, but because I thought that at the undergraduate level one could learn some techniques and ways of thought which would be of considerable value no matter what one did. In contrast, undergraduate studies in the Social Sciences at the time struck me as hardly worth while. I felt that the 'great books' could be read by oneself and, if one went to graduate school and were careful in course selection, there might be a course or two which would explain the great books.

I learned a great deal as an undergraduate in Mathematics and Physics. In particular I learned that I did not have the type of mind or talent to become either a decent mathematician or a physicist. But even though I was both a poor mathematician and a sloppy experimentalist in physics I learned a great amount about the precision of mathematics and the care needed to perform even a trivial experiment in physics. I was impressed by several of my teachers at Toronto. While I sat as an undergraduate in Professor Richard Brauer's class in algebra, I kept thinking that it was a waste of such obvious talent on a bunch of ordinary Ontario undergraduates. I was fascinated by the enthusiasm of Professor Coxeter, teaching the geometry of perfect solids. But at the same time it struck me that there were many forms of intelligence and that a great mathematician could be a totally naive citizen. As I was politically active as an undergraduate I was impressed by the different form of intelligence possessed by local politicians as contrasted with the processors of mathematics.

On graduating from Mathematics and Physics I thought for a while that I would take a higher degree in Applied Mathematics, but it was evident to me that I had neither the real interest, insight nor talent. I decided instead to take an MA in economics at Toronto. There were several positive reasons for doing so. I was involved with the CCF Party and contemplated running for parliament. But as I examined the platform concerning the electrification of North Ontario I kept asking myself, 'Where is the money going to come from?' I found the politicians to be willing to make dogmatic statements concerning items about which they knew next to nothing. This contrasted with the mathematicians who required a definition and precision for virtually everything. I decided that some formal economic theory might help. I took a course in Great Books in the Social Sciences from Harold Innis and found it to be fascinating and certainly mind stretching. Yet at the same time, while listening to his major thesis about the relationship between democracy and the oral tradition and written language and communication and the authoritarian state, I did some research on the Inca empire and found that it did not fit into the Innis thesis. I had the uneasy feeling that in the Social Sciences one merely refitted the facts to make sure that they fitted appropriately onto the Bed of Procrustes.

In my course in Economic Theory, we were required to write a book review of a major book in economics. While browsing the University of Toronto library I came across, in the newly purchased book display, in 1948, the *Theory of Games and Economic Behavior*. I read parts of it and did not really understand it. I probably wrote a relatively bad essay on it. I have no recollection whatsoever about what I said, but within a few days of finding the book, I felt that this *had to be* the way to go. Within a few weeks I decided that I wanted to go to Princeton (although I had originally contemplated MIT as well). I wrote to Princeton, saying that I wanted to study with Professor Morgenstern. I applied to both the Economics and Mathematics Departments at Princeton and the Mathematics Department had the good sense to reject my application, thereby saving all parties from potential grief.

Princeton was a sheer pleasure, far better than I suspect any drug high might offer. On the faculty I grew to like and admire Oskar Morgenstern as a great intellectual entrepreneur with great common sense and a marvelous sense of purpose. William Baumol became a friend for life and I was impressed by both the fine scholarship and the unpleasant competitive debating attitude of Jacob Viner; it seemed to me that winning the debate appeared to be more important to him than trying to grope towards whatever the truth might be. Von Neumann's intellect staggered me and gave me a feeling for what major talent looks like. When Irwin Panovsky, the great art historian, gave a lecture which I attended, I had the feeling of having seen a completely different, but nevertheless major, talent.

My intellectual stimulus came from the mathematics graduate students far more than from the economics graduate students. In particular, John McCarthy, Marvin Minsky, John Nash and Lloyd Shapley provided considerable intellectual stimulation. Nash, Shapley and I interacted on our mutual interest in game theory. A possible distinction between Nash, Shapley and myself is that they were clearly both interested in and capable of making basic contributions to the underlying mathematical theory, as is evinced by the Nash noncooperative equilibrium and the Shapley value, whereas I was primarily concerned with application to economics in particular, and the social sciences in general.

My first papers came out in 1952. I published 'A Business Cycle Model with Organized Labor Considered' in *Econometrica* (1952) where my concern was to catch the idea that the threat power of the union varied with the position of the business cycle. The next paper was both less technical and far more important. It seemed to me that the information requirements in much of economic theory were far too high. An elementary consideration of the combinatorics and costs has to limit the amount of information utilized. 'Information, Theories of Competition, and the Theory of Games' appeared in *The Journal of Political Economy* (1952).

In my discussions with Nash, who had finished his thesis on the existence of noncooperative equilibria, it was clear to me that the Cournot model of oligopoly utilized an equilibrium concept which was in essence a noncooperative equilibrium. Nash and I agreed to apply both his noncooperative and cooperative papers to a duopoly model. After my making many errors in lengthy and boring calculations, we enlisted Jim Mayberry to join us in this enterprise.

Shapley and I had worked previously on two-person cooperative games with variable threats, but stopped after Nash produced his solution. I am still not satisfied with the economic meaning of how to smooth away a continuum of equilibria to yield a unique outcome.

Although my interest in cooperative game theory was, and remains, high, it seemed to me that there was much to be done, especially in oligopoly theory, utilizing noncooperative theory. I decided that this would be the main concept employed in my thesis. I had read both Cournot and Edgeworth's *Mathematical Psychics* and they both, in different ways, suggested to me the idea of studying the behavior of solutions to games under replication. Intuitively the idea that the influence of a single individual becomes small as numbers of like, or almost like, individuals increases seemed to be commonsensical.

My thesis, which eventually became a book entitled *Strategy and Market Structure* (1959) utilized John Nash's noncooperative equilibrium, which I regarded as a considerable generalization of Cournot's equilibrium concept (although my impression in conversations with Nash was that he was not aware of Cournot's work). While working on my thesis it seemed to me that the most satisfactory link between oligopoly models and perfect competition had to involve the appropriate model of very small agents, so small that a single individual would not influence price. Shapley pointed out that to do this right would require measure theory, where one could consider a single agent as a set of measure zero, but as I knew no formal measure theory I settled for examining replicated finite sets of agents.

In the second part of my 1959 book, I wanted to take some steps towards dynamics. I concluded that an economic dynamics had to place stress on information conditions and on some basic aspects of institutional structure. In order to stress this, I entitled the second part of the book 'Mathematical Institutional Economics'.

My goal was to understand economic behavior and, to a lesser extent, political and bureaucratic strategic voting behavior. Thus, when Shapley showed me his newly developed value solution and described the class of simple games as a formal way to represent voting structures, it appeared to me that this offered an opportunity for a basic application to Political Science. We jointly wrote an article entitled 'A Method for Evaluating the Distribution of Power in a Committee System' (1954). Much to my surprise, it was accepted almost immediately by *The American Political Science Review*; to this day I do not know how or why this major journal in political science took a more or less mathematical article from two unknown graduate students, neither of whom was in Political Science.

Even as a graduate student, I was struck by the contrast between cooperative game theory, the seeds of which I regarded as already present in Edgeworth, and noncooperative theory, which was present in Cournot. I wanted to see if game theoretic methods applied to both of these works would yield limiting results related to the concept of economic competition. In the early 1950s, Shapley, Gillies and I had discussed the core of an *n*-person game in cooperative form. It seemed to me that Edgeworth's work on two-sided markets could be regarded as a precursor of a cooperative game approach to markets. While at the Center for Advanced Study in the Behavioral Sciences, I wrote an article entitled 'Edgeworth Market Games' (1959) which was finally published four years later. At the time of writing, except for some assistance by Howard Raiffa, I had no one to check my poor mathematics, and the article appeared with several false theorems. By the time it was finally published, I had realized most of my errors and had conjectured on the full significance of the core and its convergence to the competitive equilibria under broad conditions, but I knew that I was incapable of proving it. Most of my earlier errors were due to a confusion of the relationship between the sidepayment and no-sidepayment games (this linkage was eventually resolved with the analysis of the λ-transfer games). I corrected most of my errors in a publication which took several years to get published, entitled 'Extended Edgeworth Bargaining Games and Competitive Equilibrium' (1968). I posed my general conjecture to Lloyd Shapley and then to Herbert Scarf in a walk back from a seminar which he had given at Columbia University (Shapley proved, but I believe did not publish, the sidepayment results) and Scarf was able to prove the general conjecture for games without sidepayments.

Another major challenge which I perceived as a graduate student was to construct satisfactory models of the firm in a dynamic context which included the possibility of bankruptcy. I constructed a class of games which I called 'Games of Economic Survival' (1959) and managed to enlist the mathematical collaboration of Gerald Thompson in analyzing them as stochastic games with an infinite horizon and with bankruptcy represented by an absorbing barrier for the controlled random walk. At the time I did not yet see the critical importance of bankruptcy as a necessary feature in constructing any satisfactory viable dynamic theory of competition.

The few years I spent at General Electric gave me an appreciation for both macroeconomic detail and managerial practice which would have been difficult to obtain in academia. While there I became interested in the possibilities for gaming as a teaching and planning device; George Feeney was building a business game and I had talked with Richard Bellman about his game for the AMA. Even earlier, a chance meeting with Sidney Siegal in 1956 in the High Sierras had convinced me that experimental gaming was going to be of considerable importance. We began what we hoped was going to be a long collaboration; I commuted every now and then from General Electric in New York to Pennsylvania State University, but distance and Sid's early death cut our plans short. In 1960, while visiting Yale, I ran an experiment which was published as 'Some Experimental Non-Zero Sum Games with Lack of Information about the Rules' (1962). It is my belief that, to this day, the full importance of experimental gaming and games for teaching in economics has not been appreciated.

While I was at General Electric, the importance of the relationship between the assignment of joint costs and incentive systems within the firm seemed to me to be critical. I had talked about this with Harlan Mills, whom I regarded as a fine operations research practitioner. It appeared to me that one way of looking at this problem was via the Shapley value which assigned a combinatoric marginal worth to the players. I wrote up these observations in 'Incentives, Decentralized Control, the Assignment of Joint Costs and Internal Pricing' (1962).

Another of my concerns has been defense studies. In particular there is a central problem which has certainly been the object of formal study for around 40 years, but which still is by no means adequately solved. It is the meaning and credibility of threat in international relations, in economic competition and in social life.

My debt to Lloyd Shapley cannot be overstated. There are two major books on game theory published under my name which, as far as I am concerned, are joint work. But when one works with a close colleague and friend for many decades, individual choice should be respected, even if it is not understood. He chose to not appear as a coauthor, but consented to have me publish what I regard as essentially joint work. In order to fully explain our collaboration to myself, I have said, partly in jest, partly quite seriously, that I could usually tell when a paper with Lloyd was finished – I could no longer read it.

My strengths have been in conjecturing and model building, not in proof. Much of my time has been spent in finding an individual with the mathematical talent who was willing and able to understand my models and conjectures and interested enough to collaborate. The great benefits of working with Lloyd were that, not only does he have great talent in game theory, but his ability to find both logical and substantive flaws in models is considerable.

Volume one

The collection of reprints presented here covers three areas of interest. The first is devoted to my general writings in political economy and the others to two more specific topics which interested me early in my career: oligopoly theory and experimental gaming.

I was convinced from the start of my professional career that the treatment of information was critical to the understanding of economic process. This has run as a constant theme through several of my papers on theory and experimentation and has culminated in my concern for the information processing aspects of money and financial institutions and for what I regard as one of the key challenges for the behavioral sciences of the future: that is, the development of adequate models to describe decision making among individuals with limited abilities in perception, memory, computation and prediction.

Much of economic activity involves fiduciary decision making. The spenders of large sums of money are usually playing primarily with other people's money. Thus, in order to understand some basic aspects of economic decision making, it is desirable to understand how hierarchies and bureaucracies differ from individuals in the production of economic decisions.

About 25 years ago I offered my opinions on the problems and future of microeconomics in a paper entitled 'A Curmudgeon's Guide to Microeconomics' (1970). On rereading it,[1] although some of my more sanguine predictions may be running late, I believe that for the most part the development has been along the lines suggested, although the importance of computational methods and the computer were probably underestimated, even though they were advocated.

Oligopoly

For at least one brand of mathematical microeconomist, oligopoly theory, especially duopoly theory, plays the same role as the five-finger exercises for the pianist, or even worse the double acrostic for those who have nothing better to do with their time. When I saw the connection between John Nash's thesis and Cournot, I was convinced that games in strategic or extensive form provided a fruitful way to extend our knowledge of competition among the few. I believe now that I was simultaneously right in some aspects and wrong in others. The formal process models of game theory have permitted us to begin to obtain an appreciation of the importance of production, price, product variation, location, capacity, entry, inventories and other basic economic features of the firm in competition. We have also been able to gain better insight into the role of numbers of competitors. Most of my papers are addressed to these features. But it has become more and more apparent that we do not have an adequate multiperiod dynamic theory for competition among the few. The three-firm problem in economics is probably far harder analytically and more difficult to model than the three-body problem in physics. In recent years the 'New Industrial Organization' has taken over, highly laced with game theory and based on a noncooperative equilibrium theory which does not appear to be adequate to capture the dynamics. Mathematical economics and game theory are undoubtedly helpful in illustrating interesting possibilities and ways of thinking about oligopolistic competition. They are probably of value in providing expert witnesses in antitrust litigation with intellectual ammunition for their cannons. But I suspect that the future development in industrial organization will swing back more and more to better models of administrative behavior and more detailed institutional understanding of specific industries, perhaps bolstered by simulations and specific game models. An exception may be had in the construction of unified engineering, marketing and financial models which will blend these diverse aspects of the control of the firm into far richer (but computationally feasible) models than is currently the practice.

Experimental gaming

In my discussions with Sidney Siegal, Merrill Flood and others dating back to the 1950s I was (and am still) convinced of the enormous importance of gaming in economics and in business for teaching, training, planning and experimentation. Furthermore experimental games provide a natural forum for cooperative work among behavioral scientists with different backgrounds: in particular, economists, psychologists and social psychologists. Much of my own work was done in collaboration with Siegal and Gerrit Wolf. Economists working alone, especially with highly computerized games, may easily fail to appreciate the many artifacts and constraints guiding their results.

Although one can hardly classify as a formal experiment asking individuals their opinions about how they would play certain games, I have found that considerable insight concerning several basic aspects of game theory can be gleaned from doing so. In 'Cooperative Game Solutions: Australian, Indian, and U.S. Opinions' (1986) the attractiveness of the core as a solution concept is examined as its size and 'fairness' are varied.

I have been concerned with the design of simple games to illustrate problems and principles in multiperson decision making. Two games are presented here. '"So Long Sucker"' (1964) was concocted by a group of us when we were graduate students at Princeton in the early 1950s. We wanted to design a game where informal cooperation was necessary, but not sufficient to win. At some point there would appear great pressures to doublecross one's partner. 'The Dollar Auction Game' (1971) struck me as a nice illustration of the potential for mutually destructive escalation.

Except in highly structured games where the players interact through anonymous markets, experience with gaming indicates (at least to some of us) the inadequacy of the model of the well informed rational decision maker of *homo oeconomicus*. The actual beast is far more complex and subtle. The power of general equilibrium theory and the writings on the price system derives for the most part from the fortunate feature that some aspects of economics can be treated as an *asocial science*: the large numbers of agents, to a great extent, wipe out much of the sociopsychological interaction. The wise economist who wishes to experiment is well advised to appreciate how differently the social psychologist, historian, anthropologist, zoologist or biologist will look at the same phenomenon.

Volume two

The price system

My specific goal for many years (essentially from 1952 to 1984) was to understand the relationship between the competitive equilibrium price system and the various game theoretic solutions applied to economic exchange and production modeled as a game. Without the deep collaboration with Shapley, I would not have had a hope of understanding the deep relationships that are present. My first attempts involved the understanding of the core, which could be regarded as reflecting countervailing power. My program was to examine and contrast the competitive price system with the core, the von Neumann–Morgenstern stable set, the Shapley value (the nucleolus, kernel and bargaining set some years later) and the noncooperative equilibrium. Already in my two articles on the core it was clear to me that, although the price equilibrium would be in the stable set, there was no indication that the stable set would converge towards the price system. This was further illustrated in work with Shapley in 'Concepts and Theories of Pure Competition' (1967). In this publication we also were able to present a nonsymmetric closed strategic model of exchange with one group of agents as price takers, but the others as strategic Cournot quantity strategy players.

In 1967, William Lucas published his counterexample to a conjecture of von Neumann that all coalitional games would have a stable set solution. This result intrigued me, especially when, in a chance conversation, Michael Maschler mentioned that this game had a core. In particular, I wanted to know if this could possibly apply to a model of economic exchange or production formulated as a game in coalitional form. I spent several weeks trying to cook up an example. I finally patched up an example which I suspected was wrong, but I sent it to Shapley anyhow, hoping that what did happen, would happen. He found the errors, brooded over the problem and constructed an accurate example which established my hunch that the Lucas counterexample could arise from an innocent-looking economic background. We then talked further and made an extremely simple but fundamental observation. Any subgroup of agents in an n-person exchange economy can also form an exchange economy. But as we knew that the game representation of an exchange economy has a core, this implied that every one of the 2^n games which can be formed from the n players also has a core. From this observation in the paper, 'On Market Games' (1969), we developed and explored the relationship between exchange economies and totally balanced games and showed that all exchange economies could be represented as totally balanced games. Some years later, in 'Competitive Outcomes in the Cores of Market Games' (1976), we went in the opposite direction and showed that any totally balanced game would have at least one representation as an exchange economy.

While at IBM in 1962, I had conjectured that the Shapley value of an exchange economy did not necessarily lie in the core. I suggested this to Scarf and Shapley and actually was able to construct a simple and correct example. I then guessed that, even though it was not in the core, on replication it might converge to the limit point of the core. I interested Shapley in this idea and he was able to establish my conjecture. We published 'Pure Competition, Coalitional Power and Fair Division' (1969) showing that the value also converged to the same limit point as the core which could be interpreted as the competitive price system.

There are three further solution concepts, little known to the broad public, but regarded as of interest by the faithful in game theory. They are the nucleolus, the kernel and the bargaining set. As the nucleolus solution is always in the core, if the core is nonempty, the proof of convergence came as a free good, given the core convergence. Shapley and I worked on the convergence of the bargaining set and kernel solutions, but we never published a paper on it, although I provided a painstakingly calculated (and possibly inaccurate?)[2] example in my book, *A Game-Theoretic Approach to Political Economy* (1944, pp. 343–9).

Missing from the solutions noted above are the convergence results for the non-cooperative equilibrium models of the economy which should reflect individual strategic power. They are covered in the next section, for reasons which are indicated below.

A brief summary of the investigation of the functioning of the price system in an exchange economy is as follows:

- the competitive equilibrium – decentralization;
- the core – the limitations imposed by countervailing power;
- the value – an *a priori* concept of fair division;[3]
- the nucleolus – the minimization of the maximum complaint any group might have against the outcome;

- the noncooperative equilibrium – the equilibrium outcome resulting from the use of individual strategic power in the presence of consistent (self-fulfilling) expectations.

The theory of money and financial institutions

In 1961, while at the IBM Watson Laboratories, I had begun to think as deeply as I could about the role of money in the economy. I tried to obtain a satisfactory model and failed miserably. This problem dogged me for years. In the late 1960s, I made some small headway, writing and publishing 'Pecuniary Externalities: A Game Theoretic Analysis' (1971). But it did not seem to go anywhere.

In 1970–71, I took a leave from Yale and spent over a year at RAND. In late 1970, I reviewed my work of the previous years and came to the conclusion that there were two major problems of deep interest to me which were still outstanding. They were the symmetric embedding of the Cournot oligopoly model in a closed exchange model and a macroeconomic theory of money. I decided that, as I was already 44, and had been working unsuccessfully for around 10 years on the theory of money, even though it was my number one problem I should finally abandon it and concentrate on the last piece of the package of game theoretic solutions applied to understanding models of a closed exchange economy. This was the noncooperative equilibrium.

The reason why I had failed to obtain the correct formulation for the Cournot and Bertrand–Edgeworth models of competition, although Shapley and I had marched through essentially all of the cooperative solutions, was that I had failed to note a fundamental conceptual difference in the basic models required. The characteristic function or coalitional form representation of a game provides a list of how much any grouping of individuals can obtain if they all coordinate their actions. It leaves out all transactions costs; communication is implicitly free; all the problems of coalition formation, cooperation and contract are costless. The coalitions do not exist, they are 'in being'; they are mathematical artifacts which enable us to consider the combinatorics of finite *n*-person games at a high level of abstraction, using only the characteristic function. In essence, it is a context and institution free representation of multiperson collaboration. The primitive concepts utilized are players, coalitions and payoffs to coalitions.

The general equilibrium representation of economic competition is also presented at a high degree of abstraction, but not quite as high as the characteristic function. The primitive concepts behind the general equilibrium are individually owned, traded and consumed goods (and services), production processes and prices. Beyond those features there is no institutional specification and no context is supplied.

Neither the characteristic function representation of a game nor the general equilibrium model of the economy provides a specification of process. The mathematical specification of the mechanisms which serve to carry process amount to adding (albeit at a high level of abstraction) an institutional description. Essentially *the rules of the game are the carriers of process* and, in the economy, *the carriers of process are the elementary institutions*. It was an eventual understanding that the Cournot description of competition and Bertrand's diatribe against Cournot could be recast as arguments about the appropriate *extensive form* description of the game which enabled me to see why the noncooperative equilibrium models did not fit in with our previous analysis. The models had to be utterly different and required, at the least, even if at a high level of abstraction, an extensive form description of the game. An extensive form description spells out the nature of moves and information conditions in detail. It is far more engineering and physics oriented than most mathematical economic models built solely for equilibrium analysis. It deals with mechanisms. Markets are mechanisms where market prices are formed. In general, equilibrium theory market price is a primitive concept and the existence of a set of efficient market prices is proved. In a process description of the economy prices are formed in a market mechanism.

Von Neumann and Morgenstern, in their great book on the theory of games, developed three separate representations of a game, any one of which could be treated as a primitive concept: (1) the game in extensive form (2) the game in strategic form and (3) the coalitional or cooperative or characteristic function form representation. The first is the most detailed: it provides an explicit description of all potential paths of play and all information conditions. The second, the strategic form, provides a great reduction which obliterates the fine structure of moves and information and reduces the portrayal to one dealing only with sets of strategies for each individual. The formal description uses, as its primitives, players, strategies and payoffs. The cooperative form of the game goes a step further in the abstraction. It obliterates the details on strategic interaction and deals only with players' coalitions and their associated joint payoffs.

The work of Shapley and myself on the price system had utilized games in coalitional or cooperative form and the various solutions applied to this form. In order to encompass the noncooperative equilibrium solution, the model needed to be cast in either the extensive or the strategic form. Fortunately these forms coincide in the simplest case of interest, that is games with a single move for each agent, each made without information about the actions of others. For pure simplicity, I started to try to construct a model in strategic form in order to produce an intrinsically symmetric model of an exchange economy as a game in strategic form.

At this time, I reread Chamberlin's *Monopolistic Competition* and Mrs Robinson's *Imperfect Competition* to see if I might get any modeling ideas from them. A chapter heading entitled 'A World of Monopolies' from Mrs Robinson's book gave me an idea. I would try to construct a model with n traders each owning a different commodity. Thus there would be n traders and n commodities. I began by considering a strategic game in which each trader would announce price. But at this point I hit a snag. It appeared to me that a price was an exchange ratio, some quantity divided by another quantity. I could build a well-defined strategic game by selecting one of the commodities and using it as numeraire and have all other traders quote prices in terms of an exchange for this good. But if I did this, the game constructed was not intrinsically symmetric. The act of selecting the numeraire would take away the strategy set of the individual owning the commodity selected as numeraire. I fiddled with this for a while and then decided that, if I wanted to get a fully symmetric game, I seemed to be short of one degree of freedom. There appeared to be an easy way to patch up my model. Consider the model with n traders, each one a monopolist, but where there are $n + 1$ commodities and each trader holds a supply of his own good and a supply of the $n + 1$st commodity. For various technical reasons, I decided at this point that the quantity strategy model was easier to formulate and analyze than the price strategy model, so I switched to the Cournot-style strategic model where each trader offered a quantity of his good for sale and offered a quantity of the $n + 1$st good to bid for the other n goods. I then considered the replication of this model with k traders of each type. I took my half-baked model to Shapley who constructively tore it to bits in many different ways. But after each tearing up we were able to repatch and strengthen it. In particular Lloyd was quick in pointing out that, if no one offered a commodity for sale, but the $n + 1$st commodity had been bid for it, as price of the jth commodity was defined as the quantity of $n + 1$st good offered for it divided by the quantity of the jth commodity put up for sale, there would be a division by zero. We discussed this and observed that the specialist's role on the New York exchange requires that he make an 'orderly market', that is, that he has a small inventory available for sale. This can be mathematically approximated by relating an 'epsilon-defined' game to the game under consideration where we assume that there is always some small amount of 'epsilon' available in each market. This would guarantee compact payoff sets, and Lloyd thought that this might work.

I worked at a simple example and the convergence seemed to be going through. We apparently had the model that was going to complete the package for the study of the price system.

Some four to six weeks later, somewhere in early/mid-January of 1971, while walking on the beach near the end of Sunset Boulevard, it dawned on me that the $n + 1$st commodity in the model could be interpreted as money. I had been looking at the solution to my second most important problem, the Cournot convergence, and it turned out to provide the key insight into the problem I had abandoned. To this day, I am amazed that I could have worked for about two months without even considering the connection. My concerns were with symmetry and an extra degree of freedom and when I introduced the $n + 1$st commodity I did not understand what I was doing in the context of introducing the role of money in the economy.

My first published paper containing the new model was 'Commodity Money, Oligopoly, Credit and Bankruptcy in a General Equilibrium Model' (1972). I would have preferred to publish a mathematically more sophisticated joint paper, but I felt that it was important not to wait much more than a year or two, given my experience with the core. Some years later, Shapley and I published 'Trade Using One Commodity as a Means of Payment' (1977). In this paper, primarily through Lloyd's ingenuity, we were able to milk the Edgeworth Box diagram for basic exposition of a model in which all of one good was offered for sale. Pradeep Dubey and I, in a paper entitled 'The Noncooperative Equilibria of a Closed Trading Economy with Market Supply and Bidding Strategies' (1978), published the basic proof for the existence and convergence of the model where individuals can both buy and sell all commodities. A key lemma was supplied by Shapley. We originally had an elaborate example of 'wash sales' where the markets were made thicker by all traders buying and selling in the same markets, but the editor cut it out and I published it separately in my book, *A Game-Theoretic Approach to Political Economy* (1944).

During my stay at RAND, I suspected that the eventual development of the models I had begun to construct would call for modeling the economy as a set of parallel dynamic programs with a continuum of agents. I spent a certain amount of time trying to persuade Richard Bellman, who was a friend, to consider working with me on this. It took me considerable time to interest him, but before we began a serious collaboration he was tragically stricken with a brain tumor and never fully recovered from the ensuing operation. Subsequently I obtained the collaboration of Ward Whitt and he and I published a lengthy paper analyzing an infinite horizon two-person game with fiat money (Shubik and Whitt, 1973)[4]. There was a hiatus of 20 more years before I could find two colleagues to work on the infinite horizon dynamic model with stochastic elements. The paper by Karatzas, Shubik and Sudderth (1994)[5] which has appeared in *Mathematics of Operations Research*, and several subsequent papers, are based on several unpublished papers of Whitt preliminary to a projected joint paper which was never completed.

Some time in the late 1970s, I decided that the class of games being considered was sufficiently important and different from the class of market games that Shapley and I had developed and used to study the price system for these to need a name to distinguish them from the others. 'Strategic Market Games' seemed to be appropriate, as it emphasized that the agents have a strategic role.

Possibly through a combination of my interest in gaming and the poorness of my mathematics, my approach to the theory of money has been to break the analysis into many small simple parts and to try to digest them one at a time prior to reassembling them to obtain an integrated overview. Thus the first attack was concentrated on an exchange economy with a commodity money. Having managed to formulate and analyze the Cournot model, the Bertrand–Edgeworth model remained to be investigated. The article 'The Noncooperative Equilibria of a Closed Trading Economy with Market Supply and Bidding Strategies' (1978), by Pradeep Dubey and myself, contained my basic model. Dubey published the full technical version with formal proofs somewhat later in *Econometrica*.

A key concern of mine, dating back to my book, *Strategy and Market Structure* (1959), has been (and remains) the understanding of the role of bankruptcy and reorganization in an ongoing economy. In an economy with a credit structure and incomplete markets, rules covering the possibility of bankruptcy are required. For many years bankruptcy has been ignored in macroeconomic theory, or regarded as a minor institutional nuisance. My belief is that it is central to the understanding of equilibrium with complete or incomplete markets. Beginning with a model of complete markets, Charles Wilson and I were able to explore the requirement that a bankruptcy rule in an economy using fiat money be harsh enough to prevent strategic bankruptcy. In a paper entitled 'The Optimal Bankruptcy Rule in a Trading Economy using Fiat Money' (1977) we explored the relationship between the money supply and the bankruptcy penalty.

Over a decade later, Pradeep Dubey, John Geanakoplos and I examined the equilibrium analysis of an economy with bankruptcy and incomplete markets; this paper is not yet published. The fundamental contrast between the economies with complete and incomplete markets is that, in the former, equilibrium requires bankruptcy conditions harsh enough to prevent strategic election of bankruptcy, but in the latter the bankruptcy rules emerge as an important public good which set the standards of risktaking acceptable to society as a whole.

It appears to me that there is a deep analogy between the bankruptcy rules of an economy and the role of mutation among biological organisms. The more lenient the bankruptcy laws, the higher is the level of innovation and the greater the failure rate. The higher the mutation rate, the greater is the potential for adaptability of the organism, but the higher is the level of nonviable mutants.

A general criticism leveled against my strategic market game models from the start was that they were too unrealistic. My attitude is fairly simple. Actual economic life is a highly complex, but playable, game. Economic models must be simpler, but they still should be cast as playable games. The essay form or the pure equilibrium model both run the danger of containing subtle omissions which prevent them from being adequate representations to be used in the study of process. In the analysis of both the Cournot and the Bertrand–Edgeworth models, I was looking for the simplest mechanisms to form competitive prices. My idea was to look for *minimal institutions*. After one had achieved that goal and had looked at enough special cases, the next step would be to generalize by trying to axiomatize the concept of price formation and a market. I had discussed this with Pradeep Dubey and we found out that Andreu Mas-Colell had been thinking somewhat along the same lines. We got together and wrote 'Efficiency Properties of Strategic Market Games: An Axiomatic Approach' (1980).

I have described my first approach to embedding the Cournot model in a closed economic system; several years later I realized that there was a general problem concerning the number of markets which remained open. I had considered a world with $n + 1$ commodities and n markets where the $n + 1$st commodity played a special role. However, in contrasting this type of model with a general equilibrium exchange economy, one could conceive of an economy which had trading posts for every pair of goods. In this instance this would require $n(n + 1)/2$ markets. In such an economy, every commodity would serve as a 'money' in the sense that it would be a means of exchange which could be utilized in the purchase of every other commodity. In my model, individuals would offer goods in all markets or 'trading posts' and each market would clear independently. Shapley suggested a somewhat different model where the strategies were the same, but with a clearing house which summed together all the goods of the same type and then a central mechanism would calculate the prices to clear the markets for all goods. Two excellent graduate students, Sahi at Yale and Yao at UCLA, worked on the different models as part of their PhD theses. Amir, Sahi, Shubik and Yao (1990) published a paper showing the existence of equilibria in the decentralized 'trading posts' model. This paper also established the possibility that the distribution of resources could be so skewed that the usual price consistency conditions brought about by an arbitrage chain would not hold without the introduction of credit. The concept of enough money in an economy depends, not only on the market and clearing structure and the institutional needs to cover transactions, but on both the quantity of the money and its distribution.

In fact, the world does not have active markets for each pair of commodities, but the models with all markets provide a natural format to consider the endogenous selection of a money based on consideration of transactions costs and the thickness of markets.

The macroeconomic theory of money and financial institutions, after many years, is finally beginning to emerge, not merely in the work noted but also in the works of

Grandmont, Geanakoplos, Cass, Shell and others. The various individuals involved have each approached this topic from their own bents, be it temporary general equilibrium theory, or enlarged general equilibrium models with incomplete markets, or strategic market games.

Methodologically I suspect that the mathematics involved in the different approaches is coming together. Although some of us have spent a great amount of time investigating one-period models using game theory methods for a finite number of agents, many of the broad interesting questions concerning money and financial institutions lie with infinite horizon dynamic models and with overlapping generations. In my opinion, there is no unique adequate game theoretic solution concept for these models with a finite number of agents. The system dynamics is based on local optimization.

In the actual world of business and finance, some individuals have considerable strategic power in some markets. Face-to-face bargaining and contract and agency problems are of concern. Furthermore, a well-written contract may be more productive than 30 years of hard work. Thus attention to microdetail is critical. The applied game theorist, along with the lawyers and practicing financial analysts may be able to offer some insights into some of the basic features of the drawing up of contracts, but both the conceptual and mathematical difficulties to be faced in the application of *n*-person dynamic games are such that they are limited to special *ad hoc* problems.

If, however, one limits one's attention to models with considerable economic structure and with a continuum of agents, then the mathematical methods required for the study of the noncooperative equilibrium solution applied to strategic market games appear to be the same as or closely related to those employed in variants of competitive equilibrium analysis. This means that a whole body of methods is available with relatively easy modification both for proofs of the existence of equilibrium and for computational methods. The basic difference in approach, however, remains. By adhering to the playable game test of a model, the stress is on carriers of process, or minimal institutions, not on equilibrium. As the models are fully defined process models, they can be simulated or otherwise studied for behavioral solutions other than the noncooperative equilibrium.

Even if one adopts a strategic player view of the individual, I believe that both the competitive equilibrium and the perfect noncooperative equilibrium solutions with so-called ' individualistic economic maximizers' are useful but merely crude behavioral approximations for the type of socialized, capacity-constrained, limited information individuals we are. The next major advance in economic understanding will come in being able to model agents with different levels of expertise and ability to process information. The key concerns will be with expectations, inference and prediction. All of these are critical to the understanding of economic dynamics, but are essentially finessed or grossly simplified in the emphasis on equilibrium studies.

But for now, the next steps in the creation of a theory of money and financial institutions call for the modeling of the economy as a set of parallel dynamic programs with a continuum of agents, with stochastic elements, with given rules for the creation and destruction of money and credit and bankruptcy and reorganization rules specified. The process models describe the nature of the institutions. From there the next steps will involve the study of limited rationality and the use of solution concepts stressing the methods of inference used in forecasting. Details such as the financing of the float and why fiat money is the only financial instrument which causes the model of exchange to be a set of nonhomogeneous equations must be explained. The reasons why the intrinsic lack of consumer value of fiat money fixes the price level and removes the indeterminacy in the general equilibrium price level have to be spelled out. The importance of the relationship between human error in decision making and the length of time used to make the decisions remains to be explored. The thrust of my current work is in this direction.

Notes

1. I did Hicks an injustice, underestimating the importance of his comments on dynamics.
2. I suspect that no one except myself has read it and checked it, and I do not trust my own checking.
3. This is not unique for nosidepayment games. Further exposition would require a lengthy discourse on λ-transfer.
4. 'Fiat Money in an Economy with One Nondurable Good and No Credit (A Noncooperative Sequential Game)' (with W. Whitt) in *Topics in Differential Games* (ed. A. Blaquiere), Amsterdam: North Holland, 1973, pp. 401–48.
5. 'Construction of Stationary Markov Equilibria on a Strategic Market Game' (with Ioannis Karatzas and William D. Sudderth), *Mathematics of Operations Research*, **19**(4), November 1994.

PART I

THE PRICE SYSTEM

[1]

EDGEWORTH MARKET GAMES*

Martin Shubik

§1. ECONOMIC INTRODUCTION

Much mathematical work has been done in the investigation of the von Neumann and Morgenstern solution to an n-person game. However very little economic analysis has been applied to the formulation of the problems studied. In order to connect this particular type of game theory with economic analysis we will examine games whose characteristic functions are constructed in such a manner that they reflect certain economic aspects of a market.

The specific economic problem which is to be examined here is an extension of a bargaining or bilateral monopoly problem studied by Edgeworth[1]. Edgeworth considered a market which can be characterized by a two-person non-zero sum game. Each player was assumed to have a set of preferences which could be represented by a family of convex curves (not necessarily the same for each player) which at least gave an ordering to each player's preference for any combination of the two goods being traded. The player's initial fortunes were assumed to consist of (a, o) for the first and (o, b) for the second. These state, respectively that the first player has a units of the first good and o units of the second good; while the second player has o units of the first good and b units of the second.

Edgeworth selected as a solution to this trading problem a set of distributions which satisfy the conditions that at any of these distributions a player is at least as well off as he would be if he failed to trade. No distribution of the set is preferred by both players to any

* Most of the work in this paper was done at The Center for Advanced Study in the Behavioral Sciences. The writer is indebted to Roy Radnor and Howard Raiffa for their conversations on this topic.

1 Edgeworth, F. Y., Mathematical Psychics, London, 1881, pp. 37-42.

other distribution of the set. Any distribution not in the set is dominated by some distribution in the set in the sense that both players can find a distribution of the set which they both prefer to one not contained in the set. This set of distributions usually forms a curve (infinite divisibility of the goods being traded and continuous and convex indifference curves are assumed) and is called the contract curve.

The Edgeworth solution can be demonstrated easily by means of a diagram often used in economics textbooks. In Figure 1 below the initial fortune of the first player is represented by the line Oa. His preference system is represented by the family of indifference curves of which ad is one. Any point on ad for instance, represents a combination of quantities of goods of the first and second type which has the same utility to the first player as does the initial combination of (a, O).

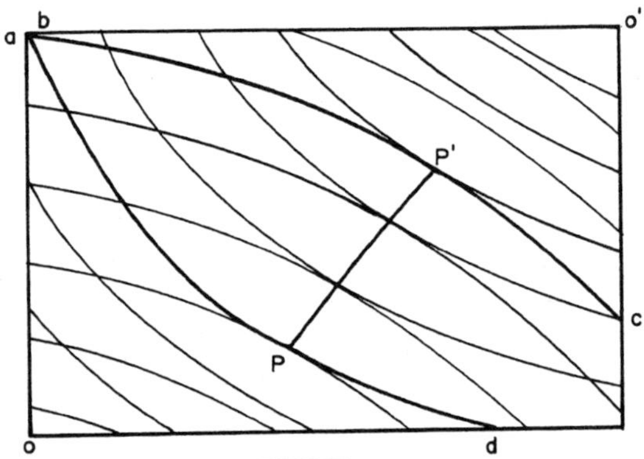

FIGURE 1

The preference system for the second player is portrayed by a second family of curves of which bc is one. The diagram is constructed in such a manner that the point O represents the state at which the first player has (O, O) and the second has (a, b), whereas at O' the first has (a, b) and the second (O, O). The Edgeworth solution or contract curve is the locus of the tangent points between the two families of indifference curves, bounded respectively by ad and bc. This is indicated by PP'. It is easy to see that this solution is very similar to that of von Neumann and Morgenstern in that it consists of a set of distributions which do not dominate each other but dominate all other distributions. The solution is not identical with that of von Neumann and Morgenstern in as much as sidepayment in the form of a third commodity or "utils" is not considered. It is not even necessary to assume that the two players have comparable utilities for the goods being traded. However Edgeworth noted that if the players had comparable utilities then there would be a single point which he called the "utilitarian point" at which trade could take place in such a manner that joint utility, i.e., the sum of the utilities, would be maximized.[2] The von Neumann and Morgenstern solution selects this point to determine the trading and then determines a range for the sidepayments. The range of the von Neumann-Morgenstern solution is bounded by the values assigned to the positions of the players prior to trading.

Edgeworth considered bargains with more than two players and concluded that if the players became numerous on both sides of the market (under certain plausible conditions) that the bargaining would be such that the "range of indeterminacy" characterized by the contract curve would narrow and in the limit a single distribution would result. This distribution would correspond to the market price under pure competition.

In this paper we examine the limiting behavior of the solutions and certain other aspects of an n-person game related to the Edgeworth model, as the number of players becomes large.

§2. THE MATHEMATICAL MODEL

An Edgeworth market game consists of a set $I = M \cup N$ of players, consisting of traders in commodity 1 (the set of players, M) and traders in commodity 2 (the set of players, N). Each player has a utility function consisting of a set of strictly convex indifference curves (i.e., an indifference map). The utility function of the ith player is denoted by $\psi_i(x, y)$ where x and y are the quantities of the first and second commodity. It is assumed that:

$$\lim_{x \to \infty} \psi_i(x, y) < \infty \quad \text{and} \quad \lim_{y \to \infty} \psi_i(x, y) < \infty$$

and that the ψ_i are twice differentiable. The ith player in the set M starts with initial resources $(a_i, 0)$ of goods 1 and 2; where the first number is the amount of the first good and the second, the amount of the second. The jth player in set N has initially $(0, b_j)$. The characteristic function for the game is:

$$v(\theta) = 0$$

$$v(S) = \max_{x_i, y_i} \left[\sum_{i \in S \cap M} \psi_i(a_i - x_i, y_i) + \sum_{j \in S \cap N} \psi_j(x_j, b_j - y_j) \right],$$

where $\Sigma x_i = \Sigma x_j$, $\Sigma y_i = \Sigma y_j$ and $x_j, y_i, \geq 0$.

In order to examine this game fully we would have to consider many special cases depending upon the shapes of the utility functions, the

[2] Ibid., pp. 42-43.

initial stocks of the traders and the number of traders. We limit ourselves to the simple case where all traders have identical utility functions $\psi_i(x, y) = \psi(x, y)$. Each trader belonging to M initially has (a, o) and each trader belonging to N has (o, b).

Because of the above restrictions we can immediately write down the explicit characteristic function for any game of this type. If there is no joint gain to be obtained from trade, the game is inessential and the value of any coalition is the sum of the individual initial utilities. If the game is essential and trade takes place, then by the symmetry and convexity, at a position of joint maximization for any group of players all players will obtain equal quantities of the same good (at this point the marginal rate of substitution between the two commodities being traded will be the same for all players trading). The characteristic function is:

$$v(\theta) = 0$$

$$v(S) = s\,\psi\left(\frac{as_m}{s}, \frac{bs_n}{s}\right)$$

where $s = |S|$, $s_m = |S \cap M|$ and $s_n = |S \cap N|$. The characteristic function for the two-person game becomes:

$$v(\theta) = 0$$
$$v(\{1\}) = \psi(a, o)$$
$$v(\{2\}) = \psi(o, b)$$
$$v(\{1, 2\}) = 2\psi\left(\frac{a}{2}, \frac{b}{2}\right) .$$

The solution to this game consists of the set of imputations of the form:

$$\left(2p\,\psi\left(\frac{a}{2}, \frac{b}{2}\right) - p\,\psi(o, b) + (1 - p)\,\psi(a, o),\right.$$
$$\left. 2(1 - p)\,\psi\left(\frac{a}{2}, \frac{b}{2}\right) + p\,\psi(o, b) - (1 - p)\psi(a, o)\right),$$
$$0 \leq p \leq 1 .$$

It can be shown that if the number of traders in one commodity is the same as the number of traders in the other, then for any size of the market there is a solution analogous to the solution to the two-person game (although such a solution will no longer be the only solution). There exists a set of imputations which form a solution such that all traders on the same side of the market always obtain an equal amount. The set of imputations is characterized by a parameter p (as in the two-person game) which can be interpreted as a market price. The range of this price remains constant for all games with the same number of players on each side of the market.

THEOREM 1. An Edgeworth market game denoted by $[n, n]$ for any size n where the number of traders in one commodity is the same as the number of traders in the other commodity and the traders have the same preferences will have a solution consisting of all imputations of the form:

$$\left(2p\,\psi\left(\frac{a}{2}, \frac{b}{2}\right) - p\psi(o, b) + (1 - p)\,\psi(a, o), \ldots, 2(1 - p)\,\psi\left(\frac{a}{2}, \frac{b}{2}\right) + p\psi(o, b) - (1 - p)\,\psi(a, o), \ldots \right), \tag{1}$$

$$0 \leq p \leq 1 .$$

PROOF. No imputation of the form of (1) can dominate another imputation of that form. A set to be effective for any imputation requires members from both M and N, i.e., trade must take place; but no imputation of the form (1) dominates another of form (1) for a set containing members from both M and N.

Any imputation not of the form (1) has a pair $\gamma_i' + \gamma_j'' < 2\psi(\frac{a}{2}, \frac{b}{2})$ where γ_i' is the amount a player $i \in M$ and γ_j'' is the amount a player $j \in N$ obtains from this imputation. If we select a p such that

$$\frac{\gamma_i' - \psi(a,o)}{2\psi\left(\frac{a}{2},\frac{b}{2}\right) - \psi(a,o) - \psi(o,b)} < p < 1 - \frac{\gamma_j'' - \psi(o,b)}{2\psi\left(\frac{a}{2},\frac{b}{2}\right) - \psi(a,o) - \psi(o,b)}$$

then the imputation of form (1) for this value of p dominates the imputation γ. This completes the proof.

Although the solutions to the Edgeworth market games do not exhibit any tendency to converge to a single point as the number of players increases, it can be shown that under certain conditions the <u>core</u> of the games approaches a single imputation in the limit. This single imputation has economic meaning either in the theory of monopoly or the theory of pure competition.

The core of a game consists of the set of imputations which are undominated in all solutions. The first class of games to be examined have as an economic anologue a monopolist selling to many customers. We consider the Edgeworth market game with one trader of the first type and examine the behavior of the core as the number of traders of the second type increases. The core approaches a single imputation at which the single trader of the first type acts as a perfectly discriminating monopolist and obtains all of the gain from trading.

THEOREM 2. In an Edgeworth market game denoted by $[1, n]$, for any n the imputation:

$$\eta = \left((n+1)\,\psi\left(\frac{a}{n+1}, \frac{nb}{n+1}\right) - n\psi(o,b),\ \overbrace{\psi(o,b),\ \psi(o,b), \ldots, \psi(o,b)}^{n} \right)$$

lies in the core. For any ϵ there exists an $n(\epsilon)$ such that for all $n \geq n(\epsilon)$ no imputation ζ with $\zeta_1 = \eta_1 - \epsilon$ lies in the core.

PROOF. No set not containing player 1 is effective for η. Consider a set containing player 1 and k players contained in N:

$$(2) \quad (n+1)\psi\left(\frac{a}{n+1}, \frac{nb}{n+1}\right) - n\psi(o,b) \geq (n-k+1)\psi\left(\frac{a}{n-k+1}, \frac{(n-k)b}{n-k+1}\right) - (n-k)\psi(o,b)$$

for any integer k, however, in any coalition smaller than I the quantity on the right-hand side of (2) is as much as can be given to player 1 if the others are to obtain at least $\psi(o, b)$ each. Hence no set containing player 1 is effective for the imputation η except I. Given I as the only effective set for η there exists no imputation which dominates η which is therefore in the core.

LEMMA. For any arbitrarily small ϵ there always exists an $n(\epsilon)$ large enough such that

$$|v(I) - v(I - \{j\}) - \psi(o, b)| < \frac{\epsilon}{n} \quad \text{for all} \quad n \geq n(\epsilon)$$

where $\{j\}$ is a player belonging to N.

PROOF. Expand $n\ |\ v(I) - v(I - \{j\}) - \psi(o, b)|$ by Taylor series about (o, b).

$$v(I) = (n+1)\,\psi\left(\frac{a}{n+1}, \frac{nb}{n+1}\right) = (n+1)\psi\left(o + \frac{a}{n+1}, b - \frac{b}{n+1}\right)$$

$$= (n+1)\psi(o, b) + a\psi_1 - b\psi_2 + \frac{1}{2}\left\{\frac{a^2}{n+1}\psi_{1,1} - 2\frac{ab}{n+1}\psi_{1,2} + \frac{b}{n+1}\psi_{2,2}\right\}$$

$$+ R_1 \; .$$

$$v(I - \{j\}) = n\psi\left(\frac{a}{n}\,\frac{(n-1)b}{n}\right) = n\psi\left(o + \frac{a}{n}, b - \frac{b}{n}\right)$$

$$= n\psi(o, b) + a\psi_1 - b\psi_2 + \frac{1}{2}\left\{\frac{a^2}{n}\psi_{1,1} - 2\frac{ab}{n}\psi_{1,2} + \frac{b^2}{n}\psi_{2,2}\right\}$$

$$+ R_2 .$$

$$n|v(I) - v(I - \{j\}) - \psi(o, b)| \leq \frac{1}{2}\left|\frac{na^2}{n+1}\psi_{1,1} - a^2\psi_{1,1}\right|$$

$$+ \left|\frac{nab}{n+1}\psi_{1,2} - ab\psi_{1,2}\right| + \frac{1}{2}\left|\frac{nb^2}{n+1}\psi_{2,2} - b^2\psi_{2,2}\right| + |R_1 - R_2|$$

as $n \longrightarrow \infty$ all terms on the right can be made arbitrarily small.

Now select an ϵ, then for all games of size n where $n \geq n(\epsilon)$ we consider any imputation of the form:

$$\gamma = \left\{v(I) - n\psi(o, b) - \epsilon, \psi(o, b) + \mu_2, \psi(o, b) + \mu_3, \ldots \psi(o, b) + \mu_{n+1}\right\}$$

where

$$\sum_{j=2}^{n+1} u_j = \epsilon \quad .$$

There will be at least one player $j \in N$ who obtains $\psi(o, b) + \mu_j$ where $\mu_j \geq \epsilon/n$, then by excluding this player, the coalition $I - \{j\}$ will be effective for γ. Say $v(I - \{j\}) - [v(I) - \psi(o, b - \frac{\epsilon}{n}] = \delta$, then the imputation π obtained by adding $\frac{\delta}{n}$ to all components of the imputation γ, except the jth component, dominates γ. Hence γ cannot be in the core.

THEOREM 3. In an Edgeworth market game denoted by $[m, n]$ such that $m = km'$, $n = kn'$ and

$$\psi\left(\frac{m'a}{m'+n'}, \frac{n'b}{m'+n'}\right) = \max\, \psi\left(\frac{s_m a}{s}, \frac{s_n b}{s}\right)$$

the imputation

$$\eta = \left(\frac{v(I)}{m+n}, \frac{v(I)}{m+n}, \ldots, \frac{v(I)}{m+n}\right)$$

is always in the core; and for any ϵ there exists a $k(\epsilon)$ such that for all $k \geq k(\epsilon)$ no imputation ζ with a component

$$\zeta_i < \frac{v(I)}{m+n} - \epsilon$$

is in the core, $i = 1, 2, \ldots, m + n$.

The proof of this theorem is similar to that of Theorem 2. An interpretation of the condition that

$$\psi\left(\frac{m'a}{m'+n'}, \frac{n'b}{m'+n'}\right) = \max \psi\left(\frac{s_m a}{s}, \frac{s_n b}{s}\right)$$

is obtained by examining the nature of the function

$$s\psi\left(\frac{s_m a}{s}, \frac{s_n b}{s}\right) \quad .$$

It is homogeneous in the two variables, s_m and s_n. If s_m and s_n are increased in a constant ratio the value of the optimum trade increases in a constant manner. In economic terminology there are constant returns to scale. As the ratio of s_m to s_n varies there are different returns. Figure 2 indicates the lines of constant return and

$$\psi^* = \max \psi\left(\frac{s_m a}{s}, \frac{s_n b}{s}\right) \quad .$$

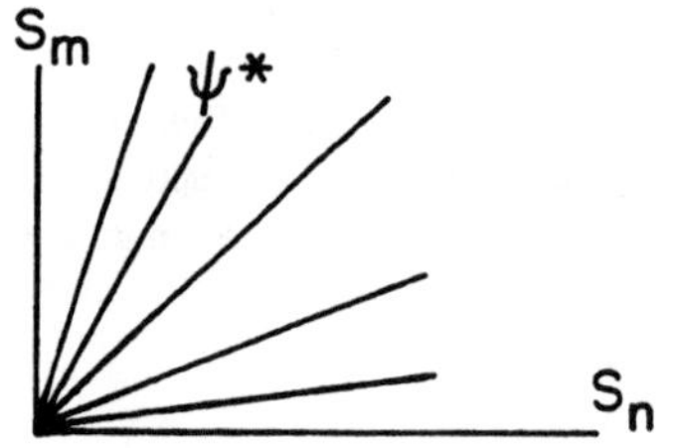

FIGURE 2

The function

$$\psi\left(\frac{s_m a}{s}, \frac{s_n b}{s}\right)$$

represents the average return to all members of S. As the ratios between the types of traders change, the average return fluctuates between the bounds of 0 and ψ^*. If we examine the behavior of the core for the limit games whose size increases in any ratio other than that which obtains at ψ^* there will always be a subset of players, as soon as the game is sufficiently large, who can obtain an average return of ψ^* in a coalition of this optimal size through which any imputation can be dominated and the core of the limit games will be empty.

Two trivial theorems can be established by considering the meaning of increasing and decreasing returns.

DEFINITION. Increasing returns are present in an economy if for at least two coalitions S and S' where $|S'| = |S| + 1$ and $S' \supset S$

$$\frac{v(S')}{|S'|} > \frac{v(S)}{|S|} .$$

Decreasing returns are present if for S and S', as above:

$$\frac{v(S')}{|S'|} < \frac{v(S)}{|S|} .$$

THEOREM 4. In an n-person economy with increasing returns everywhere, a core will always exist for all size n.

PROOF. The imputation

$$\eta = \left(\frac{v(I)}{n}, \frac{v(I)}{n}, \ldots, \frac{v(I)}{n}\right)$$

must lie in the core. As there are increasing returns everywhere, for any set $S \subset I$

$$\frac{v(S)}{|S|} < \frac{v(I)}{n} \quad \text{or} \quad v(S) < \frac{|S|}{n} v(I)$$

hence there is no effective set for η except I. As no imputation exists which dominates η, η must lie in the core.

THEOREM 5. In an n-person economy with decreasing returns for all coalitions with more than s^* members where $s^* < n$, no core exists.

PROOF. Given decreasing returns

$$\frac{v(I-\{j\})}{n-1} > \frac{v(I)}{n} \quad \text{or} \quad v(I\{j\}) > \frac{n-1}{n} v(I) \quad \text{for all } j.$$

Consider any imputation η; it will always have $n - 1$ components such that:

$$\sum_{i=1}^{n} \eta_i - \eta_j \leq \frac{n-1}{n} v(I) .$$

The coalition $v(I - \{j\})$ is effective for η and we can construct a β which dominates η using $(I - \{j\})$ as the effective set.

§3. ECONOMIC INTERPRETATION OF THE EDGEWORTH MARKET GAMES

The connection between the von Neumann and Morgenstern analysis and Edgeworth's and other economic theories of the market can now be made. The solutions to the n-person games reflect the sociological possibilities available through coalition structures. The core of a game consists of those imputations which are unaffected by any type of coalition structure. Following Edgeworth's terminology we may state that if an imputation in the core existed there would be no motivation for any group to recontract. Certain properties of increasing and decreasing returns to scale appear to be important in these games. If decreasing returns to scale are present then the core of the game is empty, i.e., all imputations are dominated by some coalition. In general this states that in an economy with decreasing returns present everywhere, under an optimal allocation of resources no matter how total product is divided there always will be some group in the economy which can form a coalition that is effective against any imputation that is suggested. This suggests that when decreasing returns are present everywhere the division of total product must depend upon the coalition structure, the social or political features of the economy under all circumstances.

In an economy of "optimum size", i.e., one in which a stage of increasing returns to scale has just given place to constant returns, the core approaches a single point at which every individual is awarded his "marginal value" productivity.[3] In this economy there is only one imputation that is not dominated by some coalition of players.

We prove in Theorem 1 that if the number of players in an Edgeworth Market game are the same on both sides of the market, then a set of imputations analogous to the contract curve[4] will be a solution no matter how many players there are in the market. Even if it happens that

[3] It may be argued that this is a trading game and there is no production taking place. This makes no real difference however; the players are able to add to joint product by their willingness to trade. In this sense we can assign them a marginal value productivity.

[4] Edgeworth, F. Y., op. cit., p. 21.

the core of the game converges to a single point as the players become numerous, this solution will not change. This feature emphasizes the difference between the von Neumann and Morgenstern solution and other economic analysis. The cooperative game analysis takes into account all possible coalition structures, thus even though there may be only one undominated imputation in the economy, there will be sets of mutually stalemating imputations which make up possible solutions given the presence of various coalitions. Edgeworth noted that if coalitions were permitted, the length of the contract curve would not shrink.[5] However, there will be many other von Neumann and Morgenstern solutions as well (only the two person game will, in general, have a single solution).

If there are increasing returns present in an economy (see Theorem 4) there will always be a set of imputations in the core. This merely reflects a basic feature of an expanding economy; the value of extra individuals is sufficiently great that there will be imputations such that no subgroup of the economy can make itself better off by refusing to cooperate.

In Theorem 2 it was demonstrated that as the number of players facing a single monopolist increases, the only undominated imputations approach the single imputation at which the monopolist is totally discriminating and obtains all the gain to be had from trade.

The core of a game is determined by technological conditions. At any imputation in the core all players obtain at least as much as they are able to enforce in any coalition. By means of n-person game analysis and an examination of the core we can give meaning to such vague ideas as "a world of monopolies". In any economy, any imputation of wealth which does not lie in the core implies that some group of individuals is profiting at the expense of another group. As the core will be empty if there are decreasing returns (see Theorem 5), in such an economy there will be no a priori "non-discriminatory" division of product.

If the core contains more than one point, then the combination of technological and coalition power is not sufficient to enforce all the possible imputations in a society. There remains an area on which the outcome (as is the case for the two person contract curve) must be determined by other considerations. If there is more than one monopolist in an economy (consider the Edgeworth games [2, n], or if there are three goods and three sets of players M, N and R, [1, 1, r]) and if all the monopolists are essential (i.e., their trade adds value to the economy) the core will not contain a single imputation, but there will be an area of indeterminacy between the shares of the monopolists.

[5] Ibid., pp. 42-43.

The games considered in Theorems 1, 2 and 3 were all highly symmetric. Little is known about the behavior of solutions for this type of trading game as the number of players increases in a non-symmetric manner. The observations about the behavior of the core, however, are quite general and do not depend on symmetry but only upon the properties of the indivisibilities and complementarities present in the economy; i.e., the nature of the super-additivity of the characteristic function.

Martin Shubik

General Electric Company

Econometrica, Vol. 34, No. 4 (October, 1966)

QUASI-CORES IN A MONETARY ECONOMY WITH NONCONVEX PREFERENCES[1]

BY L. S. SHAPLEY AND M. SHUBIK

A model of a pure exchange economy is investigated without the usual assumption of convex preference sets for the participating traders. The concept of *core*, taken from the theory of games, is applied to show that if there are sufficiently many participants, the economy as a whole will possess a solution that is sociologically stable—i.e., that cannot profitably be upset by any coalition of traders.

1. INTRODUCTION

IN HIS STUDY of bilateral monopoly, Edgeworth **[8]** introduced a solution concept, called the *contract curve*, which consists of those Pareto-optimal exchanges that are at least as attractive to each monopolist as the initial, no-trade position. He then went on to observe that if the number of entrepreneurs on each side of the market were increased, then the contract curve, or contract set, should diminish in size and converge eventually to the set of *competitive* allocations, which are those that can be derived from the initial position by direct budgetary optimization on the part of the individual traders under a fixed schedule of prices. This shrinking of the contract set depended on the possibility of *recontracting*, it being postulated that no proposed allocation of goods would be finally acceptable to the market as a whole if there were any subset of traders who, by exchanging goods only among themselves, could all do better.

The recontracting principle corresponds closely to the notion of *domination* that underlies two important solution concepts in n-person cooperative game theory, to wit, the *core* and the von Neumann-Morgenstern *solution*.[2] Indeed, if one regards the expanded Edgeworth model as a multi-person game, then the contract curve, as curtailed by recontracting, is precisely the core. The uncurtailed contract curve, on the other hand, is a solution, though not generally the only one. In the original case of bilateral monopoly, the core, the contract curve, and the unique solution all coincide.

A number of refinements and extensions of Edgeworth's convergence theorem

[1] This research is sponsored by the United States Air Force under Project RAND, Contract No. AF 49(638)-700 monitored by the Directorate of Development Plans, Deputy Chief of Staff, Research and Development, Hq USAF. Views or conclusions contained in this memorandum should not be interpreted as representing the official opinion or policy of the United States Air Force.

[2] The *core* of a game is the set of all undominated outcomes **[11, 1]**. A *solution* is any set of outcomes, mutually undominating, that collectively dominate all others **[21]**. The term "core" was introduced by Gillies and Shapley **[13, 10]** in studying properties of the von Neumann-Morgenstern solutions; the core as an independent solution concept was developed by the latter in lectures at Princeton in the fall of 1953.

have been obtained by Shubik [17], Debreu and Scarf [6, 7], Aumann [2], and Vind [19, 20], who all exploit to some degree the game-theoretic point of view. The preferences of the individual traders are usually assumed to be convex, since, as is well known, without convexity a competitive equilibrium may not exist. The present paper, on the other hand, is concerned especially with markets that possess neither a competitive equilibrium nor a core; indeed, in the presence of either one our chief theorems would be almost trivial to prove.[3]

The general tenor of our results is that nonconvexity in the preference sets is economically unimportant when the number of participating individuals is large. This may not be a new observation, but here we give it a mathematically precise expression in a fresh context.

More specifically, we show under quite general conditions that even if the true core cannot be assured, because of nonconvexity, nevertheless certain quasi-cores will always appear when the number of traders in the market is large enough. These quasi-cores are characterized by the requirement that recontracting traders must show a definite positive profit before being permitted to block a proposed final allocation. A distinction must be made ("strong" vs. "weak" ε-cores) depending on whether the ε-threshold is applied to the whole set of recontracting traders, or to each one individually. The conditions for existence are somewhat different in the two cases (Theorems 2 and 4).

There is reason to believe that the ε-cores, in a double limit process involving both ε and the number of traders, would shrink down à la Edgeworth upon the competitive allocations of a certain related economy, obtained by "concavification" of the original utility functions. In the original market these allocations, though feasible, need not be Pareto optimal. The associated prices, on the other hand, may lead under individual optimization only to infeasible allocations—that is, to mismatched supply and demand. The prices will at least be in *pseudo-equilibrium*, however, in the sense that if the excess demand for any good is positive, then another scheme of individual optimizations can be found that makes it negative, and vice versa. (Nonuniqueness of optimum demand schedules is to be expected when preferences are nonconvex). As yet we have not found sharp conditions for the convergence of the ε-cores, and our mathematical results in this paper deal only with the existence question.

Recently, new attention has been focused on the economic implications of nonconvex preferences [5, 9, 12, 15]. Unfortunately, convexity is too often assumed in economics merely because tools for handling the opposite assumption are lacking. Not the least important feature of the present investigation, we believe, lies in its contribution to the mathematical technology of nonconcave utility functions. We have already mentioned the device of concavification; we also introduce a condition of "spannability" that helps restore some of the regularity

[3] In brief, we have: convex preferences → competitive equilibrium → core → strong ε-core → weak ε-core. The implications are generally not valid in the reverse direction.

that is lost when concavity is abandoned, and which seems potentially useful beyond the immediate context of cores and competitive equilibria.

Several simple examples are discussed in Sections 3 and 11 in order to give the reader a feeling for the effects of nonconvex preferences. In an effort to keep clear the main lines of the argument, some of the drier mathematical demonstrations have been placed in appendices.

2. TWO SPECIAL ASSUMPTIONS: MONEY AND EQUAL TASTES

The mathematical results that we present are to a certain extent illustrative, rather than comprehensive, since they rest upon two fairly drastic assumptions in an otherwise very general setting. This is partly a matter of technical expedience. It is often good policy, especially in the social sciences, to work with simplified models in order to gain insight and discover techniques for tackling more complex situations. But we should emphasize that our present assumptions appear to have no direct bearing on the main subject of this investigation, namely, the effects of nonconvex preferences on the existence of cores.

First, we shall assume that in addition to the actual commodities in exchange there is a kind of "money" available, which can be separated out of individual utility functions in the following way:

$$u^i(x_1^i, ..., x_m^i, \xi^i) \equiv U^i(x_1^i, ..., x_m^i) + \lambda_i \xi^i . \tag{1}$$

Here x_j^i is the amount of the jth commodity held by the ith individual, ξ^i is his "money" level, and λ_i is a constant.[4] We assume that (1) holds to a sufficient degree of approximation for a range of values of ξ^i sufficiently wide to encompass the action of the model. Like the actual commodities, "money" is assumed perfectly divisible and freely transferable; unlike them, is not constrained by an *a priori* lower bound. This entails no loss of realism, since in practice the relation (1) will begin to fail long before the individual faces bankruptcy or a hard and fast credit ceiling.

It is worth remarking that this idealized money, though sometimes called "transferable utility," does not entail interpersonal comparisons of utility at the psychological level. Money serves merely to de-couple individual tastes. If a dollar is taken from millionaire X, then his utility is diminished by *one dollar's worth of X-utils*; if it is given to beggar Y, then the latter's utility is increased by *one dollar's worth of Y-utils*. There is no psychological comparison of "degrees of happiness." The key phrase "one dollar's worth" has meaning because of the identity (1), which is assumed separately for each individual. Admittedly, it is a poor assumption for a beggar, since it denies any income effect. It might be a good assumption

[4] The constant marginal utility for money in (1) is not of itself at issue. We could just as well assume the form $w^i \equiv f_i(U^i + \lambda_i \xi^i)$; with f_i any order-preserving transformation. It would then be merely a matter of relabeling utility levels to go from w^i to u^i.

for a millionaire, within a suitably restricted range. In any event, it is not an interpersonal assumption.[5]

The advantage of the assumption (1), in the present case, is that it permits us to reduce the game model to *characteristic-function* form (see Section 4). This is a powerful analytic and conceptual device first introduced by von Neumann and Morgenstern [**21**]. Whether our present results could be duplicated without its aid is not yet clear.

The second special assumption is that all individuals in the exchange economy have equal tastes. Formally, this means that their utility functions u^i are interrelated by order-preserving transformations f_{ij}:

$$(2) \qquad u^i(x_1, ..., x_m, \xi) \equiv f_{ij}(u^j(x_1, ..., x_m, \xi)), \qquad \text{all } i, j.$$

Since we are free to recalibrate the individual utility scales, this means that all preferences can be expressed by means of a single function u. This kind of assumption is often made in economic modeling.

It will become clear that "equal tastes" does not symmetrize the model to the point of triviality. For one thing, the initial endowments of the traders may be different. Moreover, the solutions that one gets with nonconvex preferences are often nonsymmetric even when the initial holdings are identical—the "economics"

[5] It has long been the fashion to formulate elementary models of exchange on a barter basis. Though somewhat removed from everyday experience, such models are easy to think about and work with, and have an air of abstract generality that is appropriate to the fundamental questions that they deal with. Money is, after all, a very complex phenomenon with many nonelementary ramifications. Build a theory of barter economics first, we are urged, and let "money" appear, if at all, only in the guise of a commodity in exchange.

In game-theoretic models, however, which are usually an order of magnitude more complex than the associated mechanistic models, money paradoxically is often a simplifying factor. The expediency of the "barter-first" approach is therefore called into question. Money, the de-coupler, smooths away many of the difficulties that arise in dealing simultaneously with many independently motivated sources of strategic decision. Both the models and their solutions can be more readily formulated, analyzed, and interpreted if a suitably idealized money is present.

The authors feel that this is not merely a formal theoretical consideration, but that a monetary de-coupler is a very basic feature of human economic behavior in bargaining situations, and that when both goods and information are freely transferable it would be difficult in fact, hence unrealistic in a model, to prohibit the adoption of some such comparator and carrier of value. To treat money as just another commodity, whether this is technically expedient or not, only conceals the unique and remarkable role that it plays in real bargaining processes.

In this view, an elementary model of an exchange economy could quite properly include an idealized "money" like (1) from the start, and leave to later elaborations the important but less than vital questions of transfer restrictions, income effects, individual spending ceilings, credit costs, etc.

A reader unable to accept this view is, of course, free to regard our results as pertaining only to special economies in which one of the commodities, in plentiful supply, happens to enter the utility functions in the manner (1). What we are trying to suggest, however, is that the hypothesis of a "transferable utility" is not only expedient, in this context, but methodologically respectable as well.

of the situation can drive identical traders to behave differently. This effect will be illustrated in the next section.

The advantage of assuming (2), in the present work, is that it allows us to write down explicitly certain optimizing allocations, beginning at (17) below. This simplifies the mathematical arguments in several places, but similar proofs of similar theorems could be carried out, we believe, without this aid.

The combined effect of our two assumptions is to permit us to postulate a single utility function for goods and money of the form

(3) $$u(x_1, \ldots, x_m, \xi) \equiv U(x_1, \ldots, x_m) + \xi,$$

applicable to all traders in the economy.

3. AN EXAMPLE

The following example is intended to give some idea of the role played by money and equal tastes in a market with nonconvex preferences.[6] The model is highly symmetric, having n identical traders and two identical goods, but the competitive and core solutions, when they exist, will require nonsymmetric behavior on the part of the traders.

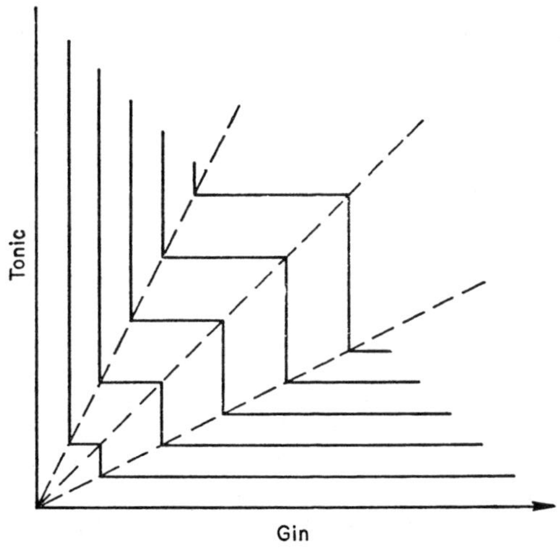

FIGURE 1.
The commodity space

The nonconvexity in the preference sets (see Figure 1) comes from preferred ratios in the consumption process. For a numerical utility function, we may take

(4) $$U(x) = \max[\min(2x_1, x_2), \min(x_1, 2x_2)].$$

[6] This example received brief mention in [7]. Some other simple examples are examined in Section 11 below.

The traders are assumed to start with one unit of each good apiece.

For example, the goods might be gin and tonic. Each trader is indifferent as between weak drinks (1 to 2) and strong drinks (2 to 1), but he will not take both, and rejects intermediate (or more extreme) concoctions.

In looking for a competitive equilibrium, we first observe that *unequal* prices will never work, whether or not there is money available, since the traders would all want to buy the cheaper good and sell the other. With equal prices, however, a competitive allocation can sometimes be reached. In fact, if n is *even*, each man can trade 1/3 units of one good for an equal amount of the other, and end up with an efficient bundle—either (4/3, 2/3) or (2/3, 4/3). The result is worth 4/3 to everyone. Of course, it may require cooperative action or a determined hostess to decide who is to get which drink!

But if n is *odd*, there is no series of exchanges (at equal prices) that gives everyone an efficient ratio, and without money there can be no competitive equilibrium. Rather, we have the pseudo-equilibrium situation described in the Introduction, in which the excess demand for either commodity can be construed as either positive or negative, but not zero.

Introducing money (3) changes the picture somewhat. In the *even* case the competitive prices and final utility levels remain as before, but the allocative possibilities are opened up. Because of the first degree homogeneity of (4), we can let some of the traders liquidate their holdings for cash, provided that others spend an equal amount to increase their consumption and all stick to efficient ratios. In the *odd* case, this same option works to circumvent the previous difficulty in matching supply and demand, and it is easy to find feasible competitive allocations. For example, one man might sell out completely, leaving an even number of traders to divide the goods efficiently. The final payoff, as in the even case, will be exactly 4/3 to each trader. Thus, we see that it is possible for money to overcome the "ill" effects of nonconvexity.

The homogeneity of (4) was crucial to this result, however. Had we assumed diminishing returns to scale, then money would no longer "save" the odd case. To see this, let us modify the utility function to be the square root of the function in (4). Then the equilibrium prices for each good would have to be (as it turns out) precisely $\sqrt{1/12}$. What is important is not this number, but the fact that each trader would have only two ways to optimize, neither way involving any net money transfer. In short, money would be irrelevant and there would be no competitive solution.

What of the cores? Returning to the original function (4), and assuming $n=2$, one can easily determine the Edgeworth contract curve for the no-money case. It proves to be a rather spectacular "curve": four triangular regions arranged in a ring. The core, its image in the utility space, is the bent line *QPR* shown in Figure 2. It includes the competitive payoff *P*, and is included in the set of Pareto-optimal payoffs *SPT*. The introduction of money is not easily shown in the commodity space because of the added dimension, but in the utility space the effect is simply to move the Pareto set out to the straight line with slope −1 passing through the point P, of which the segment $Q'PR'$ constitutes the core.

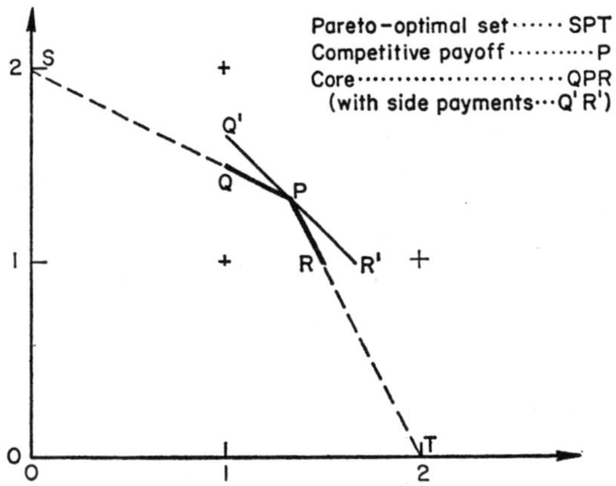

FIGURE 2.
The utility space

For $n>2$ the recontracting principle takes over with a vengeance: the shrinkage is immediate and total. The cores actually contain no points other than the competitive payoffs. That is, when n is odd and side payments are not permitted, there is no core at all; in all other cases the core is a single point. This drastic curtailment of the core is rather atypical and may be ascribed to the simple polyhedral form chosen for the indifference map.

These assertions about the cores are not meant to be obvious. The proofs, given in Appendix 1, may be of interest to the reader wishing experience with the techniques of core analysis.[7]

4. GAMES AND CORES

The *characteristic function* of a game **[21, 14]** is designed to express the optimum result or results obtainable by each coalition S of players, regardless of the actions of the players outside S. In the transferable-utility case,[8] it is a real-valued set

[7] The present example provides no instance of a core without a competitive equilibrium; such cases, however, are easily constructed.

[8] Without transferable utility, the characteristic function is *set-valued* and can be defined in two ways: $v(S)$ represents either (α) the set of payoff vectors that S can surely achieve, or (β) the set of payoff vectors that S cannot be prevented from achieving. The two theories that emerge are not generally equivalent because there is no minimax theorem for coalitions in the absence of transferable utility. (See Aumann and Peleg **[4]**.)

function $v(S)$, arbitrary except for the condition $v(O)=0$ and for the property of *superadditivity*:

(5) $$v(S \cup T) \geqslant v(S)+v(T), \qquad \text{if } S \cap T=O .$$

The outcome of the game is conveniently summarized in a *payoff vector*, α, with components measured in utility units; clearly, we must have

(6) $$\sum_N \alpha_i \leqslant v(N) ,$$

where N denotes the set of all players. A payoff vector is called an *imputation* if it satisfies two further requirements:

(7) $$\sum_N \alpha_i \geqslant v(N)$$

and

(8) $$\alpha_i \geqslant v(\{i\}) \qquad \text{for all } i \in N.$$

These correspond respectively to *Pareto optimality* and *individual rationality.* Conditions (6), (7), and (8) are always consistent because (5) implies that $v(N) \geqslant \Sigma v(\{i\})$.

The more stringent requirement of *group rationality*:

(9) $$\sum_S \alpha_i \geqslant v(S) \qquad \text{for all } S \subset N ,$$

includes both (7) and (8). The set of payoff vectors thereby delimited, if any, is called the *core*. Since (6) and (9) may well be inconsistent, however, the core need not exist.

Speaking informally and intuitively, a coreless game ought to be more competitive and harder to stabilize than one with a core. Indeed, in any game, an observed outcome falling outside the core would seem to admit just two interpretations: (a) it is a *transient event* in some dynamic process, or (b) it is evidence of a *social structure* among the players that inhibits some coalitions from developing their full potential. On the other hand, an observed outcome *in* the core tells us nothing about the organization of society; the core is sociologically neutral **[16]**.

The mathematical results in this paper depend on the device of enlarging the core by a small amount. Two related concepts will be needed. If ε is a small positive number, we define the *strong ε-core* as the set of payoff vectors α satisfying

(10) $$\sum_S \alpha_i \geqslant v(S)-\varepsilon, \qquad \text{for all } S \subset N .$$

We define the *weak ε-core* as the set of payoff vectors satisfying

(11) $$\sum_S \alpha_i \geqslant v(S)-s\varepsilon, \qquad \text{for all } S \subset N ,$$

where s denotes the number of elements of S. It is easy to verify the following chain of set-inclusions:

$$\text{weak } \varepsilon\text{-core} \supset \text{strong } \varepsilon\text{-core} \supset \text{weak } \frac{\varepsilon}{n}\text{-core} \supset \ldots \supset \text{core},$$

where n is the number of players in the game.

These quasi-cores are not merely technical devices, looking toward an eventual convergence theorem. They provide a way, for example, of taking into account the costs of coalition-forming. Under the weak definition, the costs would depend on the size of the coalition; in the strong case there would be a fixed charge. Alternatively, we might regard the organization costs as negligible, or already included in $v(s)$, but view the ε or $s\varepsilon$ as a threshold, below which the blocking maneuver implicit in (9)—the actual exercise of "group rationality"—is not considered worth the trouble.

5. THE MARKET MODEL

Let there be m different commodities and t_0 different *types* of traders, distinguished by the stocks of goods they hold at the beginning of the trading session. The initial endowment of a player of type "t" will be denoted by a vector

$$a^t = (a_1^t, \ldots, a_m^t).$$

If S is a set of players, and if s_t denotes the number of players in S of type "t," then the aggregate initial endowment of S may be written as follows:

$$a(\mathbf{s}) = \sum_{t=1}^{t_0} s_t a^t,$$

where $\mathbf{s}$, called the *profile* of S, is an abbreviation for the integer vector $(s_1, \ldots, s_{t_0})$. The total supply of goods in the game is then $a(\mathbf{n})$, where $\mathbf{n}$ is the profile of the set N of all players.

At the conclusion of trading, the players hold bundles x^i, $i \in N$, which must account for the total quantities initially present in the market. Thus we have

$$(12) \qquad x^i \in E_+^m, \quad \text{and} \quad \sum_N x^i = a(\mathbf{n}),$$

where E_+^m denotes the closed positive orthant of Euclidean m-space. Subject to these constraints, all final allocations are assumed possible. (In particular, the outcome is not assumed to be symmetric as between players of the same type.) In addition, there may be direct transfers of money among the players. Thus, if U is the common utility measure for goods (see (3)), the possible final *payoffs* will take the form

$$\alpha_i = U(x^i) - \pi_i, \qquad \text{all } i \in N,$$

subject to (12) and $\Sigma_N \pi_i = 0$.

By symmetry, the characteristic function depends only on the profiles of the coalitions. Since internal money transfers will cancel, we have

$$(13) \qquad v_U(\mathbf{s}) = \sup_{y^\nu} \sum_{\nu=1}^{s} U(y^\nu), \quad \text{subject to} \begin{cases} \text{all } y^\nu \in E_+^m, \\ \Sigma y^\nu = a(\mathbf{s}), \end{cases}$$

where s denotes the sum of the s_t. Players outside the coalition do not affect this value, since they can neither force nor be forced into dealings with the coalition members.[9]

To ensure that the "sup" in (13) is finite, we shall assume from the outset that U is bounded above on compact subsets of E_+^m or, equivalently, that there is a continuous function K such that $U(x) \leqslant K(x)$ for all $x \in E_+^m$. (Compare condition (15) below.)

Let $\Gamma_U(\mathbf{n})$ designate the game we have just defined. The subscript "U" will serve to distinguish it from an auxiliary game to be introduced in Section 7.

6. THE COMPETITIVE SOLUTION

An allocation $\{x^{*i}\}$ satisfying (12) is called *competitive* if there exists a price vector $p = (p_1, \ldots, p_m)$ such that, for each individual $i \in N$, the bundle x^{*i} maximizes the expression

$$(14) \qquad U(x^i) - p \cdot (x^i - a^i) .$$

The numbers p_j are called *equilibrium prices*. Observe that there are no "budget" conditions $p \cdot (x^i - a^i) = 0$, since our money enters directly into the complete utility function (3). Had we approached this definition via a budgeted money-of-account (taking our present "valuable" money to be the $m+1$ commodity), then the usual definition would give us equilibrium prices $p'_1, \ldots, p'_{m+1}$, related to the above by $p_j = p'_j / p'_{m+1}$. Thus, the actual numbers p_j are significant, not just their ratios, as is evident from the form of (14).

Let α_i^* denote the maximum of (14) for a given equilibrium price vector p^*. Then it is a simple matter to show that the vector α^*, which we shall refer to as a *competitive imputation*, is in the core of the game $\Gamma_U(\mathbf{n})$. Indeed, we have at once

$$\sum_N \alpha_i^* = \sum_N U(x^{*i}) \leqslant v_U(\mathbf{n}) ,$$

verifying the feasibility requirement (6). To verify the core inequalities (9), we fix $S \subset N$ and $\varepsilon > 0$, and, using (13), find bundles $y^i \in E_+^m$ such that

$$\sum_S U(y^i) \geqslant v_U(\mathbf{s}) - \varepsilon \quad \text{and} \quad \sum_S y^i = a(\mathbf{s}) .$$

[9] Hence, questions regarding the credibility and proper evaluation of threats do not arise in this case, though they play an important role in the general theory of cooperative games.

Since α^* maximizes, we have

$$\alpha_i^* \geqslant U(y^i) - p^* \cdot (y^i - a^i) \qquad \text{for each } i \in S.$$

Summing, we obtain

$$\sum_S \alpha_i^* \geqslant \sum_S U(y^i) \geqslant v_U(\mathbf{s}) - \varepsilon \, .$$

Since this is valid for arbitrarily small ε, (9) follows, completing the proof.

THEOREM 1: *Every competitive imputation is in the core.*

We remark that this result does not depend on equal tastes, since in the proof just given each $U(\cdot)$ could be replaced by $U^i(\cdot)$. We should also emphasize that nothing has been said about the *existence* of a competitive solution.

7. CONCAVIFICATION OF THE UTILITY FUNCTION

For our theorem on weak ε-cores we shall impose almost no conditions on the function U. It need not be concave, continuous, or monotonic, and it may be either bounded or unbounded. We shall require, however, that its asymptotic growth be no more than linear, and that it be bounded from below on all compact subsets of E^m_+. To this end, we assume the existence of a *linear* function L_0 and a *continuous* function K_0 such that the inequalities

$$(15) \qquad K_0(x) \leqslant U(x) \leqslant L_0(x)$$

hold for all x in the commodity space E^m_+.

Let us now define a function C on E^m_+ as follows:

$$(16) \qquad C(x) = \sup \sum_{h=1}^{m+1} \lambda_h U(y^h), \qquad \text{subject to} \begin{cases} \lambda_h \geqslant 0, & \Sigma \lambda_h \;\; = 1, \\ y^h \in E^m_+, & \Sigma \lambda_h y^h = x. \end{cases}$$

The finiteness of this "sup" is ensured by (15); indeed, we have $C(x) \leqslant L_0(x)$. The function C is concave, it majorizes U, and it is the least such function.[10] We may remark that C is continuous at every interior point x of E^m_+ and possesses a linear support there, i.e., a linear function L such that $L \geqslant C$ and $L(x) = C(x)$. If the "sup" in (16) is actually achieved for all x, we shall say that U is *spannable* (see Section 10).

We intend to use C as an artificial utility function in defining a *concave majorant game* $\Gamma_C(\mathbf{n})$, identical in every other respect to the game $\Gamma_U(\mathbf{n})$ previously defined. The characteristic function v_C has a simple explicit form. Since players have identical, concave utilities (hence *convex* preferences), a coalition achieves maximum profit by dividing its total endownment equally among its members, and we have

[10] The use of $m + 1$ spanning points is sufficient to "concavify" any linearly bounded function on E^m. Use of a larger number in (16) would not affect the definition of C.

(17) $$v_C(\mathbf{s}) = sC\left(\frac{a(\mathbf{s})}{s}\right).$$

We see that $v_C(\mathbf{s})$, like $a(\mathbf{s})$, is homogeneous of degree one:

(18) $$v_C(k\mathbf{s}) = kv_C(\mathbf{s}), \qquad (k=0, 1, 2, ...,)$$

indicating constant returns to scale to a uniformly expanding coalition in the artificial game $\Gamma_C(\mathbf{n})$. For the true game $\Gamma_U(\mathbf{n})$, on the other hand, we do not have homogeneity in general, but only the inequality

(19) $$v_U(k\mathbf{s}) \geqslant kv_U(\mathbf{s}), \qquad (k=0, 1, 2, ...,)$$

a consequence of superadditivity (5). Our results on the existence of ε-cores will hinge on showing that $v_U(\mathbf{s})$ is nevertheless "almost" homogeneous, in a sense made specific in the lemmas accompanying Theorems 2 and 4.

8. EXISTENCE OF THE WEAK ε-CORE

We consider exchange economies based on utility functions U satisfying (15).

THEOREM 2: *For every profile* $\mathbf{n} = (n_1, ..., n_{t_0})$, *and for every* $\varepsilon > 0$, *there exists a constant* k_0 *such that all games* $\Gamma_U(k\mathbf{n})$ *with* $k \geqslant k_0$ *possess weak* ε*-cores.*

LEMMA ON WEAK ε-HOMOGENEITY OF v_U: *For every profile* $\mathbf{s}$ *and for every* $\varepsilon > 0$, *there exists a constant* $k_0 = k_0(\mathbf{s}, \varepsilon)$ *such that*

(20) $$v_C(\mathbf{s}) - \varepsilon \leqslant \frac{1}{k} v_U(k\mathbf{s}) \leqslant v_C(\mathbf{s})$$

holds for all $k \geqslant k_0$.

Proof: Fix $\mathbf{s}$ and ε, and let $x^* = a(\mathbf{s})/s$. Then, by (17),

$$v_C(\mathbf{s}) = sC(x^*).$$

Using (16), find a convex representation $x^* = \Sigma \lambda_h y^h$ such that

(21) $$C(x^*) \leqslant \sum_{h=1}^{m+1} \lambda_h U(y^h) + \frac{\varepsilon}{2s}.$$

Let l_h denote the greatest integer in $ks\lambda_h$, $h = 1, ..., m+1$. Then

(22) $$\Sigma l_h y^h \leqslant ks\, \Sigma \lambda_h y^h = ksx^* = a(k\mathbf{s}).$$

Thus, in a coalition with profile $k\mathbf{s}$ it is possible to assign the bundle y^1 to the first l_1 players, y^2 to the next l_2 players, and so on. If "$<$" holds in (22), there will be goods left over after this allotment, but there will also be at least one player left over, too. Allotting the excess goods equally among the extra players gives them each an allocation y^* that lies within the convex hull of the $\{y^h\}$. Since there are at most m extra players, we can write down an upper bound for the amount that such an allotment of excess might cause to be deducted from the total coalition utility, namely:

$$B = m \cdot |\min_{y^*} K_0(y^*)|, \qquad \text{subject to } y^* \in \text{convex hull of } \{y^h\}.$$

(This is the only use of the function K_0 postulated in (15).) The important fact about this bound is that it is independent of k.

We have thus described a feasible allocation, whose value to the coalition is at least $\Sigma l_h U(y^h) - B$. Thus

$$v_U(k\mathbf{s}) \geqslant \Sigma l_h U(y^h) - B .$$

Applying (21) and the definitions of l_h and x^*, we obtain

$$\begin{aligned} v_U(k\mathbf{s}) &\geqslant ks \;\Sigma\lambda_h U(y^h) + \Sigma(l_h - ks\lambda_h)U(y^h) - B \\ &\geqslant ks\; C(x^*) - \frac{k\varepsilon}{2} + \Sigma(l_h - ks\lambda_h)U(y^h) - B \\ &\geqslant kv_C(\mathbf{s}) - \frac{k\varepsilon}{2} - \Sigma\,|\,U(y^h)\,| - B \\ &= kv_C(\mathbf{s}) - \frac{k\varepsilon}{2} - \frac{k_0\,\varepsilon}{2}, \end{aligned}$$

where

$$k_0 = \frac{2}{\varepsilon}(\Sigma\,|\,U(y^h)\,| + B) \geqslant 0 .$$

Then $k \geqslant k_0$ implies that

$$v_U(k\mathbf{s}) \geqslant kv_C(\mathbf{s}) - k\varepsilon ,$$

giving us one side of (20). The other side is a consequence of (18) and the general inequality $v_U \leqslant v_C$. This completes the proof of the lemma.

Proof of Theorem 2: Let α be the payoff vector associated with a competitive allocation of the concave game $\Gamma_C(\mathbf{n})$. For example, let $x^* = a(\mathbf{n})/n$ and take

$$\alpha_i = C(x^*) - p^* \cdot (x^* - a^i), \qquad \text{all } i \in N ,$$

where p^* is the gradient of any linear support function $L^* \geqslant C$ with $L^*(x^*) = C(x^*)$.[11] By Theorem 1, α is in the core of $\Gamma_C(\mathbf{n})$. Moreover, for every k, the k-fold replication of α is the payoff vector of a competitive allocation of the larger game $\Gamma_C(k\mathbf{n})$ and lies in *its* core. Denote this k-fold replication (a vector with kn components) by $\alpha^{(k)}$. We shall now construct a nearby imputation $\beta^{(k)}$ of the game $\Gamma_U(k\mathbf{n})$.

Denote the difference $v_C(k\mathbf{n}) - v_U(k\mathbf{n})$ by g; clearly $g \geqslant 0$. Choose an arbitrary

[11] Regardless of equal tastes, it is a simple matter to find the competitive solution(s) of an exchange economy with money when utilities are concave.

n-vector γ whose components sum to g/k and satisfy

$$0 \leqslant \gamma_i \leqslant \alpha_i - v_U(\mathbf{i}),$$

where $\mathbf{i}$ denotes the profile of the 1-player set $\{i\}$. This is possible because of the two inequalities:

$$g/k \leqslant v_C(\mathbf{n}) - v_U(\mathbf{n}) \leqslant \Sigma_i(\alpha_i - v_U(\mathbf{i}))\,,$$

and

$$0 \leqslant v_C(\mathbf{i}) - v_U(\mathbf{i}) \leqslant \alpha_i - v_U(\mathbf{i})\,.$$

(The first follows from (18), (19), (7) applied to $\Gamma_C(\mathbf{n})$, and the superadditivity of v_U; the second is a consequence of (8) applied to $\Gamma_C(\mathbf{n})$.) Let $\gamma^{(k)}$ be the k-fold replication of γ and define

(23) $$\beta^{(k)} = \alpha^{(k)} - \gamma^{(k)}\,.$$

This is the desired imputation of $\Gamma_U(k\mathbf{n})$. We wish to show that it lies in the weak ε-core, in the sense of (11), whenever k is greater than the constant $k_0 = k_0(\mathbf{n}, \varepsilon)$ provided by the lemma.

Consider, therefore, an arbitrary subset S of the set of all kn players. Note first that

(24) $$\sum_S \alpha_i^{(k)} \geqslant v_C(\mathbf{s}) \geqslant v_U(\mathbf{s})\,,$$

since $\alpha^{(k)}$ is in the core of $\Gamma_C(k\mathbf{n})$ and $C \geqslant U$. Also we have

(25) $$\sum_S \gamma_i^{(k)} \leqslant s\,\frac{g}{k} = s v_C(\mathbf{n}) - \frac{s}{k} v_U(k\mathbf{n}) \leqslant s\varepsilon\,,$$

since each $\gamma_i \leqslant g/k$ and $k \geqslant k_0(\mathbf{n}, \varepsilon)$. Combining (23), (24), and (25) now gives the desired result:

(26) $$\sum_S \beta_i^{(k)} \geqslant v_U(\mathbf{s}) - s\varepsilon\,.$$

This completes the proof of Theorem 2.

We may observe that $\gamma \to 0$ as $\varepsilon \to 0$, so that the weak ε-cores can be said in a certain sense to possess at least one limit point—namely, the imputation α replicated an infinite number of times. Let us state this more precisely. Given any competitive payoff α of the concave game $\Gamma_C(\mathbf{n})$, and any $\delta > 0$, then for all sufficiently small $\varepsilon > 0$ and for all $k \geqslant k_0(\mathbf{n}, \varepsilon)$ there is an n-vector at a distance less than δ from α, whose k-fold replication is in the weak ε-core of $\Gamma_U(k\mathbf{n})$.

It further appears, in analogy with results of Debreu and Scarf [7], that the above does not hold when α is not a competitive payoff of $\Gamma_C(\mathbf{n})$. This would mean that the weak ε-cores converge (in the above sense) to exactly the set of competitive payoffs of the concavified game. But we have not carried through a proof of this convergence, and it is possible that some regularity assumptions on U might be required.

9. FURTHER CONDITIONS ON U

The extremely weak condition (15) that we have so far imposed upon the utility function will need reinforcement before a result analogous to Theorem 2 can be obtained for strong ε-cores; we shall require "spannability" and a certain amount of differentiability for U. Our primary purpose, of course, is the study of the effects of nonconcavity, not the wholesale abandonment of regularity assumptions of all kinds. Our policy of keeping the hypotheses as general as possible, however, serves to make clear precisely what our results do and do not depend upon.

We shall postulate, then, that U is *radially differentiable* and *spannable*. The first assumption means that U is differentiable along all rays in E_+^m emanating from the origin.[12] This is considerably weaker (if $m>1$) than assuming the existence of first partial derivatives $\partial U/\partial x_i$.[13] We do not demand that the radial derivatives be continuous, or, indeed, that U itself be continuous.

The second assumption means (see Section 7) that there exists a concave function $C \geqslant U$ such that for each $x \in E_+^m$, there are $m+1$ (or fewer) points $y^h \in E_+^m$, and weights $\lambda_h \geqslant 0$, such that

$$\Sigma\lambda_h = 1, \quad \Sigma\lambda_h y^h = x, \quad \text{and} \quad \Sigma\lambda_h U(y^h) = C(x) .$$

This clearly implies that U is bounded above by a linear function, as previously assumed (right side of (15)). The matching assumption that U is bounded below by a continuous function (left side of (15)) is not implied, and will in fact not be needed.

10. DISCUSSION OF SPANNABILITY

The notion of spannability may be unfamiliar to many readers, but it appears to be quite fundamental to any investigation of the relaxation of convexity conditions. Hence a short digression is in order, linking this notion to other analytic conditions. Some simple examples will follow in the next section.

Let U be called *sublinear* if for every linear function L with positive coefficients, the difference $U-L$ has a finite upper bound. For example, logarithmic (Bernoullian) utility is sublinear; also, any bounded utility function is sublinear.

THEOREM 3: *If U is continuous, sublinear, and strictly increasing, then U is spannable.*

This theorem (proved in Appendix 2) is not at all sharp. For example, it is apparent from the proof that "strictly increasing" is hardly necessary; a certain very weak *insatiability* condition would suffice instead, as follows: For each $x \in E_+^m$ and each $j=1, 2, \ldots, m$, there is a $y \in E_+^m$ differing from x only in the jth component such that $y_j > x_j$ and $U(y) > U(x)$.

[12] We shall actually need the existence of radial derivatives only at points where U and C are equal.

[13] Figure 1 provides an example.

Another extension of Theorem 3 results from observing that if U is spannable and L is linear, then $U+L$ is spannable, despite the fact that adding L can destroy both sublinearity and monotonicity (or insatiability).

11. SOME EXAMPLES

Let us now consider a very simple example, having just one commodity and one type of player, in order to show how spannability and differentiability will be crucial to the existence of strong ε-cores in the limit. Let $U(x)=[x/2]$, i.e., the

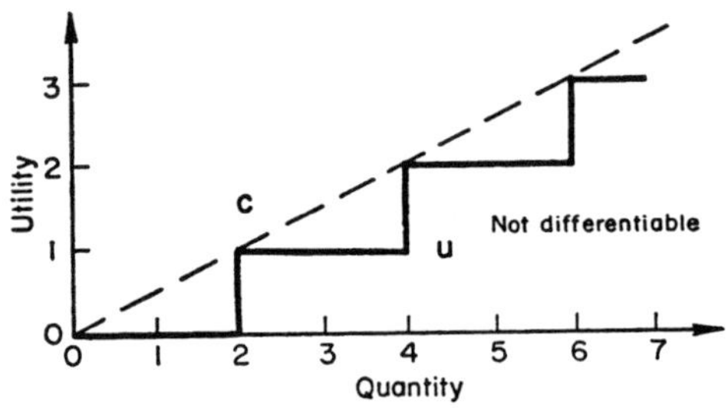

FIGURE 3

greatest integer less than or equal to $x/2$, and let all the initial endowments be 1 unit. Here $C(x)\equiv x/2$, and U is spannable, but not (radially) differentiable. (See Figure 3.) For a coalition with s members we have

$$v_U(s)=[s/2] \quad \text{and} \quad v_C(s)=s/2.$$

If the game happens to have an *odd* number of players, say $n=2r+1$, then in any Pareto-optimal payoff vector the most-favored player receives at least r/n units. The $2r$ least-favored players therefore receive at most $r-r/n$. This must be compared with the amount r that they can obtain in coalition. The difference, r/n, converges not to 0, but to 1/2 as $r\to\infty$. Hence strong ε-cores do not exist for large, odd n and small ε.

Now let us change U to spoil spannability, but at the same time make U differentiable (Figure 4).[14] One can then verify that

$$(27) \qquad v_U(s)=U(s)=\left[\frac{s-1}{2}\right], \qquad (s=1, 2, \ldots,) .$$

In other words, a coalition can do no better than allot all its goods to one player.[15] If there happens to be an *even* number of players in this game, say $n=2r$, then the

[14] The precise form of the curved parts of U within the small squares is immaterial.

[15] Proof: If the coalition allots in integer units, U might as well be the lower step-function in the figure (dotted lines). If fractional shares are used, the result can be no better than discarding one unit and then using the upper step-function. In either case the value (27) is best possible.

least-favored set of $n-1$ players will always get $(r-1)-(r-1)/n$ or less, compared with the amount $r-1$ they can obtain in coalition. Again, the difference goes to 1/2 as $r \to \infty$, and strong ε-cores do not exist for large, even n and small ε.

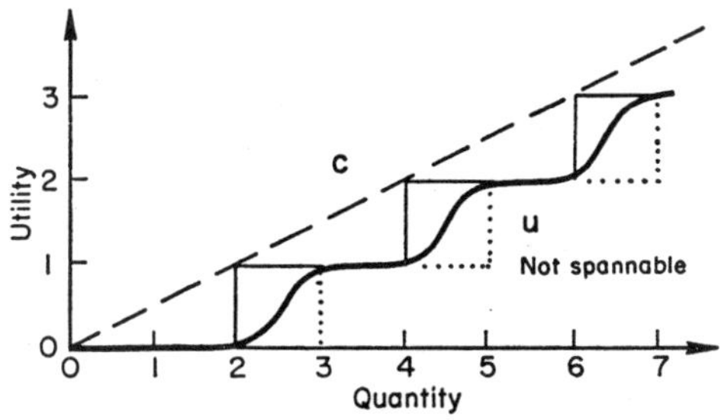

FIGURE 4

Finally, let us restore spannability, as in Figure 5, taking care to make U differentiable at the points of contact with C. A coalition with $n=2r+1$ members will now be able to make an allotment consisting of $2+1/r$ units to r of its members and nothing to the other $r+1$ members, and thereby receive a total utility of

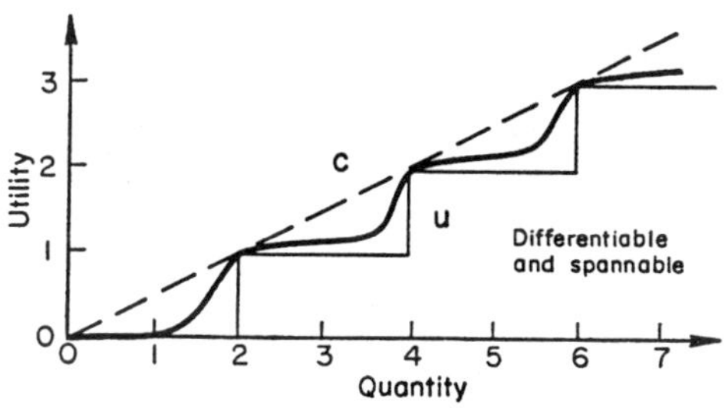

FIGURE 5

$r(1+1/2r-O(1/r^2))$. Thus we have $v_U(n)=n/2-O(1/n)$. This is also valid for even n, trivially. For any n, then, the imputation that gives equal shares to all will assign to each s-player set an amount $(s/n)v_U(n)=s/2-O(1/n)$. This must be compared with the amount, at most $s/2$, that they can obtain in coalition. In this case the difference *does* go to zero, and we have strong ε-cores in the limit as $n \to \infty$, for arbitrarily small ε.

12. EXISTENCE OF THE STRONG ε-CORE

We now consider exchange economies based on a utility function U that is radially differentiable and spannable.

THEOREM 4: *For every profile* $\mathbf{n}$ *and for every* $\varepsilon>0$, *there is a constant* k_0 *such that every game* $\Gamma_U(k\mathbf{n})$ *with* $k \geqslant k_0$ *has a strong* ε-*core.*

LEMMA ON STRONG ε-HOMOGENEITY OF v_U: *For every profile* **s** *and for every* $\varepsilon>0$ *there exists a constant* k_0 *such that*

$$(28)\qquad v_C(\mathbf{s}) - \frac{\varepsilon}{k} \leqslant \frac{1}{k}\, v_U(k\mathbf{s}) \leqslant v_C(\mathbf{s})$$

holds for all $k \geqslant k_0$ *(compare* (20)).

Proof: Fix **s** and ε and let $x^* = a(\mathbf{s})/s$. Then

$$(29)\qquad v_C(\mathbf{s}) = sC(x^*)\,.$$

Using the spannability of U, we can find a convex representation $x^* = \Sigma\lambda_h y^h$ such that $C(x^*) = \Sigma\lambda_h U(y^h)$. Given k, we wish to "move" the points y^h slightly to make the coefficients λ_h integer multiples of $1/ks$. The technical details of this maneuver have been relegated to Appendix 3; the result is a new convex representation $x^* = \Sigma\mu_h z^h$ for each k, with the property that for each h, $ks\mu_h$ is an integer, and

$$(30)\qquad ks\,\big|\,U(z^h) - L^*(z^h)\,\big| \to 0 \text{ as } k \to \infty\,,$$

where L^* is any linear function supporting C at x^*. (This result makes essential use of the radial differentiability of U.)

The new representation is a feasible allocation for any coalition with profile $k\mathbf{s}$, and we therefore have $v_U(k\mathbf{s}) \geqslant \Sigma ks\mu_h U(z^h)$. Using (30), we have for sufficiently large k, $v_U(k\mathbf{s}) \geqslant ks\ \Sigma\mu_h L^*(z^h) - \varepsilon$. The right side is equal to $ksL^*(x^*) - \varepsilon$, by linearity. This in turn is equal to $ksC(x^*) - \varepsilon$. Applying (29) gives the desired result: $v_U(k\mathbf{s}) \geqslant kv_C(\mathbf{s}) - \varepsilon$. The other inequality in (28) is immediate by (18). This completes the proof of the lemma.

The proof of the theorem itself proceeds exactly like that of Theorem 2 in Section 8 but uses the more powerful lemma that we just have established. Note that in order to ensure that the constructed imputation $\beta^{(k)}$ is in the strong ε-core, we must use "ε/n" for "ε" in applying the lemma. The last line of the proof (compare (26) in Section 8) will then read as follows:

$$\sum_S \beta_i^{(k)} \geqslant v_U(\mathbf{s}) - \frac{s\varepsilon}{kn} \geqslant v_U(\mathbf{s}) - \varepsilon\,.$$

The remarks at the end of Section 8, concerning the limiting behavior of weak ε-cores, apply equally to the present case.

13. CONCLUSION

The core is a more general economic concept than the competitive allocation in that it is possible to have markets with a core but no competitive equilibrium, but not conversely. This can happen whether the absence of a competitive equilibrium is due to nonconvex preferences or to other factors not considered in this paper, such as nonconvex production sets or external production economies. Although no pure price mechanism will clear the market, there may nevertheless exist an imputation of wealth—a core payoff—that is stable against both individual and joint action by the participants. Achieving this imputation may involve taxes, transfer payments, or multiple prices.

There are also cases of interest where the core itself is vacuous. This means, in terms of Edgeworth's mechanism, that there will always be a group wishing to recontract. To resolve the inherent instability of this situation, we must resort to introducing social, cultural, or institutional restraints. The theory of games offers several solution concepts that provide frameworks for the systematic introduction of these factors; among the more promising are the von Neumann-Morgenstern "solutions" [**21**], the Luce "ψ-stable pairs" [**14**], and the Aumann-Maschler "bargaining sets" [**3**].

Another tool that seems potentially useful in this connection is the quasi-core, as developed in the present paper. It can easily be shown that ε-cores (both weak and strong) always exist if ε is large enough. The sociological factor involved here can be interpreted as an organizational cost prerequisite to cooperative action, proportional to the parameter ε. Theorems 2 and 4 of this paper indicate that even if ε is small, the quasi-cores will exist when the market is large enough. Of course, when ε-cores exist for small values of ε, it is not unlikely that the core itself exists as well, making the market fully stable against recontracting. But even without a true core, the profit to be gained from recontracting out of an ε-core would be small and a near stability can be achieved.

By passing to the limit with these quasi-cores, or by direct concavification of the finite model, one can hopefully define a price structure to take the place of the missing competitive equilibrium. These pseudo-equilibria might repay further study. Perhaps (for example) one could find cases where they are the limits of convergent tâtonnement processes [**18**] when no true equilibrium exists.

The RAND Corporation
and
Yale University

APPENDIX 1

Determination of Cores in the Example of Section 3

Generalizing (4), let the utility function be given by

$$U(x) = \max[\min(ax_1, x_2), \quad \min(x_1, ax_2)], \tag{31}$$

where $a > 1$. Set $A = 2a/(1 + a)$, and note that $a > A > 1$.

Suppose first that money side payments are permitted. Then it is easily verified that any set of s

players, $2 \leqslant s \leqslant n$, can achieve a combined payoff of sA and no more. Thus, the payoff vector P that assigns A to each player is undominated and lies in the core. Any other payoff vector will assign *less* than $(n-1)A$ to the $n-1$ least-favored players; hence, if $n \geqslant 3$, that set of players can block it. Thus the core consists of the single point P. When $n = 2$, however, blocking only occurs if one player is assigned less than 1, or if the two together get less than $2A$. Thus, the core is the line segment joining the points $(1, 2A-1)$ and $(2A-1, 1)$ (Q' and R' in Fig. 2).

Next, let side payments be prohibited, and let $\alpha = (\alpha_1, ..., \alpha_n)$ be an undominated payoff vector. We may assume that

$$(32) \qquad \alpha_1 \leqslant \alpha_2 \leqslant ... \leqslant \alpha_n .$$

Now, the coalition $\{1, 2\}$ can divide its assets as follows:

$$(t, t/a) \text{ to } 1, \quad (2-t, 2-t/a) \text{ to } 2 ,$$

and thereby achieve the payoffs $\beta_1 = t$, $\beta_2 = 2 - t/a$ for any t between 0 and A (i.e., any point on the line from S to P in Fig. 2). To avoid domination of α by such β, we must have

$$(33) \qquad \alpha_1 \geqslant 2a - a\alpha_2 .$$

Combined with (32), this entails

$$(34) \qquad \alpha_2 \geqslant \frac{2a}{a+1} = A .$$

We also have, of course,

$$(35) \qquad \alpha_1 \geqslant 1 .$$

Let $\{(x_1^i, x_2^i)\}$ be an allocation that yields α. Let p be the number of indices $i > 1$ such that $x_1^i \leqslant x_2^i$, and let $q = n - 1 - p$. We note from (31) that $x_1^i \leqslant x_2^i$ implies that $\alpha_i = \min(ax_1^i, x_2^i)$, and that $x_1^i > x_2^i$ implies that $\alpha_i = \min(x_1^i, ax_2^i)$. Hence

$$p\alpha_2/a + q\alpha_2 \leqslant \sum_2^n x_1^i = n - x_1^1 ,$$

by (32). Hence

$$x_1^1 \leqslant n - (p/a + q)\alpha_2 = B_1 .$$

Similarly

$$x_2^1 \leqslant n - (p + q/a)\alpha_2 = B_2 .$$

Without loss of generality, we can assume that $p \geqslant q$. Then $B_1 \geqslant B_2$, and we have

$$(36) \qquad \alpha_1 \leqslant \min(B_1, aB_2) .$$

Case (i): Suppose $p = q$. Then $n = 2p + 1$, and

$$(37) \qquad 1 \leqslant \alpha_1 \leqslant B_1 = (2p+1) - p\left(\frac{1}{a} + 1\right)\alpha_2 \qquad \text{(by (35) and (36))}$$

$$\leqslant 2p + 1 - p\frac{a+1}{a}A \qquad \text{(by (34))}$$

$$= 1.$$

Hence there is equality throughout (37), and $\alpha_1 = 1$ and $\alpha_2 = A$. But this violates (33). Hence we cannot have $p = q$.

Case (ii): Suppose $p \geqslant q+2$. Then

$$\begin{aligned} 1 \leqslant \alpha_1 \leqslant aB_2 &= (p+q+1)a - (pa+q)\alpha_2 \qquad \text{(by (35) and (36))} \\ &\leqslant a + p(a-aA) + q(a-A) \qquad \text{(by (34))} \\ &\leqslant a + p(a-aA) + (p-2)(a-A) = 2A - a \\ &= 1 - \frac{(a-1)^2}{a+1} < 1 . \end{aligned}$$

Hence we cannot have $p \geqslant q+2$. There remains...

Case (iii): Suppose $p = q+1$. Then $n = 2p$, and we have

$$2a - a\alpha_2 \leqslant \alpha_1 \leqslant aB_2 = 2pa - (pa+p-1)\alpha_2 ,$$

from (33) and (36). Hence

(38) $$(p-1)(a+1)\alpha_2 \leqslant 2(p-1)a .$$

Now if $p > 1$, this gives $\alpha_2 \leqslant A$; hence, by (34), $\alpha_2 = A$, and we have equality in (38). But in deriving (38) we made essential use of (33) and most of (32); therefore, equality must prevail in these places as well, and we find that all the α_i are equal to A. In other words, the only possible candidate for the core (if $p > 1$) is the vector $\alpha = (A, A, ..., A)$. It is easily verified that this vector is in fact a feasible payoff and is undominated; hence, we have a one-point core as claimed. Note that n in this case is even; for n odd there is no core.

If $p = 1$ (i.e., if $n = 2$), then (38) is no restriction, and it is easy to show that the core consists of the bent line from $(1, 2 - 1/a)$ to (A, A) to $(2 - 1/a, 1)$ (QPR in Figure 2).

APPENDIX 2

Proof of Theorem 3

Define C as in (16), take x^* interior to E^m_+, and let L be a linear support to C at x^*. Then L is strictly increasing in each x_j, and so is $L/2$. By sublinearity, we can find a function L' parallel to $L/2$ such that $L' \geqslant U + c$, where c is a preassigned positive constant. Let R denote the region of E^m_+ in which $L \leqslant L'$; clearly R is compact and contains x^*. We now wish to consider convex representations of x^* that "almost" achieve the value $C(x^*)$, in the sense of the "sup" in (16). In order to distinguish between vertices lying within R and those outside, the representations will be written in the following way:

(39) $$x^* = \alpha y + \bar{\alpha} z = \alpha \sum \lambda_h y^h + \bar{\alpha} \sum \mu_k z^k, \qquad y^h \in R,\ z^k \notin R .$$

Here $\bar{\alpha}$ denotes $1 - \alpha$ and is understood to be 0 if there are not points of the second type in the representation, i.e., if z is not well defined. (Note that y is always well defined; this follows from the fact that x is outside the convex set $E^m_+ - R$.) Given $\varepsilon > 0$, by (16) we can find a representation satisfying

$$\alpha \sum \lambda_h U(y^h) + \bar{\alpha} \sum \mu_k U(z^k) \geqslant C(x^*) - \varepsilon .$$

Hence we have

$$\alpha L(y) + \bar{\alpha} L'(z) - \bar{\alpha} c \geqslant L(x^*) - \varepsilon ,$$

or, from (39) and the linearity of L, $\varepsilon \geqslant \bar{\alpha}[L(z) - L'(z) + c]$. Since $[L(z) - L'(z) + c] \geqslant c > 0$ for $z \in E^m_+ - R$, we see that $\bar{\alpha} \to 0$ as $\varepsilon \to 0$. But since L' and $L/2$ are parallel, the expression in brackets is of the form $L'(z) + c'$, with a new constant c'. Thus, even though $\|z\|$ may be unbounded, we nevertheless have $\bar{\alpha}\|z\| \to 0$ and $\bar{\alpha} L'(z) \to 0$, as $\varepsilon \to 0$. Hence, $y = (1/\alpha)(z^* - \bar{\alpha} z) \to x^*$ as $\varepsilon \to 0$, and

$$\begin{aligned} \sum \lambda_h U(y^h) &\geqslant \frac{1}{\alpha}[C(x^*) - \varepsilon - \bar{\alpha} \sum \mu_k U(z^k)] \\ &\geqslant \frac{1}{\alpha}[C(x^*) - \varepsilon - \bar{\alpha} L'(z)] \\ &\to C(x^*) . \end{aligned}$$

Since the y^h are restricted to the compact region R and since U is assumed to be continuous, there exists a limiting representation $x^* = \Sigma\, \lambda_h y^h$ with $\Sigma\, \lambda_h U(y^h) = C(x^*)$. Hence C is spanned by U at the arbitrary interior point x^*.

If x^* is not interior to E^m_+, this argument is not directly valid, since a linear support L may not exist. But we can then reduce the dimension of the problem without affecting either the definition of $C(x^*)$ or the hypotheses of continuity, sublinearity, and strict monotonicity. In the reduced problem, x^* will be interior.

APPENDIX 3

Derivation of (30) in Section 12

LEMMA: *Let L^* be a support to C at x^*. Let there be a convex representation $x^* = \Sigma\, \lambda_h y^h$ such that $\Sigma\, \lambda_h U(y^h) = C(x^*)$. Assume that U possesses a radial derivative at each y^h. Let s be a fixed integer and k a variable integer. Then there exist convex representations $x^* = \Sigma\, \mu_h z^h$, depending on k, such that for each h, $ks\mu_h$ is an integer and*

(30) $$ks \mid U(z^h) - L^*(z^h) \mid \to 0 \qquad as\ k \to \infty .$$

Proof: Assume first that all λ_h are positive. For any k, we can find nonnegative integers l_h with sum ks and such that $\mid ks\lambda_h - l_h \mid \leqslant 1$. Now define:

$$\mu_h = l_h/ks \quad \text{and} \quad z^h = \frac{\lambda_h}{\mu_h} y^h .$$

For k sufficiently large, the μ_h will be positive and all the statements in the lemma are obviously satisfied, except for (30). To verify the latter, note that L^* is tangent to U at each y^h, and that z^h approaches y^h along the ray Oy^h. (If $y^h = 0$ and there is no ray, then (30) is trivial.) Since the derivative of U at y^h exists along that ray, we have

$$\frac{\mid L^*(z^h) - U(z^h) \mid}{\| z^h - y^h \|} \to 0, \qquad \text{as } k \to \infty .$$

(We are concerned, of course, only with values of k for which $z^h \neq y^h$.) But, for large k,

$$\| z^h - y^h \| = \left| \frac{\lambda_h - \mu_h}{\mu_h} \right| \cdot \| y^h \|$$
$$\leqslant \frac{1/ks}{\lambda_h - 1/ks} \| y^h \| .$$

Hence $ks \| z^h - y^h \|$ goes to zero and (30) follows.

If some of the λ_h are zero, we can set the corresponding $\mu_h = 0$ and $z^h = y^h$, and proceed as above, with h restricted to the indices for which λ_h is positive.

REFERENCES

[1] AUMANN, R. J.: "The Core of a Cooperative Game without Side Payments," *Transactions of the American Mathematical Society*, Vol. 98, 1961, pp. 539–552.
[2] ———: "Markets with a Continuum of Traders," *Econometrica*, Vol. 32, 1964, pp. 39–50.
[3] AUMANN, R. J., AND M. MASCHLER: "The Bargaining Set for Cooperative Games," *Annals of Mathematics Studies*, No. 52, Princeton University Press, Princeton, New Jersey, 1963, pp. 443–476.

[4] Aumann, R. J., and B. Peleg: "von Neumann-Morgenstern Solutions to Cooperative Games without Side Payments," *Bulletin of the American Mathematical Society*, Vol. 66, 1960, pp. 173–179.

[5] Bator, F. M.: "On Convexity, Efficiency, and Markets," *Journal of Political Economy*, Vol. 69, 1961, pp. 480–483; "Rejoinder," *ibid.*, Vol. 69, 1961, p. 489.

[6] Debreu, G.: "On a Theorem of Scarf," *Review of Economic Studies*, Vol. 30, 1963, pp. 177–180.

[7] Debreu, G., and H. Scarf: "A Limit Theorem on the Core of an Economy," *International Economic Review*, Vol. 4, 1963, pp. 235–246.

[8] Edgeworth, F. Y.: *Mathematical Psychics*, C. Kegan Paul, London, 1881.

[9] Farrell, M. J.: "The Convexity Assumption in the Theory of Competitive Markets," *Journal of Political Economy*, Vol. 67, 1959, pp. 377–391; "A Reply," *ibid.*, Vol. 69, 1961, pp. 484–489; "Rejoinder," *ibid.* Vol. 69, 1961, p. 493.

[10] Gillies, D. B.: "Some Theorems on n-Person Games," Ph.D. Thesis, Princeton University, 1953.

[11] ———: "Solutions to General Non-zero-sum Games," *Annals of Mathematics Studies*, No. 40, Princeton University Press, Princeton, New Jersey, 1959, pp. 47–85.

[12] Koopmans, T. C.: "Convexity Assumptions, Allocative Efficiency, and Competitive Equilibrium," *Journal of Political Economy*, Vol. 69, 1961, pp. 478–479.

[13] Kuhn, H. W.: *The Theory of n-person Games* (report of an informal conference held at Princeton University, March 20–21, 1953), Princeton University, 1953.

[14] Luce, R. D., and Howard Raiffa: *Games and Decisions: Introduction and Critical Survey*, John Wiley & Sons, Inc., New York, 1957.

[15] Rothenberg, J.: "Nonconvexity, Aggregation, and Pareto Optimality," *Journal of Political Economy*, Vol. 68, 1960, pp. 435–468; "Comments on Nonconvexity," *ibid.*, Vol. 69, 1961, pp. 490–492.

[16] Shapley, L. S., and M. Shubik: "Concepts and Theories of Pure Competition," *Essays in Mathematical Economics: In Honor of Oskar Morgenstern*, Princeton University Press, Princeton, New Jersey, 1967, pp. 63–79.

[17] Shubik, M.: "Edgeworth Market Games," *Annals of Mathematics Studies*, No. 40, Princeton University Press, Princeton, New Jersey, 1959, pp. 267–278.

[18] Uzawa, H.: "Walras' Tâtonnement in the Theory of Exchange," *Review of Economic Studies*, Vol. 27, 1960, pp. 182–194.

[19] Vind, K.: "Edgeworth-Allocations in an Exchange Economy with Many Traders," *International Economic Review*, Vol. 5, 1964, pp. 165–177.

[20] ———: "A Theorem on the Core of an Economy," *Review of Economic Studies*, Vol. 32, 1965, pp. 47–48.

[21] von Neumann, J., and Oskar Morgenstern: *Theory of Games and Economic Behavior* Princeton University Press, Princeton, New Jersey, 1944, 1947, 1953.

[3]

Concepts and Theories of Pure Competition

By L. S. SHAPLEY and MARTIN SHUBIK*

1. INTRODUCTION

Underlying much of basic economic theory concerning the formation of prices and the operation of markets is the idea that if the number of individuals trading in all markets is sufficiently large, and if there are no institutional bounds or opportunities for cooperative arrangements between groups, then a competititve price system will emerge. Each individual may proceed to maximize his own welfare, utilizing only his knowledge of the price levels and of his own desires and assets, with the overall result that all markets will be cleared and the resulting imputation of goods and services will meet certain broad requirements of optimality.

At the other end of the spectrum, Edgeworth [1], and many others since, have observed that in bargaining among a *few* individuals (of more or less comparable strength), there will be a whole range of outcomes that are optimal, in the appropriate sense, and economic conditions alone will not determine a specific outcome.

A third possibility is a situation in which individuals of one type are few in number, or are organized in some manner, while the others are not. In this case it has been observed that the few can employ "monopolistic practices" against the others, and that the resultant outcome will be determinate but not optimal. Institutionally, such a situation may arise when a few firms confront many customers in a market; the natural setting for this type of analysis is within the framework of the theory of oligopoly.

An interpretation and unification of these three different viewpoints is presented here in terms of the theory of games. The results described are based primarily upon the joint findings of Shapley and Shubik [2], but also upon recent papers by Shubik [3], Shapley [4], Scarf [5], Aumann [6], and Debreu [7].

2. BILATERAL MONOPOLY

2.1. THE EDGEWORTH MODEL. For simplicity we consider a market with two traders, trading in two commodities. The initial holding of the first trader is $(a, 0)$ and that of the second is $(0, b)$. Let their preferences be represented by two families of continuously differentiable, convex indifference curves, denoted by $\psi(x, y)$ and $\varphi(x, y)$, respectively, where x and y are the amounts held of the first and second commodity. By superimposing the two families of curves, with coordinates oppositely oriented, we obtain

* The RAND Corporation and Yale University.

the familiar "Edgeworth box" [1], as illustrated in Figure 1. Any point in the box represents a jointly attainable trade and will have coordinates $(a - x, b - y)$ for the first trader and (x, y) for the second. The point R represents the initial position of both players prior to trading. The first trader's zero point is at O', and his initial holdings are represented by the vector $O'R$, which has length a. His goal is to carry out trades that move the outcome in a "south-west" direction as far as possible, i.e., away from

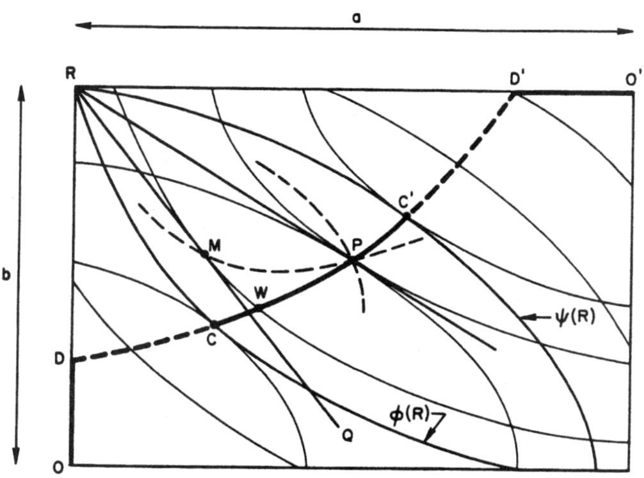

FIGURE 1

O'. The zero point for the second trader is at O, and OR represents his initial stock of b units of the second commodity. He wishes to trade in a manner that moves the outcome in a "north-east" direction as far as he can.

There are three basic models, or "games," which can be formulated in bilateral monopoly to illustrate three strategically very different trading procedures; we shall denote them by the symbols $(1, 1)_0$, $(1, 1)_1$, and $(1, 1)_2$. In Figure 1 they will lead to the outcomes indicated by the point P, the point M, and the curve CC', respectively.

2.2. THE GAME $(1, 1)_0$: THE COMPETITIVE EQUILIBRIUM. The essence of the economic model of pure competition is that all participants act as *price-takers*. One assumes that in some manner or other a schedule of prices has been established in the market, and that each individual takes note of these prices and buys or sells accordingly. He does not actively control or influence the prices. When the number of participants in the market is large, it is usually assumed either implicitly or explicitly that a dynamic market mechanism produces the prices, but in our "game," which at present has just two players, we shall unrealistically assume that the omniscient referee performs his calculations and names an appropriate set of prices, then acts as a clearing house for all transactions. By the rules of the game, the players are strategically constrained to act as price-takers. The subscript on the symbol $(1, 1)_0$ is meant to convey that neither of the two traders has sufficient freedom of strategy to manipulate price.

We know from basic economic theory [8] that when consumer's tastes are independent and can be represented by convex indifference sets (as is the case here), then a *competitive equilibrium* will always exist. This is illustrated in Figure 1 by the equilibrium price line *RP* and the competitive allocation point *P*. In this two-commodity example, the slope of the line *RP* gives the relative price of one commodity in terms of the other. If prices in this ratio are announced by the referee, then trading will continue until the point *P* is reached. No further gain can be made by either side because of the tangency between the price line and each traders' indifference curve. This so-called "competitive" solution exists under quite general conditions, but it is not necessarily unique.[1]

The point *P* is *Pareto optimal*, which means that, given the final distribution of resources, it is not possible to improve the welfare of any one individual without decreasing the welfare of another. The extended curve $ODD'O'$ describes the full Pareto-optimal set in Figure 1. (The interior portion DD' is the locus of the points of tangency between the two sets of indifference curves.) Any point *off* this curve can be dominated by a point *on* it. For instance, *M* and *W* in Figure 1 are indifferent to the second trader but *W* is preferred to *M* by the first trader. Hence *M* is not Pareto optimal since the welfare of at least one trader can be improved without damage to the other.

2.3. THE GAME $(1, 1)_1$: THE MONOPOLISTIC SOLUTION. The second case we consider is $(1, 1)_1$, where one of the traders, say the first, is strategically empowered to name the price, while the other is restricted to maximizing his own welfare, taking price as given. Any ray through the initial point *R* represents a set of (relative) prices. The object of the first trader will be to select a ray such that the final trade is as favorable as possible to himself, knowing that the second player will trade up to the point of tangency between the price ray and his family of indifference curves. This is indicated in Figure 1 for the ray *RQ* at the point *M*. A monopolistic trading curve may be drawn—the curve *MP* in the figure—which is the locus of points of tangency between the price rays and the indifference curves of the second

[1] Somewhat unfortunately, the term "competitive" has been attached by established usage to just that solution concept, out of the three or four that we shall be considering, that has the least to do with game theory. Paradoxically, the competitive equilibrium involves no interplay between the "competitors," and might better be described as an "administered-price" solution.

trader. (There will be a similar "response curve" for the first player, intersecting the other at *P*.) The optimum for the first player is in fact the point *M*, where his indifference curve is tangent to the second player's response curve.

We see that this "monopolistic" solution is not Pareto optimal, and that it favors the price-naming monopolist as compared to the solution *P* of the previous case.

2.4. THE GAME $(1, 1)_2$: THE CONTRACT CURVE. In the third case, denoted by $(1, 1)_2$, we assume that both players are strategically capable of naming price and may negotiate in any manner they choose. There is no specification of the dynamic process—the particular "rules of the game"—beyond noting that any bargaining or haggling is permitted. The solution propounded by Edgeworth to this model of bilateral monopoly was the *contract curve* CC' in Figure 1. This is not the entire Pareto-optimal set, but just that part of it where neither side can force a more favorable distribution by a refusal to trade. The two indifference curves through *R* provide the bounds *C* and C'.

This Edgeworth solution does not yield a unique prediction of the imputation of resources; it merely indicates a range. The price-taking model $(1, 1)_0$, on the other hand, did produce a determinate, Pareto-optimal outcome; the latter was located on the contract curve but it was obtained by imposing somewhat unrealistic restrictions on the trading possibilities, considering the small number of traders.

3. MARKETS WITH MANY TRADERS. THE GAME $(n, n)_0$

Implicitly or explicitly in all models of trading, assumptions have to be made concerning the institutional nature of the market. In the broad sweep of economics when many individuals are involved, we expect markets to be more or less insensitive to minor institutional differences, but when two individuals engage in face to face bargaining we suspect that personality, cultural factors, psychological details, the fine structure of moves and timing, etc., all play a major role.

This being the case, it is entirely appropriate to use the "game" $(1, 1)_2$ when we deal with bilateral monopoly. When we discuss markets with many traders on each side, however, we often tend to use something like $(m, n)_0$ as the model under investigation. In other words, we effectively assume that in markets with many participants on all sides, the individual is constrained to act as a price-taker. There are many good reasons for this, such as the cost of communication, the lack of time to talk to everyone, and other organizational factors that drive toward impersonal mechanisms of trade.

On the other hand, a sociologist or an anthropologist might point out that in spite of numerical size and communication problems, patterns can exist and persist in a society that rely on the overt or covert coordination of many individuals. There may be stable configurations involving the compliance of large groupings or coalitions within the society. Furthermore, both long-run socioeconomic considerations and a study of information processes indicate that although many costs and much expenditure of time and energy may be incurred in creating a set of intricate coalitions *ab initio*, yet given their existence, little effort need be spent in maintaining them. In view of these considerations, it will be worth investigating the characteristic properties of games of the form $(m, n)_1$ and $(m, n)_2$, as well as $(m, n)_0$.

For simplicity in the sequel, we shall not only restrict our attention to bilateral markets, but we shall assume $m = n$. Traders of the same type are assumed to have identical preferences and initial holdings. Under these assumptions, the discussion of $(n, n)_0$ becomes especially easy.

Indeed, in going from $(1, 1)_0$ to $(n, n)_0$, the form of the solution remains the same, since the competitive equilibrium point (illustrated for the two-person case as the point P in Figure 1) has a direct $2n$-dimensional analogue. In fact, the outcome will be the same (under our highly symmetric assumptions) as though n games of the form $(1, 1)_0$ were being conducted independently and simultaneously. As n becomes large, the competitive equilibrium point, from the viewpoint of the economist, will be much more "reasonable" as a solution than it was in the case of bilateral monopoly, for the reasons already given.

4. COOPERATIVE GAMES. THE GAME $(n, n)_2$

4.1. Cores and solutions. Before we proceed to $(n, n)_1$ or $(n, n)_2$, a brief digression into the subject of cooperative n-person game theory will be necessary. For ease and clarity the exposition will be mostly in terms of games where the players have transferable, measurable, and comparable utilities. (It is as though the players all attach the same worth to money and have a constant marginal utility for it.) It must be emphasized that the results we shall give concerning $(n, n)_2$ are independent of these utility restrictions.

In order to explain the concepts of *cooperative solution* and *core*, we must also define what is meant by the following terms: the *characteristic function* of a game, an *imputation*, an *effective set* of players, and *domination* of one imputation by another.

The *characteristic function* specifies the worth that a coalition can achieve if they limit their trades strictly to themselves. Mathematically it is a function $v(S)$ defined on sets of players S, with the properties

$$v(\theta) = 0,$$

$$v(S \cup T) \geq v(S) + v(T), \quad \text{whenever } S \cap T = \theta.$$

The first condition merely states that the amount achievable by the null set is nothing. The second condition is the fundamental economic property of superadditivity: if two separate groups having commerce only amongst themselves are joined together, the resultant group is at least as effective as were the two independent groups. Beyond these two conditions there is nothing more than can be said *a priori* about a characteristic function.

If we denote the set of all players in a game by N, then $v(N)$ specifies the total amount that the whole group can obtain by cooperation.[2] A reasonable form of "cooperative" behavior would be for the players to agree to maximize jointly, and then to decide how the proceeds are to be apportioned, or "imputed." We define an *imputation* α to be a division of the proceeds from the jointly optimal play of the game among all the n players:

$$\alpha = (\alpha_1, \alpha_2, \alpha_3, \ldots, \alpha_n),$$

where

$$\alpha \geq v(\bar{i}) \quad \text{and} \quad \sum_{i=1}^{n} \alpha_i = v(N).$$

The condition $\alpha_i \geq v(\bar{i})$ embodies the principle that no individual will ever consent to a division that yields him less than he could obtain by acting by himself. It is often convenient to normalize the individual scales so that $v(\bar{i}) = 0$.

A set of players is said to be *effective* for an imputation if by themselves they can obtain at least as much as they are assigned in that imputation. Symbolically, S is effective for α if and only if

$$v(S) \geq \sum_{i \in S} \alpha_i.$$

If ">" rather than "=" holds, we shall say that S is *strictly effective.*

An imputation α *dominates* an imputation β if there exists an effective set S for α such that for all members of S, $\alpha_i > \beta_i$. Following the notation of von Neumann and Morgenstern [9], we write

$$\alpha \succ\!\!\dashv \beta.$$

In other words, if a set S of players is in a position to obtain by independent action the amounts that they are offered in the imputation α, and if, when they compare the amounts offered in α to the amounts offered in β, all of them prefer the former, then α dominates β. There is a potential coalition that prefers α to β and is in a position to do something about it. Note that S is necessarily *strictly* effective for β, the dominated imputation.

[2] Their utilities being transferable, this is properly represented by a single number, which denotes maximum obtainable welfare. If utilities were not transferable, $v(N)$ would instead have to represent the Pareto-optimal surface, and similarly for smaller coalitions. (See [6], [4].)

Finally, we may define two "solution" concepts. The *core* of an n-person game is the set of undominated imputations, if any. A von Neumann-Morgenstern *solution*, on the other hand, consists of a set of imputations which do not dominate each other but which collectively dominate all alternative imputations. There is at most one core, but there may be many solutions. All solutions contain the core, if it exists.

4.2. Some examples. A series of simple, three-person games will illustrate these concepts. Consider first the game in which any player acting by himself obtains nothing, but any pair of players acting in concert can demand three units to share between them, while all three players in coalition are also awarded three. The characteristic function of this game is

$$
\begin{aligned}
&v(\theta) = 0,\\
&v(\bar{1}) = v(\bar{2}) = v(\bar{3}) = 0,\\
&v(\bar{1}\bar{2}) = v(\bar{1}\bar{3}) = v(\bar{2}\bar{3}) = 3,\\
&v(\bar{1}\bar{2}\bar{3}) = 3,
\end{aligned}
$$

where $\bar{1}\bar{2}$ means "the set consisting of players 1 and 2," etc.

We may represent the imputations in this game by triangular coordinates, as shown in Figure 2. The vertices P_1, P_2, P_3 represent the imputations (3, 0, 0), (0, 3, 0), and (0, 0, 3), respectively. The point $\omega = (1, 1, 1)$ is the center of the triangle. Consider the two imputations $\alpha = (1.9, 0, 1.1)$ and $\beta = (0, 1.5, 1.5)$. The set is $\bar{2}\bar{3}$ effective for β, and furthermore both 2 and 3 are better off in β than in α. Hence $\beta \in\!\dashv \alpha$.

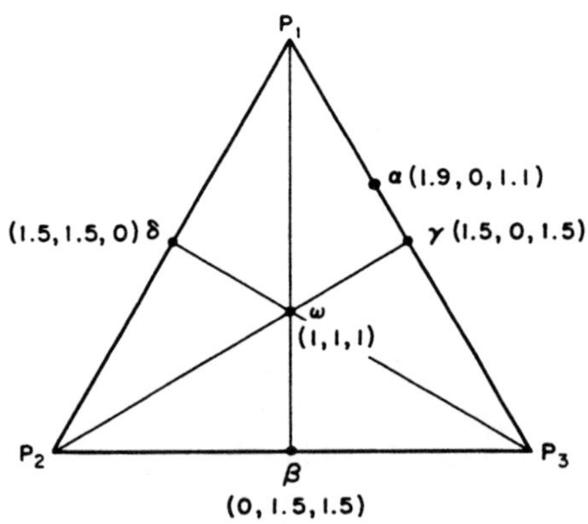

Figure 2

The trio of imputations β, γ, and δ forms a solution set to this particular game. Any other imputation gives two of the players less than 1.5 apiece and thus is dominated by one of these three imputations, but the three do not dominate each other. (There are other solution sets which we need not discuss.) *This game has no core*, since the imputations β, γ and δ, dominating all the rest, are themselves dominated by others. For example, the imputation α, which was dominated by β, in turn dominates δ via the effective set $\overline{1}\overline{3}$. Note that domination is not a transitive relation: $\beta \succ \alpha$ and $\alpha \succ \delta$ do not entail $\beta \succ \delta$.

We now consider three closely related games differing from the previous one only in what the two-person coalitions obtain. In the first variant we have

$$v(\overline{1}\overline{2}) = v(\overline{1}\overline{3}) = v(\overline{2}\overline{3}) = 0.^{3}$$

In this case, *all* imputations are in the core. The only set of players that is effective, for most imputations, is the three-person set; however, this is useless for domination since, on examining the distribution of welfare from the viewpoint of all three players, we see that if one player prefers one of two imputations, then at least one of the other players will prefer the other, the sum of the allotments being constant. In fact, it suffices to point out that no set of players is *strictly* effective for any imputation—hence there is no domination. The core is therefore as large as possible, and is also the unique von Neumann-Morgenstern solution.

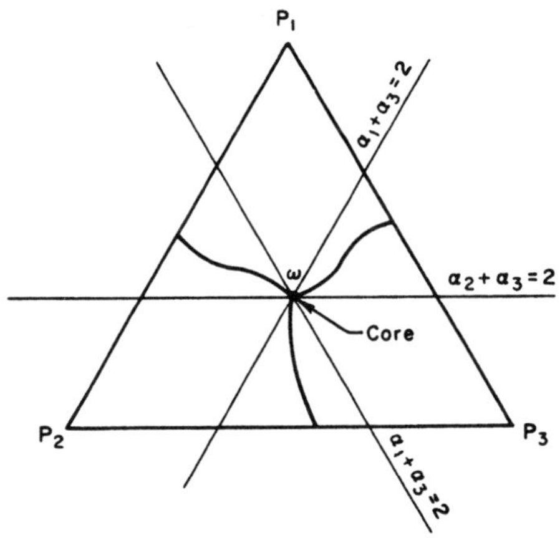

FIGURE 3

[3] This all-or-nothing type of characteristic function, like the previous one, is associated more with political than economic processes [10]. The previous game was a majority-take-all situation; the present one is a veto situation, since if one member wishes to be the "dog in the manger," he can prevent the others from obtaining any payoff. In economics such extremes—called *simple games*—are not typical. We shall presently consider variants in which the two-person coalitions obtain intermediate amounts, reflecting the more usual situation in which any new adherent to a coalition means added possibilities for profit.

In our third example we assume

$$v(\overline{1}\overline{2}) = v(\overline{1}\overline{3}) = v(\overline{2}\overline{3}) = 2.$$

As shown in Figure 3, the lines which describe the amount obtainable by each coalition of two players intersect in a single point, the imputation ω with coordinates (1, 1, 1). This is the only undominated imputation of the game, and thus constitutes a single-point core. Since ω fails to dominate the three small triangles adjoining it in the diagram, however, it is not a von Neumann-Morgenstern solution by itself. To get a solution we must add some more or less arbitrary curves, as shown, traversing the three triangular regions (see [9], pp. 550-554).

In the final variant, we assume that the two-person coalitions are only half as profitable as in the preceding example. That is, we have

$$v(\overline{1}\overline{2}) = v(\overline{1}\overline{3}) = v(\overline{2}\overline{3}) = 1.$$

The lines indicating the ranges of effectiveness of these coalitions are spread apart, as shown in Figure 4, revealing a large, hexagonal core. All imputations in that area are undominated. As in the second example, this core is the unique solution.

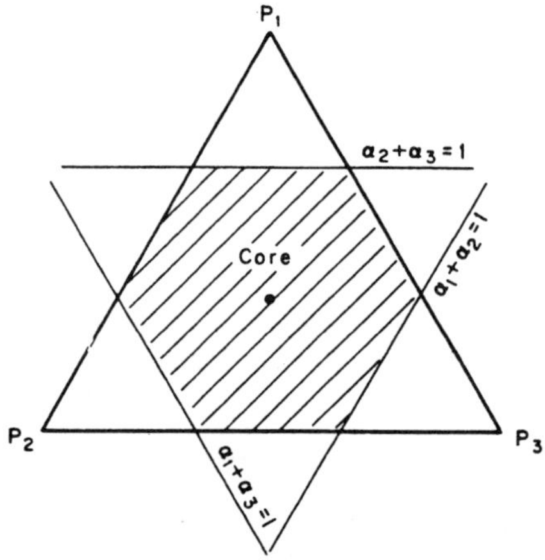

FIGURE 4

A superficial examination of these four examples suggests a relationship between the size of the core and the "fatness" of the coalitions in a game, i.e., how much they can promise their members per capita as compared to the per capita amount available in the whole game. In all four instances, the latter amount was $v(\overline{1}\overline{2}\overline{3})/3 = 1$. Denote $v(\overline{i}\overline{j})/2$ by f_2. In the first game, f_2 was 1.5, which is greater than 1, and there was no core. In the third game, f_2 was exactly 1, and the core was a single point. In the fourth game,

f_2 was 1/2, and there was a large core, while in the second game, f_2 was 0, and every imputation was in the core.

Of course, in a less symmetric situation, this principle would not reveal itself in such a clean-cut manner. However, a general rule of thumb seems to persist: the more power there is in the hands of the middle-sized groups, the more narrowly circumscribed is the range of outcomes of the co-operative game. This rule, vague as it is, applies to solutions as well as cores.

When we do not permit the transfer of utility, we can no longer talk about the amount attainable by a group of players as a single number; nevertheless, cores [6] and solutions [11] can still be defined, and the idea of the per-capita gain in one coalition being larger than in another can still be utilized in a vectorial sense.

4.3. THE EDGEWORTH MARKET GAME, $(n, n)_2$. Let us return to the Edgeworth bilateral-monopoly game $(1, 1)_2$. Even without the simplifying assumption of transferability of utility, it is not hard to see that the Edgeworth contract curve is the "cooperative solution," in the spirit of the von Neumann-Morgenstern definition, but with weaker utility assumptions [11]. It is also the core. The reason is simply that with just two players, there are no coalitions of intermediate size, between the individual and the whole group, and hence no domination occurs between outcomes that are both Pareto optimal and individually rational.

It can be shown that the whole contract curve (considered in the higher dimensions) will remain as a cooperative solution to the game $(n, n)_2$ for any number of traders. However, it will no longer be the core if there are more than two players, and there will be many other von Neumann-Morgenstern-type solutions, which may in general be quite difficult to compute (see [4]).

It has been shown that as the number of players increases, the core shrinks down upon the competitive equilibrium point (or set of points) [3, 5, 12]. Hence, if we regard the core as our concept of cooperative solution, the game $(1, 1)_2$ has the contract curve as its solution, while the game $(n, n)_2$, as n grows large, has an increasingly determinate solution consisting of a small neighborhood of the competitive equilibrium.

Figure 5 illustrates the shrinking of the core as the number of players is increased. It is similar to Figure 1 of Edgeworth ([1], pp. 20–25). The line RP is the competitive price ray on the exchange diagram for $2n$ traders, consisting of n of each type. The arc CC' is the range of the two-person contract curve, and I is a typical point on that curve. The shaded area between I and I_1 indicates a domain that is preferred to I by a player of the first type; similarly the shaded area between I and I_2 is preferred to I by a player of the second type. These areas are bounded by the two indifference curves which are tangent at I; each area includes a portion of the price ray RQ only if I is *not* the competitive allocation P. We shall describe the conditions under which I (or more precisely, the imputation associated with I) is dominated by the outcome of some trade among a subset of the players.

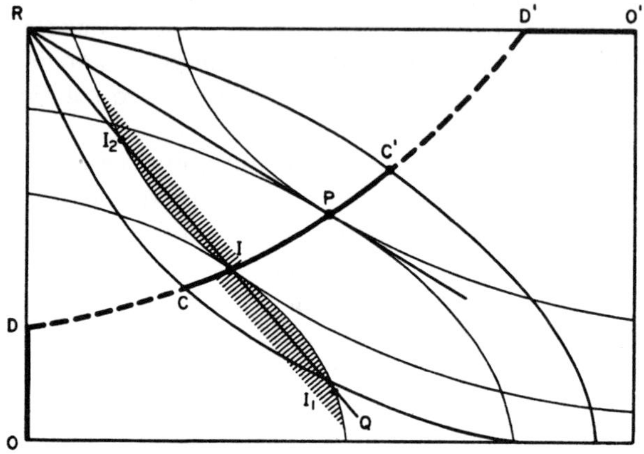

FIGURE 5

When only two traders enter into contract, exchanging x units of the first good for y of the second, the final distribution of assets between them is given by $(a - x, y)$ and $(x, b - y)$. This can be represented by a single point on the Edgeworth diagram. Suppose that a coalition of k traders of the first type and m traders of the second type forms, and suppose that traders of the same type decide to divide their gains equally. Together the coalition controls ka units of the first commodity and mb of the second. After trade, a player of the first type will have $\left(a - \frac{x}{k}, \frac{y}{k}\right)$, and a player of the second type will have $\left(\frac{x}{m}, b - \frac{y}{m}\right)$, for some x, y. This outcome can be represented on the Edgeworth diagram by two points, corresponding to the two types of players, which fall on the line RQ, one on either side of the contract curve. If $k = m$, the two points coincide at I. If the ratio of k to m is only slightly different from unity, it will be possible for the point corresponding to each type of player to lie in the shaded region in which that type of player gains, in comparison to the distribution offered at I. The imputation corresponding to I will therefore be dominated. As n grows larger, the available ratios k/m become denser around 1, and the portion of the contract curve that escapes domination shrinks down to an arbitrarily small neighborhood of the competitive equilibrium point. (If the latter is not unique, the convergence will be to the set of competitive equilibria.)

It is remarkable that such widely different sets of modeling assumptions as were used in $(n, n)_0$ and $(n, n)_2$ should lead, in the limit, to the same solution. In the one case, the participants operate on a minimum of information, they have essentially no strategic initiative, and they are prevented from cooperating or even interacting with their fellows. In the other case, information is freely available, exchanges of goods may be made in all possible ways, without regard for prices, and collusion not only is permitted but is in fact essential to the maintenance of stability.

Viewed as a limit of cores, the competitive equilibrium for large n is seen to be "sociologically neutral." No coalition is effective against it. It is not a von Neumann-Morgenstern-type solution by itself, even in the limit, since it does not dominate the other points on the contract curve; nevertheless, every such solution will include the competitive imputation. Thus, without any dynamic assumptions regarding prices or other mechanisms of the market place, and without special hypotheses concerning costs of communication, information, and so forth, the competitive equilibrium plays an important and distinctive role purely on the ground of social stability.

5. NONCOOPERATIVE GAMES. THE COURNOT OLIGOPOLY GAME

5.1. NONCOOPERATIVE EQUILIBRIUM POINTS. An approach highly different from that of cooperative game theory, and connected less directly to welfare considerations than to problems of oligopoly and control of industry, is the idea of the *noncooperative* solution to a game. The spirit of much of the classical discussion of oligopoly behavior and equilibrium in monopolistic competition has been to regard the firms as powerful and their customers as weak. This conforms to our commonsense notions of the oligopolistic market place. The automobile or tobacco companies, for example, have some form of control over output, prices, brands, and so forth, but their customers are, for the most part, price-takers.

Both the classical and many of the more recent works in oligopoly theory have made use of "open" models, in the sense that they investigate the behavior of a group of firms or an industry, taking the behavior of customers and suppliers as given. Underlying the writings of Cournot [13], Bertrand [14], Edgeworth [15], Hotelling [16], Stackelberg [17], Chamberlin [18], and others, is the concept of the attainment of a noncooperative equilibrium by the firms in competition. On the mathematical side, Nash [19] was first to develop the basic idea of noncooperative solution in the abstract framework of the theory of games. The crucial concept in the noncooperative theory is that of "equilibrium point," which may be defined as follows:

Consider a game with n players. Let player i have a class of possible "strategies" S_i, and let s_i denote a particular strategy belonging to the class. The payoff to the ith player is denoted by $P_i(s_1, s_2, s_3, \ldots, s_n)$, a function of the strategies of all the players. A strategy vector $(\bar{s}_1, \bar{s}_2, \bar{s}_3, \ldots, \bar{s}_n)$ is said to constitute an *equilibrium point* if, for all i, the function

$$P_i(\bar{s}_1, \bar{s}_2, \ldots, \bar{s}_{i-1}, s_i, \bar{s}_{i+1}, \ldots, \bar{s}_n)$$

is maximized by setting $s_i = \bar{s}_i$. In other words, a set of strategies, one for each player, forms an equilibrium point if each player, knowing the strategies of all others, will not be motivated to change. If the equilibrium point is unique, as it proves to be in many of the classical economic models, it is termed the "noncooperative solution" to the game.

5.2 THE COURNOT MODEL. An early ecónomic example of noncooperative equilibrium was presented by Cournot [13]. Suppose that two firms are constrained to name amounts q_1, q_2 offered for sale. These quantities are their strategic variables. Furthermore, suppose that there is a market mechanism of some sort which selects a price that exactly clears the market, say $p = D(q_1 + q_2)$. Let the cost functions of the two firms be $C_1(q_1)$ and $C_2(q_2)$, respectively. The payoffs are then

$$P_1 = q_1 D(q_1 + q_2) - C_1(q_1),$$

$$P_2 = q_2 D(q_1 + q_2) - C_2(q_2)$$

respectively. An equilibrium point will be a pair of strategies $(\bar{q}_1, \bar{q}_2)$ such that

$$P_1(q_1, \bar{q}_2) \text{ is maximized at } q_1 = \bar{q}_1,$$

$$P_2(\bar{q}_1, q_2) \text{ is maximized at } q_2 = \bar{q}_2.$$

This is the "Cournot" solution [20].

In the game described above, the welfare of the customers was included only implicitly, through the workings of the demand relation; it is thus an open model. Considering only the welfare of the two duopolists, it is easy to see that the noncooperative equilibrium is not Pareto optimal in the open model. They could both improve matters by joint action, restricting the amounts offered to be sold.

It is nevertheless conceivable that the noncooperative equilibrium might be Pareto optimal in the closed model, in which the welfares of the customers are also taken into account; however, even this is not the case in general. We already have a simple example of this in the game $(1, 1)_1$, which can be viewed as a monopolist dealing with a compliant, price-taking customer. In this case, quantity naming is usually equivalent to price naming. Referring to Figure 6, we see that if the monopolist (origin at 0′) offers the amount e to be sold, for example, the price mechanism

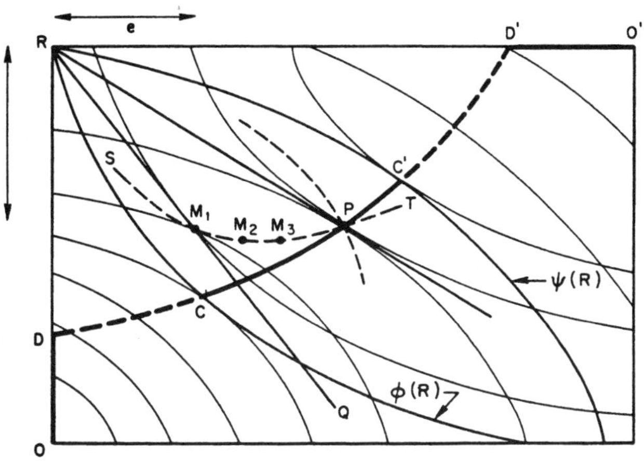

FIGURE 6

will determine the price ray RQ that just clears the market when the other player optimizes, selecting the point M, where the ray intersects his response curve ST. Under suitable assumptions, this derived price ray will be unique, for any value of e, giving the monopolist effective control over prices.

Returning to our point about optimality, we note that the best outcome from the viewpoint of the monopolist is the point M_1, which is clearly not Pareto optimal in the closed model since it lies off the contract curve. In this instance of monopolistic exploitation, all could have their welfare improved by negotiation.

5.3. THE GAME $(n, n)_1$. In order to define the game $(n, n)_1$, we must specify the market mechanism more fully, making explicit the scope of strategic choice. There are many ways in which this can be done. We might assume that the firms playing the role of oligopolists are in a position to name price. This is the manner in which Bertrand and Edgeworth handled oligopoly. Though complicated, it is possible to specify a mechanism whereby each oligopolist names both a price and a limit to the amount he is willing to trade. In order to complete the specification, the nature of the market-clearing process, in the presence of possibly different prices, would also have to be made explicit.

An easier model to define, involving fewer arbitrary institutional assumptions, is the analogue to Cournot's approach, in which the oligopolists are free to specify the *amounts* to be offered for trade, but not the prices. A mechanism then sums together all the offers and calculates the prices that will exactly clear the market. The traders of the second type are constrained to be price-takers, as before. We shall adopt this quantity-naming model as the most natural generalization of the game $(1, 1)_1$, in which, as already noted, price naming and quantity naming are essentially equivalent.

Let us assume that the "response curve" (*ST* in Figure 6) of the second player in the two-person game $(1, 1)_1$ can be expressed as a single-valued function $y = r(x)$, where (x, y) is that player's final holding, after trade. In addition, we shall assume that $r(x)$ has a bounded derivative. (The purpose of these assumptions is to exclude certain difficulties related to the possibility of different prices yielding the same demand.) Then, in the game $(n, n)_1$, the price that clears the market is given by

$$p = \frac{b - r(\bar{q})}{\bar{q}},$$

where $\bar{q} = \sum_i q_i/n$ is the average quantity offered of the first good. The final holdings will be

$$(a - q_i, pq_i),\ i = 1, 2, \ldots, n,$$

for the players of the first type, and

$$(\bar{q}, b - p\bar{q})$$

for each player of the second type. The noncooperative equilibrium is found by requiring the function $\psi(a - q_i, pq_i)$ to be a maximum with respect to q_i, for each i.

This maximum will generally not be found at the point M_1 in Figure 6, where $r(x)$ is tangent to the family ψ. The reason is that when a player varies his offer q_i, the *average* offer $\bar{q}$, which is what determines the price, varies by only $1/n$ as much. In fact, the oligopolists will tend to offer larger quantities than they would if they were in monopolistic collusion, and the noncooperative equilibrium outcome M_n will be displaced along the response curve in the direction of the competitive equilibrium outcome P. (See Fig. 6.)

To see why M_n converges to P in the limit, we merely set $\partial\psi(a - q_i, pq_i)/\partial q_i = 0$, and obtain

$$p = \frac{\psi_x}{\psi_y} - q_i\frac{\partial p}{\partial q_i} = \frac{\psi_x}{\psi_y} + \frac{q_i}{n\bar{q}}(r' + p),$$

a relation which must hold at M_n. But r' is bounded by assumption, as is q_i, and it can be shown that $\bar{q}$ does not go to zero; therefore

$$p \to \frac{\psi_x}{\psi_y} \quad \text{as} \quad n \to \infty.$$

This implies that M_∞ lies on the response curve of the players of the first type as well as of those of the second type. Hence $M_\infty = P$.[4]

The intuitive idea behind this argument is that when n becomes large, the effect of one individual on the price structure becomes negligible; i.e., $\partial p/\partial q_i \to 0$.

6. CONCLUSION

We began with three different models of bilateral monopoly, based on the same data. Three different solution concepts were advanced, each appropriate to one of the models, and three qualitatively different outcomes were predicted. As the number of traders on both sides of the market was increased, the three solutions merged into one as regards predicted outcome, but their *rationales* remained quite distinct. A relationship was thereby demonstrated among (1) administered price stability (the *competitive equilibrium*), (2) noncollusive oligopolistic exploitation (the *equilibrium point* of a noncooperative game), and (3) unrestricted bargaining between coalitions (the *core* of a cooperative game).

Each model embodied radically different assumptions concerning the strategies and information available to the participants. A scrutiny of the differences would reveal many places where sociological and institutional assumptions might be slipped in, perhaps inadvertently, in the construction of models of markets.

REFERENCES

[1] Edgeworth, F. Y., *Mathematical Psychics*, Kegan Paul, London, 1881.

[2] Shapley, L. S., and M. Shubik, manuscript in progress on applications of n-person game theory.

[3] Shubik, Martin, "Edgeworth Market Games," in: *Contributions to the Theory of Games*, Vol. 4, Annals of Mathematics Study No. 40, Princeton, Princeton University Press, 1959, pp. 267–278.

[4] Shapley, L. S., "The Solutions of a Symmetric Market Game," in: *Contributions to the Theory of Games*, Vol. 4, Annals of Mathematics Study No. 40, Princeton, Princeton University Press, 1959, pp. 145–162.

[5] Scarf, Herbert, "An Analysis of Markets with a Large Number of Participants," *Recent Advances in Game Theory* (proceedings of a conference held at Princeton University, October 4–6, 1961), Princeton University, 1962, pp. 127–155.

[6] Aumann, R. J., "The Core of a Cooperative Game without Side Payments," *Trans. Amer. Math. Soc.*, Vol. 98, 1961, pp. 539–552.

[7] Debreu, Gerard, *On a Theorem of Scarf*, Cowles Foundation Discussion Paper No. 130, November, 1961.

[4] More precisely, since uniqueness is not assured, any limit point of any sequence $\{M_n\}$ of noncooperative equilibria is a competitive allocation. Our assumptions regarding $r(x)$ are essential to this result.

[8] Debreu, Gerard, *Theory of Value, An Axiomatic Analysis of Economic Equilibrium*, Cowles Foundation Monograph 17, New York, John Wiley & Sons, Inc. 1959.
[9] von Neumann, J. and O. Morgenstern, *Theory of Games and Economic Behavior*, Princeton, Princeton University Press, 3rd Ed., 1953.
[10] Shapley, L. S., "Simple Games: An Outline of the Descriptive Theory," *Behavioral Science*, Vol. 7, 1962, pp. 59–66.
[11] Shapley, L. S., and M. Shubik, "Solutions of *n*-person Games with Ordinal Utilities" (abstract) *Econometrica*, Vol. 21, 1953, pp. 348–349.
[12] Shapley. L. S., and M. Shubik, *The Core of an Economy with Nonconvex Preferences*, The RAND Corporation, RM–3518, February, 1963.
[13] Cournot, Augustin A., *Researches into the Mathematical Principles of the Theory of Wealth*, New York, Macmillan and Co., 1897, pp. 79–80, 84.
[14] Bertrand, J., "Théorie Mathématique de la Richesse Sociale" (review), *Journal des Savants*, Paris, September 1883, pp. 499–509.
[15] Edgeworth, F. Y., *Papers Relating to Political Economy*, I. Macmillan and Co., London, 1925, pp. 111–142, 118, 120.
[16] Hotelling, H., "Stability in Competition," *The Economic Journal*, March 1929, p. 41.
[17] Stackelberg, H. von, *Marktform und Gleichgewicht*, Berlin, Julius Springer, 1934.
[18] Chamberlain, Edward H., *The Theory of Monopolistic Competition*, Cambridge, Harvard University Press, 6th Ed., 1950.
[19] Nash, J. F., Jr., "Non-Cooperative Games," *Annals of Mathematics*, Vol. 54, September 1951, pp. 286–295.
[20] Mayberry, J. P., J. F. Nash, and M. Shubik, "A Comparison of Treatments of a Duopoly Situation," *Econometrica*, Vol. 21, 1953, pp. 141–154.

[4]

Estratto da «Metroeconomica»
Vol. XX - Fasc. III
Settembre-Dicembre 1968

EXTENDED EDGEWORTH BARGAINING GAMES AND COMPETITIVE EQUILIBRIUM (*) (**)

by Martin Shubik, New-Haven

1. - Introduction

There has been little, if any, utilization of the von Neumann and Morgenstern concept of solution to an n-person non-constant sum game to economic problems. In this paper an attempt will be made to link up this solution concept with welfare problems and further to give an interpretation relating it to the properties of equilibrium and stability.

The description of an n-person game is given in terms of its *characteristic function*. As the models dealt with below are primarily based upon extensions of Edgeworth's two-person bargaining model this is used to provide the first example of a characteristic function.

Suppose that two bargainers have initial fortunes consisting respectively of $(a, 0)$ and $(0, b)$, where the first number of the pair is the amount of the first two commodities being traded which is possessed by the bargainer. The second number is the amount of the second commodity possessed by the bargainer. Suppose that both bargainers possess a preference system which can be represented by a family of convex curves for each of them. Let them be $\varphi_1(x, y)$ and $\varphi_2(x, y)$. Suppose that the bargainers were able to compare utilities. The characteristic function below describes this game:

$$v(\Theta) = 0$$
$$v(\{1\}) = \varphi_1(a, 0)$$
$$v(\{2\}) = \varphi_2(0, b)$$
$$v(\{1, 2\}) = \max_{x_1, y_1} [\varphi_1(a-x_1, y_1) + \varphi_2(x_1, b - y_1)]$$

The $v(\Theta)$ merely states that the value of a coalition with no members is zero. The next two values given are the amounts that each player can obtain for himself by individual action. Each bargainer can guarantee for himself a value no lower than that obtained by refusing to trade. The last value given is that for the largest amount that

(*) *Research undertaken by the Cowles Commission for Research in Economics under Task NR-047-006 with the Office of Naval Research.*

(**) Editorial note: This article was accepted by Metroeconomica in 1963. The delay in its pubblication is due to the reorganisation of our Journal.

the bargainers can obtain together. On the assumption that utilities are comparable this will in general be a uniquely defined amount. Edgeworth calls this the « utilitarian point ». If trading is to be worth while, the sum of the utilities to be obtained by trade should be larger than the sum of the utilities obtained without trade. Thus in the bargaining situation:

$$v(\{1,2\}) > v(\{1\}) + v(\{2\}).$$

The characteristic function is a super-additive set function. It is defined, i.e. , it takes on a value for every possible set of players in a game. In other words, it gives a value for every coalition. For example, if the game has four players the characteristic function will have 2^4 or 16 values; one for the no-member coalition, four one man coalitions, six coalitions of two, four coalitions of three and one coalition of four.

In a game that is worth playing there must be at least two sets of players who can improve their gain by cooperating. Thus in an *essential game* there will be at least two coalitions without members in common, say M and N such that $v(M \cup N) > v(M) + v(N)$. It is furthermore assumed that no coalition will ever obtain less than can be obtained by smaller coalitions consisting of the same individuals. Hence always for any two sets of players M and N with no member in common, $v(M \cup N) \geq v(M) + v(N)$. The characteristic function is super-additive as the amount to be obtained by a coalition is always at least as much as the sum of the amounts obtained by the same players acting in smaller coalitions.

The characteristic function provides a useful description for complementarity and substitutability relations among indivisible objects, hence the sets can equally well be considered to be groupings of specialized machines as well as coalitions of bargainers.

The amounts that can be obtained by coalitions are given in the values of the characteristic function. In general we will want to know the amount that each individual player receives after the bargaining has taken place and side-payments have been agreed upon. Let x_i be the amount obtained by the ith player. We denote by x the n-tuple of numbers $(x_1, x_2, \dots, x_n)$ which indicates the amount obtained by each player according to some scheme of division. The first condition we impose upon these numbers is that for all i

$$v(\{i\}) \leq x_i \tag{1}$$

This states that under any scheme of division no player will obtain less than he can obtain by acting unilaterally.

A second condition to be imposed is that of Pareto Optimality (or even more strongly joint maximality). Let us denote by I the set consisting of all of the players. In this instance they number

n. $v(I)$ stands for the amount that the players can obtain from the game by acting as one large coalition. This is the most that they can obtain under any circumstances, thus it is not possible for the sum of the payments to all players to exceed this amount. If the players behave optimally it should equal this amount.

$$(2) \qquad \sum_{i \text{ in } I} x_i = v(I) .$$

An n-tuple which meets the above conditions is referred to as an *imputation* from a game with the characteristic function v. The von Neumann and Morgenstern solution concept is addressed to the selection of certain sets of imputations as solutions to an n-person game.

The selection of the sets of imputations in a solution depends upon the relationship of *domination*. An imputation $x = (x_1, x_2, \ldots, x_n)$ dominates an imputation $y = (y_1, y_2, \ldots, y_n)$ with respect to a coalition of players S if (on the assumption that S contains some players)

$$(3) \qquad v(S) \geq \sum_{i \text{ in } S} x_i$$

and

$$(4) \qquad x_i > y_i \text{ for every } i \text{ belonging to } S .$$

In words, for x to dominate y, the set of players S must be in a position to obtain at least the amount they are suggesting for themselves in x. Furthermore, everyone in the coalition S strictly prefers x to y. The set of players S is called an *effective set.*

It is possible for an imputation x to dominate an imputation y and vice versa (with respect to different effective sets). Furthermore, the relation is not transitive. x may dominate y, which dominates z, which in turn dominates x. If we consider voting agreements, political power and horse-trading arrangements over appropriations, these possibilities are not strange. The stability of an imputation depends upon the presence of a group of individuals with something better to offer rather than upon an atomistic reaction to some price parameters.

A *solution* to a game, in the sense of von Neumann and Morgenstern, consists of a set of imputations B such that:

If x and y are imputations contained in B, then neither dominates the other.

If there exists an imputation z which is not contained in B then there is at least one imputation belonging to B which dominates z.

The meaning of this solution concept can be demonstrated with reference to the Edgeworth bilateral monopoly model. Before this is done, we can view the important distinction between the Edgeworth treatment and the von Neumann and Morgenstern treatment in terms of games with and without side payments. Figure 1 illustrates a different representation of the familiar *contract curve* shown in Figure 2.

— 302 —

If the utilities of the players were comparable it would then be meaningful to find a point where they were jointly maximized. This is denoted by *U* in this diagram. Furthermore, if there were a supply of «utiles» or a «money» commodity for which the players had a linear

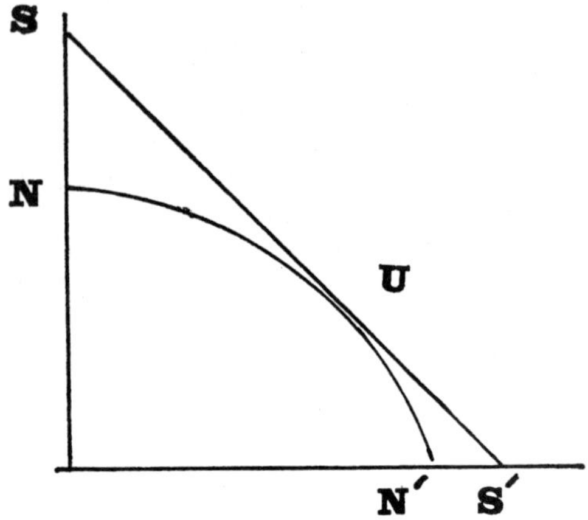

Figure 1.

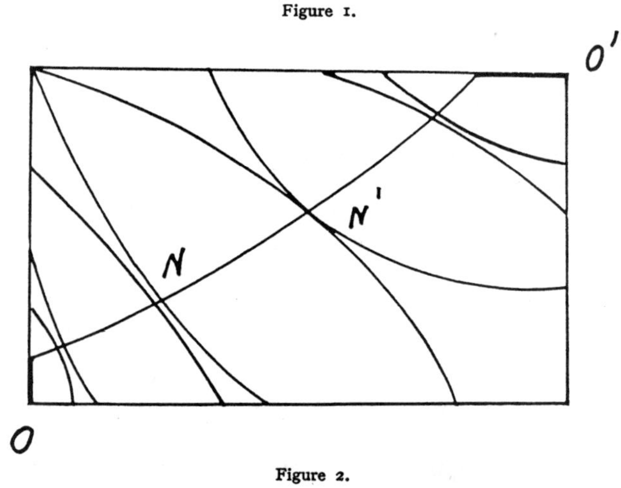

Figure 2.

utility, then they could jointly maximize regardless of the asymetry involved in doing so, and then settle up by means of side payments along the 45° line denoted by *SS′*. This is what happens in the von Neumann and Morgenstern theory. Thus their model of the Edgeworth two-person, two-commodity trade is in fact a two-person three-commodity trade model with the third good being utiles.

If side-payments are forbidden, then the outcome must be restricted to the curve *NN′* or the contract curve (which is part of the Pareto optimal surface). This is the « best » that the players can do if they are unable to settle up outside of the game. An immediate analogy can be made in terms of cartel behavior. The curve *NN'* represents the best that a pair of firms can obtain if they are not in a position to make cash side-payments after having adjusted their outputs to equalize marginal costs in both plants (in this instance the assumption that there is a third good in the form of money which can be treated as transferable utility appears to be fairly reasonable).

We can modify the von Neumann and Morgenstern assumptions to rule out side-payments in which case the best that a coalition of the two players can do is to obtain some point on the contract curve. When there are only two players it is obvious that by forbidding side-payments and modifying the von Neumann and Morgenstern solutions accordingly, then it is also not necessary to assume the comparability of utilities.

The solution to the bargain in the sense of Edgeworth (or von Neumann and Morgenstern without side-payments) is the set of imputations on the contract curve such that for any imputation (x_1, x_2), $x_1 \geq \psi_1(a, 0)$ and $x_2 \geq \psi_2(0, b)$. This is proved by showing that any imputation on this curve does not dominate any other imputation on the curve. Consider two imputations (x_1, x_2) and (y_1, y_2). If $x_1 > y_1$ then $x_2 < y_2$, hence the two person coalition is not an effective set. Furthermore, for any of these imputations excluding the end points of the contract curve, condition (3) is violated. For the end points if condition (3) is not violated condition (4) will be. This implies that no single player forms an effective set for any of the imputations on the contract curve.

Using a completely similar argument, but assuming side-payments with comparable utilities it can be shown that the von Neumann and Morgenstern solution is the « extended contract curve » *SS'* of Figure 1, as has been previously noted.

2. - Some general problems

In dealing with models of bargaining and welfare there are four problems, which though related, are distinct. They involve the implications of the assumptions concerning:

(1) The measurability of the utility of an individual
(2) The comparability of utility
(3) The transferability of utility (i. e., side-payments)
(4) The importance of the role of coalitions of players.

The fourth assumption is utterly independent of the others. Although it fast becomes mathematically complex, it has been shown that even on the assumption of ordinal non-comparable utility scales it is possible to define a solution concept broadly analagous to that of von Neumann and Morgenstern taking into account the role of coalitions [1]. Thus even rejecting the first three assumptions, there is still a need and a possibility for a theory of coalitions.

The price-parameter or competitive equilibrium theory is a theory which assumes no measurability, comparability or transferability of utility and furthermore assumes no role for coalitions. The fact that at equilibrium the market is cleared implies no side-payments.

It is remarkable that under relatively general conditions [2] a competitive equilibrium point exists on the Pareto optimal surface. Beyond the observation that it is on this surface, there is no particular ethical content which distinguishes this point from any other on the surface. In fact it may not even be unique. There is, however, a strong *administrative argument* for the competitive equilibrium. It is simple to administer in the sense that it is utterly decentralized, there is no bargaining or discussion between any individuals. However instead of being an institution-free concept of economic solution, it should be more correctly viewed as one which calls for a very special institutional set-up in which man as an economic animal takes great care to purify himself from contact, collusion or collaboration with his fellows.

The recent concern for the stability of competitive equilibria [3] illustrates the degree of specialty of the assumptions even further. Differential or difference equations must be introduced which depend in some manner upon the excess supply and demand functions [4]. The selection of the specific forms is more or less « dealer's choice ». In his series of counter-examples Scarf has shown [5] that even under some of the « most natural » choices the equilibria may be unstable. Intuitively this appears to be a disturbingly undesired result. Yet even the most casual empiricism seems to indicate that whatever stability there is in an economy tends to owe its strength to enforcement by groups or coalitions. For example, the instability of an agricultural market appears to be aggravated by the lack of coalitions.

Any concept of stability which, in my opinion, appears to be natural to an economic system should be related to the coalition structures potentially able to disturb it. The stability of an imputation should be inversely related to the number of *effective sets* of players that exist for that imputation. This imposes many more conditions

(1) Shapley L. S. and M. Shubik: *Solutions of n-person games with ordinal utilities* (Abstract), « Econometrica », 21, pp. 348-349, 1953.

(2) See for instance: Arrow J. K., and G. Debreu: *Existence of an Equilibrium for a Competitive Economy*, « Econometrica », 22, 1954, pp. 265-290; Wald A. *On Some Systems of Equations of Mathematical Economics*, « Econometrica », 19, pp. 368-403, 1951.

(3) Arrow K. J. and L. Hurwicz: *On the Stability of the Competitive Equilibrium*, « Econometrica », 26, pp. 522-552, 1958.

(4) *Ibid.*, p. 525.

(5) Scarf H.:*Some Examples of Global Instability of the Competitive Equilibrium*, « *CFDP No.* 79, » October, 1959.

than are imposed by the conditions under which the stability of a price-parameter equilibrium has been examined.

The von Neumann and Morgenstern solution (as it stands, or modified to rule out side-payments) to an n-person game is a socio-economic concept. The economic bounds are reflected in the utility functions and assets of the participants. However the stability of the imputation of wealth is examined in terms of the sociological conditions reflected by the power of groups.

The remainder of this paper is devoted to illustrating the properties of the von Neumann and Morgenstern solution and associating them with the concept of equilibrium by examining an extension of the Edgeworth bargaining model. The proofs and initial discussion are given in terms of games with comparable transferable utilities; however it will be seen that at least for the relatively simple games under discussion the results generalize immediately to the case where no side-payments are permitted.

3. - Edgeworth Market Games

Edgeworth examined bargains with more than two players and concluded that if the players became numerous on both sides of the market (under certain conditions) the bargaining would be such that the range of indeterminacy characterized by the contract curve would narrow until in the limit a single point would be reached which would correspond to the market price under pure competition.

An Edgeworth market game consists of a set $I = M \cup N$ of players where the set of players M can be regarded as traders with initial assets of commodity 1 and the set of players N are traders with initial assets of commodity 2. Each player has a utility function consisting of a set of strictly convex indifference curves (i. e., an indifference map). The utility function of the ith player is denoted by $\Psi_i(x, y)$ where x and y are the quantities of the first and second commodity. It is assumed that:

$$\lim_{x \to \infty} \Psi_i(x, y) < \infty \text{ and } \lim_{y \to \infty} \Psi_i(x, y) < \infty$$

and that the Ψ_i are twice differentiable. The ith player in the set M starts with initial resources $(a_i, 0)$ of goods 1 and 2; where the first number is the amount of the first good and the second, the amount of the second. The jth player in set N has initially $(0, b_j)$. The characteristic function for the game is:

$$v(\Theta) = 0$$

$$v(S) = \max_{x_i, y_i} \left[\sum_{i \varepsilon S \Omega M} \Psi_i(a_i - x_i, y_i) + \sum_{j \varepsilon S \Omega N} \Psi_j(x_j, b_j - y_j) \right],$$

where $\Sigma x_i = \Sigma x_j$, $\Sigma y_i = \Sigma y_j$ and $x_j, y_i, \geq 0$.

In order to examine this game fully we would have to consider many special cases depending upon the shapes of the utility functions, the initial stocks of the traders and the number of traders. We limit ourselves to the simple case where all traders have identical utility functions $\Psi_i(x, y) = \Psi(x, y)$. Each trader belonging to M initially has $(a, 0)$ and each trader belonging to N has $(0, b)$.

Because of the above restrictions we can immediately write down the explicit characteristic function for any game of this type. If there is no joint gain to be obtained from trade, the game is inessential and the value of any coalition is the sum of the individual initial utilities. If the game is essential and trade takes place, then by the symmetry and convexity, at a position of joint maximization for any group of players all players will obtain equal quantities of the same good (at this point the marginal rate of substitution between the two commodities being traded will be the same for all players trading). The characteristic function is:

$$v(\Theta) = 0$$

$$v(S) = s\,\psi\left(\frac{as_m}{s}, \frac{bs_n}{s}\right)$$

where $s = |S|$, $s_m = |S \cap M|$ and $s_n = |S \cap N|$. The characteristic function for the two-person game becomes:

$$v(\Theta) = 0$$
$$v(\{1\}) = \Psi(a, 0)$$
$$v(\{2\}) = \Psi(0, b)$$
$$v(\{1, 2\}) = 2\,\psi\left(\frac{a}{2}, \frac{b}{2}\right).$$

The solution to this game consists of the set of imputations of the form:

$$\left(2p\,\psi\left(\frac{a}{2}, \frac{b}{2}\right) - p\,\psi(0, b) + (1 - p)\,\psi(a, 0),\right.$$

$$\left.2(1 - p)\,\psi\left(\frac{a}{2}, \frac{b}{2}\right) + p\,\psi(0, b) - (1 - p)\,\psi(a, 0)\right)$$

$$0 \leq p \leq 1.$$

It can be shown that if the number of traders in one commodity is the same as the number of traders in the other, then for any size of the market there is a solution analogous to the solution to the two-person game (although such a solution will no longer be the only solution).

There exists a set of imputations which form a solution such that all traders on the same side of the market always obtain an equal amount. The set of imputations is characterized by a parameter p (as in the two-person game) which can be interpreted as a market price. The range of this price remains constant for all games with the same number of players on each side of the market.

THEOREM 1. – An Edgeworth market game denoted by $[n, n]$ for any size n where the number of traders in one commodity is the same as the number of traders in the other commodity and the traders have the same preferences will have a solution consisting of all imputations of the form:

$$\left(2p\,\psi\left(\frac{a}{2},\frac{b}{2}\right) - p\,\psi(0, b) + (1-p)\,\psi(a, 0), \ldots, 2(1-p)\,\psi\left(\frac{a}{2},\frac{b}{2}\right)\right.$$

$$\left. + p\Psi(0, b) - (1-p)\Psi(a, 0), \ldots\right), \tag{1}$$

$$0 \leq p \leq 1.$$

PROOF. No imputation of the form of (1) can dominate another imputation of that form. A set to be effective for any imputation requires members from both M and N, i.e., trade must take place; but no imputation of the form (1) dominates another of form (1) for a set containing members from both M and N.

Any imputation not of the form (1) has a pair $\gamma'_i + \gamma''_j < 2\Psi\left(\frac{a}{2}, \frac{b}{2}\right)$ where γ'_i is the amount a player $i \in M$ and γ''_j is the amount a player $j \in N$ obtains from this imputation. If we select a p such that

$$\frac{\gamma'_i - \psi(a, 0)}{2\,\psi\left(\frac{a}{2},\frac{b}{2}\right) - \psi(a, 0) - \psi(0, b)} < p < 1 - \frac{\gamma''_j - \psi(0, b)}{2\,\psi\left(\frac{a}{2},\frac{b}{2}\right) - \psi(a, 0) - \psi(0, b)}$$

then the imputation of form (1) for this value of p dominates the imputation γ. This completes the proof.

We have shown that a line in $2n$ dimensions which is the direct analogue of the contract curve is a solution to the $2n$ person game. Furthermore the range of this solution is the same as in the range in the two person game. This however is *not* the only solution to the $2n$ person gane, although the contract curve is the only solution to the two person game. It appears that there will be many sets of imputations which satisfy the stability conditions. Thus as is indicated below there can be solutions in which subsets of players may fare relatively poorly as compared to others with the same initial economic strength.

— 308 —

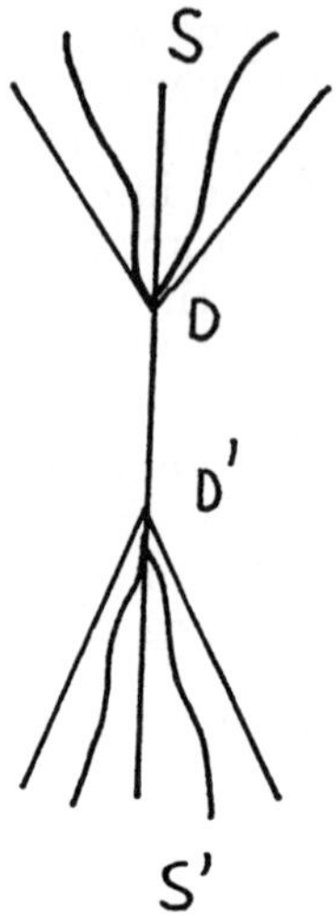

Figure 3.

The vertical line in Figure 3 must be imagined as the contract curve in *2n* dimensions. The lines departing from it represent sections of other solutions which include only part of the contract curve as members of the stable set. These may have very general shapes. No attemps is made here to calculate the totality of solutions which exist. For our purposes their *intersection* is of prime concern. In Figure 3 all the solutions have a common segment DD'. This segment is part of the contract curve, but for $n > 1$ in any of the *2n* person games the segment DD' will not be the full length of the contract curve SS'. Under the appropriate conditions as the number of bargainers becomes large, we will show that the (multidimensional) segment DD' converges to a single point.

The intersection of all of the solutions, to a game is known as its *core*. We give it an economic interpretation upon observing that it consists of a set of imputations which are always undominated under every circumstance. There is no effective set of players for any of these imputations. In economic terms, these imputations are stable against the economic power of any coalition. In the two person bilateral monopoly there is only one solution, hence the whole contract curve is the core of the game. The core appears to be determined by the technological conditions. Any imputation within the core gives to all players at least as much as they are able to enforce in any coalition. In any economy any imputation of wealth outside of the core implies that some group of individuals is profiting at the expense of another group.

The core to a game need not always exist. In particular this may hold true in an economy with increasing returns. This implies that in such an economy no division of product can be made which is stable against all coalitions.

4. - CORES AND COMPETITIVE EQUILIBRIA

The proofs of the following two theorems have appeared elsewhere (6) hence they will not be reproduced here. The discussion is limited to the economic interpretation of the results.

We consider the Edgeworth market game with one trader of the first type and examine the behavior of the core as the number of traders of the second type increases. The core approaches a single imputation at which the solitary trader of the first type acts as a perfectly discriminating monopolist and obtains all the gain from trading. The theorem may be stated as follows:

THEOREM 2. In an Edgeworth market game denoted by $[1, n]$, for any n the imputation:

$$\eta = \left((n+1)\,\psi\left(\frac{a}{n+1}, \frac{nb}{n+1}\right) - n\,\psi(0, b),\ \psi(0, b),\ \psi(0, b),\ \ldots,\ \psi(0, b)\right)$$

lies in the core. For any ε there exits an $n(\varepsilon)$ such that for all $n \geq n(\varepsilon)$ no imputation ζ with $\zeta_1 = \eta_1 - \varepsilon$ lies in the core.

In Theorem 3 a very specialized condition is introduced which results in the core of the game converging to the competitive equilibrium point of the no side-payments game.

THEOREM 3. In an Edgeworth market game denoted by $[m, n]$ such that $m = km'$, $n = kn'$ and

$$\psi\left(\frac{m'\,a}{m'+n'}, \frac{n'\,b}{m'+n'}\right) = max\ \psi\left(\frac{s_m\,a}{s}, \frac{s_n\,b}{s}\right)$$

the imputation

$$\eta = \left(\frac{v(I)}{m+n}, \frac{v(I)}{m+n}, \ldots, \frac{v(I)}{m+n}\right)$$

is always in the core; and for any ε there exists a $k(\varepsilon)$ such that for all $k \geq k(\varepsilon)$ no imputation ζ with a component

$$\zeta_1 < \frac{v(I)}{m+n} - \varepsilon$$

is in the core, $i = 1, \varepsilon, \ldots, m+n$.

(6) SHUBIK M.: *Edgeworth Market Games*, in Tucker, A. W., and R. D. LUCE, (eds.), *Contributions to the Theory of Games* V. IV (Princeton: Princeton University Press, pp. 267-278, 1959.

The proof of this theorem is similar to that of Theorem 2. An interpretation of the condition that

$$\psi\left(\frac{m' a}{m' + n'}, \frac{n' b}{m' + n'}\right) = max\ \psi\left(\frac{s_m a}{s}, \frac{s_n b}{s}\right)$$

is obtained by examining the nature of the function

$$s\ \psi\left(\frac{s_m a}{s}, \frac{s_n b}{s}\right).$$

It is homogeneous in the two variabies, s_m and s_n. If s_m and s_n are increased in a constant ratio the value of the optimum trade increases in a constant manner. In economic terminology there are constant returns to scale.

The meaning of the optimum ratio can be seen for this rather specialized symmetric game by observing from the form of the characteristic function that in the process of obtaining the joint maximum each individual has as his final stock p units of a and (l-p) units of b where $0 \leq p \leq 1$ depending upon the ratios of the different types of players. There will be an optimum size for p or in other words there will be an optimal coalition.

This size will be determined by the size of the initial stock of the players and the shape of the indifference maps. If the size of the game increases with the players remaining in the ratio given by p, then the core converges to the single point which is the point of tangency between the hyperplane representing the imputation space in the side-payment game and the Pareto optimal surface in the no side-payment situation. But this point also has the property of a competitive equilibrium. Given the initial stocks of the players there exists a price which clears the market and brings them to the Pareto optimal surface.

In this instance we have the property that the competitive equilibrium is the only undominated imputation in the limiting game.

A simple example of the optimum ratio for traders is given if we suppose that all members of the society maintain that the best proportion between the amount of tea and sugar used is 2 ounces of the first with 13 ounces of the second. If all players of the first type come to the market with 2 ounces of tea, and all of the second come with 6 ounces of sugar, then the optimal proportions between the players is 6 of the first type to 13 of the second.

There are two theorems concerning the core of a game and increasing and decreasing returns. These results were obtained after the first submission of this article and in various forms are due to the work of Aumann, Scarf, Shapley and Shubik.

THEOREM 4. In an n-person economy with increasing returns a core may not exist.

THEOREM 5. In an n-person economy with decreasing returns the core always exists.

5. - CONCLUSIONS.

One of the major reason why it has been so hard for economists to even examine the relevance of the von Neumann and Morgenstern theory for the cooperative solution of games is that in their book very little attention was paid to constructing games with an economic content. In order to do so the characteristic function of the game must be obtained from economic considerations such as the initial assets of individuals, production functions and preference relations. Having given the characteristic function economic content it then appears to be reasonable to ask if the solutions appear to bear any relationship to the body of economic knowledge.

Although in the applications of welfare, utility comparisons are made daily and transfer payments are based upon them and the power of the coalitions involved; the pills of comparable and transfetable utility are hard to swallow. This still does not rule out attempting to introduce the important role of coalitions into any solution concept. The neat formulation of the characteristic function with a single value for each coalition must be abandoned and replaced by a « Pareto sub-surface » which indicates the restraints upon the payoffs between the players. Edgeworth was aware of the role of coalitions and pointed out that if they were permitted, the contract curve would not shrink as players became numerous (7). Even without giving the formal definition for a solution to a game without transferable utilities, in the $2n$ person games described, making use of symmetry it is easy to see that the full contract curve is still a solution.

What does a unique point core mean? Is it a pure coincidence that the core and the equilibrium point coincided in the symmetric game discussed above? A unique point core means that there is only one possibile imputation of wealth in that society which yields as much or more to all participants than every conceivable subgroup considered as a whole can offer to its members.

In Section 4 an example of an optimal combination of 6 tea traders with 13 sugar traders was given. The goods are infinitely divisible, the preferences of the players may be assumed to be continuous, however the players are not divisible. Thus although the price parameter or competitive equilibrium solution for the games $(6n, 13n)$ will always be the same, in the (6, 13) market game the

---

(7) EDGEWORTH F. Y.: *Mathematical Psychics*, (London: KEGAN PAUL, 1881), pp. 42-43.

coalitions are at loggerheads between any point on the contract curve. This is because of the fewness and indivisibility of the players. For a large *n* , there is always the possibility of finding a « more reasonable » optimal group of traders, thus the relative importance of the indivisibility caused by the presence of subgroups of optimum size diminishes. The possibility that large groups may still act together is reflected in the multiplicity of solutions; the narrowing of the core reflects the lessening effect of fewness as more combinations of players obtain power over the imputations.

6. - Post script.

Since 1961 it has been shown by Scarf and Debreu [8] that the convergence of the core to the competitive equilibrium takes place without transferable utility. The equilibrium is not unique [8]. Recently Lucas has shown that the von Neumann-Morgenstern solution does not necessarily always exist [9], this raises once more some fundamental questions concerning the nature and meaning of social stability.

---

(8) Debreu G. - Scarf H. « A limit theorem on the core of on Economy » *International Economic Review* 1963 volume 4 n. 3 pp. 235-246.

(9) Lucas W. F. « A Game with no solution » *RM-5518 PR* October 67.

[5]

JOURNAL OF ECONOMIC THEORY, 1, 00–00 (1969)

On Market Games[1]

LLOYD S. SHAPLEY

The RAND Corporation, Santa Monica, California 90406

AND

MARTIN SHUBIK[2]

Department of Administrative Sciences, Yale University, New Haven, Connecticut 06520

The "market games"—games that derive from an exchange economy in which the traders have continuous concave monetary utility functions, are shown to be the same as the "totally balanced games"—games which with all their subgames possess cores. (The core of a game is the set of outcomes that no coalition can profitably block.) The coincidence of these two classes of games is established with the aid of explicit transformations that generate a game from a market and vice versa. It is further shown that any game with a core has the same *solutions*, in the von Neumann–Morgenstern sense, as some totally balanced game. Thus, a market may be found that reproduces the solution behavior of any game that has a core. In particular, using a recent result of Lucas (see RM–5518–PR), a ten-trader ten-commodity market is described that has no solution.

1. INTRODUCTION

Recent discovery of n-person games in the classical theory which either possess no solutions [*8, 9*], or have unusually restricted classes of solutions [*6, 7, 10, 18*], has raised the question of whether these games are mere mathematical curiosities or whether they could actually arise in applica-

[1] This research is supported by the United States Air Force under Project RAND —Contract No. F44620–67–C–0045—monitored by the Directorate of Operational Requirements and Development Plans, Deputy Chief of Staff, Research and Development, HQ USAF. RAND Memoranda are subject to critical review procedures at the research department and corporate levels. Views and conclusions expressed herein are nevertheless the primary responsibility of the author, and should not be interpreted as representing the official opinion or policy of the United States Air Force or of The RAND Corporation.

This memorandum extends to economic models some recently discovered counter examples in n-person theory. It is part of a continuing investigation by the authors of the application of game theory to economics.

[2] Dr. Shubik, a RAND consultant, is Professor of the Economics of Organization at Yale University.

tion. Since the most notable applications of n-person game theory to date have been to economic models of exchange, or exchange and production [*3*, *13*, *15*, *16*, *19–23*], the question may be put in a more concrete form: Are there markets, or other basic economic systems, that when interpreted as n-person games give rise to the newly-discovered counterexamples? If so, can they be distinguished from the ordinary run of market models, on some economic, heuristic, or even formal grounds —i.e. do they give any advance warning of their peculiar solution properties?

These questions stimulated the present investigation. The answers are "bad news": Yes, the games can arise in economics; No, there are no outwardly distinguishing features. In reaching these conclusions, however, we were led to a positive result: a surprisingly simple mathematical criterion that tells precisely which games can arise from economic models of exchange (with money). In fact, this criterion identifies a very fundamental class of games, called "totally balanced", whose further study seems merited quite apart from any consideration of solution abnormalities. Of technical interest, our derivation of the basic properties of these games and their solutions makes a substantial application of the recently-developed theory of *balanced sets* [*1*, *2*, *11*, *17*], as well as of the older work of Gillies on *domination-equivalence* [*4*, *5*].

In the present note we confine our attention to the classical theory "with side payments" [*23*]; this corresponds in the economic interpretation to the assumption that an ideal money, free from income effects or transfer costs, is available.[3] We further restrict ourselves to exchange economies, without explicit production or consumption processes, in which the commodities are finite in number and perfectly divisible and transferable, and in which the traders, also finite in number, are motivated only by their own final holdings of goods and money, their utility functions being continuous and concave and additive in the money term.

For our immediate purpose, these strictures do not matter, since the anomalous games are already attainable within the limited class of ideal markets considered. But for our larger purpose—that of initiating a systematic study of "market games" as distinct from games in general—some relaxation may be desirable, particularly with regard to money. The propects for significant generalizations in this direction appear good, and we intend to pursue them in subsequent work.

1.1 Outline of the Contents

The notions of game, core, and balanced set are reviewed. A game is called "balanced" if it has a core, and "totally balanced" if all of the

[3] For a discussion of this assumption, see [*20*], pp. 807–808.

subgames obtained by restricting the set of players have cores as well (Section 2).

A "market" is defined as an exchange economy with money, in which the traders have utility functions that are continuous and concave. The method of passing from a market to its "market game" is described. The market games with n traders form a closed convex cone in the space of all n-person games with side payments. Every market game not only has a core, but is totally balanced (Section 3).

A canonical market form—the "direct market"—is introduced, in which the commodities are in effect the traders themselves, made infinitely divisible, and the utility functions are all the same and are homogeneous of degree one. The method of passing from any game to its direct market is described, the utilities being based upon the optimal assignment of "fractional players" to the various coalitional activities. The "cover" of a game is defined as the market game of its direct market; the cover is at least as profitable to all coalitions as the original game. Every totally balanced game is its own cover, and hence is a market game. This shows that *the class of market games and the class of totally balanced games are the same*. Moreover, every market is game-theoretically equivalent to a direct market (Section 4).

The notions of imputation, domination, and solution are reviewed. Games are "*d*-equivalent" if they have identical domination relations on identical imputation spaces. They therefore have identical solutions (or lack of solutions), and their cores, if any, are the same. It is shown that *every balanced game is* d-*equivalent to a totally balanced game*. Hence, for every game with a core, there is a market that has precisely the same set of solutions (Section 5).

Using Lucas's solutionless game [*8*, *9*], a direct market is constructed that has ten traders and ten commodities (plus money), and that has no solution. Another version in the form of a production economy is also presented. Several other examples of market games with unusual solution properties are mentioned, and in one case, where the solutions contain arbitrary components, the utility function is worked out explicitly (Section 6).

2. GAMES AND CORES

For the purpose of this note, a *game* is an ordered pair $(N; v)$, where N is a finite set [the players] and v is a function from the subsets of N [coalitions] to the reals satisfying $v(O) = 0$, called the *characteristic function*. A *payoff vector* for $(N; v)$ is a point α in the $|N|$-dimensional vector space E^N whose coordinates α^i are indexed by the elements of N. If $\alpha \in E^N$ and $S \subseteq N$, we shall write $\alpha(S)$ as an abbreviation for $\sum_{i \in S} \alpha^i$.

The *core* of $(N; v)$ is the set of all payoff vectors α, if any, such that

$$\alpha(S) \geqq v(S), \quad \text{all } S \subseteq N, \tag{2-1}$$

and

$$\alpha(N) = v(N). \tag{2-2}$$

If no such α exists, we shall say that $(N; v)$ *has no core*. (Thus, in this usage, the core may be nonexistent, but is never empty.)

2.1 Balanced Sets of Coalitions

A *balanced set* $\mathscr{B}$ is defined to be a collection of subsets S of N with the property that there exist positive numbers γ_S, $S \in \mathscr{B}$, called "weights", such that for each $i \in N$ we have

$$\sum_{\substack{S \in \mathscr{B} \\ i \in S}} \gamma_S = 1. \tag{2-3}$$

If all $\gamma_S = 1$, we have a partition of N; thus, balanced sets may be regarded as generalized partitions.

For example, if $N = \overline{1234}$, then $\{\overline{12}, \overline{13}, \overline{14}, \overline{234}\}$ is a balanced set, by virtue of the weights 1/3, 1/3, 1/3, 2/3.

A game $(N; v)$ is called *balanced* if

$$\sum_{S \in \mathscr{B}} \gamma_S v(S) \leqq v(N) \tag{2-4}$$

holds for every balanced set $\mathscr{B}$ with weights $\{\gamma_S\}$.[4]

Theorem 1. *A game has a core if and only if it is balanced.*

This is proved in [*17*]. In Scarf's generalization to games without transferable utility [*11*], all balanced games have cores, but some games with cores are not balanced. If our present results can be generalized in this direction, we conjecture that it will be the balance property, rather than the core property, that plays the central role.

2.2 Totally Balanced Games

By a *subgame* of $(N; v)$ we shall mean a game $(R; v)$ with $O \subset R \subseteq N$. Here v is the same function, but implicitly restricted to the domain consisting of the subsets of R. A game will be said to be *totally balanced* if all of its subgames are balanced. In other words, all subgames of a totally balanced game have cores.

[4] These conditions are heavily redundant; it suffices to assert (2-4) for the *minimal* balanced sets $\mathfrak{B}$ (which moreover have unique weights). In the case of a superadditive game, only the minimal balanced sets that contain no disjoint elements are needed. (See [*17*].)

Not all balanced games are totally balanced. For example, let $N = \overline{1234}$ and define $v(S) = 0, 0, 1, 2$ for $|S| = 0, 1, 3, 4$ respectively, and, for $|S| = 2$:

$$v(\overline{12}) = v(\overline{13}) = v(\overline{23}) = 1$$
$$v(\overline{14}) = v(\overline{24}) = v(\overline{34}) = 0.$$

This game has a core, including the vector $(\frac{1}{2}, \frac{1}{2}, \frac{1}{2}, \frac{1}{2})$ among others. But it is not totally balanced, since the subgame $(\overline{123}; v)$ has no core.

3. MARKETS AND MARKET GAMES

For the purpose of this note, a *market* is a special mathematical model, denoted by the symbol (T, G, A, U). Here T is a finite set [the traders]; G is the nonnegative orthant of a finite-dimensional vector space [the commodity space]; $A = \{a^i: i \in T\}$ is an indexed collection of points in G [the initial endowments]; and $U = \{u^i: i \in T\}$ is an indexed collection of continuous, concave functions from G to the reals [the utility functions]. When we wish to indicate that $u^i \equiv u$, all $i \in T$ [the special case of "equal tastes"], we shall sometimes denote the market by the more specific symbol $(T, G, A, \{u\})$.

If S is any subset of T, an indexed collection $X^S = \{x^i: i \in S\} \subset G$ such that $\Sigma_S x^i = \Sigma_S a^i$ will be called a *feasible S-allocation* of the market (T, G, A, U).

A market (T, G, A, U) can be used to "generate" a game $(N; v)$ in a natural way. We set $N = T$, and define v by

$$v(S) = \max_{X^S} \sum_{i \in S} u^i(x^i), \quad \text{all } S \subseteq N, \tag{3-1}$$

where the maximum runs over all feasible S-allocations. Any game that can be generated in this way from some market is called a *market game*.[5]

In the special case of identical utility functions $u^i \equiv u$, we have

$$v(S) = |S| u(\textstyle\sum_S a^i / |S|), \quad \text{all } S \subseteq N; \tag{3-2}$$

this is a simple consequence of concavity. In the still more special case where u is homogeneous of degree 1, we have simply

$$v(S) = u(\textstyle\sum_S a^i), \quad \text{all } S \subseteq N. \tag{3-3}$$

3.1 Some Elementary Properties

The following two theorems are of a routine nature; they show that the property of being a market game is invariant under "strategic equivalence", and that the set of all market games on N forms a convex cone in the $(2^{|N|} - 1)$-dimensional space of all games on N.

[5] For examples, see [*13, 15, 16, 20, 21, 22*]. The abstract definition of "market game" proposed in [*13*] is not equivalent to the present one, however.

THEOREM 2. *If $(N; v)$ is a market game, if $\lambda \geqq 0$, and if c is an additive set function on N, then $(N; \lambda v + c)$ is a market game.*

Proof. We need merely take any market that generates $(N; v)$ and replace each utility function $u^i(x)$ by $\lambda u^i(x) + c(\{i\})$. Q.E.D.

THEOREM 3. *If $(N; v')$ and $(N; v'')$ are market games, then $(N, v' + v'')$ is a market game.*

Proof. Let (N, G', A', U') and (N, G'', A'', U'') be markets that generate $(N; v')$ and $(N; v'')$ respectively. We shall superimpose these two markets, keeping the two sets of commodities distinct. Specifically, let G be the set of all ordered pairs (x', x'') of points from G' and G'' respectively; let A be the set of pairs (a'^i, a''^i) of correspondingly-indexed elements of A' and A''; and let U be the set of sums:

$$u^i((x', x'')) = u'^i(x') + u''^i(x'')$$

of correspondingly-indexed elements of U' and U''. One can then verify without difficulty that the elements of U are continuous and concave on the domain G (which is a nonnegative orthant in its own right), so that (N, G, A, U) is a market. Finally, one can verify without difficulty that (N, G, A, U) generates the game $(N; v' + v'')$. Q.E.D.

3.2 THE CORE THEOREM

THEOREM 4. *Every market game has a core.*

This theorem is well known, and has been generalized well beyond the limited class of markets we are now considering. Nevertheless we shall give two proofs, both short, for the sake of the insights they provide. In the first, we in effect determine a competitive equilibrium for the generating market (a simple matter when there is transferable utility), and then show that the competitive payoff vector lies in the core. In the second proof, we show directly that the game is balanced, and then apply Theorem 1.

Proof 1. Let $(N; v)$ be a market game and let (N, G, A, U) be a market that generates it. Let $B = \{b^i : i \in N\}$ be a feasible N-allocation that achieves the value $v(N)$ in (3-1) for $S = N$. The maximization in (3-1) ensures the existence of a vector p [competitive prices—but possibly negative!] such that for each $i \in N$, the expression

$$u^i(x^i) - p \cdot (x^i - a^i), \quad x^i \in G, \tag{3-4}$$

is maximized at $x^i = b^i$. Define the payoff vector β by

$$\beta^i = u^i(b^i) - p \cdot (b^i - a^i);$$

we assert that β is in the core. Indeed, let S be any nonempty subset of N, and let Y^S be a feasible S-allocation that achieves the maximum in (3-1), so that $v(S) = \sum_S u^i(y^i)$. Since b^i maximizes (3-4), we have

$$\beta^i \geqq u^i(y^i) - p \cdot (y^i - a^i).$$

Summing over $i \in S$, we obtain

$$\beta(S) \geqq \textstyle\sum_S u^i(y^i) - p \cdot 0 = v(S),$$

as required by (2-1). Moreover, if $S = N$ we may take $Y^S = B$ and obtain $\beta(N) = v(N)$, as required by (2-2). Q.E.D.

Proof 2. Let (N, G, A, U) be a generating market for $(N; v)$, and, for each $S \subseteq N$, let $Y^S = \{y_S : i \in S\}$ be a maximizing S-allocation in (3-1). Let $\mathscr{B}$ be balanced, with weights $\{\gamma_S^i : S \in \mathscr{B}\}$. Then we have

$$\sum_{S \in \mathscr{B}} \gamma_S v(S) = \sum_{S \in \mathscr{B}} \sum_{i \in S} \gamma_S u^i(y_S^i) = \sum_{i \in N} \sum_{\substack{S \in \mathscr{B} \\ S \ni i}} \gamma_S u^i(y_S^i).$$

Now define

$$z^i = \sum_{\substack{S \in \mathscr{B} \\ S \ni i}} \gamma_S y_S^i \in G, \quad \text{all } i \in N.$$

Note that z^i is a center of gravity of the points y_S, by virtue of (2-3). Hence, by concavity,

$$\sum_{S \in \mathscr{B}} \gamma_S v(S) \leqq \sum_{i \in N} u^i(z^i). \tag{3-5}$$

But $Z = \{z^i : i \in N\}$ is a feasible N-allocation, since

$$\sum_{i \in N} z^i = \sum_{S \in \mathscr{B}} \sum_{i \in S} \gamma_S y_S^i = \sum_{S \in \mathscr{B}} \gamma_S \sum_{i \in S} a^i = \sum_{i \in N} a^i.$$

Hence the right side of (3-5) is $\leqq v(N)$, and we conclude from (2-4) that the game is balanced and from Theorem 1 that it has a core. Q.E.D.

COROLLARY. *Every market game is totally balanced.*

Proof. If $(N; v)$ is generated by the market (N, G, A, U), and if $O \subset R \subseteq N$, then we may define a market (R, G, A', U'), where A' and U' come from A and U by simply omitting all a^i and u^i for i not in R. This market clearly generates the game $(R; v)$. Hence $(R; v)$ is balanced.

Q.E.D.

Our next objective will be to prove the converse of this corollary—i.e. that every totally balanced game is a market game.

4. DIRECT MARKETS

A special class of markets, called *direct markets*, will play an important role in the sequel. They have the form

$$(T, E_+^T, I^T, \{u\}).$$

where u is homogeneous of degree 1 as well as concave and continuous. Here E_+^T denotes the nonnegative orthant of the vector space E^T with coordinates indexed by the members of T, and I^T denotes the collection of unit vectors of E^T—in effect, the identity matrix on T.

Thus, in a direct market, each trader starts with one unit of a personal commodity [e.g. his time, his labor, his participation, "himself"]. When it is brought together with other personal commodities, we may imagine that some desirable state of affairs is created, having a total value to the traders that is independent (because of homogeneity and equal tastes) of how they distribute the benefits.

Let e^S denote the vector in E^N in which $e_i^S = 1$ or 0 according as $i \in S$ or $i \notin S$; geometrically, these vectors represent the vertices of the unit cube in E_+^N. Then the characteristic function of the market game generated by a direct market can be put into a very simple form:

$$v(S) = u(e^S), \quad \text{all } S \subseteq N \tag{4-1}$$

(compare (3-3)). Note that only finitely many commodity bundles are involved in this expression.

4.1 The Direct Market Generated by a Game

Thus far we used markets to generate games. We now go the reverse route, associating with any game (not necessarily a market game) a certain "market of coalitions". Specifically, we shall say that the game $(N; v)$ "generates" the direct market $(N, E_+^N, I^N, \{u\})$, with u given by

$$u(x) = \max_{\{\gamma_S\}} \sum_{S \subseteq N} \gamma_S v(S), \quad \text{all } x \in E_+^N, \tag{4-2}$$

maximized over all sets of nonnegative γ_S satisfying

$$\sum_{S \ni i} \gamma_S = x_i, \quad \text{all } i \in N. \tag{4-3}$$

To explain this market,[6] we may imagine that each coalition S has an activity $\mathscr{A}_S$ that can earn $v(S)$ dollars if all the members of S participate fully. More generally, it earns $\gamma_S v(S)$ dollars if each member of S devotes the fraction γ_S of "himself" to $\mathscr{A}_S$. The maximization in (4-2) is then nothing but an optimal assignment of activity levels γ_S to the various $\mathscr{A}_S$'s, subject to the condition (4-3) that each player, i, distribute exactly the amount x_i of "himself" among his activities, including of course the "solo" activity $\mathscr{A}_{\{i\}}$.

The utility function defined by (4-2) is obviously homogeneous of degree 1, as required for a direct market. But before we can claim to

[6] The essence of this model was suggested by D. Cantor and M. Maschler (private correspondence, 1962).

have defined a market, let alone a direct market, we must also establish that (4-2) is continuous and concave. Continuity gives no trouble. To show concavity, it suffices (with homogeneity) to prove that

$$u(x)+u(y) \leqq u(x+y), \quad \text{all } x, y \in E^N_+.$$

This is not difficult. By definition, there exist sets of nonnegative coefficients $\{\gamma_S\}$ and $\{\delta_S\}$ such that

$$u(s) = \sum_{S \subseteq N} \gamma_S v(S), \qquad u(y) = \sum_{S \subseteq N} \delta_S v(S);$$

and

$$\sum_{S \ni i} \gamma_S = x_i, \qquad \sum_{S \ni i} \delta_S = y_i, \qquad \text{all } i \in N.$$

Hence $\{\gamma_S + \delta_S\}$ is admissible for $x+y$, and (4-2) yields

$$u(x+y) \geqq \sum (\gamma_S + \delta_S) v(S) = u(x)+u(y),$$

as required.

4.2 The Cover of a Game

We shall now use the direct market generated by a game $(N; v)$ to generate in turn a new game $(N; \bar{v})$—schematically:

arbitrary game ⟶ direct market ⟶ market game.

We shall call $(N; \bar{v})$ the *cover* of $(N; v)$.

Combining (4-1) with (4-2) and (4-3), we obtain the following relation beteen v and $\bar{v}$:

$$\bar{v}(R) = \max_{\{\gamma_S\}} \sum_{S \subseteq R} \gamma_S v(S), \quad \text{all } R \subseteq N \tag{4-4}$$

maximized over $\gamma_S \geqq 0$ such that

$$\sum_{\substack{S \subseteq R \\ S \ni i}} \gamma_S = 1, \quad \text{all } i \in R. \tag{4-5}$$

Note that we could have taken (4-4), (4-5) as the definition of "cover", bypassing the intermediate market. Indeed, the cover of a game proves to be a useful mathematical concept quite apart from the present economic application.

We see immediately that

$$\bar{v}(R) \geqq v(R), \quad \text{all } R \subseteq N, \tag{4-6}$$

since one of the admissible choices for $\{\gamma_S\}$ in (4-4) is to take $\gamma_R = 1$ and all other $\gamma_S = 0$. Moreover, the equality cannot always hold in (4-6); indeed, $\bar{v}$ comes from a market game while v was arbitrary. Thus, the mapping $v \to \bar{v}$ takes an arbitrary characteristic function and, by perhaps increasing some values, turns it into the characteristic function of a market game.

LEMMA 1. *If $(N; v)$ has a core, then $\bar{v}(N) = v(N)$, and conversely.*

Proof. Let α be in the core of $(N; v)$. Then

$$\begin{aligned}\bar{v}(N) &= \max_{\{\gamma_S\}} \sum_{S \subseteq N} \gamma_S v(S) \\ &\leqq \max_{\{\gamma_S\}} \sum_{S \subseteq N} \gamma_S \alpha(S) = \max_{\{\gamma_S\}} \sum_{i \in N} \alpha^i \sum_{S \ni i} \gamma_S \\ &= \max_{\{\gamma_S\}} \sum_{i \in N} \alpha^i = \alpha(N) \\ &= v(N),\end{aligned}$$

the successive lines being justified by (4-4), (2-1), (4-5), and (2-2). In view of (4-6) we therefore have $\bar{v}(N) = v(N)$.

Conversely, if $(N; v)$ has no core, then (2-4) fails for some balanced set $\mathscr{B}$ with weights $\{\gamma_S\}$. Defining $\gamma_S = 0$ for $S \notin \mathscr{B}$, we see that (4-5) holds (for $R = N$). Then (4-4) and the denial of (2-4) give us

$$\bar{v}(N) \geqq \sum_{S \subseteq N} \gamma_S v(S) = \sum_{S \in \mathscr{B}} \gamma_S v(S) > v(N);$$

Hence $\bar{v}(N) \neq v(N)$. Q.E.D.

LEMMA 2. *A totally balanced game is equal to its cover.*

Proof. Let $(N; \bar{v})$ be the cover of $(N; v)$, and let $0 \subset R \subseteq N$. Then it is clear from the definitions that the cover of $(R; v)$ is $(R; \bar{v})$. But if $(N; v)$ is totally balanced, then $(R; v)$ has a core and $\bar{v}(R) = v(R)$ by Lemma 1. Hence $\bar{v} = v$. Q.E.D.

THEOREM 5. *A game is a market game if and only if it is totally balanced.*

Proof. We proved earlier (corollary to Theorem 4) that market games are totally balanced. We have just now shown that totally balanced games are equal to their covers, which are market games. Q.E.D.

4.3 EQUIVALENCE OF MARKETS

There is one more result of some heuristic interest that we can extract from the present discussion, before entering the realm of solution theory. This time, we follow the scheme:

arbitrary market ———→ market game ———→ direct market.

Let us call two markets *game-theoretically equivalent* if they generate the same market game. Then the two markets in the above scheme are equivalent in this way, since the cover of the game in the middle is just the market game of the market on the right, and these two games are equal by Lemma 2. This proves

THEOREM 6. *Every market is game-theoretically equivalent to a direct market.*

5. SOLUTIONS

An *imputation* for a game $(N; v)$ is a payoff vector α that satisfies

$$\alpha(N) = v(N) \tag{5-1}$$

and

$$\alpha^i \geqq v(\{i\}), \quad \text{all } i \in N. \tag{5-2}$$

A comparison with (2-1) and (2-2) shows that the imputation set is certainly not empty if the game has a core.[7]

Classical solution theory [*23*] rests upon a relation of "domination" between imputations. If α and β are imputations for $(N; v)$, then α is said to *dominate* β (written $\alpha \text{ — } \beta$) if there is some nonempty subset S of N such that

$$\alpha^i > \beta^i, \quad \text{all } i \in S, \tag{5-3}$$

and

$$\alpha(S) \leqq v(S). \tag{5-4}$$

A *solution* of $(N; v)$ is defined to be any set of imputations, mutually undominating, that collectively dominate all other imputations. Our only concern with this definition, technically, is to observe that it depends only on the concepts of imputation and domination; any further information conveyed by the characteristic function is disregarded.

The core is also closely dependent on these concepts. In fact, the core, when it exists, is precisely the set of undominated imputations. The converse is not universally true—there are some games that have undominated imputations but no core.[8] We can rule this out, however, by imposing the very weak condition:

$$v(S) + \sum_{N-S} v(i) \leqq v(N), \quad \text{all } S \subseteq N, \tag{5-5}$$

which is satisfied by all games likely to be met in practice.[9]

5.1 Domination-Equivalence

Two games will be called *d-equivalent* (domination-equivalent) if they have the same imputation sets and the same domination relations on them. It follows that *d*-equivalent games have precisely the same solutions, or lack of solutions. Also, if they have cores, they have the same cores;

[7] Some approaches to solution theory omit (5-2), relying on the solution concept itself to impose whatever "individual rationality" the situation may demand [*5*, *12*]. This modification in the definition of solution would make little difference to our present discussion, except for eliminating the fussy condition (5-5). In particular, Theorem 7 and all of Section 6 would remain correct as written.

[8] We are indebted to Mr. E. Kohlberg for this observation.

[9] Thus, (5-5) is implied by either superadditivity or balancedness, but is weaker than both. For a game in normalized form, i.e. with $v(i) \equiv 0$, it merely states that no coalition is worth more than N.

moreover, within the class of games satisfying (5-5) the property of being balanced is preserved under *d*-equivalence. However, the property of being totally balanced is not so preserved, as the following lemma reveals.

LEMMA 3. *Every balanced game is d-equivalent to its cover.*

Proof. Let $(N; v)$ be balanced. By Lemma 1, $\bar{v}(N) = v(N)$ and by (4-4), $\bar{v}(\{i\}) = v(\{i\})$; hence the two games have the same imputations. Denote the respective domination relations by — and —′. By (4-6) and (5-4) we see at once that the latter is, if anything, stronger than the former—i.e. α — β implies α —′ β. It remains to prove the converse.

Assume, *per contra*, that α and β are imputations satisfying α —′ β but not α — β. Then for some nonempty subset R of N we have

$$\alpha^i > \beta^i, \quad \text{all } i \in R,$$

and

$$\alpha(R) \leqq \bar{v}(R). \tag{5-6}$$

To avoid α — β we must have

$$\alpha(S) > v(S) \tag{5-7}$$

for *all* S, $O \subset S \subseteq R$. Referring to the definition of $\bar{v}$, we see that there are nonnegative weights γ_S, $S \subseteq R$, such that

$$\bar{v}(R) = \sum_{S \subseteq R} \gamma_S v(S)$$

and

$$\sum_{\substack{S \subseteq R \\ S \ni i}} \gamma_S = i, \quad \text{all } i \in R.$$

Hence, using (5-7)

$$\bar{v}(R) < \sum_{S \subseteq R} \gamma_S \alpha(S) = \alpha(R).$$

The strict inequality here contradicts (5-6). Q.E.D.

By a "solution" of a market, we shall mean a solution of the associated market game.

THEOREM 7. *If $(N; v)$ is any balanced game whatever, then there is a market that has precisely the same solutions as $(N; v)$.*

Proof. The main work has been done in Lemma 3. Indeed, let $(N, E_+^T, I^T, \{u\})$ be the direct market generated by $(N; v)$. Then the solutions of this market are the solutions of $(N; \bar{v})$, which by the lemma is *d*-equivalent to $(N; v)$ and hence has the same solutions. Q.E.D.

5.2 A TECHNICAL REMARK

The notion of *d*-equivalence is essentially due to Gillies [*4*, *5*], though he works with a broader definition of imputation, not tied to the characteristic function by (5-1). He defines a *vital* coalition as one that achieves some domination that no other coalition can achieve, and shows that two games are d-equivalent (in the present sense) if and only if they have (i) the same imputation sets, (ii) the same vital coalitions, and (iii) the same v-values on their vital coalitions.

A necessary (but not sufficient) condition for a coalition to be vital is that it cannot be partitioned into proper subsets, the sum of whose v-values equals or exceeds its own v-value. Sufficiency would require the generalized partitioning provided by balanced sets.

Given a game $(N; v)$, we can define its "least superadditive majorant" $(N; \tilde{v})$ by

$$\tilde{v}(S) = \max \textstyle\sum_h v(S_h), \tag{5-8}$$

the maximization running over all partitions $\{S_h\}$ of S. (Compare (4-4), (4-5).) It can be shown that $\tilde{v}(N) = v(N)$ if and only if $(N; v)$ has a core (cf. Lemma 1 above), in which case the two games are d-equivalent. Thus, every game with a core is d-equivalent to a superadditive game.

However, as Gillies observes, d-equivalence can also hold nontrivially among superadditive games. That is, it may be possible to push the v-value of some nonvital coalition *higher* than the value demanded by superadditivity, without making the coalition vital.[10] We are using the full power of this observation, since the cover $\bar{v}$ can be thought of as the "greatest d-equivalent majorant" of v. Thus, $v \leqq \tilde{v} \leqq \bar{v}$, and all three may be different.

6. EXAMPLES

Lucas's 10-person game [*8*, *9*] with no solution has players $N = \overline{1234567890}$ and the following characteristic function:

$$\left.\begin{aligned}
v(\overline{12}) = v(\overline{34}) = v(\overline{56}) = v(\overline{78}) = v(\overline{90}) = 1 \\
v(\overline{137}) = v(\overline{139}) = v(\overline{157}) = v(\overline{159}) = v(\overline{357}) = v(\overline{359}) = 2 \\
v(\overline{1479}) = v(\overline{2579}) = v(\overline{3679}) = 2 \\
v(\overline{1379}) = v(\overline{1579}) = v(\overline{3579}) = 3 \\
v(\overline{13579}) = 4 \\
v(N) = 5, \text{ and} \\
v(S) = 0, \text{ all other } S \subseteq N.
\end{aligned}\right\} \tag{6-1}$$

The game has a core, containing among others the imputation that gives each "odd" player 1.[11] It is not superadditive (for example, $v(\overline{12}) + v(\overline{34}) > v(\overline{1234})$); however it is d-equivalent to its least superadditive majorant $(N, \tilde{v})$, which can be calculated without difficulty,

[10] For example, at the end of Sec. 2, $v(\overline{123})$ may be increased from 1 to 3/2 without making 123 vital.

[11] The full core is a five-dimensional polyhedron, having vertices $e^S : S = \overline{13579}$, $\overline{23579}$, $\overline{14579}$, $\overline{13679}$, $\overline{13589}$, and $\overline{13570}$.

using (5-8). Moreover, one can verify that the latter is totally balanced, i.e. that $\tilde{v} = \bar{v}$ in this case. Thus, $(N, \tilde{v})$, defined by (6-1) and (5-8), is a market game with no solution.

The corresponding *market* with no solution, provided by Theorem 7, has ten traders and ten commodities, plus money. The traders have identical continuous concave homogeneous utilities $u(x)$, which may be calculated by applying (4-2), (4-3) to (6-1). Note that positive weights γ_S need be considered only for the eighteen vital coalitions and the ten singletons.[12] Of course, this is not the only utility function that works, since only a finite set of its values are actually used (cf. (4-1)).

6.1 A Production Model

Perhaps the most straightforward economic realization of Lucas's game is in the form of a production economy. (Compare the "activity" description in Section 4.1.) The production possibilities are generated by 18 specific processes, which produce the same consumer good (at constant

TABLE 1

Inputs										Output
x_1	x_2	x_3	x_4	x_5	x_6	x_7	x_8	x_9	x_{10}	x_{11}
1	1									1
		1	1							1
				1	1					1
						1	1			1
								1	1	1
1		1				1				2
1		1						1		2
1				1		1				2
1				1				1		2
		1		1		1				2
		1		1				1		2
1			1			1		1		2
	1			1		1		1		2
		1			1	1		1		2
1		1				1		1		3
1				1		1		1		3
		1		1		1		1		3
1		1		1		1		1		4

[12] The singleton weights are needed as slack variables, because we used "=" in (4-3) instead of "$\leqq$".

returns to scale) out of various combinations of the raw materials (see Table I). Each entrepreneur starts with one unit of the correspondingly indexed raw material. The utility is simply the consumer good: $u(x) \equiv x_{11}$; hence it is not necessary to postulate a separate money.

This type of construction is perfectly general: a production model can be set up in a similar fashion for any other game in characteristic function form, one activity being required for each vital coalition. The market game generated by such a model will be the cover of the original game, and will have the same core and solutions provided that the original game was balanced.

6.2 Other Examples

Lucas [*6*, *7*, *10*] gives several examples of games in which the solution is unique but does not coincide with the core. In [*10*] he also describes a symmetric 8-person game, very similar to the above 10-person game, that has an infinity of solutions but none that treats the symmetric players symmetrically. Shapley [*18*] describes a 20-person game, of the same general type, every one of whose many solutions consists of the core, which is a straight line, plus an infinity of mutually disjoint closed sets that intersect the core in a dense point-set of the first category. A common feature of all these "pathological" examples is the existence of a core; hence, by Theorem 7, they are d-equivalent to market games that have the same solution behavior.

We close with another "pathological" example, of an older vintage [*14*], which because of its simple form leads to a direct market with utilities that we can write down explicitly. The game has players $N = \overline{123\ldots n}$, with $n \geqq 4$, and its characteristic function is given by

$$\left.\begin{array}{c} v(N-\{1\}) = v(N-\{2\}) = v(N-\{3\}) = v(N) = 1 \\ v(S) = 0, \quad \text{all other coalitions } S. \end{array}\right\} \qquad (6\text{-}2)$$

Thus, to win anything requires the participation of a majority of $\overline{123}$, plus all of the "veto" players $4, \ldots, n$. The core is the set of all imputations α that satisfy $\alpha_1 = \alpha_2 = \alpha_3 = 0$. It is easily verified that the game is totally balanced: $v = \bar{v}$. There are many solutions; but the remarkable feature of the game is a certain subclass of solutions, as follows:

Let B_e denote the set of imputation α that satisfy $\alpha_1 = 0, \alpha_2 = \alpha_3 = e > 0$. Thus, B_e is a $(n-3)$-dimensional closed convex subset of the imputation space. In [*14*] it was shown that one may start with any closed subset of B_e whatever, and extend it to a solution of the game by adding only imputations that are at least $e/2$ distant from B_e.[13] The arbitrary starting

[13] The metric used here is $\rho(\alpha, \beta) = \max_i |\alpha_i - \beta_i|$. Our present claim entails a slight change in the construction given in [*14*], which merely keeps the rest of the solution away from the arbitrary subset of B_e, rather than from B_e itself.

set remains a distinct, isolated portion of the full solution. For example, if $n = 4$ (the simplest case), an arbitrary closed set of points on a certain line can be used.

To determine the direct market of this game, we apply (4-2), (4-3) to (6-2) and obtain the utility function

$$u(x) = \max_{\{\gamma^i\}} (\gamma^1 + \gamma^2 + \gamma^3),$$

maximized subject to

$$\left.\begin{array}{c} \gamma^1 \geqq 0, \qquad \gamma^2 \geqq 0, \qquad \gamma^3 \geqq 0; \\ \gamma^2 + \gamma^3 \leqq x_1, \qquad \gamma^1 + \gamma^3 \leqq x_2, \qquad \gamma^1 + \gamma^2 \leqq x_3; \\ \gamma^1 + \gamma^2 + \gamma^3 \leqq x_i, \qquad i = 4, \ldots, n; \end{array}\right\}$$

where γ^i abbreviates $\gamma_{N-\{i\}}$. This reduces to the closed form:

$$u(x) = \min \left[x_1 + x_2, x_1 + x_3, x_2 + x_3, \frac{x_1 + x_2 + x_3}{2}, x_4, \ldots, x_n \right]. \quad (6\text{-}3)$$

We see that u is the envelope-from-below of $n+1$ very simple linear functions.

Thus, an n-trader n-commodity market having the solutions containing arbitrary components, as described above, is obtained by giving the ith trader one unit of the ith commodity, $i = 1, \ldots, n$, and assigning them all the utility function (6-3).

REFERENCES

1. Bondareva, O. N., Nekotorye primeneniia metodov linejnogo programmirovaniia k teorii kooperativnykh igr (Some applications of linear programming methods to the theory of cooperative games). *Problemy Kibernetiki* **10** (1963), 119–139.
2. Charnes, A. and Kortanek, K. O., On balanced sets, cores, and linear programming. *Cahiers du Centre d'Etudes de Recherche Opérationelle* **9** (1967), 32–43.
3. Debreu, G. and Scarf, H., A limit theorem on the core of an economy. *International Economic Review* **4** (1963), 235–246.
4. Gillies, D. B., Some theorems on n-person games. Ph.D. Thesis, Princeton University, June 1953.
5. Gillies, D. B., Solutions to general non-zero-sum games. *Ann. Math. Study* **40** (1959), 47–85.
6. Lucas, W. F., A counterexample in game theory. *Management Science* **13** (1967), 766–767.
7. Lucas, W. F., *Games with Unique Solutions which are Nonconvex*. The RAND Corporation, RM–5363–PR, May 1967.
8. Lucas, W. F., A game with no solution. *Bull. Am. Math. Soc.* **74** (1968), 237–239; also The RAND Corporation, RM–5518–PR, November 1967.
9. Lucas, W. F., *The Proof that a Game may not have a Solution*. The RAND Corporation, RM–5543–PR, January 1968 (to appear in *Trans. Am. Math. Soc.*).
10. Lucas, W. F., *On Solutions for n-Person Games*. The RAND Corporation, RM–5567–PR, January 1968.

11. SCARF, H., The core of an n-person game. *Econometrica* **35** (1967), 50–69.

12. SHAPLEY, L. S., *Notes on the n-Person Game—*III: *Some Variants of the von Neumann–Morgenstern Definition of Solution.* The RAND Corporation, RM–817, April 1952.

13. SHAPLEY, L. S., *Markets as Cooperative Games.* The RAND Corporation, P–629, March 1955.

14. SHAPLEY, L. S., A solution containing an arbitrary closed component. *Ann. Math. Study* **40** (1959), 87–93; also The RAND Corporation, RM–1005, December 1952, and P–888, July 1956.

15. SHAPLEY, L. S., The solutions of a symmetric market game. *Ann. Math. Study* **40** (1959), 145–162; also The RAND Corporation, P–1392, June 1958.

16. SHAPLEY, L. S., *Values of Large Games—*VII: *A General Exchange Economy with Money.* The RAND Corporation, RM–4248, December 1964.

17. SHAPLEY, L. S., On balanced sets and cores. *Naval Research Logistics Quarterly* **14** (1967), 453–560; also The RAND Corporation, RM–4601–PR, June 1965.

18. SHAPLEY, L. S., *Notes on n-Person Games—*VIII: *A Game with Infinitely "Flaky" Solutions.* The RAND Corporation, RM–5481–PR, 1968.

19. SHAPLEY, L. S. and SHUBIK, M., "Concepts and theories of pure competition." *Essays in Mathematical Economics: In Honor of Oskar Morgenstern*, Martin Shubik (ed.). Princeton University Press, Princeton, New Jersey, 1967, 63–79; also The RAND Corporation, RM–3553–PR, May 1963.

20. SHAPLEY, L. S. and SHUBIK, M., Quasi-cores in a monetary economy with non-convex preferences. *Econometrica* **34** (1966), 805–827; also The RAND Corporation, RM–3518–1, October 1965.

21. SHAPLEY, L. S. and SHUBIK, M., *Pure Competition, Coalitional Power, and Fair Division.* The RAND Corporation, RM–4917–1–PR, March 1967 (to appear in *International Economic Review*).

22. SHUBIK, M., Edgeworth market games. *Ann. Math. Study* **40** (1959), 267–278.

23. VON NEUMANN, J. and MORGENSTERN, O., *Theory of Games and Economic Behavior.* Princeton University Press, Princeton, New Jersey, 1944, 1947, 1953.

RECEIVED: October 8, 1968.

[6]

On the Core of an Economic System with Externalities

By LLOYD S. SHAPLEY AND MARTIN SHUBIK*

It is well known that if a productive socio-economic system generates external economies or diseconomies, then the existence of prices that will maintain an efficient or Pareto optimal distribution of goods and services is not in general assured [2]. Hence, a social incentive may exist for other methods of allocation. The mathematical theory of games provides some tools for the study of such systems. In this note we shall be concerned especially with the social stability of cooperative agreements that impute the costs or benefits associated with such externalities, using the key notion of the core of an n-person cooperative game. We shall also make some remarks on the possibility of taxation or subsidization schemes to restore optimality to the competitive market.

We recall that an external economy or diseconomy is a production or consumption activity, or by-product thereof, which affects the welfare of outsiders regardless of their desires. Thus, smoke, noise, or other pollution may represent external diseconomies, extending far beyond the group responsible for them; while landscaping or fireproofing a place of business, for example, or founding a cultural center or a shopping center, may create external economies for others. One classical example has an apple grower and a beekeeper in adjacent fields; the blossoms with their nectar help the bees make more honey, while the pollination service increases the yield of fruit—thus, two external economies appear in the same setting, see J. E. Meade [5]. Certain "public goods," as roads, parks, police, schools, can also sometimes be viewed as external economies, though other considerations intrude[1] as seen in M. Shubik [14].

The distinctive property of an external, as opposed to a normal economic good, is that of *appropriability.* Externalities are not owned, in the ordinary sense. An externality in production usually comes in the form of a joint product, or by-product, that is not appropriated by the producer but is transmitted by noneconomic means to other individuals. The smoke is not kept by the factory, or even counted in its output statistics, but is carried by the winds to the shirts and lungs of the passerby or the walls and furnishings of the nearby homes. The orchardist does not appropriate and sell the nectar, nor the apiarist the pollination service.

We shall limit our discussion to externalities in production, because they are easier to define and quantify than those arising on the consumption side. Our results could perhaps be made to apply, at least at the formal level, to both cases; but consumption externalities entail interlinked consumer preferences and involve socio-psychological motivations like envy, status, conformity, or even pity. Such considerations are remote from the substance of our present inquiry.

I. *The Core*

In recent work in the application of game theory to economics it has been shown that if a static model of exchange and production admits an efficient decentralizing price system, and if one views the model as an n-person cooperative game, then there will be a set of socially stable distributions, called the *core,* which cannot be overthrown by any coalition acting in its own interests. The core is related to, but not identical with, the "competitive" equilibrium (or equilibria) arising out of the price system.[2] In fact, there is a basic theorem that asserts that the competitive distributions are always included in

* The RAND Corporation and Yale University.

[1] In particular, the notion of a corporate "public" or "state" which can own things, and which may have an independent set of values.

[2] Paradoxically, the competitive equilibrium is *not* a "game" solution, under either the cooperative or non-cooperative approaches to the theory of games.

the core, [3], [10], [12]. The core generally has more in it, however, and in any case it is defined without reference to prices.

Economic models aside, the core of an abstract n-person game may be defined as the set of feasible outcomes, if any, that enjoy the property of Group Rationality (*GR*), for all groups of players. This means that no coalition, acting independently, can get more for itself than any outcome in the core provides. ("More," for a coalition, means more for each member of the coalition.) Thus, *GR* includes Individual Rationality (*IR*), by way of the one-player sets, as well Pareto Optimality (*PO*), by way of the all-player set. While *IR* and *PO* can be achieved quite generally in an n-person game, *GR* cannot, and there is no assurance that a core will exist.

A simple example will help to illustrate. Suppose a donor promises to distribute \$3 among three contestants, but only if they agree unanimously on the shares. However, if only two can agree, he will distribute a smaller prize between them, with nothing to the third. We consider two versions: (1) the smaller prize is \$1.50 and (2) it is \$2.50.

The two versions can be summed up conveniently by "characteristic functions," which are set-functions that state the monetary worth of each possible coalition:[3]

First game	*Second game*
$v(\bar{1}) = v(\bar{2}) = v(\bar{3}) = 0$	$v(\bar{1}) = v(\bar{2}) = v(\bar{3}) = 0$
$v(\overline{12}) = v(\overline{13}) = v(\overline{23})$	$v(\overline{12}) = v(\overline{13}) = v(\overline{23})$
$= 1.5$	$= 2.5$
$v(\overline{123}) = 3$	$v(\overline{123}) = 3$

In either game, a *PO* imputation of wealth is a division of \$3, i.e., a vector $\alpha = (\alpha_1, \alpha_2, \alpha_3)$ such that $\alpha_1 \geqq 0$, $\alpha_2 \geqq 0$, $\alpha_3 \geqq 0$, and $\Sigma_1^3 \alpha_i = 3$. A *GR* imputation would demand, in addition, that

$$\alpha_1 + \alpha_2 \geqq v(\overline{12}),\ \alpha_1 + \alpha_3 \geqq v(\overline{13}),$$
$$\alpha_2 + \alpha_3 \geqq v(\overline{23}).$$

In the first game there is a sizeable core, because there are many ways of dividing \$3 so that any two shares add to at least \$1.50. Examples are (1, 1, 1) or 1, 1.4, 0.6) or (1.5, 1.5, 0). But in the second game there is no core at all; we cannot arrange matters so that every pair gets at least \$2.50.[4]

Of these two games, the second, lacking a core, could never arise from an economic model having an efficient competitive equilibrium. The first game, on the other hand, can indeed arise in this way. In fact, we have shown elsewhere [12] that every "totally balanced" characteristic function, of which the first game is an example,[5] is derivable from a simple exchange economy.

For the rest of this paper we shall assume that the production and consumption (i.e., utility) functions meet suitable conditions that would, *except for the presence of externalities of production*,[6] assure the existence of a *PO* competitive equilibrium. Does a core nevertheless exist? We shall see that the answer is yes if there are only external economies, but no (or rather, not necessarily) if external diseconomies are present.

II. *External Economies*

We first consider the case where the external factors have only beneficial effects. To demonstrate the existence of a core is quite straightforward—so much so that we

[3] For brevity, coalitions are indicated by a bar, written over the indices of the members, without punctuation or spacing; thus "$\overline{123}$" replaces the more conventional "[1, 2, 3]" or "[P_1, P_2, P_3]."

[4] A simple proof: Suppose α were in the core. Then we would have

$$\begin{aligned} \alpha_1 + \alpha_2 \quad\ \ &\geqq 2.5 \\ \alpha_1 \quad\ \ + \alpha_3 &\geqq 2.5 \\ \alpha_2 + \alpha_3 &\geqq 2.5. \end{aligned}$$

But adding and dividing by 2 yields $\alpha_1+\alpha_2+\alpha_3 \geqq 3.75$, which is infeasible.

[5] A game with a numerical characteristic function is "balanced" if it has a core, and "totally balanced" if all its subgames have cores as well. See [12].

[6] That is, external factors that arise out of the production processes. Their impact may be felt either in other production processes or directly in the utility functions.

can dispense with most of the formal trappings.

Let us note in advance that the proof does not depend on the existence of a characteristic functional representation, which would entail the availability of "*u*-money," or something closely akin to it,[7] and which would sharply limit the scope of the result. Indeed, *GR* is basically an *ordinal* condition (like *PO* and *IR* which it generalizes) and requires neither measurability nor transferability of the individual utilities.[8]

Now for the proof. Let Γ denote a productive economic system having k external economies and no diseconomies. Let n denote the total number of individuals (i.e., households or firms). We construct an associated, artificial system Γ' without externalities, as follows: Every external product in Γ is replaced by n separately-tradable "labelled" products in Γ', each one "addressed" to a particular individual. These new commodities (there are nk of them), though freely transferable, have value only to their addressees; in fact, each one enters its addressee's utility or production function in Γ' in exactly the same way as its original counterpart did in Γ.[9] Similarly, the new commodities are produced in Γ' exactly as their counterparts were in Γ. That is, where in Γ one unit of an external good would be produced, in Γ' one unit of *each* of the n associated "labelled" goods is forthcoming, from the same inputs. But in the latter case, the producer appropriates all of his outputs, and trades them as normal commodities.

How are the cores of Γ and Γ' related? How effective are the potential "blocking" coalitions in the two models? Let X denote a distribution that is attainable in Γ' by a subeconomy if it breaks off and ignores the possibilities of trade with the rest. If X is efficient, i.e., *PO* with respect to the capabilities of the subeconomy, then we might as well assume that all "labelled" goods are handed over to their addressees whenever possible, since they have no value to anyone else. But then it follows that X can also be attained by the corresponding subeconomy of the original system Γ, since in that model the corresponding external factors are delivered automatically to their beneficiaries. If production activity outside the subeconomy bestows further involuntary benefits, so much the better. We conclude that the original system Γ is *at least as favorable* to any coalition as is the associated system Γ'.[10]

On the other hand, a disaffected coalition cannot rely on receiving outside benefits even in the Γ model, in assessing the prospects of a separatist subeconomy. The only external economies it can count on enjoying are those that can be internalized—i.e., produced within the coalition. But these benefits can be obtained in the Γ' model as well, via ordinary trading within the coalition. Thus, insofar as determining the core is concerned, Γ is *no more favorable* than Γ' to any coalition. Thus, the two models are equivalent for the purpose of our proof, and their cores are the same.[11]

Our starting assumption was that except for the externalities, sufficient conditions for the existence of an efficient competitive equilibrium are satisfied by the production and utility functions. In passing from Γ to Γ' we have not changed this; hence Γ', which by definition is free from externalities, has a *PO* competitive equilibrium. It follows from the basic theorem mentioned near the beginning of Section I that Γ' has a core. Hence Γ too has a core, as was to be demonstrated.

[7] This term, introduced in [13], denotes an idealized money that is freely transferable, is present in sufficient supply, and enters each utility function as a separable, linear term. It is useful in economic game theory in much the way that an "ideal gas" is useful in thermodynamics.

[8] See [1], [6], and [9]. A distinction is made in [1] between two kinds of core ("alpha" and "beta"). This makes no difference in the case of external economies, but the distinction can be significant in the case of a "directed" external diseconomy, without money.

[9] Some of these new goods may be completely valueless, since we do not assume that everyone actually benefits from every externality.

[10] In characteristic function terms we would have $v(S) > v'(S)$ for all S. In the next paragraph the reverse inequality is established.

[11] In other respects the two models, though usefully related, are not equivalent; see Sec. IV.

III. *External Diseconomies*

Our results regarding external diseconomies are on the negative side, and are best displayed by means of examples.

The Garbage Game. Each player has a bag of garbage which he must dump in someone's yard. The utility of having b bags dumped in one's yard is $-b$.

The diseconomy here is "directed" rather than "diffuse"; however, we can eliminate the problem of directed threats by assuming that a coalition can "take care of its own," shifting the burden either by money side-payments or by redistributing garbage within the coalition. It therefore makes sense to use a characteristic function, and we have

$$v(S) = \begin{cases} -(n - s) & \text{if } s < n, \\ -n & \text{if } s = n. \end{cases}$$

Here n is the number of players and S is any subset of players of size s. The all-player set is in a special position: it alone cannot avoid "fouling its own nest."

To see whether there is a core, we proceed as in Section I. Let α be any GR payoff vector (distribution of utility). For S of size $s = n-1$ we have

$$\sum_{i \in S} \alpha_i \geqq v(S) = -1.$$

There are n such inequalities; adding them yields

$$(n - 1) \sum_{i=1}^{n} \alpha_i \geqq -n.$$

But feasibility requires that

$$\sum_{i=1}^{n} \alpha_i = v(\text{all-player set}) = -n.$$

Eliminating α, we obtain

$$(n - 1)(-n) \geqq -n;$$

i.e., $n^2 \leqq 2n$, or $n \leqq 2$. We conclude that if there are more than two players, there is no core. No agreement on garbage disposal can satisfy every group of players to the point where they cannot, by violating it, be sure of doing better.

We shall see presently that the directability of the diseconomy in this example was not solely responsible for the lack of a core. First, however, we illustrate a common type of nondirected diseconomy, which happens to give rise to a very large core.

The Lake. There are n factories around a lake. It costs an amount B for a factory to treat its wastes before discharging them into the lake. It costs an amount uC for a factory to purify its own water supply, if u is the number of factories that do *not* treat their wastes. We assume that $C<B<nC$.

Here IR sets an upper bound of nC on what a single factory should consent to spend, while PO says that all factories together should pay no more than nB. The potential social gain from cooperative action is therefore $n(nC-B)$ perhaps a substantial amount.

Assuming that u-money is available for side payments, we have

$$v(S) = \begin{cases} -snC & \text{if } s \leqq B/C \\ -snC + s(sC - B) & \text{if } s \geqq B/C \end{cases}$$

where s denotes the size of S. For smaller sets, $v(S)$ is merely an additive function, so there is no incentive to form coalitions. But for larger sets, $v(S)$ exhibits increasing returns to scale; in fact, each new adherent to a coalition brings in more than the last, and cooperation becomes increasingly profitable. Two effects combine to make this true: (1) a larger coalition can control at the source a more significant fraction of the pollution, and (2) a larger coalition includes more of the beneficiaries of such control. In short, it can afford to internalize the diseconomy.

Technically, we have here a *convex* game [4], [8].[12] There is a core, but it is so large that it tells us little about how to assign the costs of pollution control. In fact, among the outcomes in the core are some that assess several factories the maximum cost permitted by

[12] A formal definition of convexity is that

$$v(S) - v(S - i) \leqq v(T) - v(T - i)$$

whenever $i \in S \subset T$.

IR, while the other factories enjoy all the benefits. Although such a control arrangement is patently "unfair," the victims' threat to secede does not suffice to block it.[13]

It should not be inferred from this example that large cores are the rule in this type of application. For example, if we make a minor change in the intake purification cost formula, from uC to $max[(u-1)C, 0]$, then the core collapses to a single point, representing the control arrangement that has everyone pay just for treating his own effluent.

Our third example again has a diffuse diseconomy.

The Smelting Game. One "load" of ore plus one "load" of coke make one "load" of iron plus one "cloud" of smoke. At the start, n_1 players have loads of ore and n_2 have loads of coke. At the end, each player's payoff is the amount of iron he has, reduced by K (a small constant) times the amount of smoke in the air.

There is enough symmetry in this game so that all that matters is the ***profile*** of a coalition, i.e., its composition by player-types. In fact, if S contains s_1 and s_2 of the "ore" and "coke" types respectively, then we have

$$v(S) = \begin{cases} min(s_1, s_2) - Ks[min\,(s_1, s_2) \\ \qquad + min\,(n_1 - s_1, n_2 - s_2)], \\ \qquad\qquad \text{if } s \leqq 1/K, \\ -Ks\, min\,(n_1 - s_1, n_2 - s_2), \\ \qquad\qquad \text{if } s \geqq 1/K, \end{cases}$$

where $s = s_1 + s_2$. Again, there are two cases. The larger coalitions are disinclined to go into production, while the smaller coalitions, with less to lose from the smoke, are not so inhibited.

Although the diseconomy is diffuse, as in the last example, this game is not convex, and whether or not a core exists depends in a rather complicated way on the values of the three parameters n_1, n_2, and K.

To illustrate, we have set $n_1 = n_2 = n/2$ and $K = 1/10$. Table 1 shows v as a function of s_1 and s_2 for small values of n; for convenience all entries have been multiplied by 10.[14]

The four-person game has a core; indeed any payoff distribution of the form $(a, a;$ $6\text{-}a, 6\text{-}a)$ with $1 \leqq a \leqq 5$ is unblockable. The only operative constraints are that profile (1, 1) must get at least 6 points, and profiles (1, 2) and (2, 1) at least 7. Thus, the core is a line segment.

The six-person game also has a core, but it contains only the single distribution (2, 2, 2; 2, 2, 2). Any other feasible distribution would give some five-person set less than the 10 points it can claim under *GR*. The eight-person game has no core at all, because every seven-person set can demand 9, whereas an average of only 1 per capita is available in the game as a whole. Note that all of the games thus far are so small that full production is still socially desirable, despite the smoke.

If we continue the series, we find that the core remains in eclipse for $n = 10$, 12, 14, and 16. Although production is no longer socially desirable, some subeconomies can still attain a positive payoff on their own, and would therefore block any agreement that would shut them down. At $n = 18$ and beyond this

[13] Other game-theoretic solution principles, notably the *value* [11] and the *nucleolus* [7], could be invoked to provide more determinate guidelines, but we shall not discuss them here.

Of course, this example is so symmetrical that it is obviously "fair" that each factory take care of its own effluent. But we could easily desymmetrize the model, for example, by introducing unequal cost coefficients $[B_i, C_i]$ for the different factories, or unequal outputs of waste matter. The game would still be convex, and the core excessively large, but the "proper" or "fair" imputation of costs is no longer so obvious.

[14] For example, when $n = 8$ (four players of each type), a coalition with profile (2, 0) is worth only -4 points, since it can produce no iron yet must breathe the others' smoke (2 "clouds" times 2 sufferers times —10K equals -4). Nevertheless, out of coalition they would be even worse off, since the *IR* limit is -3 per capita. Including people of the other type would rapidly strengthen the coalition; reading across the table we see that the first would add 5 points, the second 3, the third 1, and the fourth 3.

Note that the data in the table all come from the top line of the formula for $v(S)$; i.e., we are still only considering "small" coalitions, with $s \leqq 1/K$.

TABLE 1

$n=4$

	$s_2=0$	1	2
$s_1=0$	0	−1	0
1	−1	6	7
2	0	7	12

$n=6$

	$s_2=0$	1	2	3
$s_1=0$	0	−2	−2	0
1	−2	4	4	6
2	−2	4	8	10
3	0	6	10	12

$n=8$

	$s_2=0$	1	2	3	4
$s_1=0$	0	−3	−4	−3	0
1	−3	2	1	2	5
2	−4	1	4	5	8
3	−3	2	5	6	9
4	0	5	8	9	8

is no longer true, and the core reappears. But it consists only of the zero vector, representing a general agreement to stop all production.

When $n_1 \neq n_2$ it can be shown that the zero vector is the only possible core; the smaller games, in which production is socially desirable, are all coreless.

To sum up, in considering three simple examples representing rudimentary types of external diseconomy, we have found a wide diversity of core behavior. Perhaps further work will uncover classifications that will permit general existence, nonexistence, or convergence theorems to be proved for restricted classes of models. Until then, the core is still an investigative tool, for case-by-case application, that can give important insights into the prospective stability of cooperative internalization of external diseconomies.

IV. *Taxes and Subsidies*

We return briefly to the discussion of external economies. In Section II we proved the existence of the core with the aid of an artificiality, the associated model Γ′, in which the external products were neatly packaged addressed and transmitted to their beneficiaries, through the normal channels of trade. This artificial model always has a competitive equilibrium, however, and it is interesting to attempt an interpretation for it in the original model, Γ.

Indeed, fixing on a particular set of competitive prices for Γ′ (in case the equilibrium is not unique) we proceed as follows: Let each normal good in Γ be assigned the same price as in Γ′. Let each person in Γ who benefits from an external good be *taxed*, i.e., made to pay for it, at the price of the corresponding "labelled" good in Γ′. Finally, let each producer of an external economy in Γ be *subsidized*, i.e., paid for the product, at a rate equal to the sum of the prices of its *n* "labelled" counterparts in Γ′. These taxes and subsidies are of course stated in terms of the same, valueless accounting money used in the sale and purchase of normal goods in Γ, and each individual is required to take his tax and subsidy payments into account in balancing his budget.

Because of the direct relationship between Γ and Γ′, the above prescription of prices, taxes, and subsidies constitutes a kind of generalized competitive equilibrium for Γ. Among its virtues is that it is in the core, i.e., no faction will be tempted to split off and "go it alone," even though they can thereby escape the taxes. Among its drawbacks (both administrative and philosophical) is the fact that each beneficiary of an external economy would in general have to be taxed differently, not only for a different amount but at a different rate, depending on how much enjoyment he personally derives from the benefits thrust upon him.

In the diseconomy case, there is no comparable "associated" model and therefore no assurance that a tax-and-subsidy scheme can be found that enjoys the same combination of both competitive and collusive stability.[15]

[15] Further modeling problems, of high game-theoretic content, are encountered when one attempts to include taxation policy among the strategic options for society, when joint social-political goods are involved. We defer these questions to another occasion [13].

REFERENCES

1. R. J. AUMANN, and B. PELEG, "Von Neuman-Morgenstern Solutions to Cooperative Games Without Side Payments," *Bull. Amer. Math. Soc.* 1960, *66*, 173–79.
2. F. M. BATOR, "The Anatomy of Market Failure," *Quart. J. Econ.*, Aug. 1958, *72*, 351–79.
3. G. DEBREU, and H. SCARF, "A Limit Theorem on the Core of an Economy," *Int. Econ. Rev.* Sept. 1963, *4*, 235–36.
4. M. MASCHLER, B. PELEG, and L. S. SHAPLEY, *The Kernel and Bargaining Set for Convex Games*, The RAND Corporation, RM-5372-PR, August 1967.
5. J. E. MEADE, "External Economies and Diseconomies in a Competitive Situation," *Econ. J.* Mar. 1952, *62*, 54–67.
6. H. SCARF, "The Core of an *N* Person Game," *Econometrica*, Jan. 1967, *35*, 50–69.
7. D. SCHMEIDLER, *The Nucleolus of a Characteristic Function Game*, Research Memorandum RM23, Mathematics Department, Hebrew University, Jerusalem, October 1966.
8. L. S. SHAPLEY, *Notes on N Person Games VII: Cores of Convex Games*, The RAND Corporation, RM-4571-PR, October 1965.
9. ———, and M. SHUBIK, "Solutions of *n*-person games with ordinal utilities" (abstract), *Econometrica*, 1953, *21*, 348–49.
10. ———, and ———, "Quasi-cores in a monetary economy with nonconvex preferences," *Econometrica* Oct. 1966, *34*, 805–827.
11. ———, and ———, "*Pure Competition, Coalitional Power, and Fair Division* (to appear in *Int. Econ. Rev.*).
12. ———, and ———, "*On Market Games*" (to appear in *J. Econ. Theor. 1*, 1969.)
13. ———, and ———, *Competition, Welfare, and the Theory of Games* (book in progress).
14. ———, *Notes on the Taxonomy of Problems Concerning Public Goods*, Cowles Foundation Discussion Paper No. 208, Yale University, April 1966.

[7]

INTERNATIONAL ECONOMIC REVIEW
Vol. 10, No. 3, October, 1969

PURE COMPETITION, COALITIONAL POWER, AND FAIR DIVISION*

BY LLOYD S. SHAPLEY AND MARTIN SHUBIK[1]

1. INTRODUCTION

THIS PAPER IS CONCERNED with the conceptual foundations of the theory of economic competition, and with the effect thereon of a new solution concept borrowed from the mathematical theory of games.

In Section 2, we consider three basic principles of distribution in an economic society—pure competition, coalitional power, and fair division—and show how they lead to three different ideas of what might constitute the "solution" of a mathematical model of the marketplace. Two of these solutions, the *competitive equilibrium* and the *core*, have been found to be intimately related, despite sharp differences in heuristic interpretation [6, 4, 25, 21]. Our present purpose is to introduce to economic analysis the third of these solutions, the *value of the game*, and to compare and contrast it with the other two.

Like the core, the value solution presupposes that the market is a collusive, multi-person game. The value solution, however, looks for a unique, equitable compromise among all opposing interests, whereas the core merely delimits a "no man's land" between unyielding coalitions. The competitive equilibrium, for its part, recognizes no collusion at all. But when the number of traders is large, it can nevertheless be shown under rather general conditions that all three varieties of solution come into agreement, predicting the same outcome, but for different reasons.[2]

In Sections 3, 4, and 5, which may be read independently of one another, we consider some typical applications. In Section 3, numerical solutions are determined for a simple symmetric market model involving complementary goods (see [18]). In Section 4, the mutual convergence of value, core, and competitive equilibrium is demonstrated for a general class of Edgeworth market games (see [24, 20]). In Section 5, a particular Edgeworth model is viewed as a game without transferable utility and is solved explicitly as a function of market size. This exercise represents the maiden voyage of a new definition of the value, and the technical argument is accordingly spelled out in some detail.

* Manuscript received May 23, 1966, revised March 20, 1967.

[1] This research was sponsored by the United States Air Force under Project RAND, Contract No. F44620-67-C-0045, monitored by the Directorate of Operational Requirements and Development Plans, Deputy Chief of Staff, Research and Development, Hq USAF. Views or conclusions contained in this paper should not be interpreted as representing the official opinion or policy of the United States Air Force.

[2] Agreement in the limit, for a restricted class of oligopolistic models, has been established for yet another game-theoretic solution: the Nash *noncooperative equilibrium point* (see [21]).

The main definitions of the formal theory are summarized in the Appendix.

2. THREE APPROACHES

Underlying traditional writings and utterances on the virtues of competition and the "invisible hand" of the open marketplace are several fundamentally different approaches to the question of distribution and trade in an economic society. The three that will concern us in this essay are, in brief: (1) *pure competition*: individuals interact only through an impersonal market, which provides them with information and receives their separate decisions; (2) *coalitional power*: individuals join forces and coordinate decisions when it is to their advantage, constrained ultimately only by the countervailing power of opposing coalitions; (3) *fair division*: gains in welfare arising out of economic activity are distributed equitably among all participants, in accordance with their contributions to the economy.

These three conceptual points of departure, seemingly at odds, may or may not lead to conflicting results in application. The mathematical theory of games provides a formal frame of reference within which these ideas may be analyzed, compared, and, to a certain extent, integrated.

2.1. *The competitive equilibrium.* The classical theory of pure competition has demonstrated that, under certain general conditions concerning consumer preferences and production possibilities, there will exist a schedule of equilibrium prices for the goods and services in the economy [**27, 10, 2**]. If each individual maximizes his own welfare, on the assumption that these prices will prevail and without spending more than he receives, then demand and supply will just balance. The resulting "competitive" allocation of goods and services has the property of *Pareto optimality*: no redistribution could benefit anyone without hurting someone else.

The beauty of this kind of price mechanism, which uses money only as a bookkeeping device, is that it replaces a complicated, joint maximization problem by a set of simple, individual maximization problems, the solutions of which all synchronize to give an optimal result. It is a decentralized decision system with a (weak) welfare property.

A drawback, however, is the absence in the model of a method of generating the equilibrium prices that determine the pattern of trade. One might postulate a central authority, which not only calculates and publishes the prices but also selects a particular price schedule if there should happen to be more than one with the equilibrium property. The informational requirements of such an agency, however, would seem to vitiate any subsequent advantages of decentralization. Alternatively, one might hope that competitive prices would arise out of some dynamic adjustment process, steered by tentative bids and offers from the individual traders. Unfortunately, convergence and stability are not ensured without added assumptions about the economy that exclude many cases of interest [**1, 14, 12, 13**].

These difficulties are closely related to the possibility of nonuniqueness of the competitive solution, a phenomenon that can occur without visible peculiarities in the data of the model. It must be admitted, however, that even without uniqueness, the set of competitive allocations (arising from a given initial allocation) is generally a very substantial narrowing of the set of all Pareto optima.

2.2. *The core.* In order to discuss the question of coalitional power, we turn to a game-theoretic solution known as the core. In a general n-person cooperative game, the core may be described as the set of final imputations of wealth that are not "coalitionally dominated" by other possible divisions of welfare. In other words, for an imputation to belong to the core, no group of participants should be able, through collusion, to ensure an outcome of the game that every member of the group would prefer to the given imputation.[3]

The core concept may be regarded as an extension of the notion of Pareto optimality, taking into account the possibility of independent optimization by subsets of the economy, as well as by the economy as a whole. It may also be regarded as an extension, to groups, of the individualistic principle that says that a man will not accept any redistribution of wealth that worsens his initial position, unless compelled to do so. Thus, the core is both Pareto optimal and individually rational.

In many general n-person games the core is void.[4] In these cases the intermediate coalitions—between the one-person sets and the all-player set—are too strong and cannot all be satisfied at once. On the other hand, it has been shown [**6**, **23**] that in an economy in which competitive prices exist, the core is not void; indeed, it contains all competitive allocations. This means, somewhat remarkably, that the decentralized "pure competition" equilibrium cannot be upset by collusion among any subset of traders, even though they violate all the rules of pricing, communication, and trading on which the competitive equilibrium in principle depends.

Even more remarkably, when the number of participants in the economy is increased in a suitably homogeneous manner, the core can be shown to shrink, until in the limit only the competitive solution is left [**24**, **7**, **6**]. Note that prices play no part in the definition of the core. Nevertheless, in an economy with a large number of participants, the device of giving free rein to coalitions has the effect of generating competitive prices.

When there are only a few participants, the core may be large. For example, in trading between just two individuals, the core must consist of the entire Edgeworth contract curve, there being no intermediate coalitions to raise objections. It is also worth noting that the core can and does exist

[3] A simple three-person money-sharing game will illustrate this. Suppose the players are told that (1) they may have \$1 if they can agree how to divide it; (2) failing this, any two of them may exclude the third and get 80¢, again provided they agree on the division; (3) failing this, all get nothing. This game has no core, for no matter how the \$1 might be divided, some two players will get less than 80¢ together.

Now change the payoff rules to make the coalitions of two worth 50¢, instead of 80¢. Then there is a core, consisting of all allocations of \$1 that give every *pair* of players at least 50¢; for example (50¢, 0, 50¢), (15¢, 40¢, 45¢), etc.

[4] For example, all essential zero-sum games in the sense of von Neumann and Morgenstern have empty cores [**26**, (280)].

in some economic models in which there is no competitive solution because of the disequilibrating effect of such features as joint production costs or nonconvex or interdependent preferences [23, 22].

2.3. *The value.* The idea of "fair division" in a socio-economic situation calls for the application of yet another solution concept from the theory of games—the value. This solution seeks to evaluate each player's position in the game *a priori*, taking into account both his own strategic opportunities and his bargaining position with respect to gains attainable through collaboration.

The value can be defined most easily when a common measure of utility exists, together with a vehicle, such as money or credit, which permits utility to be transferred freely among the players. In the absence of such transferability, a value can still be defined, but intrinsic rates of utility comparison between the players must simultaneously be derived from the strategic and bargaining possibilities of the game itself, and multiple solutions are possible.

Intuitively speaking, the value solution seeks to impute the proceeds of total cooperation among the participants in a way that takes fair account of each person's contribution to each possible cooperative venture. In calculating the value of an economic game, one must determine the *marginal worth* of an individual to every subset of other individuals and form an average.[5] Thus, an ability to measure the economic "worth" of a set of individuals is presupposed. If a money (in the sense of the preceding paragraph) is available, then utilities can be measured on a common scale, and the worth of a coalition can be defined as the maximum combined wealth that the coalition can achieve for its members by its own efforts. These worths, determined for every possible subset of players, comprise the *characteristic function* [26], from which the value of the game can be computed by a formula (see Appendix A.1).

Historically, the formula for the value was first derived from postulates of symmetry, Pareto optimality, additivity, and—most crucially—the value's sole dependence on the kind of information conveyed in the characteristic function [17]. Harsanyi, working constructively from a model of the bargaining process, and later Selten, working deductively from postulates on the move and payoff structure of the game, arrived at a value definition that applies the same mathematical formula to a modified characteristic function [8, 15, 16].[6] In the present economic models without externalities, the modifi-

[5] We illustrate with a variant of the previous example. Avoiding total symmetry, we now assume that pairs $\overline{12}$, $\overline{13}$, and $\overline{23}$ can divide 50¢, 50¢, and 80¢, respectively, failing a general agreement to divide \$1. Player 1 is in the weakest position. His marginal worth to the coalition $\overline{123}$ is just 20¢; to $\overline{12}$ and $\overline{13}$ he is worth 50¢; and to the "coalition" $\overline{1}$, consisting of himself alone, he contributes nothing. Averaging these numbers (giving equal weight to each size of coalition, not to each coalition) we find Player 1's value to be 23⅓¢. A similar calculation gives 38⅓¢ to Player 2 and the same to Player 3. This imputation happens to be outside the core, since the coalition $\overline{23}$ can do better alone.

[6] Use of the classical characteristic function [26] implies in effect that a coalition

cation proves to be irrelevant; the classical characteristic function yields the Harsanyi-Selten value and is simpler to work with.

In general, the value may lie outside the core. That is, there may be coalitions that could seize by their own efforts more than they are allotted in the value.[7] Indeed, the value exists even when the core is empty.

When no "money" is available, evaluation of the game must proceed indirectly. We *assume* that utility is transferable, and then try to arrange matters so that the value imputation does not, in the end, require any net transfer of utility. This seems at first too much to arrange, but there is some freedom of maneuver. Since the individual utility scales are not related *a priori* to each other, or to a common monetary unit, we are free to adjust them separately, before making utility transferable. (Equivalently, we could leave the individual units alone but permit transfer only at prescribed rates of exchange, as in an international money market.) This freedom to rescale (or to prescribe rates of exchange) proves to be just enough to ensure the existence of a value that is feasible—i.e., one that can be achieved with no net transfer of utility. Thus the value solution, in the "no-transfer" theory, can be defined as the set of all feasible "transfer" values.[8]

This is not the place for a formal axiomatic justification of this definition. However, we may remark that its plausibility rests on a form of the "independence of irrelevant alternatives" principle, namely, the assertion that if the solution-with-transfers-permitted can be achieved without transfers, then it must remain a solution when transfers are forbidden. Of course, it could be objected that the *possibility* of transfers may well influence the outcome even in those cases where the ultimate net transfer is nil.

We shall sketch another view of the value definition which may be of interest. It depends on the observation that when a particular outcome of an n-person game is adopted, this implies not only that an interpersonal comparison of utilities has taken place, but implies it in two distinct ways. The first way relates the given outcome to other possible outcomes on the Pareto surface, basing the comparison on the existence of tradeoffs that were available but did not occur. The other way relates the given outcome to the initial positions (and strategic potentials) of the contestants, using it to refer something about the relative intensities of their desires. The first method infers the individuals' weights in the measuring of social welfare; the second their weights in the sharing of social profit. Our present value concept embodies a *principle of equivalence* between these two methods of inter-

---

always expects the worst so far as actions of outside players are concerned. In the present economic context, this "worst" is simply a boycott, involving no special costs to the outside players. In other contexts, the most damaging threats might be so costly to make that they should be discounted in determining what a coalition is worth. It is therefore advisable, in a general value theory based on a characteristic function, to recast the definition in terms of "optimum threats," in the sense of Nash [**11**].

[7] See footnote 5.

[8] Appendix A.3 has the formal definition. The key idea of using implicitly-determined utility exchange ratios was adapted from a proposal of Harsanyi [**9**].

personal comparison: the values of the game are just those outcomes for which the two sets of derived weights coincide.

Thus, under this definition, a value of the game represents a kind of equilibrium. It is as though the players introduced a money-of-account as an aid to rational bargaining, but with the prices of the different people's "units" chosen so that when the books are closed all accounts are miraculously in balance. The game itself is called upon to provide an intrinsic, "equitable" comparison among the personal utility units. As previously noted, the comparison factors are not always unique.

A close affinity between the value of the game and the competitive equilibrium is suggested by the money-of-account analogy. They are not the same, however, as can be seen from the fact that the former makes essential use of cardinal utility while the latter does not. A nonlinear, order-preserving transformation of the utilities will generally change the value, but not the competitive equilibrium nor, for that matter, the core.

In the remainder of this paper we apply the value theory to three different models, correlating the value solution in each case with the core and competitive solutions. The three applications are independent of one another. In the Appendix we have assembled some of the important formal definitions.

3. A SYMMETRIC MARKET GAME

Our first model was chosen for the contrast it provides between the value and the other solutions under discussion. We shall find that the value gives an intuitively more satisfactory measure of the "equities" of the situation while avoiding a violent discontinuity exhibited by both the competitve equilibrium and the core. Also, exploiting that discontinuity, we shall obtain a simple example of convergence to *different* limits, by the value and the other solutions, when the set of traders is expanded linearly but not homogeneously.

The model can be formulated in terms of gloves. Each player starts with one glove—either right- or left-handed, and the players may trade them, or buy or sell them for money, without restriction. At the end of the game, an assembled pair is worth $1 to whoever holds it. For example, there might be an outside market that would pay that price.[9]

The characteristic function of the game, which states the dollar potential of each coalition S, is given by the equation

$$v(S) = \min(|S \cap R|, |S \cap L|) . \tag{1}$$

Here R and L are the original sets of owners of right- and left-handed gloves, respectively, and the notation "$|X|$" means the number of elements of the set X.

Equation (1) expresses a rudimentary form of complementarity between economic units of different types. Traders on the same side of the market stand in the position of perfect substitutes; traders on opposite sides, perfect complements. A further discussion of this characteristic function will be found in [18].

[9] An alternative formulation, without money, is given at the end of the section.

3.1. *The competitive equilibrium and core.* The three kinds of solutions discussed in Section 2 will now be determined. Let $r = |R|$ and $l = |L|$, and suppose first that $r < l$. Then the equilibrium price of left-handed gloves in a competitive market is zero, and the members of R can acquire complete pairs at no cost. The unique competitive imputation is therefore

$$\omega_i = \$1 \quad \text{for} \quad i \in R\,, \qquad \omega_j = \$0 \quad \text{for} \quad j \in L\,. \tag{2}$$

The case $r > l$ is just the opposite. In the transition case, $r = l$, the equilibrium prices are not unique; we know only that the sum of the two prices must be \$1. A continuum of competitive imputations results, as follows:

$$\begin{aligned} \omega_i(p) &= \$p \quad &\text{for} \quad i \in R\,, \\ \omega_j(p) &= \$(1-p) \quad &\text{for} \quad j \in L\,, \end{aligned} \qquad (0 \leqq p \leqq 1)\,. \tag{3}$$

The core of the game necessarily contains all competitive imputations. In this case, it happens to contain no other imputations. Indeed, in any *non-competitive* imputation the least-favored member of R and the least-favored member of L get less than \$1 combined, and can therefore form a blocking coalition. Thus the core is also given by (2) (or its opposite) or (3).

3.2. *The value.* To calculate the value, we shall make use of the "random order" version of the definition (see Appendix A.1), in which an imputation is built up one player at a time by awarding each player the increment that he brings to the coalition consisting of his predecessors. The value of the game is equal to the average of these imputations, over all possible orderings of players [17].

Let $\phi_{\text{right}}(r, l)$ denote the *sum* of the values to the r members of R. If we consider separately those orderings that end with a member of R (probability $r/(r+l)$), and those orderings that end with a member of L (probability $l/(r+l)$), we obtain the following difference equation:

$$\begin{aligned} \phi_{\text{right}}(r, l) = {} & \frac{r}{r+l}[\phi_{\text{right}}(r-1, l) + v(R \cup L) - v(R' \cup L)] \\ & + \frac{l}{r+l}[\phi_{\text{right}}(r, l-1)]\,, \end{aligned} \tag{4}$$

where R' is any set satisfying $R' \subset R$, $|R'| = r - 1$. If we assume that $r > l$ (thereby eliminating the "v" terms in (4)), then the relevant boundary conditions are

$$\phi_{\text{right}}(r, 0) = 0 \quad \text{and} \quad \phi_{\text{right}}(r, r) = \frac{r}{2}\,, \qquad \text{all } r\,,$$

and the solution of the difference equation for $r \geqq l$ is

$$\phi_{\text{right}}(r, l) = \frac{r}{2} - \frac{r-l}{2} \sum_{k=0}^{l} \frac{r!\, l!}{(r+k)!\,(l-k)!}\,.$$

(The reader may verify this by direct substitution in (4).) This amount is divided equally among the members of R. With the aid of Pareto optimality:

$\phi_{\text{right}} + \phi_{\text{left}} = v(R \cup L) = l$, the values to members of L are easily determined. The complete value imputaion for $r \geqq l$ is

(5)
$$\phi_i = \frac{1}{2} - \frac{r-l}{2r} \sum_{k=0}^{l} \frac{r!\,l!}{(r+k)!\,(l-k)!} \quad \text{for} \quad i \in R\,,$$
$$\phi_j = \frac{1}{2} + \frac{r-l}{2l} \sum_{k=1}^{l} \frac{r!\,l!}{(r+k)!\,(l-k)!} \quad \text{for} \quad j \in L\,.$$

The case $r \leqq l$ is symmetrical. Table 1 gives an idea of how these equations behave for small numbers of traders.

TABLE 1

VALUE TO A MEMBER OF R

r \ l	0	1	2	3	4	5	6	7	8
1	0	0.500	0.667	0.750	0.800	0.833	0.857	0.875	0.889
2	0	0.167	0.500	0.650	0.733	0.786	0.822	0.847	0.867
3	0	0.083	0.233	0.500	0.638	0.720	0.774	0.811	0.838
4	0	0.050	0.133	0.272	0.500	0.629	0.710	0.764	0.802
5	0	0.033	0.086	0.168	0.297	0.500	0.622	0.701	0.755
6	0	0.024	0.060	0.113	0.194	0.315	0.500	0.616	0.693
7	0	0.018	0.044	0.081	0.135	0.214	0.330	0.500	0.610
8	0	0.014	0.033	0.061	0.099	0.153	0.230	0.341	0.500

The value solution definitely favors the "short" side of the market, individually and collectively. For example, if $l < r$, the members of L, with *less* than half the population, get *more* than half the total profit, which is $v(R \cup L) = l$. On the other hand, the "long" side of the market is not totally defeated, as in the other solutions discussed. The value of the game is less abruptly sensitive to the balance between supply and demand than the competitive equilibrium and the core, since it gives some credit for the bargaining position of the group in oversupply.[10] It is not strange that the competitive solution, with its decentralized outlook, fails to recognize collusive bargaining power, but it is a little surprising that the core—a cooperative-game concept—misses it as well.

3.3. *Asymptotic behavior.* In Figure 1 we show the effect of altering the ratio of trader types, holding the *size* of the market fixed. In the second

[10] For example, if $r = l + 1$, then the members of R, faced with total defeat under pure competition, might select two of their number to withdraw from the market, thus turning the tables on L. This behavior would not be Pareto optimal, since only $l - 1$ pairs of gloves could be formed, but the threat is credible enough and might well raise the price for right-handed gloves.

The reader will recognize this as a standard price-support tactic in situations where collusion is possible. Of course, the value of the game does not directly consider such details of process, but it recognizes and measures the coalition potentials that make such maneuvers effective. For further discussion, see [**19**].

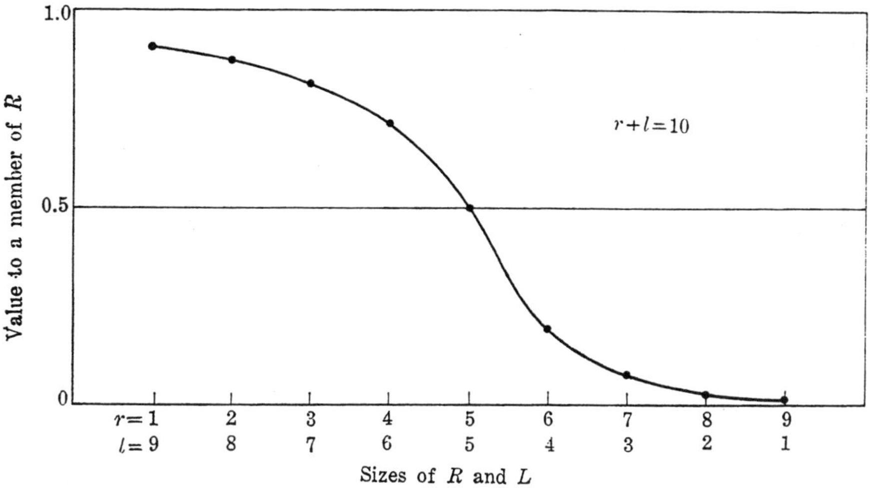

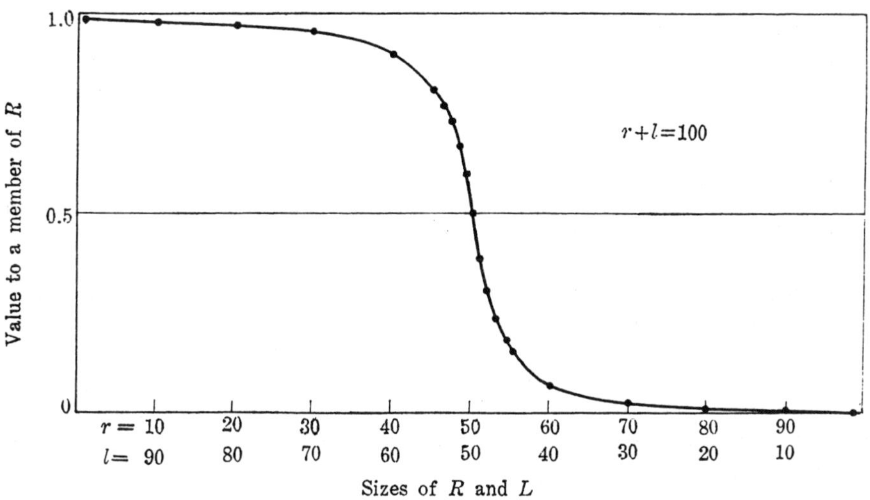

FIGURE 1

VALUE AS A FUNCTION OF COMPOSITION

graph, with ten times as many traders, the slope of the curve in the vicinity of the transition case is noticeably steeper. In the limit, the curve approaches the ⅂_-shape associated with the core and competitive solutions.

The effect of increasing the size of the market while holding the *ratio* of trader types fixed is illustrated in Table 2, for the ratio 2 : 1. We see that owning a right glove is worth $16\frac{2}{3}$¢ when there are three traders, but less than 6¢ if there are 30 traders and less than 1¢ if there are 300. The

TABLE 2

VALUE SOLUTIONS FOR TRADER TYPES IN RATIO 2:1

r	l	ϕ_i	ϕ_j	ϕ_{right}	ϕ_{left}
2	1	0.1667	0.6667	0.333	0.667
4	2	0.1333	0.7333	0.533	1.467
6	3	0.1131	0.7738	0.679	2.321
8	4	0.0990	0.8020	0.792	3.208
10	5	0.0884	0.8232	0.884	4.116
20	10	0.0589	0.8822	1.178	8.822
200	100	0.0092	0.9816	1.842	98.158
as $r = 2l \to \infty$		$\frac{1}{l} - \frac{9}{l^2}$	$1 - \frac{2}{l} + \frac{18}{l^2}$	$2 - \frac{18}{l}$	$l - 2 + \frac{18}{l}$

general asymptotic formulas for fixed ratio $a:b$, with $a > b$, are

$$\phi_i(ak, bk) = \frac{b}{(a-b)^2 k} - \frac{3b(a+b)}{(a-b)^4 k^2} + O(k^{-3}) \qquad \text{for } i \in R ,$$

$$\phi_j(ak, bk) = 1 - \frac{a}{(a-b)^2 k} + \frac{3a(a+b)}{(a-b)^4 k^2} + O(k^{-3}) \qquad \text{for } j \in L .$$

(We omit the derivation.) The rate of convergence of the value to the competitive imputation is of the order of $1/k$. The same rate has been observed in other examples and may be presumed to be typical for homogeneously expanding markets.

Another asymptotic result is available. If we let the market expand with the *difference* $d = r - l$ held constant, we obtain the following estimates:[11]

$$\phi_i(r, r-d) = \frac{1}{2} - \frac{d}{2}\sqrt{\frac{\pi}{r}} + O(r^{-1}) \qquad \text{for } i \in R ,$$

$$\phi_j(r, r-d) = \frac{1}{2} + \frac{d}{2}\sqrt{\frac{\pi}{r}} + O(r^{-1}) \qquad \text{for } j \in L .$$

In other words, if traders are added in equal numbers on both sides of the market, the distinction between "long" and "short" disappears in the limit, and all get equal shares of the profit. This provides an example (if $d \neq 0$) of a linearly, nonhomogeneously expanding market in which the value and the competitive imputation do *not* tend to the same limit.

3.4. *Concluding remarks.* Still another solution concept—the von Neumann-Morgenstern stable sets [26]—has been applied to the symmetric market game (1) by one of the authors [18]. It was found that in the transition case, $r = l$, the unique stable set is the core itself, in other words, the straight line of imputations defined by (3) above. In the other cases, $r > l$ and $r < l$, there

[11] The method of derivation, in brief, is to multiply the summands in (5) by $(r+l)!/r!l!$, to obtain truncated sums of binomial coefficients that can be estimated with the aid of Stirling's formula.

are infinitely many stable sets, each one a continuous curve of imputations emanating from the one-point core and possessing certain monotonicity properties. Thus, a "price-like" parameter can always be identified, but most of the stable sets will involve some kind of price discrimination—i.e., unequal treatment of traders of the same type.

A nonsymmetric generalization of the present example, typified by Böhm-Bawerk's celebrated horse market [5], has also been considered by one of the authors [**19**]. Again, it was found that the traders that are priced out of the market receive consideration in the value of the game, but not in the core or competitive equilibrium. The stable sets are again continuous, monotonic curves.

A final remark: We could have avoided the use of money in this example if we had made the commodities continuously divisible and given the traders identical (cardinal) utility functions:

$$u^i(x, y) = \min(x, y), \qquad \text{all } i \in N . \tag{6}$$

Money would then be superfluous, since the relevant utility transfers could be accomplished (on a constant-sum basis) by transferring bundles containing equal amounts of the two commodities.

4. A CONVERGENCE THEOREM

A proof of the convergence of the value to the competitive solution was given in [**20**] for a general class of replicated markets with money, as formulated in Appendix A.2. Here we shall focus on a two-sided market with equal tastes, in order to display the main sequence of ideas of that more general proof while avoiding certain secondary complications. Along the way, we shall prove the convergence of the core to the competitive solution.

The two types of traders are distinguished only by their initial commodity bundles, which are, respectively, $(a, 0)$ and $(0, b)$. Let there be km traders of the first type and kn of the second; here k is to be regarded as variable, m and n fixed. The *relative composition* (μ, ν) of the market will thus remain fixed at $\mu = m/(m + n)$, $\nu = 1 - \mu = n/(m + n)$.

All traders are assumed to have the same concave and twice-differentiable utility function of the separable form $u(x, y) + \xi$, where ξ is the net change from the initial money level. The characteristic function of the game will depend only on the numbers s, t of traders of each type in a coalition and will be written $v(s, t)$. Because of equal tastes, total utility for any coalition is maximized by an equal sharing of goods, and we have at once

$$v(s, t) = (s + t)u(\sigma a, \tau b) , \tag{7}$$

where (σ, t) is the relative composition of the coalition, thus: $\sigma = s/(s + t), \tau = 1 - \sigma = t/(s + t)$. Note that v is homogeneous of the first degree: $v(\lambda s, \lambda t) = \lambda v(s, t)$; this holds regardless of the homogeneity of u.

4.1. *The competitive solution.* Pareto optimality can be achieved by allocating $(\mu a, \nu b)$ to everyone, followed by an arbitrary money transfer. In order to support this goods allocation, the competitive prices must be

$$\pi_x = \frac{\partial u(\mu a, \nu b)}{\partial x} \qquad \text{(first good)},$$

$$\pi_y = \frac{\partial u(\mu a, \nu b)}{\partial y} \qquad \text{(second good)}.$$

The competitive payoffs are therefore given by

$$(8) \qquad \begin{aligned} \omega_1 &= u(\mu a, \nu b) + \nu a \pi_x - \nu b \pi_y \qquad \text{(first type)},\\ \omega_2 &= u(\mu a, \nu b) + \mu b \pi_y - \mu a \pi_x \qquad \text{(second type)}. \end{aligned}$$

In these expressions, the first term is the utility of the final holding, the second is the payment received for selling off part of the initial holding, and the third is the money spent on buying the other good.

Now there may be competitive *allocations* other than $(\mu a, \nu b)$, since u need not be strictly concave. But the competitive *imputation* ω is unique, as are the prices.[12] Note also that ω_1 and ω_2 are independent of k. As we change the size of the market, the competitive solution remains fixed.

4.2. *The core.* Next, let us examine the behavior of the core. Expanding u in (7) in a Taylor's series about $u(\mu a, \nu b)$, we have

$$v(s, t) = (s + t)\,[u(\mu a, \nu b) + (\sigma - \mu)a\pi_x + (\tau - \nu)b\pi_y + O((\sigma - \mu)^2)]\,.$$

Using (8) and the relation $\sigma - \mu = \nu - \tau = \sigma\nu - \tau\mu$, we have

$$(9) \qquad v(s, t) = s\omega_1 + t\omega_2 + (s + t)O((\sigma - \mu)^2)\,.$$

Moreover, by concavity, the remainder term is either zero or negative. Hence $v(s, t) \leqq s\omega_1 + t\omega_2$, and the competitive imputation satisfies every coalition and is an element of the core.

Now let α be any Pareto-optimal imputation, i.e., one with total payoff $v(km, kn)$. If α is not symmetric, i.e., if α gives unequal payoffs to some pair of traders of the same type, and if $k > 1$, then the m worst-treated traders of the first type and the n worst-treated traders of the second type must together get *less* than $v(m, n)$. Hence they can block α. It follows that if $k > 1$, the core is confined to the one dimensional set P of *symmetric Pareto-optimal imputations.* We may parametrize this set by distance from ω, thus: $P = \{\alpha(c) \mid -\infty < c < \infty\}$, where

$$\begin{aligned} \alpha_1(c) &= \omega_1 + c/\mu \qquad \text{(first type)},\\ \alpha_2(c) &= \omega_2 - c/\nu \qquad \text{(second type)}. \end{aligned}$$

As we have seen, $\alpha(0) = \omega$ is in the core. We shall not trouble to determine the exact upper and lower bounds for c in the core, which depend somewhat irregularly on m, n, and k. But the convergence of these bounds to zero, and hence the convergence of the core to the competitive solution, can be shown quite easily.

In fact, let Q be a coalition having $km + kn - 1$ members, lacking only

[12] Differentiability is important here. For example, the utility function (6) in Section 3 is not differentiable at $x = y$, and the nonuniqueness of the competitive prices and imputations for the case $r = l$ (see (3)) is the direct result.

one trader of the first type. Then $\alpha(c)$ awards Q the amount

$$v(km, kn) - \omega_1 - c/\mu .$$

If we use (9) to estimate the characteristic function of Q, we obtain

$$v(km - 1, kn) = v(km, kn) - \omega_1 + O(1/k) .$$

Thus, if c is positive, and if k is sufficiently large, then Q can block $\alpha(c)$. Similarly, if c is negative and k large enough, then a coalition lacking just one trader of the second type can block $\alpha(c)$. In the limit, only the competitive imputation $\alpha(0)$ remains unblocked.

4.3. *The value.* Now let us examine the behavior of the value. For any trader, it represents his expected marginal worth to a coalition chosen at random (see Appendix A.1). We may express this (for a trader of the first type) as follows:

$$\phi_1(k) = E\{D_1(s, t)\} , \tag{10}$$

where E is an averaging operator and D_1 is the finite difference

$$D_1(s, t) = v(s + 1, t) - v(s, t) .$$

The precise form of E could be stated, but it is not relevant here. Indeed, any method of averaging the increments D_1 that will sustain the "almost all" statements in the paragraph following will suffice. The convergence theorem is therefore valid for a whole class of "values" that might be defined.

Now let equation (7) be regarded as defining a function $v(s, t)$ for all positive real numbers s and t; like u it is twice differentiable. Using homogeneity, a simple Taylor's expansion, and (7), we have

$$\begin{aligned} D_1(s, t) &= (s + t)\left[v\left(\sigma + \frac{1}{s+t}, \tau\right) - v(\sigma, \tau)\right] \\ &= \frac{\partial v(\sigma, \tau)}{\partial \sigma} + O\left(\frac{1}{s+t}\right) \\ &= \frac{\partial v(s, t)}{\partial s} + O\left(\frac{1}{s+t}\right) \\ &= \frac{\partial}{\partial s}[(s + t)u(\sigma a, \tau b)] + O\left(\frac{1}{s+t}\right) \\ &= u(\sigma a, \tau b) + \tau a\frac{\partial u(\sigma a, \tau b)}{\partial x} - \tau b\frac{\partial u(\sigma a, \tau b)}{\partial y} + O\left(\frac{1}{s+t}\right). \end{aligned}$$

The last line closely resembles the formula for ω_1 in (8). Indeed, if $s + t$ is large, and if (σ, τ) is close to (μ, ν), then $D_1(s, t)$ will approach the competitive payoff ω_1, as required. But if k is large enough, then "almost all" coalitions will be large, and "almost all" coalitions will have compositions approximating (μ, ν). The latter will be recognized as a form of the law of large numbers. The precise statement is as follows: Given any $\varepsilon > 0$, a number k_ε can be chosen so large that for any $k \geqq k_\varepsilon$ a randomly chosen coalition will, with probability at least $1 - \varepsilon$, have size $s + t$ large enough and composition (σ, τ)

near enough to (μ, ν) to ensure that $|D_1(s, t) - \omega_1| \leqq \varepsilon$. Hence we may write

$$\phi_1(k) = (1 - \varepsilon)(\omega_1 + \delta(k)) + \varepsilon C_1(k)\,, \qquad \text{all } k \geqq k_\varepsilon\,, \tag{11}$$

where $|\delta(k)| \leqq \varepsilon$, and $C_1(k)$ is the appropriately weighted average of the $D_1(s, t)$ for the "exceptional" coalitions—i.e., those occurring with probability ε that may be too small or have compositions too far from (μ, ν).

Let us consider the implications of (11). If $C_1(k)$ were known to be bounded, we could conclude that $\phi_1(k) \to \omega_1$, and we would be through. A lower bound for $C_1(k)$, namely, $v(1, 0)$, follows at once from the definition of D_1 and the superadditivity of the characteristic function. An upper bound for $C_1(k)$ cannot be deduced so directly, however, unless restrictions are imposed on the behavior of $u(x, y)$ near the boundaries of the positive quadrant. But there is a trick that takes us around this difficulty. We observe that both the value and the competitive imputation are Pareto optimal; hence we have

$$m\phi_1(k) + n\phi_2(k) = m\omega_1 + n\omega_2 \qquad \text{for all } k\,. \tag{12}$$

The lower bound on $C_1(k)$ tells us that $\liminf \phi_1(k) \geqq \omega_1$, i.e., that no limit point (finite or infinite) of the sequence $\{\phi_1(k)\}$ is less than ω_1. Hence, by (12), we have $\limsup \phi_2(k) \leqq \omega_2$. To complete the proof, we merely return to (10) and repeat the whole argument with types 1 and 2 interchanged, obtaining $\limsup \phi_1(k) \leqq \omega_1$ and $\liminf \phi_2(k) \geqq \omega_2$. In this way we finally establish the convergence of the value imputation $\phi(k)$ to the competitive imputation ω.

To sum up the essential idea of the proof: The partial derivatives of the characteristic function $v(s, t)$, show that the marginal value of a player to a *large, nearly balanced* coalition is substantially equal to his competitive payoff. However, if the economy is big enough, almost all coalitions are large and nearly balanced.

5. A MARKET WITHOUT TRANSFERABLE UTILITY

In order to illustrate the application of the theory of games without transferable utility, we shall present a simple two-sided Edgeworth market in which both the core and the value can be determined explicitly, but in which there is not so much symmetry that the solutions are uninteresting. In addition to providing working experience with the new value concept, the example will illustrate several theoretical points.

Let there be n traders on each side of the market. Let there be two goods in trade, but no money or credit. Let the initial holdings be

$$(1, 0) \quad \text{(first type)}\,,$$
$$(0, 1) \quad \text{(second type)}\,,$$

and let the (cardinal) utility functions be

$$\begin{aligned} u_1(x, y) &= \sqrt{xy} && \text{(first type)}\,, \\ u_2(x, y) &= \sqrt{x^2 + y^2 + 23xy} && \text{(second type)}\,. \end{aligned} \tag{13}$$

The number 23 is only a convenience; all that really matters is that these functions are concave, homogeneous of degree 1, and symmetric in the two goods.

Figure 2 is the "Edgeworth box" for this market [7, 24]. The origin O represents the allocation that gives $(0, 0)$ to type 2, and hence $(1, 1)$ to type 1. The opposite corner C' is the "origin" for type 1, and R is the initial or no-trade point. The segment CC' is Edgeworth's *contract curve* for the case of two traders (i.e., $n = 1$). The unique competitive allocation ω, which gives all traders $(1/2, 1/2)$, is represented by the point W; this is independent of n. (The point V represents the value allocation for $n = 1$, which will be determined later.)

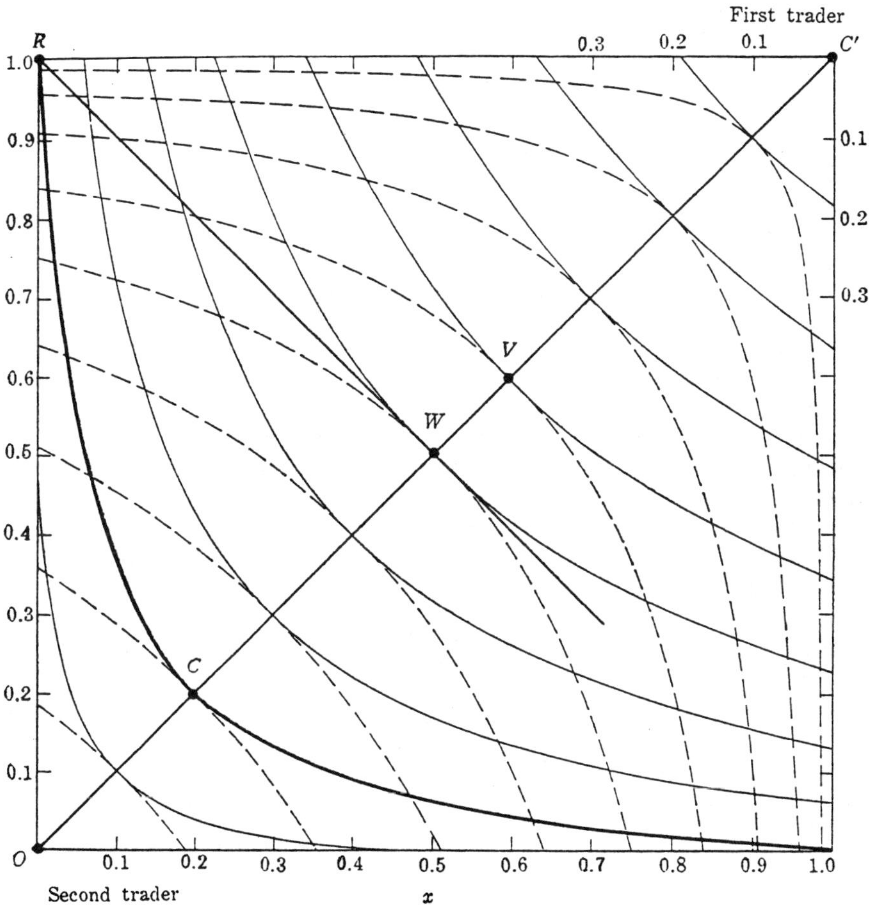

FIGURE 2
THE EDGEWORTH BOX

We emphasize that for $n > 1$, the Edgeworth box serves to represent only the symmetric allocations—those in which traders of the same type are treated alike.

5.1. *The core.* We first discuss the core. For $n = 1$ the core comprises the whole contract curve CC'. More precisely, the core is the image of CC' in the utility space under the mapping (13), as shown in Figure 3.

For $n > 1$, we first observe that all imputations in the core must be symmetric, since any nonsymmetric imputation can be profitably blocked by a two-man coalition consisting of one least-favored trader of each type. We may therefore transfer our attention from the $2n$-dimensional space of all imputations to the two-dimensional subspace of symmetric imputations.

Because of the homogeneity of (13), the Pareto-optimal symmetric imputations lie along a straight line (OC' in Figure 3); its equation is

$$5u_1 + u_2 = 5 \; . \tag{14}$$

The core, for each n, is a subset of this line. It remains to discover which points on the line can be blocked. As in the model of Section 4, the most efficient blockers are coalitions that have almost, but not quite, the same relative composition as the market as a whole.

To verify this, let $r(S)$ denote the ratio of first types to second types in an arbitrary coalition S, and let $(1 - t, 5t), 0 \leqq t \leqq 1$, be an arbitrary point on OC', representing the symmetric imputation α_t. Then a routine calculation reveals that in order for S to block α_t it is necessary and sufficient that $r(S)$ lie strictly between 1 and a certain critical ratio r_t, given by

$$r_t = \frac{t(t^2 - 0.04)}{(1 - t)(t - t^2 - 0.04)} \; .$$

Note that $r_{1/2} = 1$, showing that the competitive imputation $\alpha_{1/2}$ cannot be blocked, since there is no number strictly between 1 and 1. However, any other α_t will be blocked by some coalition if n is sufficiently large.

For each n, the coalitions of size $2n - 1$ provide the best type-composition ratios available, namely $(n - 1)/n$ and $n/(n - 1)$. Setting r_t equal to these numbers in turn, and solving for t, gives us the endpoints of the core. We have done this numerically for several values of n, as shown in Table 3 and Figure 3. Asymptotically (last line of Table 3), the length of the core varies inversely with the size of the market.

TABLE 3

ENDPOINTS OF THE CORE

n	C		C'	
1	0.800	1.000	0.000	5.000
2	0.577	2.116	0.421	2.897
3	0.546	2.272	0.453	2.733
4	0.533	2.337	0.467	2.665
5	0.525	2.374	0.474	2.628
10	0.512	2.440	0.488	2.560
as $n \to \infty$	$\frac{1}{2} + \frac{21}{184n}$	$\frac{5}{2} - \frac{105}{184n}$	$\frac{1}{2} - \frac{21}{184n}$	$\frac{5}{2} + \frac{105}{184n}$

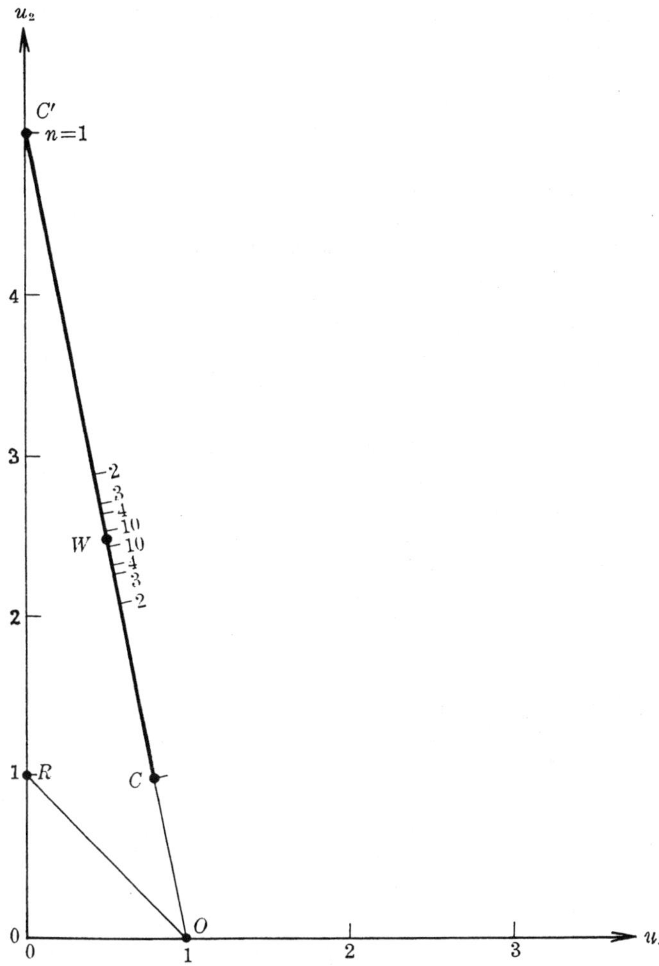

FIGURE 3

THE UTILITY SPACE (SYMMETRIC PAYOFFS ONLY), SHOWING THE SHRINKING OF THE CORE TO THE COMPETITIVE PAYOFF[13]

5.2. *The value.* We now turn to the value. The technique, as explained in Section 2.3, is to introduce a set of "weights" $\{\lambda_i\}$, nonnegative and not all zero, on which to base hypothetical exchanges of utility among the traders. Thus, if α is a payoff vector attainable in the market, we assume that $\alpha + \beta$ is also attainable for any β such that $\sum \lambda_i \beta_i = 0$. Given the weights, we are

[13] We have used the same letters to denote *allocation* points in the (x, y)-space and the corresponding *payoff* points (imputations) in the (u_1, u_2)-space.

in a position to calculate the "λ-transfer value" of the game (Appendix A.3). This value is in general infeasible, i.e., not attainable by commodity transfers alone. We must try to find weights that yield a feasible λ-transfer value; this will be, by definition, a *value* of the original market without transferable utility.

We shall first dispose of the possibility of a nonsymmetric solution—one that gives different payoffs to traders of the same type. This would require unequal weights, since the value formula itself is symmetric. Suppose therefore that $\lambda_j < \lambda_k$ where j and k are traders of the same type, and consider how the λ-transfer value might be attained. Regardless of utility transfers, the allocation of goods must maximize the weighted sum of utilities. Any goods of positive utility that j could transfer to k would increase this sum because of the following comparison:

$$(15) \qquad \begin{aligned} \lambda_k u(x^j + x^k, y^j + y^k) &\geqq \lambda_k[u(x^j, y^j) + u(x^k, y^k)] \\ &\geqq \lambda_j u(x^j, y^j) + \lambda_k u(x^k, y^k)\,, \end{aligned}$$

with strict inequality on the second line if $u(x^j, y^j) > 0$. Hence j's share of the *goods* allocation must be worthless. But his *payoff* in the λ-transfer value is easily seen to be positive, even if $\lambda_j = 0$.[14] It follows that the λ-transfer value can be attained only with the aid of a utility transfer, making it infeasible for the original game. We conclude that only equal weights need be considered for traders of the same type.

With symmetry established, the *λ-transfer characteristic function* (see Appendix A.3) can be written in the form $v^\lambda(s, t)$, where the pair of integers (s, t) is the type-composition of the coalition in question. A simple fact about this function will be useful.

LEMMA. *If $st \neq 0$, then $v^\lambda(s, t) = v^\lambda(t, s)$.*

PROOF. The inequality (15), which depends on the concavity and homogeneity of (13), shows that any coalition can attain its maximum λ-weighted total utility while concentrating all its goods in the hands of at most two traders—one of each type. For a coalition of composition (s, t), with neither s nor t equal to 0, there will be an optimal allocation that gives, say, (x, y) to one man of type 1, $(s - x, t - y)$ to one man of type 2, and nothing to the rest. This allocation is worth an amount $A = v^\lambda(s, t)$ to the coalition. Now consider a second coalition, of composition (t, s). A possible allocation is (y, x) to one man of type 1, $(t - y, s - x)$ to one man of type 2, and nothing to the rest. Since the utility functions are symmetric in the commodities, this must be worth the same amount, A, to the second coalition. Hence $v^\lambda(t, s) \geqq A = v^\lambda(s, t)$. Repeating the argument, we obtain $v^\lambda(t, s) = v^\lambda(s, t)$, completing the proof of the lemma.

Let us now consider the value formula in its "orderings" version (see Appendix A.1). The λ-transfer value ϕ_i^λ to a typical trader of type 1 can be

[14] The reason: He makes a positive λ-weighted contribution to any coalition in which at least one trader is of the opposite type and at least one trader has weight greater than 0.

expressed as a linear function of all the $v^\lambda(s, t)$, where s and t are integers ranging from 0 to n. Consider how the coefficient of a particular $v^\lambda(s, t)$ is formed. There will be positive contributions from coalitions containing i, and negative contributions from coalitions not containing i. For the positive part, we must count the number of orderings that put i in position "$s + t$" and put exactly $s - 1$ other type 1 traders in positions preceding "$s + t$." This number is

$$\binom{n-1}{s-1}\binom{n}{t}(s + t - 1)!\,(2n - s - t)!\ ,$$

or zero if $s = 0$. Similarly, for the negative part, we must have i in position "$s + t + 1$," and exactly s other type 1 traders in positions before "$s + t + 1$." The number of orderings that do this is

$$\binom{n-1}{s}\binom{n}{t}(s + t)!\,(2n - s - t - 1)!\ ,$$

or zero if $s = n$. The desired coefficient of $v^\lambda(s, t)$ is the difference of these two numbers divided by the total number of orderings, which is $(2n)!$. This reduces to

$$\binom{n}{s}\binom{n}{t}\frac{(s + t - 1)!\,(2n - s - t - 1)!\,(s - t)}{(2n)!} \tag{16}$$

or

$$-\frac{1}{2n} \qquad \text{if } s = t = 0\ ,$$

or

$$\frac{1}{2n} \qquad \text{if } s = t = n\ .$$

Since the coefficient (16) is *anti*symmetric in s and t, the lemma permits us to cancel the bulk of the terms in the value formula. It is this cancellation that makes the whole calculation manageable. We are left with

$$\begin{aligned}\phi_i^\lambda = \frac{1}{2n}v^\lambda(n, n) + \sum_{s=1}^{n}\binom{n}{s}\frac{(s - 1)!\,(2n - s - 1)!\,s}{(2n)!}v^\lambda(s, 0)\\ - \sum_{t=1}^{n}\binom{n}{t}\frac{(t - 1)!\,(2n - t - 1)!\,t}{(2n)!}v^\lambda(0, t) - \frac{1}{2n}v^\lambda(0, 0)\ .\end{aligned}$$

Homogeneity of v^λ reduces this to

$$\phi_i^\lambda = \frac{1}{2}v^\lambda(1, 1) + \sum_{s=1}^{n}\binom{n}{s}\frac{s!\,(2n - s - 1)!\,s}{(2n)!}[v^\lambda(1, 0) - v^\lambda(0, 1)]\ .$$

This, in turn, can be reduced to the very simple expression,[15]

[15] The reduction:

$$\sum_{s=1}^{n}\binom{n}{s}\frac{s!(2n - s - 1)!s}{(2n)!} = \frac{n!n!}{(2n)!}\sum_{s=1}^{n}\frac{(2n - s - 1)![n - (n - s)]}{(n - s)!n!}$$

$$\phi_i^\lambda = \frac{1}{2} v^\lambda(1, 1) + \frac{1}{n+1} [v^\lambda(1, 0) - v^\lambda(0, 1)] .$$

A symmetrical argument yields

$$\phi_j^\lambda = \frac{1}{2} v^\lambda(1, 1) - \frac{1}{n+1} [v^\lambda(1, 0) - v^\lambda(0, 1)] ,$$

where j is a typical trader of type 2.

Thus far only the concavity, symmetry, and first-degree homogeneity of the utility functions have played a role. We now refer to the particular functions (13) and make the simple determinations

$$v^\lambda(1, 0) = 0 ,$$
$$v^\lambda(0, 1) = \lambda_j ,$$
$$v^\lambda(1, 1) = \max(\lambda_i, 5\lambda_j) .$$

Inserting these values, we obtain

$$(17) \qquad \begin{aligned} \phi_i^\lambda &= \frac{1}{2} \max(\lambda_i, 5\lambda_j) - \frac{1}{n+1} \lambda_j \qquad \text{(first type)} , \\ \phi_j^\lambda &= \frac{1}{2} \max(\lambda_i, 5\lambda_j) + \frac{1}{n+1} \lambda_j \qquad \text{(second type)} . \end{aligned}$$

This is the λ-transfer value. The question is: For what choices of λ_i, λ_j is it feasible?

Clearly, only the ratio between λ_i and λ_j is significant. The situation is illustrated in Figure 4 for the case $n = 1$. In general, let α be any feasible, symmetric imputation. Then $5\alpha_i + \alpha_j \leqq 5$ (cf. (14)). But if α is to be a value of the game, we must have $\lambda_i \alpha_i = \phi_i^\lambda$ and $\lambda_j \alpha_j = \phi_j^\lambda$ for some choice of λ_i, λ_j. Hence the following inequality must be satisfied:

$$(18) \qquad 5\lambda_j \phi_i^\lambda + \lambda_i \phi_j^\lambda \leqq 5\lambda_i \lambda_j ,$$

with λ_i and λ_j nonnegative and not both zero. If we substitute (17) into (18), we find that $\lambda_i \leqq 5\lambda_j$ implies

$$\left(\frac{5}{2} - \frac{1}{n+1}\right)(\lambda_i - 5\lambda_j) \geqq 0 ,$$

while $\lambda_i \geqq 5\lambda_j$ implies

$$\left(\frac{5}{2}\lambda_i + \frac{1}{n+1}\lambda_j\right)(\lambda_i - 5\lambda_j) \leqq 0 .$$

It follows that the only solution is $\lambda_i = 5\lambda_j$; this conclusion is independent of

---

$$= \frac{n!n!}{(2n)!}\left[\sum_{s=1}^{n} \binom{2n-s-1}{n-1} - \sum_{s=1}^{n-1} \binom{2n-s-1}{n}\right]$$
$$= \frac{n!n!}{(2n)!}\left[\binom{2n-1}{n} - \binom{2n-1}{n+1}\right] = \frac{1}{n+1} ,$$

using the identity $\binom{r}{r} + \binom{r+1}{r} + \cdots + \binom{R}{r} = \binom{R+1}{r+1}$.

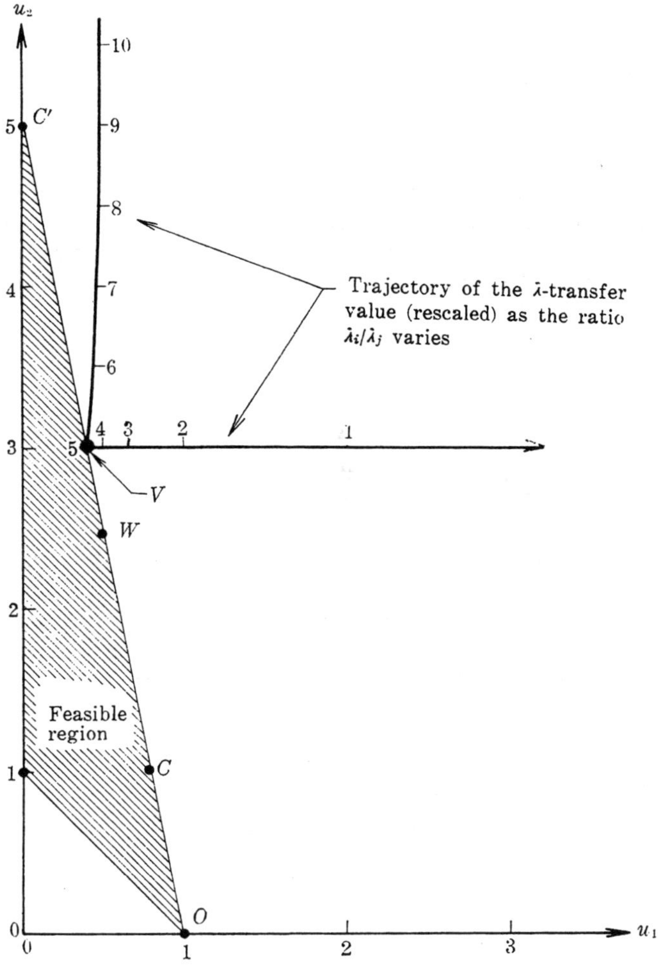

FIGURE 4
FEASIBILITY OF THE λ-TRANSFER VALUE ($n = 1$)

n.[16] Hence the value of the game is unique and is given by

$$\phi_i = \frac{1}{2} - \frac{1}{5(n+1)} \qquad \text{(first type)},$$

$$\phi_j = \frac{5}{2} + \frac{1}{n+1} \qquad \text{(second type)}.$$

Table 4 is intended for comparison with Table 3. In Figure 5 all three

[16] This 5 : 1 ratio, valid for all n, could have been deduced several pages ago, by an argument similar to the one establishing symmetry of λ. However, the λ-transfer value computation is not helped (in this case) by knowing λ in advance.

TABLE 4
THE VALUE

n	V	
1	0.400	3.000
2	0.433	2.833
3	0.450	2.750
4	0.460	2.700
5	0.467	2.667
10	0.482	2.591
as $n \to \infty$	$\frac{1}{2} - \frac{1}{5n}$	$\frac{5}{2} + \frac{1}{n}$

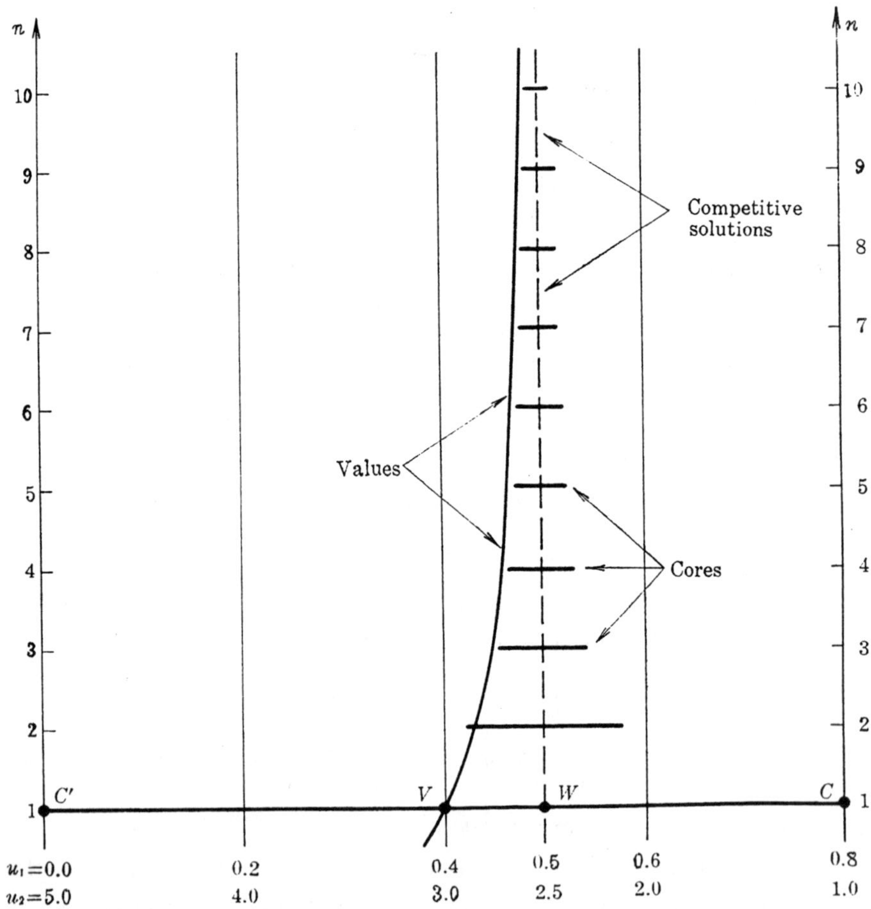

FIGURE 5
CONVERGENCE OF SOLUTIONS

kinds of solutions are shown together, as functions of market size. We see that the value is always more favorable to traders of type 2 than the competitive solution, and that it begins inside the core but moves outside at $n = 3$ because of its somewhat slower rate of convergence. These comparisons should not be interpreted too broadly, however, since as indicated at the end of Section 2 the no-transfer value is essentially a cardinal concept, while the other solutions are not. If we tamper with our example, applying nonlinear transformations to the traders' utilities, we can alter all such qualitative features. Indeed, for any fixed n, a pair of differentiable order-preserving concavity-preserving transformations for (13) can be found that place the value point V at any designated spot in the interior of the contract curve CC', without affecting the other solutions.

The RAND Corporation and Yale University, U.S.A.

APPENDIX

A.1. C-GAMES WITH SIDE PAYMENTS[1]

N: the set of *players*.

v: the *characteristic function*: $v(S)$ for each $S \subseteqq N$ is a real number representing the monetary worth of the coalition S. Ordinarily v is *superadditive*, in the sense that if $S \cap T = O$ then $v(S) + v(T) \leqq v(S \cup T)$.

A: the set of *imputations* α (feasible utility vectors), characterized by

$$\sum_{i \in N} \alpha_i \leqq v(N) .$$

C: the *core*, defined as the set of imputations $\alpha \in A$ that satisfy every coalition:

$$\sum_{i \in S} \alpha_i \geqq v(S) , \qquad \text{all } S \in N .$$

ϕ: the *value*, an imputation defined by

$$\phi_i = \sum_{S \ni i} \frac{(s-1)!\,(n-s)!}{n!} [v(S) - v(S - \{i\})] ,$$

where s, n denote the number of elements in S, N, respectively. An equivalent definition, often easier to work with, is

$$\phi_i = \frac{1}{n!} \sum_{\omega \in \Omega} [v(P_{\omega,i} \cup \{i\}) - v(P_{\omega,i})] ,$$

where Ω is the set of all orderings of N and $P_{\omega,i}$ is the set of predecessors of player i in the ordering ω.

A.2. AN EXCHANGE ECONOMY WITH MONEY[2]

N: the set of *traders*.

[1] See [17]. The term "c-game" is short for "game that is adequately represented by its characteristic function."

[2] See [20, 23].

x: an *allocation* of the m goods among the n traders:

$$x = \{x^i \mid i \in N\}, \quad \text{where} \quad x^i = (x_1^i, \cdots, x_m^i).$$

Thus, x_j^i is the amount of good j held by trader i.

U^i: the *utility function* of trader i, assumed to be of the form:

$$U^i(x^i, \xi^i) \equiv u^i(x^i) + \xi^i,$$

where u^i is concave and differentiable and ξ^i represents the net change from the initial money position.

a: the *initial allocation*. We assume every good to be present: $\sum_N a^i > 0$.

b: an *optimal allocation*, characterized by

$$\sum_N u^i(b^i) = \max_x \sum_N u^i(x^i),$$

the maximization subject to $x \geqq 0$ and $\sum_N x^i = \sum_N a^i$. It is unique if the u^i are strictly concave.

π: the *competitive price vector*, defined by

$$\pi_j \begin{Bmatrix} = \\ \geqq \end{Bmatrix} \partial u^i(b^i)/\partial x_j^i \quad \text{if} \quad b_j^i \begin{Bmatrix} > \\ = \end{Bmatrix} 0,$$

where b is any optimal allocation.

ω: the *competitive imputation*, defined by

$$\omega_i = u^i(b^i) + \pi \cdot (a^i - b^i),$$

where b is any optimal allocation. Uniqueness of π and ω follows from the differentiability of the u^i.

v: the *characteristic function*, defined by

$$v(S) = \max_x \sum_S u^i(x^i),$$

the maximization subject to $x \geqq 0$ and $\sum_S x^i = \sum_S a^i$. The *core* and *value* can now be defined as in Appendix A.1.

Replication: Let k identical economies be regarded as a single economy, having kn traders of n different types. The competitive price vector of the enlarged market is just π, while the competitive imputation is just the kn-dimensional vector $(\omega, \cdots, \omega)$ (k times). The characteristic function of the enlarged market is homogeneous of the first degree, in the sense that $v(hS) = hv(S)$, where hS denotes a coalition having exactly h times as many traders of each type as S. Unlike the competitive solution, the core and the value depend on k. However, both converge to the competitive imputation as $k \to \infty$.

A.3. *C*-GAMES WITHOUT SIDE PAYMENTS[3]

N: The set of *players*.

V: the generalized *characteristic function*: $V(S)$ is a compact, convex, nonempty set of s-dimensional vectors, representing the feasible utility vectors for the coalition $S \subseteqq N$. Ordinarily V is *superadditive*,

[3] See [3].

in the sense that if $S \cap T = O$ and if $\alpha \in V(S)$, $\beta \in V(T)$, then the $s + t$-dimensional vector (α, β) (with coordinates properly identified) is in $V(S \cup T)$.

A: the set of *imputations*: $A = V(N)$.

C: the *core*, defined as the set of imputations $\alpha \in A$ that satisfy every coalition, i.e., it is never the case that $\beta \in V(S)$ and $\beta_i > \alpha_i$, all $i \in S$.

λ: a *scaling vector* $(\lambda_i \mid i \in N)$. The components must be nonnegative and not all zero.

v^λ: the *λ-transfer characteristic function*, defined by

$$v^\lambda(S) = \max_{\alpha \in V(S)} \sum_{i \in S} \lambda_i \alpha_i \,.$$

ϕ^λ: the *λ-transfer value*, defined by

$$\phi_i^\lambda = \sum_{S \ni i} \frac{(s-1)!\,(n-s)!}{n!} [v^\lambda(S) - v^\lambda(S - \{i\})] \,.$$

Φ: the set of *values* of the game, comprising all $\phi \in A$ for which there exists a scaling vector λ such that $\lambda_i \phi_i = \phi_i^\lambda$, all $i \in N$.

REFERENCES

[1] ARROW, K. J., H. D. BLOCK, and L. HURWICZ, "On the Stability of the Competitive Equilibrium, II," *Econometrica*, XXVII (January, 1959), 82–109.

[2] ARROW, K. J. and G. DEBREU, "Existence of an Equilibrium for a Competitive Economy," *Econometrica*, XXII (July, 1954), 265–90.

[3] AUMANN, R. J., "The Core of a Cooperative Game Without Side Payments," *Transactions of the American Mathematical Society*, XCVIII (March, 1961), 539–52.

[4] ———, "Markets with a Continuum of Traders," *Econometrica*, XXXII (January, 1964), 39–50.

[5] BÖHM-BAWERK, E. von, *Positive Theory of Capital*, translated by William Smart (New York: G. E. Stechert & Company, 1923).

[6] DEBREU, G. and H. SCARF, "A Limit Theorem on the Core of an Economy," *International Economic Review*, IV (September, 1963), 235–46.

[7] EDGEWORTH, F. Y., *Mathematical Psychics* (London: C. Kegan Paul, 1881).

[8] HARSANYI, J. C., "A Bargaining Model for the Cooperative *n*-Person Game," *Annals of Mathematics Studies*, XL (Princeton: Princeton University Press, 1959), 325–56.

[9] ———, "A Simplified Bargaining Model for the *n*-Person Cooperative Game," *International Economic Review*, IV (May, 1963), 194–220.

[10] MCKENZIE, L. W., "On the Existence of General Equilibrium for a Competitive Market," *Econometrica*, XXVII (January, 1959), 54–71.

[11] NASH, J. F., "Two-Person Cooperative Games," *Econometrica*, XXI (January, 1953), 128–40.

[12] NEGISHI, T., "The Stability of a Competitive Economy: A Survey Article," *Econometrica*, XXX (October, 1962), 635–69.

[13] NIKAIDO, H., "Generalized Gross Substitutability and Extremization," *Annals of Mathematics Studies*, LII (1964), 55–68.

[14] SCARF, H., "Some Examples of Global Instability of the Competitive Equilibrium," *International Economic Review*, I (September, 1960), 157–72.

[15] SELTEN, R., "Bewertung Strategischer Spiele," *Zeitschrift für die gesamte Staatswissenschaft*, CXVI (1960), 221-82.

[16] ———, "Valuation of *n*-Person Games," *Annals of Mathematics Studies*, LII (1964), 577-626.

[17] SHAPLEY, L. S., "A Value for *n*-Person Games," *Annals of Mathematics Studies*, XXVIII (1953), 307-17.

[18] ———, "The Solutions of a Symmetric Market Game," *Annals of Mathematics Studies*, XL (1959), 145-62.

[19] ———, *Values of Large Games, V: An 18-Person Market Game*, The RAND Corporation, RM-2860-PR (November, 1961).

[20] ———, *Values of Large Games, VII: A General Exchange Economy with Money*, The RAND Corporation, RM-4248-PR (December, 1964).

[21] ——— and M. SHUBIK, "Concepts and Theories of Pure Competition," in M. Shubik, ed., *Essays in Mathematical Economics: In Honor of Oskar Morgenstern* (Princeton: Princeton University Press, 1967), 63-79.

[22] ——— and M. SHUBIK, "Ownership and the Production Function," *Quarterly Journal of Economics*, LXXXI (February, 1967), 88-111.

[23] ——— and M. SHUBIK, "Quasi-Cores in a Monetary Economy with Nonconvex Preferences," *Econometrica*, XXXIV (October, 1966), 805-27.

[24] SHUBIK, M., "Edgeworth Market Games," *Annals of Mathematics Studies*, XL (1959), 267-78.

[25] VIND, K., "Edgeworth Allocations in an Exchange Economy with Many Traders," *International Economic Review*, V (May, 1964), 165-77.

[26] von NEUMANN, J. and O. MORGENSTERN, *Theory of Games and Economic Behavior* (Princeton: Princeton University Press, 1944).

[27] WALRAS, L., *Elements of Pure Economics*, translated by W. Jaffe (London: Allen and Unwin, 1954).

[8]

Int. Journal of Game Theory, Vol. 4, Issue 4, page 229–237, Physica-Verlag, Vienna.

Competitive Outcomes in the Cores of Market Games[1])

By *L.S. Shapley*, Santa Monica[2]) and *M. Shubik*, New Haven[3])

Abstract: The competitive outcomes of an economic system are known, under quite general conditions, always to lie in the core of the associated cooperative game. It is shown here that every "market game" (i.e., one that arises from an exchange economy with money) can be represented by a "direct market" whose competitive outcomes completely fill up the core. It is also shown that it can be represented by a market having any given core outcome as its *unique* competitive outcome, or, more generally, having any given compact convex subset of the core as its full set of competitive outcomes.

1. Introduction

In a previous paper [*Shapley* and *Shubik*, 1969], the authors introduced a class of cooperative n-person games in characteristic-function form, called "market games," which come from trading economies in which the traders measure utility in money. This class of games was shown to coincide with the class of "totally balanced" games, i.e., games that have nonempty cores and all of whose subgames have nonempty cores as well. In this paper we shall compare the cores of market games with the competitive equilibria of the markets that they come from. We first consider the "direct market" of a market game, and discover that its competitive outcomes fill the entire core (Theorem 1). We then take an arbitrary point in the core and construct a market (actually a class of markets) which generates the given market game and which has the given core point as its *only* competitive outcome (Theorem 2). A modification of this construction yields an arbitrary closed convex subset of the core as the set of competitive outcomes.

Extensions of these results to games and markets without money (i.e., without transferable utility) will be considered in a subsequent paper.

---

[1]) Supported by Grant GS–31253, National Science Foundation. The authors are grateful to *Bezalel Peleg* for some helpful discussions of this work during the recent International Workshop on Basic Problems of Game Theory at Bad Salzuflen, West Germany, and to *Robert Shishko* for suggesting some improvements in the presentation.

[2]) Dr. *L.S. Shapley*, The Rand Corporation, Santa Monica, California, 90406.

[3]) Prof. *M. Shubik*, The Cowles Foundation, Yalc University, New Haven, Connecticut, 06520.

2. Games and Markets

The reader is referred to *Shapley* and *Shubik* [1969] for a more extended discussion of the matters reviewed in this section.

A *game*, for our present purpose, will consist of a finite set N and a real-valued set function v, the latter defined on all the subsets of N and satisfying $v(\emptyset) = 0$. In the standard interpretation, N represents the set of players, and $v(S)$ represents the "worth" of S, i.e., the total amount of utility that the members of S can secure if they form a coalition and play the game without help from the other players.

Outcomes of the game are expressed as N-tuples of utility: $\alpha = \{\alpha^i : i \in N\}$, measured in some common monetary unit and called *payoff vectors*. A payoff vector α is said to be "feasible" if $\Sigma_{i \in N}\, \alpha^i \leqslant v(N)$; "efficient" if $\Sigma_{i \in N}\, \alpha^i = v(N)$; "individually rational" if $\alpha^i \geqslant v(\{i\})$ for each $i \in N$; and "coalitionally rational" if $\Sigma_{i \in S}\, \alpha^i \geqslant v(S)$ for each $S \subseteq N$. The set of feasible, coalitionally rational payoff vectors is called the *core* of the game; thus, α is the core if and only if

$$\alpha \cdot e^S \geqslant v(S), \text{ all } S \subset N, \text{ and } \alpha \cdot e^N = v(N), \tag{1}$$

where e^S denotes the N-vector having $e_i^S = 1$ if $i \in S$ and $e_i^S = 0$ if $i \in N - S$. Geometrically, the core is a compact convex polyhedron, possibly empty.

It is well known (see *Shapley* [1967]) that nonemptiness of the core is equivalent to the game being *balanced*, in the sense that

$$\sum_{S \subseteq N} \gamma_S v(S) \leqslant v(N) \tag{2}$$

holds for every set of weights $\gamma_S \geqslant 0$, $\quad S \subseteq N$ such that

$$\sum_{S \subseteq N} \gamma_S e^S = e^N. \tag{3}$$

A balanced game v is said to be *totally balanced* if all of its subgames, obtained by restricting v to the subsets of R, $R \subset N$, are also balanced.

A *market*, for our present purpose, will consist of a finite set N of "traders"; a finite set M of "commodities"; an $|M|$-dimensional euclidean orthant E_+^M of "bundles"; and, for each $i \in N$, an *initial bundle* $a^i \in E_+^M$ and a continuous concave *utility function* u^i from E_+^M to the reals. In the interpretation (see *Shapley* and *Shubik* [1966], pp. 807–808) utility is understood to be measured in units of money, and the traders may not only exchange the commodities M as initially supplied to them, but they may also transfer money in any amount. The *final payoff*, which the traders seek to maximize, is therefore found for each trader by adding his net gain of money to the utility of his final bundle of commodities.

A market generates a game in a natural way. First, for any $\emptyset \subset S \subseteq N$, let us define an *S-allocation* to be an indexed set $x^S = \{x^i : i \in S\}$ of bundles in E_+^M. An S-allocation is said to be *feasible* if

$$\sum_{i \in S} x^i = \sum_{i \in S} a^i. \tag{4}$$

To generate the game, we set $v(\emptyset) = 0$ and, for each $\emptyset \subset S \subseteq N$ define

$$v(S) = \max \sum_{i \in S} u^i(x^i), \tag{5}$$

letting the maximum run over all feasible S-allocations x^S. Any v that can be represented in this way is called a *market game*.

By the "core" of a market we shall mean the core of the game it generates. Since market games are always balanced (indeed, the market games are precisely the totally balanced games), all markets have nonempty cores.

Obviously, many markets may generate the same game, while being dissimilar in other respects. In *Shapley* and *Shubik* [1969], a canonical representative for each such class of "game-theoretically equivalent" markets was introduced. We called it the *direct market* of the game, because of its simple form and the fact that the players themselves are, in a sense, the commodities being bought and sold.

To define the direct market for any game v, we first put the commodities M into one-to-one correspondence with the players N. The initial allocation is then given by $a^i = e^{\{i\}}, i \in N$. The traders all have "equal tastes," i.e., identical utility functions, $u^i \equiv u$, given by

$$u(x) = \max \sum_{S \subseteq N} \gamma_S v(S), \tag{6}$$

the maximum running over all $\{\gamma_S \geqslant 0 : S \subseteq N\}$ satisfying[4])

$$\sum_{S \subseteq N} \gamma_S e^S = x. \tag{7}$$

An intuitive explanation of this market can be given in terms of group activities or implicit production processes: associate with each coalition S an activity that "earns" the amount $v(S)$ if all members of S participate fully. For details, the reader is again referred to *Shapley* and *Shubik* [1969]. The technical justification for this construction, however, lies in the fact that if we start with a totally balanced set function v, then *taking the market game of the direct market* gives us v back again. On the other hand, if we start with a v that is not totally balanced, then we get back the so-called "cover" of v, which is the least totally balanced set function that is greater than or equal to v for all S.

From (6) and (7) and the definition of "balance" (i.e., (2), (3), relativized to R), one can verify that the utility function of the direct market of a totally balanced game satisfies

$$u(e^R) = v(R), \qquad \text{all } R \subseteq N. \tag{8}$$

---

[4]) At least one such set of weights exists. For example, take $\gamma_{\{i\}} = x_i, i \in N$; all other $\gamma_S = 0$.

This shows that u may be regarded as an *extension* of the set function ν to the domain of "fuzzy" sets, i.e., coalitions whose members may participate at fractional levels of intensity[5]).

3. The Competitive Equilibrium

The so-called "competitive" solution is not a game theory concept, but is based on the notion of an imposed schedule of prices which, if accepted by all the members of the economy, will make it possible to balance supply and demand in each commodity, "clearing the market" to everyone's satisfaction. In our present setting, we must remember that money (or "transferable utility") is implicitly one of the commodities in exchange, so that the ith trader's complete utility function has the form

$$U^i(x^i, \xi^i) = u^i(x^i) + \xi^i. \tag{9}$$

Here ξ^i denotes i's final money balance. If we wish to follow the classical definition of "competition" we must keep this new commodity explicitly in view. A typical price schedule can then be written as

$$(\pi, 1) = (\pi_1, \ldots, \pi_{|M|}, 1), \tag{10}$$

where we have set the price of money at 1, as a convenient normalization. These prices serve to evaluate everything, including money, in terms of some new accounting unit.

Acting "competitively" in the face of (10), the ith trader will seek to maximize (9) subject to the "budget" constraint

$$\pi \cdot x^i + \xi^i = \pi \cdot a^i + \xi_0^i, \tag{11}$$

which equates the inflow and outflow of the accounting unit. Here ξ_0^i denotes i's initial money balance. On the assumption of freely transferable utility, ξ^i is an unrestricted variable, so we may solve (11) and eliminate ξ^i from (9). Trader i's goal can now be restated: to maximize

$$u^i(x^i) + \xi_0^i + \pi \cdot (a^i - x^i) \tag{12}$$

where x^i is now chosen unrestrictedly from E_+^M. For the price schedule $(\pi, 1)$ to be in competitive equilibrium, there must exist a set of maximizing choices by the different traders that fit together to form a feasible N-allocation, since only then can the market be cleared to everyone's satisfaction.

---

[5]) This extension from the vertices of the unit N-cube to all of E_+^N is continuous, concave, and positively homogeneous of degree 1. Such an extension exists only when ν is totally balanced; see *Shapley* and *Shubik* [1969], Eq. 4–1, etc. It may be contrasted with *Owen's multilinear* extension [1972], which is always possible, and which is a more appropriate extension for studying the "value" solution concept. See also *Aubin* [1974] and *Aumann* and *Shapley* [1974], Ch. IV, esp. p. 166.

By this roundabout path[6]) we have arrived at the desired definition. A *competitive solution* in our model is an ordered pair (π, z^N), where π is an arbitrary N-vector of prices[7]) and z^N is a feasible N-allocation, such that

$$u^i(z^i) - \pi \cdot z^i = \max_{x^i \in E_+^M} [u^i(x^i) - \pi \cdot x^i], \qquad \text{all } i \in N. \tag{13}$$

In other words, each trader maximizes his "trading profit." Note that we have omitted the terms ξ_0^i and $\pi \cdot a^i$ appearing in (12), as they are irrelevant to the maximization problem.

Moving to the payoff space, we shall call a vector α *competitive* if it arises from a competitive solution (π, z^N), thus:

$$\alpha^i = u^i(z^i) - \pi \cdot (z^i - a^i), \qquad \text{all } i \in N. \tag{14}$$

It is not hard to establish that the competitive allocations are just those that maximize *total* utility, $\Sigma_{i \in N} u^i(x^i)$. It follows that the set of all competitive payoff vectors of a market is nonempty, compact, and convex; moreover, it is a subset of the core. To show the latter, suppose that α is competitive but not in the core. Then $v(S) > \alpha \cdot e^S$ for some $S \subset N$. This means, by (5), that there would be a feasible S-allocation x^S such that

$$\sum_{i \in S} u^i(x^i) > \sum_{i \in S} (u^i(z^i) - \pi \cdot (z^i - a^i)).$$

But $\Sigma_{i \in S} \pi \cdot a^i = \Sigma_{i \in S} \pi \cdot x^i$, by (4), and so we would have

$$\sum_{i \in S} (u^i(x^i) - \pi \cdot x^i) > \sum_{i \in S} (u^i(z^i) - \pi \cdot z^i).$$

This contradicts the maximization (13). Hence each competitive payoff vector is a core vector. The reverse is not generally true, however; indeed, the competitive solution is often unique whereas the core is typically a set of $|N| - 1$ dimensions[8]). The following theorem shows, however, that *in a direct market* the set of competitive vectors and the core coincide.

---

[6]) We could, of course, have stated and justified the following definition directly, keeping the transferable utility hidden beneath the notational surface. We wished, however, to establish a firm connection with the standard definition of competitive equilibrium, and to clear up the "mystery" of the missing budget constraint in (13).

[7]) We have not been assuming that utilities are nondecreasing, and so we do not assume here that prices are nonnegative. This approach also entails using "=" rather than "$\leqslant$" in the budget condition (11). Note also that our prices π_j are not just ratios, but have meaningful magnitudes.

[8]) A sufficient condition for uniqueness of the competitive payoff is that the functions u^i all be differentiable and that at least one competitive allocation be strictly positive. A sufficient condition for the core to be full dimensional is that all the inequalities (2) (except the trivial case where $\gamma_N = 1$) be satisfied strictly.

Theorem 1: Every payoff vector in the core of a game is competitive in the direct market of that game.

Proof: Let α be in the core of v[9]). The idea will be to show that α itself can be used as a competitive price vector for the direct market of the game. By (8) and (1) we have

$$u(e^N) = v(N) = \alpha \cdot e^N. \tag{15}$$

In other words, the "value" of the total supply of goods e^N is the same in utility terms as it is when computed using α as a price vector.

Next, take an arbitrary bundle $x \in E_+^M$, and let $\{\gamma_S : S \subseteq N\}$ be any set of nonnegative coefficients satisfying (7). Then, by (1) and (7),

$$\sum_{S \subseteq N} \gamma_S v(S) \leqslant \sum_{S \subseteq N} \gamma_S (\alpha \cdot e^S) = \alpha \cdot \sum_{S \subseteq N} \gamma_S e^S = \alpha \cdot x.$$

Hence, by (6),

$$u(x) \leqslant \alpha \cdot x. \tag{16}$$

We can now show that α is a competitive payoff vector. Define prices by $\pi_i = \alpha^i$, $i \in N$. At these prices, a trader trying to choose x^i to maximize his "trading profit" $u(x^i) - \pi \cdot x^i$, as in (13), will find that he can't make it positive, because of (16), but that he can make it zero by choosing x^i to be the bundle e^N, because of (15)[10]). By the homogeneity of u, any fraction f^i of that bundle also yields a trading profit of zero. So we can construct a competitive solution (π, z^N) by taking $z^i = f^i e^N$, where the f^i are any nonnegative numbers that sum to 1. Moreover, (π, z^N) yields the desired payoff vector α, since we have (see 14))

$$u(z^i) - \pi \cdot z^i + \pi \cdot a^i = 0 + \pi_i = \alpha^i$$

for each $i \in N$. This completes the proof of Theorem 1.

Theorem 1 tells us that every point in the core of a market game is competitive for at least one of the associated markets–namely, the direct market. The next theorem refines this result, by showing that for each core point there are associated markets for which *only* that point is competitive.

Theorem 2: Among the markets that generate a given totally balanced game, there is at least one that has any given core point as its unique competitive payoff vector.

---

[9]) If v is not balanced the core is empty, so the theorem is vacuously true. If v is balanced but not totally balanced, the core coincides with that of the cover of v (see *Shapley* and *Shubik* [1969], Lemma 3), and hence we could assume v to be totally balanced.

[10]) Indeed, if he could make it positive he could make it arbitrarily large, and no maximum would exist. But this is a special consequence of homogeneity; in general with just concave utilities a positive trading profit is quite possible at competitive equilibrium.

Proof: Let v be totally balanced, let d be a real number, and define the set function v_d by

$$v_d(S) = v(S), \quad \text{all } S \subset N, \text{ and } v_d(N) = v(N) + d. \tag{17}$$

The game v_d is obviously totally balanced if $d \geqslant 0$. Let u_d denote the utility function for the direct market of v_d (see (6), (7)). By (8) we have

$$u_d(e^S) = v_d(S), \qquad \text{all } S \subseteq N. \tag{18}$$

Let α be an arbitrary core point of v, and define the function $u_{d,\alpha}$ by

$$u_{d,\alpha}(x) = \min(u_d(x), \alpha \cdot x). \tag{19}$$

This is continuous and concave, and so it can serve as the utility function of a market with equal tastes, using the same commodity space and initial bundles as the direct market of v. We shall show that for any positive value of d this market has the properties claimed in the statement of the theorem, namely, (a) that its market game is v and (b) that its unique competitive payoff vector is α.

(a) Since we have equal tastes and homogeneity of degree 1, the market game w generated by $u_{d,\alpha}$ is given by

$$w(S) = u_{d,\alpha}\left(\sum_{i \in S} a^i\right) = u_{d,\alpha}(e^S) \tag{20}$$

(compare *Shapley* and *Shubik* [1969], Eq. (3–3). By (19) and (18), this means that

$$w(S) = \min(v_d(S), \alpha \cdot e^S).$$

By (17) and (1), we see that this minimum is equal to $v(S)$ both when $S \neq N$ and when $S = N$. Hence $w = v$ as claimed.

(b) As noted previously, the competitive solution maximizes total utility. Each competitive price vector will therefore be the gradient of a linear support to $u_{d,\alpha}(x)$ at $x = e^N$, since $u_{d,\alpha}$ is homogeneous of degree 1 and concave. But when d is positive, $u_{d,\alpha}(x)$ coincides with the linear function $\alpha \cdot x$ in at least a small neighborhood of $x = e^N$, by (19), because u_d is continuous and

$$u_d(e^N) = v_d(N) > v(N) = \alpha \cdot e^N.$$

Hence $\alpha \cdot x$ is the only support at $x = e^N$, and so the unique competitive prices are $\pi_i = \alpha^i$, $i \in N$. As we saw in the proof of Theorem 1, these yield the payoff vector α. This completes the proof of Theorem 2.

By a simple extension of this proof, a market can be constructed having any given closed convex subset of the core as the set of competitive payoff vectors. Indeed, it is only necessary to define

$$u_{d,A}(x) = \min_{\alpha \in A} u_{d,\alpha}(x),$$

where A is the desired convex set, and proceed as above. (We omit the details.) When A equals the core, the market obtained in this way is independent of the parameter d, and in fact reduces to the direct market; thus, this more general construction unified Theorems 1 and 2.

4. Concluding Remarks

The relation between the core and the competitive equilibrium can be viewed in terms of the information that is lost in passing from a market to the game that it generates. In the first place, all details concerning the commodities and their distribution among the traders are suppressed, since the analysis of the market *game* takes place in the utility or payoff space, not the allocation space. In the second place, the game actually takes cognizance of only a finite number of the possible outcomes of the market process, i.e., the best result for each coalition. Most of the detailed preference information contained in the utility functions is ignored, as may be seen clearly in Eqs. (8) or (20) above, where the utilities are evaluated only at the vertices of a cube.

This loss of information suggests that the core is a rather blunt solution concept, and accounts for the many-to-one correspondence between markets and their market games. It is not surprising that we were able to find plenty of markets (without even looking beyond the special type of markets where the commodities are identified with the traders) having just the competitive outcomes that we needed for our proofs.

In a subsequent note we shall consider the competitive equilibria of markets *without* money, using the framework set forth in the work of *Billera* and *Bixby* [1973, 1974] and *Billera* [1974]. Here too, although the space of games is far richer (the analog of the function v being set-valued), there is a great loss of information and a similar many-one relationship between markets and games. It turns out, however, that the locus of competitive payoffs is *not* the entire core of the game, but only a certain "inner core." This inner core can also be characterized game-theoretically, without explicit reference to any economic model. It should be observed, however, that our method of analysis depends heavily on the assumption of concave utility functions, which in turn require cardinal utilities (see *Billera*'s comments [1974, pp. 129–130]). A radically new approach may be required before the analogous characterization of "ordinal" market games and their competitive solutions can be obtained.

References

Aubin, J.-P.: Coeur et valeur des jeux flous à paiement lateraux, C.R. Acad. Sc. Paris **279**, 891–894, 1974.

Aumann, R.J., and *L.S. Shapley*: Values of Non-Atomic Games, Princeton 1974.

Billera, L.J.: On games without side payments arising from a general class of markets, J. Math. Econ. 1, 129–139, 1974.

Billera, L.J., and *R.E. Bixby*: A characterization of polyhedral market games, Int. J. Game Theory 2, 253–261, 1973.

– : Market representations of *n*-person games, Bull. Amer. Math. Soc. **80**, 522–526, 1974.
Owen, G.: Multilinear extensions of games, Management Science **18**, P64–P79, 1972.
Shapley, L.S.: On balanced sets and cores, Nav. Res. Log. Q. **14**, 453–460, 1967.
Shapley, L.S., and *M. Shubik*: Quasi-cores in a monetary economy with nonconvex preferences, Econometrica **34**, 805–827, 1966.
– : On market games, J. Econ. Theory **1**, 9–25, 1969.

Received May, 1975

PART II

THE THEORY OF MONEY AND FINANCIAL INSTITUTIONS

[9]

Pecuniary Externalities: A Game Theoretic Analysis

By MARTIN SHUBIK*

When one firm or an industry expands, its actions may influence the cost of factors to another firm. The change in the cost of factors represents an externality to the firm which finds its input prices changed. The firm is not the master of its own destiny. An action by an outside agent has influenced its economic well being.

I. Pecuniary Externalities and Markets

Why should a pecuniary externality, i.e., an externality that is manifested in a change in prices, be different from a physical externality such as the production of smoke or other pollution? Modern economic analysis provides the answer that the change in prices is merely a reflection of the maximization process in a competitive market embodied in a general equilibrium system. The apparent externality shows up as a feedback from one market to another in a comparative statics explanation of adjustment to change.

The paradox of the pecuniary externality is caused by the nature of the partial equilibrium analysis of the firm. A real or a pecuniary externality appear to be the same when viewed from the viewpoint of the individual firm. When embedded in a general equilibrium context the differences emerge.

In our attempt to set up the appropriate economic model we find that the paradox of the pecuniary externality is closely related to the implicit assumption of the existence of markets rather than the assumption of the emergence of markets as part of the economic process.

The paradox does not even appear if, instead of viewing the behavior of the individual firm or customer as a mere mechanistic adjustment to a price system, it is viewed as the acts of a player in an economic game.

* Professor, economics of organization, Yale University, on leave to the RAND Corporation. Research undertaken by the Cowles Commission for Research in Economics under Contract Nonr-3055(01) with the Office of Naval Research.

Instead of assuming the existence of a price system, we deduce its existence as the limit of the core of an economy with many players (see Shubik). Although the concept of the *core* has gained wide recognition in the past few years, it may still be unfamiliar to some readers. For this reason a definition is supplied in the footnotes.[1]

F. Y. Edgeworth first suggested approaching the formation of the price system via the contracting and recontracting of groups in a market. He indicated the limits of a bargain between traders by means of a contract curve and indicated that as the number of traders is increased the contract curve shrinks and in the limit (with many traders) the price system emerges. In my 1959 article, I noted that the Edgeworth contract curve was the same as the game theoretic concept of the core of a game. I treated a special case; Herbert Scarf treated the general case and proved the convergence of the core to the competitive equilibrium price in a general context. Gerard Debreu worked with Scarf in shortening and strengthening the proof.

The reason for the paradox of the pecuniary externality and why it does not appear when we approach the price system via the core can be explained by means of a simple

[1] The core is composed of that set of imputations of wealth on the Pareto optimal surface which cannot be blocked by any coalition. An imputation can be blocked if there exists a coalition that could obtain more for all of its members without cooperating with the others. The following simple game illustrates the core. Suppose players 1, 2, 3 can together obtain \$30. The Pareto optimal surface consists of a set of imputations of the form (a_1, a_2, a_3) where $\sum_{i=1}^{3} a_i = 30$. Suppose that any individual alone can obtain nothing and that any pair can obtain \$15. The core consists of the imputations satisfying the conditions $a_1 \geq 0$, $a_2 \geq 0$, $a_3 \geq 0$, $a_1+a_2 \geq 15$, $a_1+a_3 \geq 15$, $a_2+a_3 \geq 15$, $a_1+a_2+a_3 \geq 30$, and $\sum_{i=1}^{3} a_i = 30$. It is easy to observe, for example, that the imputations (8, 9, 13) or (10, 10, 10) are in the core.

Not all games have cores. In the example above, suppose that any pair could obtain \$25 then three of the inequalities would become $a_1+a_2 \geq 25$, $a_1+a_3 \geq 25$, $a_2+a_3 \geq 25$ and no set of (a_1, a_2, a_3) adding to 30 can exist.

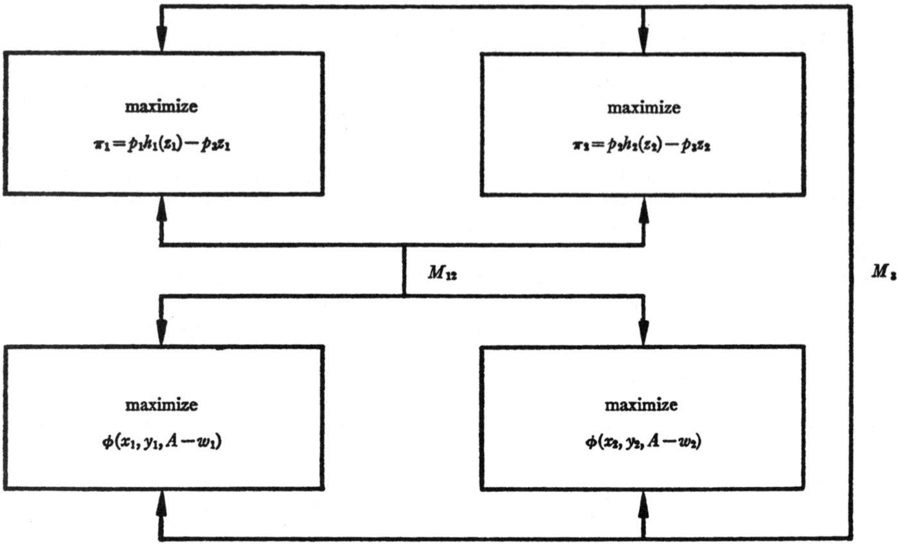

FIGURE 1

example of a closed economy with two industries, customers, two final products, and labor.

Figure 1 shows the general equilibrium market model and Figure 2 shows a closed economy modeled as a game. They are modeled for two firms and two customers. For simplicity, we assume that raw materials are free inputs, hence the profits of firms 1 and 2 can be expressed as:

$$\Pi_1 = p_1h_1(z_1) - p_3z_1$$

and

$$\Pi_2 = p_2h_2(z_2) - p_3z_2$$

where $x = h_1(z_1)$ and $y = h_2(z_2)$ are the production functions for the two consumer goods. The distributions of the consumer goods to the customers are $x_1 + x_2 = x$ and $y_1 + y_2 = y$, where the subscripts to x and y

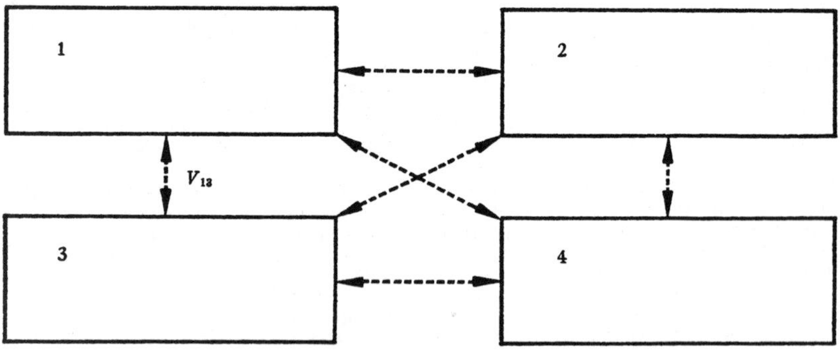

FIGURE 2

denote customers. The amount of labor bought equals the amount sold, i.e., $z_1+z_2=w_1+w_2$. The prices of the first and second consumer goods are p_1 and p_2; the price of labor is p_3. For simplicity we assume that the utility functions of the consumers are the same, ϕ, and their initial endowments are also the same $(0, 0, A)$.

The lines and arrows marked by M_{12} show the presence of a market for the consumer goods (or two interlinked markets). The market structure aggregates and interlinks firms and customers so that any individual trades with an impersonal market rather than indulges in personal face-to-face bargaining and trading. Both firms and customers are assumed to face prices and they maximize accordingly. The lines marked by M_3 denote the labor market.

If we wished we could introduce one more set of lines representing the payments of dividends $D_1+D_2=\Pi_1+\Pi_2$ back to the consumers. For the purpose at hand this is an unnecessary complication. Furthermore if the production functions are homogeneous of order 1 (i.e., constant returns) then profits will be zero and no dividends will be paid.[2]

Figure 2 shows the two firms (as players 1 and 2) and the two customers (as players 3 and 4). The dotted lines indicate that anyone can deal with anyone directly. There is no imposed market structure or price system.

In particular the two firms and customers can form 2^4-1 or fifteen different trading groups. The trades that these different groups can achieve can be denoted by a characterizing function, (see my forthcoming book with Lloyd Shapley).

$V(\{i\})=0$ is the set of outcomes that a coalition of one individual i can obtain if he deals with nobody else. This amount can be assigned the value 0. It is the status quo payoff. There are four one-person coalitions.

$V(\{i, j\})$ stands for the set of outcomes that a coalition of two individuals can obtain.

$V(\{1, 2\})=V(\{3, 4\})=0$. If the firms or the customers try to trade only among themselves they cannot improve over the status quo.

$V(\{1, 3\})=V(\{1, 4\})=$ the Edgeworth contract curve between the first firm and a single customer (where the customers have the same tastes and endowments).

$V(\{2, 3\})=V(\{2, 4\})=$ the Edgeworth contract curve between the second firm and a single customer.

There are six two-person coalitions possible. Similarly there are four three-person coalitions: $V(\{1, 2, 3\})$, $V(\{1, 2, 4\})$, $V(\{1, 3, 4\})$, and $V(\{2, 3, 4\})$; and there is one four-person coalition whose trading possibilities are characterized by $V(\{1, 2, 3, 4\})$.

II. Orthogonal Coalition Games and Decentralization

Despite its frightening name, the concept of an orthogonal coalition game is relatively straightforward and is the key to the analysis of externalities. The basic idea is that in such a game, once the participants have chosen sides, those excluded from a specific coalition have no influence on the fate of the coalition. This is highly related to the view of open trading in a market. When two traders agree upon a bargain, for all intents and purposes they do not care what the rest of the market is doing. This does *not* mean that the availability of other traders has no influence on the market; however, no one is forced to deal with anyone else. The device of recontracting may be regarded as a comparative statics way of checking trading possibilities; however (as is the case in most actual market places), a contract will be honored by both parties and does not depend upon third parties.

When we view the market with two firms and customers, as described in Section I, as one in which any coalition is feasible then as soon as you have signed your contract with some set of participants the remainder of the market is irrelevant.[3] The essence of an

[2] There is a further complication relating to production. Are the production functions individually owned or can any factory produce anything if it has the inputs? Here we assume the former. This might happen if each owner had a nonmarketable talent or capital good used in the process. A fully satisfactory answer to this question is not attempted here as it is not central to discussion of monetary externalities although it calls for care in treating capacity constraints, nonmarket goods and dividends.

[3] It is possible that, for unspecified reasons, the outsiders might wish to give away extra goods to you,

orthogonal coalition game in general and of a market game in particular is "doing your own thing."

Consider the coalition {1, 3} which can obtain $V(\{1, 3\})$. This is indicated in Figure 2 by V_{13}, and leaves the first firm and customer utterly independent of the second firm and customer. This is not possible in the setup described in Figure 1. The other firm and customer are always interlinked to everyone else through the impersonal market where a price to one is a price to all simultaneously. The market in Figure 1 forces a feedback upon all participants whether they like it or not. They cannot cut out and form their own community. The market in Figure 2 allows for any grouping anyone wants. Once the groups have been decided upon they are strategically isolated. They have no further interaction with each other. In other words they are completely decentralized.

If an organization can be described as an orthogonal coalition game, this means that it is possible to decentralize it in such a complete manner that the performance of any subdivision will depend only upon itself.[4]

Suppose that we were to increase the number of customers and the number of firms of type 1 and type 2, the market we have described is merely a special instance of the market treated by Debreu and Scarf. As the number of firms and customers increases the core shrinks and a price system emerges.

III. Partial Equilibrium Analysis, Comparative Statics and Externalities

The full force of the paradox of pecuniary externalities comes in when we discuss changes in supply costs to an industry considered in isolation; where the changes have been brought about by a shift elsewhere in the economy. For example, suppose that an industry in the same district has had a change in technology which calls for a different type of labor input. The effect of this change may feed back on the labor costs of the first industry.

Following the description of the two models in Section I we can see immediately the effect of the different treatments. Suppose that there was a shift in technology so that the production function for firm 2 becomes $h_2^*(z_2)$ instead of $h_2(z_2)$. Solving the general equilibrium system indicated in Figure 1 a new set of prices will emerge. In particular the cost of the labor supply to Figure 1 will have changed. Viewed from the position of the head of the first firm looking only at his own market, an external force over which he has no control has affected his costs. The reason for this is because we have implicitly assumed that he has to deal via the overall impersonal price system of the mass market whether he likes it or not.

Let us now consider this change in terms of the model in Figure 2. The only values of the characterizing function that have changed are those involving firm 2. If your group was not dealing with firm 2 before its technological breakthrough, it does not have to deal with it now. If it chooses not to, then it will earn the same amount as before.

The effect of the technological change in firm 2's production has not gone unnoticed. It has increased the value of eight out of the fifteen coalitions. By doing so it has made certain coalitions more attractive to some players than they were before and has changed the pattern of worthwhile recontracts. If we were to take the new characterizing function and add more players, the core will now converge to a new limit which is the new price system that takes into account the change in technology.

IV. Markets, Economic Reality, Pecuniary and Physical Externalities

Suppose that exogenous change in the production process of firm 2 were such that it involves a physical externality. For example it switches to a new process which may be labor saving to itself but which dumps smog on the consumers and on the other firm. Returning to Figure 2 and to the characterizing function; we would have to modify the figure by introducing solid interlinkages

---

whether you like it or not. However, if strangers bearing gifts turn up you do not have to take them.

[4] This does not solve the problem of what is the optimal decentralization for the organization as a whole. The answer for a market appears to be: "give up the doing-your-own-thing level of decentralization. Put in a price system in the central office and announce the prices to the otherwise decentralized divisions."

between firm 2 and the three others. Regardless of the market organization they cannot avoid the connection to firm 2. The game is no longer an orthogonal coalition game; we cannot even easily state what a coalition can obtain without first finding out what firm 2 intends to do. It is no longer possible to do your own thing. The characterizing function is not particularly helpful in describing the structure of the threats and interference that the firm with the physical externality can export to all of the other players.

When we try to patch up the characterizing function to take into account the actions of firm 2 we find that the core of the new game may not exist (see Shapley and Shubik 1969). Furthermore if the core were to exist there is no guarantee that it converges as more firms and customers are added. No price system emerges.

In terms of conventional analysis where we *assume* the existence of prices, if we try to solve the general equilibrium system with physical externalities present, although it may be possible to determine prices which clear the markets, they no longer call forth a Pareto optimal outcome.

Returning to Figure 1 we would have to patch up all production functions and the consumer preferences to reflect the role of the smog as an extra commodity. The smog is not an economic good and will not appear with a price, nor can it be traded or avoided. Independent maximizing behavior based upon a price system which does not include a method for pricing and controlling the distribution of the extra commodity will be optimal only by pure coincidence.

In an economy such as ours, markets exist and, at least in some areas, the price system is prevalent. Although the argument in this article has attempted to show that by adopting a more basic approach to trade, based upon the core of an economy and all coalition structures, the phenomenon of pecuniary externalities poses no problem; yet we know that our institutions have markets and we do not deduce their existence *de novo* by calculating the effect of billions of coalitions.

If we assume that firms *must* trade through organized markets, then an analysis of partial equilibrium competition shows that a pecuniary externality may be just as real to a competitor as is smog being dumped on his factory.

If the individual large firm has strategic control over at least part of the environment, its management will not care (unless they are socially motivated) whether costs are going up because a new industry in the area is bidding up the price of engineers or causing costs to rise because it is now more expensive to clean polluted water for the industrial use that the firm has for it. In either case the strategy of interest to the oligopolistic firm is to block the entry of the new industry or to extract a tribute from it to pay for the additional costs it will incur. The firm per se does not care whether (as in the first case) the extra costs are caused by the functioning of a technically efficient price system correctly (from a social point of view) increasing the cost of skilled labor to it; or whether costs are increasing due to a socially undesirable externality which has not been accounted for.

In summary there are three phenomena which must be separated when comparing pecuniary and physical externalities. We may view the whole economy as a mechanistic functioning of the price system. The single price-taking producer will regard a change in costs to him via the price system as bad as a change via a physical externality. However, he is constrained (by being a price taker) to do the right thing when the cost change merely reflects a feedback in the general equilibrium process. He will almost always do the wrong thing when the cost change is caused by a physical externality.

When we view the whole economy as a cooperative game with all coalitions possible, the paradox of the pecuniary externality disappears and is shown to vanish into the overall maximization problem where exogenous change is correctly reflected in the change in the values of some of the outcomes shown in the characterizing function.

When we consider the possibility that an individual firm may be an oligopolist with market power, it is not constrained "to do the right thing" when its costs go up, no matter how they go up. Hence in the sense of an oligopolistic general equilibrium, an individual firm may be expected to use its power to noncooperatively fight any type of perceived externality.

REFERENCES

G. Debreu and H. Scarf, "A Limit Theorem on the Core of an Economy," *Int. Econ. Rev.*, Sept. 1963, *4*, 235–46.

F. Y. Edgeworth, *Mathematical Psychics*, London 1881, pp. 20–25.

L. S. Shapley and M. Shubik, *Competition Welfare and the Theory of Games*, forthcoming, ch. 6.

——— **and** ———, "On the Core of an Economic System with Externalities," *Amer. Econ. Rev.*, Sept. 1969, *59*, 678–84.

M. Shubik, "Edgeworth Market Games," in A. W. Tucker and R. D. Luce, eds., *Contributions to the Theory of Games*, Princeton 1959.

[10]

Reprinted for private circulation from Western Economic Journal
Vol. X, No. 4, Dec. 1972 • Copyright 1972 • Printed in U.S.A.

COMMODITY MONEY, OLIGOPOLY, CREDIT AND BANKRUPTCY IN A GENERAL EQUILIBRIUM MODEL

MARTIN SHUBIK*
Yale University

General equilibrium theory presents a description of the functioning of the price system that is essentially both static and nonstrategic. Oligopoly theory has been presented as strategic, but with few exceptions models of oligopolistic behavior have been open or one-sided. Most frequently the stress has been upon the oligopolistic competitors as firms with the customers as passive price taking automata, best represented by demand functions.

From Cournot, through Chamberlin the predominant solution concept used in models of oligopoly has been that of the noncooperative equilibrium point [4]. The noncooperative solution has been employed primarily with the strategic variable being quantity offered or price offered by the firms.

Although much of the verbal treatment of the noncooperative models has been phrased in dynamic or quasidynamic terms, the mathematical formulation of the noncooperative equilibrium can be given as a static "one-shot" system to be solved simultaneously. In other words, the market may be represented by a game in strategic form where all players simultaneously announce their strategies.

There is little doubt that an adequate model of oligopoly calls for an explicitly dynamic treatment; however, before attempting to develop dynamics we may consider a problem that arises in statics even though the model that must be built is not "realistic."

Why is it that it is so easy to formulate a static symmetric noncooperative game as a representation of an open oligopolistic system; but it does not appear to be either easy or "natural" to formulate a static symmetric noncooperative game as a representation of a closed economic system?

In the remainder of this paper it is shown that the answer to this question is closely related to the role of money in an economy, and this in turn provides a link from statics to dynamics.

I. ON CLOSED AND OPEN SYSTEMS

Before starting to construct an explicit model, the distinction between

*This research was supported by the Office of Naval Research. The research was also partially supported by a grant from the Ford Foundation. The author is deeply indebted to Lloyd Shapley for his help and insights given during years of joint work and many conversations and calculations on all aspects of applying game theory to economics. Miscalculations, assertions and errors are my own.

open and closed models is made.

The Edgeworth bilateral model of trade is a closed nonmonetary system. Goods flow within the system and after trade the change in the amount of goods in the system is zero. This is shown in Figure 1. The Walrasian system may be regarded as this type of system.

Figure 1

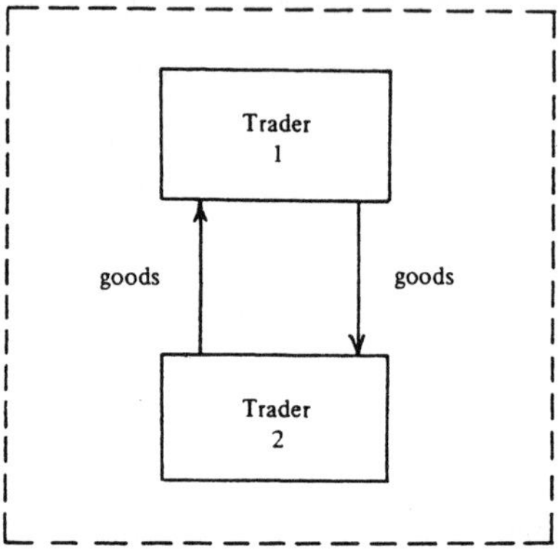

The Cournot duopoly model is an open market model involving money. After trade has taken place *neither* the amount of goods *nor* the amount of money in the system is conserved. Goods flow out into "the market" and money flows in from the market. This is shown in Figure 2. Virtually all models of oligopoly are of this type.

Figure 2

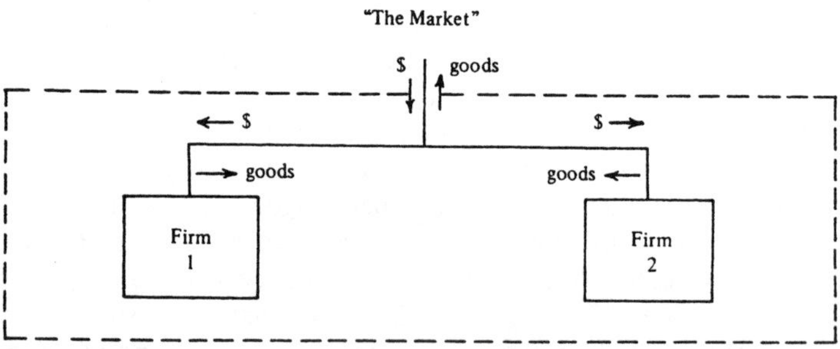

A general equilibrium system that trades in money must conserve *both* goods and money. Thus the flow diagram for a two-trader market becomes that shown in Figure 3. We may also wish to consider systems that are "open" or "closed" in time. This involves specifying interperiod balances.

Figure 3

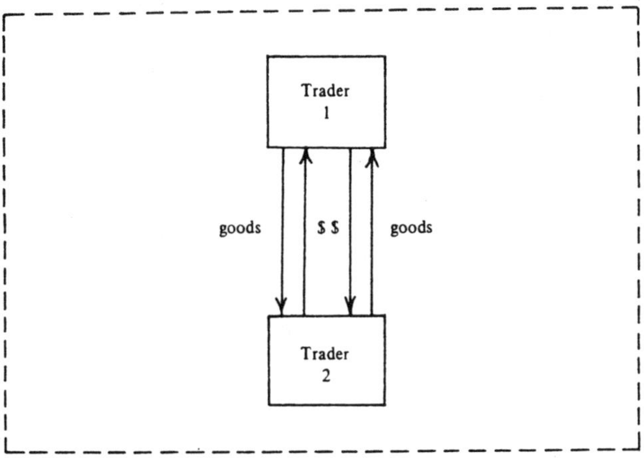

II. THE DIFFICULTIES WITH THE EDGEWORTH BOX

The simplest and most attractive device for illustrating the properties of the price system is the Edgeworth box. Given two traders with initial endowments of $(A_1, 0)$ and $(0, A_2)$ and preferences that can be represented by utility functions $\phi(x_1, y_1)$ and $\psi(x_2, y_2)$. The familiar Edgeworth box diagram is drawn in Figure 4. The line *OP* indicates the price system. *P*

Figure 4

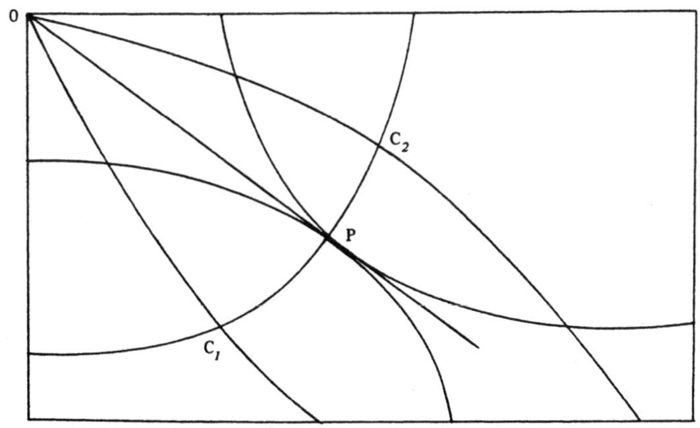

represents a point on the Pareto optimal surface that trading will reach when the price system is used. This simple model illustrating the price system for two traders, trading in two goods can be generalized for *n* traders trading in m goods [1]. In the general case it can be shown that m prices $(p_1, p_2, \ldots, p_m)$ exist such that if each trader is constrained to maximize his welfare taking these prices as given, the resultant trades will all balance and the outcome will be Pareto optimal (as is illustrated in the simpler case by P). It is well-known that the price system is homogeneous of order 0. As the prices in the general equilibrium model are essentially exchange ratios among the goods, we could use one of the goods, say the m-th, as a numeraire, fixing its price at 1. The price system becomes $(p_1/p_m, p_2/p_m, \ldots, p_{m-1}/p_m, 1)$ where the $m-1$ prices are independent.

The general equilibrium closed competitive economic system is modeled nonstrategically in the sense that each individual is *constrained* to maximize his welfare as though prices were fixed and he has no market influence.

In Figure 4 the curve joining C_1C_2 is the contract curve. This set of imputations suggested by Edgeworth as a solution to the bilateral monopoly bargaining problem also coincides with the cooperative game theory solution known as the *core*. Game theory solutions involve making the assumption that the individuals have strategic freedom and influence.

Cooperative solutions, such as the core, can be illustrated on the Edgeworth box because the strategic freedom, the dynamics, haggling and bargaining lie outside of the formal economic model. They are only implicit in the formal mathematical description. In particular no description is given of the communications or the nature of the discussion that takes place among the traders. It is assumed that after they have finished bargaining they then act jointly or "cooperatively" in selecting some outcome on curve C_1C_2.

The noncooperative solution is explicitly strategic; this is in contrast with the cooperative solutions (implicitly strategic) and the competitive equilibrium (nonstrategic). Duopoly models, which are open economic models (the customers being represented by a demand mechanism), have been studied using the noncooperative equilibrium solution where either price or quantity has been used as the strategic variable employed by the firms [8, Chs. 4 and 5]. Given that the noncooperative solution to duopoly is well-known and has been in economic literature since the time of Cournot, a natural question to ask is: can the noncooperative equilibrium solution be applied to a bilateral monopoly? Furthermore, do there exist two versions, one with quantity and the other with price as the strategic variables?

Another important question concerns the role of symmetry. The duopoly model is inherently symmetric in the roles of the two firms. The symmetry is reflected in the resultant noncooperative games. The bilateral monopoly model is also inherently symmetric. Can we construct a two-person symmetric noncooperative game that is a reasonable model of a closed bilateral monopoly?

One further question must be asked. Although the noncooperative equilibrium solutions to duopoly can be given a quasi-dynamic interpretation and in some instances some formal dynamic processes may be specified which lead to the noncooperative equilibrium, they can equally well be discussed in a static context. The existence of a noncooperative equilibrium point in a Cournot duopoly can be examined in much the same way as the existence of the competitive equilibrium can be examined. Can we obtain a noncooperative equilibrium model of a bilateral monopoly that is reasonably well protrayed in a static context?

The Price Game. We discover immediately that we are unable to formulate a satisfactory bilateral monopoly model where the traders are trading in only two commodities, as a symmetric noncooperative game with price as the strategic variable for each trader. In the closed model, prices can be regarded as ratios. If we select one commodity as the numeraire, we destroy the symmetry of the model and deprive the monopolist in that commodity of his strategic freedom. If we do not pick a numeraire, the resulting price game amounts to the game of "pick the biggest number" because there is really only one price in the market, which is determined by the ratio of the two numbers named. No matter what the other trader has named, it pays the remaining trader to name a higher number.[1]

The Quantity Game. We are not able to formulate a satisfactory symmetric bilateral monopoly model with two commodities using quantity as the strategic variable. If each names a quantity and the market is cleared, the only equilibrium is $(0,0)$. Somewhat surprisingly, this changes for bilateral duopoly. If there are two traders of each type, then each trader offers a quantity of his good to the market and the one price is determined by

$$p = (x_1 + x_2)/(y_1 + y_2) \tag{1}$$

Trader 1, for instance, will obtain in return for x_1 the amount:

$$(y_1 + y_2)(x_1)/(x_1 + x_2) \tag{2}$$

This model has at least two equilibrium points; one is $(0,0,0,0)$ or *no trade* or an *inactive market*.[2] This equilibrium is generally highly unstable;

1. When there are two traders of each type naming prices, a model can be defined; however, the difficulties in describing demand conditions [3] [8] in the resultant market call for a somewhat dynamic treatment, and dealing with them would not particularly add to illustrating the major points of this article.

2. The presence of inactive markets is undoubtedly closely related to transactions costs, especially those related to information and trust [11].

nevertheless, one needs at least two individuals on each side of the market to have a market become active. There is at least one other equilibrium with active trade, and these equilibria approach the competitive equilibria as the number of traders increases.

Suppose that there are several traders of two types trading in m commodities. For ease in exposition two traders are of the same type if each has the same initial endowment, say, $(A_1, A_2, A_3, \ldots, A_m)$ for traders of the first type and $(B_1, B_2, B_3, \ldots, B_m)$ for traders of the second type. Traders of the same type have the same preferences.

When there were only two commodities, it was fairly easy to define a simple strategy for each trader. He offers a quantity of one commodity to "the market" in return for a quantity of the other. Price emerges as is shown in (1) and the quantity he obtains is indicated in (2) (modified for more than four traders).

When there are more than two commodities, the natural description of the market is not so obvious. We may consider two different possibilities. The first calls for the existence of $m(m-1)/2$ trading posts or markets, one for each pair of goods. A trading strategy is to offer an amount of one commodity (possibly zero) for trade in each market. The prices or exchange rates are determined in a manner similar to that in (1). This market model enables us to define a noncooperative game, but it is both cumbersome and not necessarily conducive to optimal trade (for instance, it fails when there is no double coincidence of wants).

There is a far easier trading setup which involves only m markets instead of $m(m-1)/2$. This involves the utilization of some commodity in a manner that is *strategically* different from the way in which the other commodities are traded.

For notational ease, suppose that there are $m+1$ commodities in existence and that we choose to use the $m+1$st commodity as a money. Each individual may have his preferences represented by a utility function involving all $m+1$ commodities. A priori, no particular attempt is made to distinguish the "money" from any other of the commodities in trade.[3] No particular asymmetry is introduced in the domain of individual preferences; however, asymmetry is assumed in the strategic treatment of this good.

A strategy for an individual with an initial endowment $(A_1, A_2, \ldots, A_m, A_{m+1})$ consists of a set of $2m$ numbers representing a buy or sell decision in the m markets.

3. At this level no attempt is made to restrict the utilitarian or consumption properties of the commodity money; however, it is reasonable to expect that desirable properties should be suggested by the mathematical analysis.

If an individual wishes to sell, he offers an amount of the good he wishes to sell, where the amount must be less than his endowment.[4] If he wishes to buy, he offers an amount of the $m + 1$st commodity. In his buying, *if no credit is granted*, we must restrict his expenditures to a sum of no more than A_{m+1}, i.e., the amount of cash he has on hand.

The introduction of this mode of trade has done two things to the general equilibrium model. It has explicitly introduced the rudimentary institutions of markets[5] and it has introduced a *cash flow constraint* which in general is much more binding on trade than the budget constraint in general equilibrium models. In order to loosen this constraint it is necessary to introduce credit into the model of trade.

III. A WORLD OF OLIGOPOLIES

The nature of the noncooperative game and equilibria can be most easily seen if we begin with an economy consisting of n monopolists, each of whom possesses the sole supply of some commodity and has a supply of the commodity money.

Let there be n traders, each a monopolist in one commodity. There are $n + 1$ commodities in the economy. Trader i has an initial endowment given by $(0, 0, \ldots, 1, \ldots, 0, M_i)$ and has a utility function of $\phi_i(x_{i,1}, x_{i,2}, \ldots, x_{i,n+1})$.

A strategy by trader i is to offer a quantity $0 \leqslant q_{i,i} \leqslant 1$ to the market and to offer quantities of money $d_{i,j} \geqslant 0\ (j \neq i)$ for the remaining goods, where $\Sigma_{j \neq i} d_{ij} \leqslant M_i$.

The price of a commodity j is given by:

(3) $$p_j = \Sigma_{i \neq j} d_{i,j} / q_{j,j} \qquad \text{for}^6\ j = 1, \ldots, n \text{ and } j \neq i$$

(4) $$p_{n+1} = 1$$

The quantity of the j-th commodity obtained by trader i is:

(5) $$x_{i,j} = d_{i,j} q_{j,j} / \Sigma_{h \neq j} d_{h,j} \qquad \text{for}^6\ j = 1, \ldots, n \text{ and } j \neq i$$

4. In this static model no futures contracts are considered.

5. At a more technical level the economy with markets is no longer a *c*-game [5], whereas the general equilibrium model without markets is. The existence of markets and a "money" introduce pecuniary externalities [7].

6. In both of these instances, if $q_{j,j} = 0$ or $\Sigma_{h \neq j} d_{h,j} = 0$, we must define values for p_j and $x_{i,j}$. They will cause difficulties with a general investigation of the mathematical properties of this model. There is always a (highly unstable) equilibrium point at $(0, \ldots, 0)$. These conditions will be dealt with in a further projected paper with L. S. Shapley.

(6) $$x_{i,i} = 1 - q_{i,i} \text{ and}$$

(7) $$x_{i,n+1} = M_i + p_i q_{i,i} - \Sigma_{j \neq i} d_{i,j}$$

The goal of Trader i in the noncooperative game is to select

$$(d_{i,1}, d_{i,2} \ldots, d_{i,i-1}, q_{i,i}, d_{i,i+1}, \ldots, d_{i,n})$$

to maximize

$$\phi_i(x_{i,1}, x_{i,2}, \ldots, x_{i,n+1})$$

The ϕ_i are assumed to be twice differentiable with bounded first derivatives. We may write down the n^2 first-order conditions for individual maximization in the noncooperative game. Although they can be written down formally, reflection on the nature of the model when the market consists of monopolists indicates that there will exist only one highly unsatisfactory noncooperative equilibrium point where no trade takes place. If only one individual offers a commodity and obtains all of the money that the others have offered, regardless of the amount he offers, then his optimal strategy is to offer nothing. This paradoxical behavior associated with infinite price and a zero offering occurs for "the world of monopolies." When we consider that there are two or more traders of each type, this paradoxical result remains; for if all but one of the traders offer none of their goods for sale, the remaining one will also offer nothing. However, a new and more reasonable equilibrium also appears.

Suppose that there are s traders of each type in the market where $s \geqslant 2$. We may express the first-order conditions for the existence of a noncooperative equilibrium by:

(8) $$\partial\phi_i/\partial d_{i,k,j} = 0 \qquad \text{for } k = 1, \ldots, s \text{ and } i, j = 1, 2, \ldots, n \text{ but } j \neq i$$

and

(9) $$\partial\phi_i/\partial q_{i,k,i} = 0 \qquad \text{for } k = 1, \ldots, s \text{ and } i = 1, \ldots, n$$

These may be written out more fully for a trader of type 1 as:

(10) $$\partial\phi_1/\partial q_{1,k,1} = -\partial\phi_1/\partial x_1 + p_1(\partial\phi_1/\partial x_{n+1}) - q_{1,b,1}(p_1/\Sigma q_{1,b,1})(\partial\phi_1/\partial x_{n+1}) = 0 \qquad \text{for } k, b = 1, \ldots, s$$

(11) $$\partial\phi_1/\partial d_{1,k,2} = (\partial\phi_1/\partial x_2)[\Sigma_{k=1}^{s} q_{2,k,2}/\Sigma_{i=2}\Sigma_{b=1}^{s} d_{i,b,2} - d_{1,k,2}\Sigma_{k=1}^{s} q_{2,k,2}/(\Sigma_{i\neq 2}^{n}\Sigma_{b=1}^{s} d_{i,b,2})^2] - \partial\phi_1/\partial x_{n+1} = 0$$

and similarly for the others. These give us two equation types for when an individual sells one of the first n commodities or when he buys. They can be expressed as:

$$(12) \quad p_i = (\partial\phi_i/\partial x_i)/(\partial\phi_i/\partial x_{n+1})[1 - (q_{i,k,i}/\Sigma^s_{b=1} q_{i,b,i})]$$

$$(13) \quad \left\{ 1/p_j - [d_{i,k,j}\, \Sigma^s_{b=1} q_{j,b,j}/(\Sigma^n_{b \neq j} d_{h,b,j})^2] \right\} = (\partial\phi_i/\partial x_{n+1})/(\partial\phi_i/\partial x_j) \quad \text{for } i \neq j$$

Given the symmetry among the traders of a specific type we may replace the summation over a type by s times the action of an individual of that type. Thus, equation (12) becomes

$$(14) \quad p_i = [(\partial\phi_i/\partial x_i)/(\partial\phi_i/\partial x_{n+1})](1 - 1/s)$$

and (13) becomes

$$(15) \quad p_j = [(1 - d_{i,j}/s\Sigma_{h \neq j} d_{h,j})(\partial\phi_i/\partial x_j)]/(\partial\phi_i/\partial x_{n+1})$$

We know that $0 \leqslant d_{i,j} \leqslant \Sigma_{j \neq i}\, d_{i,j}$ hence

$$(16) \quad 1 \geqslant [1 - d_{i,j}/s\Sigma_{h \neq j} d_{h,j}] \geqslant (1 - 1/s)$$

Thus as s increases in size, (14) and (15) approach the form of

$$(17) \quad p_j = (\partial\phi_i/\partial x_j)/(\partial\phi_i/\partial x_{n+1}) \qquad \text{for } j = 1, 2, \ldots, n$$

These together with $p_{n+1} = 1$ can be written as:

$$(18) \quad [\partial\phi_i/(\partial x_{i,k,1}/p_1)] = [\partial\phi_i/(\partial x_{i,k,2}/p_2)] = \ldots = [\partial\phi_i/(\partial x_{i,k,n+1}/1)] \qquad i = 1, \ldots, n$$

But the equations (18) are similar in form to the conditions at a competitive equilibrium point in the $n + 1$ commodity economy. This is *suggestive* that there may exist a class of noncooperative oligopoly games in a closed economy for which the limiting (in the sense of replication) symmetric noncooperative equilibrium points coincide with the competitive equilibrium points (see Appendix).

The noncooperative equilibria will be influenced by the amount of commodity money in existence. Furthermore there must be "enough money" held by each trader or a boundary solution may result. The meaning of enough is discussed in Section IV.

The indication of the relationship between the noncooperative equilibrium and the competitive equilibrium given above for a world of one-product oligopolists appears to hold far more generally. Instead of limiting the model to individuals holding one product and "money" we may consider a market with n individuals and $m+1$ goods ($m \geq 1$) where each individual i has as his initial holdings amounts $(A_{i,1}, A_{i,2}, \ldots, A_{i,m+1})$. In this game a strategy is a $2m$-dimensional vector where an individual may decide either to buy or sell commodity j. If he sells, he offers an amount $q_{i,j} \leq A_{i,j}$. If he buys, he offers an amount of money $d_{i,j}$ where $\Sigma_j d_{i,j} \leq A_{i,m+1}$.

IV. DISCUSSION AND INTERPRETATION

Commodity Money. The commodity money serves as an imperfect but adequate strategic decoupling device. It enables each trader to announce his strategy without coordination with the others. The key property required to make it work is that by custom or law it is accepted in trade. If it has a consumption value, then the price system will be anchored in terms of it, and the distribution of an absolute amount of commodity money needed to finance trade without credit becomes an important consideration. If credit is granted, then the conservation conditions on the amount of money are relaxed. Thus, for example, there need be only a minute amount (or zero) gold in a system using gold, provided that there are sufficient credits (or gold certificates). This is discussed further below in reference to credit and bankruptcy.

When an economy runs with little or no credit and with a commodity money, some of the desirable properties needed to facilitate trade are that it is in sufficient supply and of low enough consumption worth that it is reasonable for all to be able to keep a store of it. It should be durable, and the closer it comes to having a relatively constant marginal utility, the more stable the economy becomes with respect to injections of new money. In various societies, coconuts, salt bars dried fish, bricks of tea, etc., have come reasonably close to fulfilling these requirements [2].

It is likely that credit facilities in a small region or in a nation under a common law will be easier to foster and police than in international trade. On the other hand, a commodity such as gold may by custom have gained international acceptance; hence, until international law reaches the enforcement levels of national law, there will be a predisposition in international trade to use a good commodity money.

The Markets and Dynamics. The noncooperative market described used quantity as the strategic variable. In considering the behavior of a large economy and in trying to make the connection between statics and dynamics, it appears reasonable to consider that the amount of goods offered and the amount of "money" chasing them are the determinants of prices rather than vice versa, although there are undoubtedly components of both. The basic problem in modeling trade is to be able to avoid having to specify too many details or sequences of trading, on minor variants of information conditions, etc., and to be able to justify the validity of the approximation.

We note that the noncooperative equilibrium, although it can be given a static interpretation, is an interlink to a dynamic model. The trading conditions reflect this in the cash or commodity money flow account, in the sense that the items sold do not provide spendable funds for this period. They build up the individual's supply of money for the next trading period.

Thin and Broad Markets. By introducing m markets explicitly into the model and by considering behavior as the number of traders increases, the importance of the relationship between the *volume* of trade and the thinness or broadness of a market appears. This is shown in equations (11) and (13). The volume of trade is the combined measure of the number of individuals active on each side of the economy and the amounts of goods and money being offered by each. Even in an economy with many individuals, if there are many commodities, some may be lightly traded.

Credit and Bankruptcy. In the model in Section III a *cash flow constraint* was introduced into trade. In many ways this is as unrealistic a model as is the general equilibrium model with only a *budget constraint*. The latter implies that an individual may always obtain credit in trade up to his net worth. The former calls for "cash on the barrel head." Thus, for example, if an individual has many goods, but no money, he cannot buy anything during the immediate period of trade because he must first liquidate some part of his holdings. A reasonable way to modify the overly harsh effect of the cash flow constraint is to consider the existence of a "shadow bank" or outside mechanism that will finance trade in the sense that it can grant to individuals credits which are accepted by all. We could consider that the bank limits the credit line to each trader directly or that it permits any trader to borrow any amount. We adopt the latter convention. This amounts to relaxing completely the cash flow constraint.

As the market is being viewed as a noncooperative game, what is there to prevent an individual from borrowing an enormous amount and then going bankrupt? In order to preserve an orderly market and to prevent strategically profitable bankruptcies, a bankruptcy law must be enforced which is strong enough to make strategically planned bankruptcy not worthwhile, but is lenient enough not to discourage other borrowing.

An *optimal bankruptcy law* will be one that when applied as a condition to a replicated noncooperative model of a market economy will in the limit call forth the amount of borrowing that precisely relaxes the cash flow constraints and enables the system as a whole to evolve prices that give rise to a Pareto optimal distribution. These prices will arise from a limiting noncooperative equilibrium and be the same as the competitive equilibrium.

A bankruptcy law that appears to satisfy the optimality conditions is as follows. Suppose that an individual i borrows the amount $D_{i,m+1}$. After trade he has achieved the state $\phi_i(x_{i,1}, x_{i,2}, \ldots, x_{i,m} x_{i,m+1})$. If $x_{i,m+1} \geqslant D_{i,m+1}$, he pays off his debt and his final utility is $\phi_i(x_{i,1}, x_{i,2}, \ldots, x_{i,m+1} - D_{i,m+1})$. Suppose that his debt is larger than his final holdings of the money (the $m + 1$st commodity), the bank reclaims from him by selling off other goods at market price until the debt is paid or he has no further assets.[7] In selecting the goods to garnish, the bank acts to *minimize* the trader's payoff. Thus his payoff under bankruptcy is:

$$\text{(19)} \qquad \min \phi_i(x_{i,1} - z_1, x_{i,2} - z_2, \ldots, x_{i,m} - z_m, 0)$$

subject to $\Sigma_{j=1}^{m} p_j z_j \leqslant D_{i,m+1} - x_{i,m+1}$ and $0 \leqslant z_j \leqslant x_{i,j}$.

At first glance it might appear that this law is unduly harsh; however, this is not the case, and the spirit behind it can be illustrated as follows. An individual will wilfully go bankrupt for some amount with the express purpose of maximizing his welfare on the assumption that the creditor will be able to reclaim items of less worth to the bankrupt than his gains from bankruptcy. The rule above prevents that. It makes wilful bankruptcy just unprofitable.

A harsher rule would discourage borrowing by those who were not intentionally trying to go bankrupt. Although the principle for the selection of a bankruptcy law can be clearly specified in terms of the properties of limit solutions to replicated noncooperative games, it has not yet been possible to determine the full class of laws that satisfy or to prove uniqueness.[8,9]

7. An operational difference between *insolvency* and bankruptcy needs to be noted when dynamic models are considered. In such models it appears that the ability to *service debt* is more important than to repay debt.

8. An important distinction must be made between noncooperative equilibrium solutions that require only pure strategies and involve no exogenous uncertainty and those that involve either mixed strategies or random moves by Nature (such as a crop failure). The equilibria not involving random components are invariant under ordinal transformations of the utility functions. The others are more naturally defined for von Neumann-Morgenstern cardinal utility. The bankruptcy law noted above is designed to stabilize the pure strategy equilibria. H. Markowitz has suggested that the bank try to *maximize* the individual's welfare—subject to reclaiming its loan. This is considerably more lenient than the rule above.

9. A slightly better bankruptcy law for small numbers might call for liquidation which includes the possible depression of market prices by the bankruptcy sale. There are several difficulties in formulating this.

Monetizing the Economy. In Section III the strategies for the traders were described. Individuals had the choice to offer part or all of their holdings to the markets. An alternative convention would be to require all to offer *all* of their holdings for sale and then to buy back whatever they wish. This convention has the property that all economic activity is monetized. Under this convention the accounting and price system would reflect the totality of economic activity far more closely than does the convention adopted in Section III. Furthermore, the amount of credit required to run the economy would be equal to or larger in this instance than that called for by the convention adopted.

The Amount of Money and Credit. In this model we could consider that the individuals have an initial supply of some commodity money such as gold and that the amount of this money is conserved. We could also consider initial amounts of a paper fiat money with no consumption worth. This too would have a physical existence and be conserved. In either case, if the proportions and/or amounts were not appropriate, credit would be needed.

The amount of money and credit required to enable a competitive price system to emerge can be calculated and will vary in size both with the level of monetization of trade and the strategic rules on borrowing.[10]

V. TRUST, RATIONALITY AND NONCOOPERATIVE SOLUTIONS

As the general equilibrium model of the economy is nonstrategic, it is natural for a modeler to implicitly ignore the problem of how strangers in a market economy can trade together efficiently and more or less impersonally with a minimal need for trust. When the economy is modeled as a noncooperative game, this problem must be faced.

If individuals trust only cash and there is no banking system, then although it is highly likely that optimality in trade cannot be attained without credit, at least all trades that are made are made for immediate value received by both parties. Hence, as all claims are discharged in the exchange, not only do the individuals not need to trust each other, they do not even need to consider the motivations or rationality of their trading partners.

When a banking system issues credit so that strangers can trade with virtual anonymity, the bank must either trust [6] the individuals to whom it grants credit, or it must have rules which can be enforced that make it unprofitable for individuals to fail to pay their debts. This calls for bankruptcy laws. The economic system that has institutions which grant credit and has bankruptcy laws to police the borrowers can function efficiently

10. For example, it may be possible to obtain credit to buy one type of asset and not another. You may borrow to buy real estate but not stocks, in some instances.

only if the creditors are able to assume that the debtors are rationally motivated. If the debtors do not understand that failure to repay is unprofitable, they could destroy the credit system.[11]

"Noncooperative cooperation" presumes the existence of an enforcement mechanism and the intelligence of all parties to understand that it does not pay to suffer the consequences of failing to repay credits granted.

VI. ON FIAT MONEY

The important distinction in this model between the use for money and an ordinary good comes in its strategic role. This is *not* necessarily directly related to its intrinsic consumption worth. It is not particularly directly important to the functioning of the system if the money is a commodity or a fiat money, provided that there are loan or credit facilities. The key factor is the *acceptance* by law or convention of the monetary unit and the acceptance of abstract credit. Thus, in particular, if the monetary unit accepted is gold, the bank may call its credit line "gold certificates," and actual trade will be carried on in gold and gold certificates. At the end of trade, if bankruptcy is avoided, the books will balance.

The steps in modeling required to transform the one-period trading model with a commodity money into a multiperiod model with fiat money are relatively straightforward and have been taken elsewhere [10] [12]. Some of the simpler dynamic models of trade with fiat money have been solved, and the results are presented in [12]. A three-commodity example of trade with a commodity money has been calculated and presented in an earlier paper [9].

APPENDIX

Individual maximization in a competitive market involves the individual's maximizing his welfare subject to his *budget constraint*. This is the value of his entire holdings at market prices. Thus, suppose an individual begins with $(A_1, A_2, \ldots, A_m, A_{m+1})$ and ends with $(A_1 + x_1, A_2 + x_2, \ldots, A_m + x_m, A_{m+1} + x_{m+1})$ where:

$$\begin{aligned} s_j &= x_j \text{ if } x_j \leqslant 0 \quad \text{and} \quad & d_j &= x_j \text{ if } x_j \geqslant 0 \\ &= 0 \text{ if } x_j > 0 & &= 0 \text{ if } x_j < 0 \end{aligned}$$

Then the *budget constraint* can be described as:

11. A close analogy can be drawn between economic exchange and trust and social exchange and trust. In the extended family group (uncles, cousins etc.) visits among cousins who do not particularly like or trust each other, "duty calls" on the sick and old incur social debts and credits the repayment of which is enforced by tradition and indoctrination which play a role analogous to the bankruptcy laws.

(1) $$\Sigma p_j x_j = 0 \qquad j = 1, \ldots, m+1$$

and the *cash flow constraint* as

(2) $$\Sigma p_j d_j \leqslant A_{m+1} \qquad j = 1, \ldots, m$$

(2) is far more binding than (1), as can be seen by rewriting (1) as:

(3) $$\Sigma p_j d_j = A_{m+1} + \Sigma p_j s_j$$

If trade is for cash, (2) may constrain the maximization, and a shadow price for the shortage of money emerges. By introducing sufficient credit, this shadow price becomes zero and optimal trades become feasible.

REFERENCES

1. G. Debreu, *Theory of Value*. New York 1954.
2. P. Einzig, *Primitive Money*. London 1948.
3. R. E. Levitan, "Demand in an Oligopolistic Market and the Theory of Rationing," Yorktown Heights 1964.
4. J. F. Nash, Jr., "Equilibrium Points in N-Person Games," in *Proceedings of the National Academy of Sciences of the U.S.A.*, 1950, *36*, 48-49.
5. L. S. Shapley and M. Shubik, *Competition, Welfare and the Theory of Games* (Manuscript in process).
6. M. Shubik, "Money, Trust and Equilibrium Points in Games in Extensive Form," Part VII, A Theory of Money and Financial Institutions, CFDP 331, Revised Feb. 1973, Cowles Commission for Research in Economics at Yale University.
7. ———, "Pecuniary Externalities: A Game Theoretic Analysis," *Am. Econ. Rev.*, Sept. 1971, *61*, 713-18.
8. ———, *Strategy and Market Structure*. New York 1959.
9. ———, "A Theory of Money and Financial Institutions," Part III, CFDP 324, Oct. 1971, Cowles Commission for Research in Economics at Yale University.
10. ———, "Fiat Money and Noncooperative Equilibrium in a Closed Economy," Part IV, A Theory of Money and Financial Institutions, *International Journal of Game Theory* (to appear).
11. ———, "Transactions Costs in a Market Economy," Part VIII, A Theory of Money and Financial Institutions, CFDP 336, April 1972, Cowles Commission for Research in Economics at Yale University.
12. ——— and W. Whitt, "Fiat Money in an Economy with One Nondurable Good and No Credit (A Noncooperative Sequential Game)," in A. Blaquière, ed., *Topics in Differential Games*, Amsterdam (to appear).

[11]

ON THE EIGHT BASIC UNITS OF A DYNAMIC ECONOMY CONTROLLED BY FINANCIAL INSTITUTIONS*

BY MARTIN SHUBIK
Cowles Foundation, Yale University

The elegant general equilibrium world of Arrow and Debreu has had a considerable mathematical development in the last decade. Underlying this work is an extremely parsimonious model of the economic system. In particular, only one economic actor is distinguished, the consumer who maximizes his welfare (the firm which maximizes profit is an automaton). One class of economic entities is considered. These are goods and services. There is no important operational distinction made between a good, which is durable, and a service, which is not.

It is suggested here that a more fruitful basic economic model needed to achieve a unification of micro and macroeconomic theory needs both more actors and more basic economic units. Specifically, the structure of process in a political-economy is such that even at the level of relatively abstract theory operational differences among consumers, entrepreneurs, administrators, financiers, and politicians should be discernible. Furthermore, several basic economic entities (or "basic particles") must play important discernible roles in an adequate theory. In particular, in the "real sector" physical assets should play a major role, i.e., the distinction between durable goods and consumables or services should be important. The paper sector must also be present with the roles of fiat money, ownership claims and contracts all distinguished. It is argued here that any economy can be characterized in terms of two real and six paper basic units: goods, services and six financial instruments. All other financial instruments can be obtained as mixtures of this basic set.

1. THE ACTORS AND UNITS OF TRADE IN A GENERAL EQUILIBRIUM WORLD

In the general equilibrium system several ingenious modelling simplifications were made in order to define a mathematically tractable model which casts some light on production and distribution in a timeless equilibrium.

The actors in the mathematical model are *consumers* and *firms*. At the most abstract level consumers own a vector of resources and any consumer may buy or sell any number of resources. The firms have no institutional content. They are devices which maximize profits and instantaneously flow them through to the consumers who may be deemed to be holding nonvoting common stock in these institutionless mechanisms called firms.

There is no need to distinguish between goods and services because no operational use is made of this distinction.

There is no need for any dynamic mechanism of prediction because by the trick of time dating all future commodities as different commodities which are traded now, all distinctions between spot and future markets are wiped out. By then solving the one period or normalized form of the multiperiod market simultaneously there is no need to consider any evolutionary process or disequilibrium state.

Similarly by expanding the commodity space to include commodities which have not only a time dating but also an existence attached to a contingent event,

*The research described in this paper was undertaken by a grant from the Office of Naval Research. The author is indebted to his friend and colleague Martin J. Whitman for many valuable discussions.

we can formally handle exogeneous statistical uncertainty and still solve the whole system as a one-shot market.

This whole *tour-de-force* is by no means trivial. The proof of the existence of a price system under these conditions is a substantial achievement. But possibly the price paid for the result was too high to be of use when we ask of this model how can it guide us in a world where individuals try to base their actions of today upon predictions of tomorrow. In a world where perfect trust does not exist, where most future markets do not exist and where money and many financial institutions do exist, the general equilibrium model does not come to our aid because it is static. It cannot handle the dynamics or the array of actors, goods, financial paper and institutions needed even to start to reflect the economic process.

General Equilibrium theory gets along with:

one actor	the consumer-trader-owner;
one shadow actor	a mechanism without institutional substance called the firm;
one financial paper	nonvoting stock with no other financial properties beyond the receipt of dividends;
one commodity type	this is a catch-all for goods, services, future and contingent goods and services;
no expectations	because it is static; or if one wishes to try for a dynamic interpretation then all actors have self fulfilling prophecies of the future;
one market	because all of the future including contingencies is collapsed into one period: another interpretation is perfect future markets for everything;
perfect trust	because the only constraint on the system is the budget constraint at the end: cash flow considerations are irrelevant because credit up to the final known budget constraint is always available;
no fiat money	it plays no role;
no financial institutions	they play no role.

2. The Different Actors Needed for an Adequate Theory of Economic Process

It is my belief that as is done in macroeconomic theory the distinction between a consumer and investor is worth making. In particular in a process world where information and evaluation are critical and bureaucracies abound the differentiating characteristics of the following actors need to be developed:

(1) The Consumer:	passive buyer, voter, saver who sells little beyond his services.
(2) The Businessman:	the active investors of private resources.
(3) The Administrators:	the bureaucrats and technicians who are the major carriers of routines and processes.

(4) The Financiers: the group whose function and tools include information processing, evaluation, brokerage and the control of money and financial papers.

(5) The Politicians: who are brokers of power and fidiciaries of the public sector coordinating among other factors, public administration finance and business.

A detailed discussion of these suggested categories has been given elsewhere[5]. In this paper my intention is merely to call attention to the possibility that to go from general equilibrium to a useful economic dynamics requires not only that attention be paid to information processing and frictions in the system but these require that we differentiate economic and other actors.

In particular it is my belief that economic dynamics is *financial dynamics.* The financial system is the neural network and dynamic guidance system of the real economy.

3. The Eight Basic Units

In a dynamic economy with many frictions and transactions costs, imperfect information, errors, limited trust and aggregated measures, at best there will be some limits to the extension of credit. The magic of the static general equilibrium model is not there. People need bankers, insurance agents, investment bankers and other evaluators of individual, corporate or governmental economic plans.

It does not appear to be particularly easy to guess the price system that will exist until the end of an individual's life to work out his total expected future income and to let him allocate his resources constrained only by this budget constraint.

Instead of guidance by the budget constraint the way the world appears to run is by a predominant use of spot markets with trade in money, financial paper and other contracts which permit a limited extension into the future.

The eight basic units which provide a rich enough basis for the development of an adequate microeconomic disequilibrium model of financial control are:

(1) Services;
(2) Goods;
(3) Money;
(4) Ownership claims;
(5) Service contracts;
(6) Futures;
(7) Promissory notes;
(8) Calls.

An adequate model of financial control of an economy requires not only a specification of the roles of these basic units, but also a specification of the rules, mechanisms or institutions which control the demand and supply of these items and a description of the motivation of the actors who control the institutions and manipulate the rules. The nature of the actors has been sketched in Section 2. The institutions are noted in Section 5. Some further comments on the basic units are called for.

(1) *Services*: A service may be regarded as an input which cannot be inventoried, and hence cannot be used as a security against a loan.

(2) *Goods*: For many economic purposes it may be useful to break down the category of goods into consumer, intermediate and capital goods. However, as a first cut, given the time unit under consideration possibly from the point of view of finance the key distinction between a service and a good is that the latter (in varying degrees) provides an asset or a store of value which can be used as a security against which ownership paper or loans may be issued.

(3) *Money*: Money is *the* liquid asset. One might wish to distinguish among commodity money, specie and bank money. In international trade, gold may still play an important role. However, leaving aside commodity money, the distinction among say fiat, deposits and time accounts for most questions is probably less the slight differences in liquidity than the distinction among the mechanisms or institutions generating supply.

(4) *Ownership Claims*: In a world with no uncertainty, perfect divisibility, liquidity and zero transactions costs the sale of ownership claims rather than the sale of an asset might not provide any particular economic advantage. In the world as it is, the existence of various states of indivisibility, uncertainty and ownership transfer costs make the use of ownership claims efficient and important. The main forms of this paper are claims to land, common stock, and titles to buildings and machinery.

(5) *Service Contracts*: Salary and wage contracts are probably the most important form of service contracts, although car rental, apartment rental, building cleaning, payroll processing and so forth provide many other examples. In the case of salaries and wages the only asset backing the service supply is the individual or human capital, and since indentured servitude and slavery are no longer in the rules of the game, that capital asset is of little worth as a backing for a financial instrument.[1] In contrast the fleet of cars, the hotels or the computers of various service firms are excellent items with which to obtain financing.

(6) *Futures*: A futures contract is a financial instrument which guarantees the delivery of a good at some future period. It is possible that the good cannot be delivered at its due date. Hence futures contracts (and for that matter service contracts) must be considered as not fully defined unless a complete specification is given of what happens if the economy evolves to a state where the contract cannot be honored. This involves specifying rules to cover recontracting, bankruptcy, market corners, insolvency and so forth.

(7) *Promissory Notes*: This category encompasses a large array of instruments which involve promises to pay money in the future in return for a money payment now. Thus a government bond involves present money paid by a purchaser in return for a future promise by the government to pay. Care must be taken in defining conditions on these instruments especially when loans are rolled over or perpetuities are issued.

[1]This is not strictly true when information on an individual's integrity exists. A sophisticated lender may feel that an individual with a reputation of integrity is a better risk than one with no reputation but with physical assets.

(8) *Calls*: Calls are financial instruments involving the exchange of money for an option to purchase an ownership certificate at or before some date in the future. Puts, warrants and options are related instruments.

All of the intertemporal financial instruments noted may involve some degree of risk. Thus for example an insurance policy may involve the payment of money now in return for money in the future under a set of contingencies. In a world without uncertainty a call would be a superfluous instrument. Most promissory notes and service contracts are written in a way that presupposes a low risk of contract nonfulfillment. Any change in risk short of disaster is borne by the seller of the contract.

Undoubtedly variations in the conditions of uncertainty, limitations on the length of time over which financial instruments can be issued, changes in physical transactions costs, indivisibility, data transmission and processing all provide the reasons for the production of a continuous blend of financial instruments and the continuous creation and change in financial institutions. However it appears that they can all be built up from the eight basic units. Furthermore one need only consider two time periods "NOW" and "NOW + 1." A transition matrix of the units is given in Table 1.

TABLE 1

NOW \ NOW + 1	S	G	M	O.C.	S.C.	F	P.N.	C	Nothing
Services									X
Goods		X							X
Money			X						X*
Ownership Certificates			X	X					X
Service Contracts	X		X						X
Futures Contracts		X	X						X
Promissory Notes		X	X						X
Calls			X	X					X

*The creation and destruction of money (and for that matter all paper involving future claims) is clearly far more complex than this simple transition matrix indicates, specifically in the sense that they all appear on at least two balance sheets and hence even at the simplest level involve systems properties.

Table 1 merely shows the main transformation of types and does not specify the actual transformation probabilities that exist in a world of uncertainty. Thus for example a new automobile is a good which transforms into a one year old car of varying quality depending upon the chances of use. It could even be destroyed.

Money may be withdrawn by an issuing agency. A firm can go bankrupt and hence ownership claims can be wiped out, or the owner may obtain some amount of money in the liquidation. The writer of a futures contract or a service contract may fail to meet his obligations in which case the holder is totally or partially wiped out. In the latter case he usually will obtain a money payment and/or a partial fulfillment of contract such as "all the copper you have, plus a monetary settlement for the remaining futures undelivered."

When a promissory note is honored a monetary payment will be made. If it is not honored there is a possibility that it becomes valueless. If it is secured by goods the goods may either be appropriated or sold. Not indicated in Table 1 is the possibility that a debtor's wages may be garnished in which case a claim on the proceeds from the sale of services is established.

3.1. *Where Do Taxes and Subsidies Fit?*

In the history of taxation and subsidies these items have been due to or from governmental bodies in the forms of claims on services, goods or money or combinations thereof. Tithes, tariffs, octrois, social security benefits, income taxes, sales taxes, death taxes, levies against wealth and so forth are all claims against goods, services or money and hence can be considered as mixtures of the instruments already noted.

The agencies for the issue and destruction of these claims in the United States include on the order of 70,000 Federal, state and local government bodies with taxing power. The mechanics of the creation and destruction of claims on taxes poses many accounting problems in economies where the nature of some of the claims is contingent upon certain events taking place such as death or the sale of assets.

In the modern world almost all taxes are due in money (or other paper such as certain bonds). The choice of working on the roads, fortifications or in the fields of the lord is no longer available.

3.2. *On the Unit of Time.*

How does the usefulness of this classification scheme depend upon the unit of time selected? It is my belief that due to the cyclical aspects of the world, to custom, law and human decisionmaking processes, the year is the natural unit. However from the viewpoint of the development of a theory of finance the classification scheme is not influenced in any conceptually important manner by the choice of time unit.

The major change in extending the size of the unit of time is to lump several of the more short-lived goods together with services. Thus for example if the unit of time were two months as contrasted with two minutes, ripe tomatoes would be classed together with services as items which cannot be stored to the next period. If the period were two minutes then the tomatoes can be stored.

The essential question here however is what the financial implications of the existence of goods with different lives and other properties are. The prime consideration is how suitable they are to be posted as security against the issue of various financial instruments. Ripe tomatoes are not particularly good security.

It is often important to "perform a sensitivity analysis" or to ask if a change in the time unit will have a radical and unnatural effect on the model postulated. If the continuous version of an economic model exhibits properties radically different from the finite time increment model one may suspect that something is unsatisfactory in the formulation of the problem. Here this does not appear to be the case.

3.3. *On Time, Pyramids and Complex Financial Instruments*

By permitting the rollover of short term loans and the issue of long term loans including perpetuities, by permitting the renegotiation of contracts, the delicate tuning of the control structure imposed by the past and present on the future can be achieved. This leads to a considerably different transition matrix than the one shown in Table 1.

Once one starts to consider a multiperiod financial structure which includes the exchange of one form of financial paper or futures contract for another form, it becomes necessary to correctly interpret the paradoxes of infinity such as understanding the implications of servicing but never repaying debts. Furthermore at this point the possibilities for many-tiered financing appear. It becomes possible to create chains of paper backed by paper. These reflect new possibilities for distributing risk of different quality;[2] as well as institutional peculiarities based upon social and historical phenomena.

4. The Relative Importance and Time Structure of the Basic Units

In this section some comments are given together with a few figures which are at best extremely crude and are meant only to be qualitatively suggestive for the U.S. economy rather than quantitatively meaningful.

Unless otherwise stated the figures quoted are for the year 1970 and are mostly gleaned from the 1971 *Statistical Abstract of the United States.*

Goods and services are what an economic system is about. Steak is still better eating than financial instruments. However even though the goals of an economy are primarily measured in terms of real goods and services, the control mechanism is measured or at least partially described in terms of financial instruments.

(1) *Services*: As an extremely crude approximation we may regard all wages as payments for services. Beyond that we have net income value added from the service sectors. These are shown below, based on [9].

Wages		$540.1 billion
Net income of service sectors		
Transportation	30.3	
Communication	16.3	
Finance, Insurance, Real Estate	88.5	
Services	104.4	
Wholesale, Retail	121.8	
		351.3
		891.4

[2]I purposely use the somewhat vague word "quality" to emphasize the evaluative aspect to the treatment of financial risk. Probabilities are being assessed but dimensions are also being sought out. "We understand the deal, but what factors have we left out"—hence consultants, evaluators, appraisers, insurance and reinsurance.

This does not count housewives' services but this sector is neither monetized nor accounted for. Nor does it adequately cover rentals.

The stock behind services is human capital plus around half of corporate assets plus housing.

(2) *Goods*: 1n 1968 assets were evaluated as below, based on [10]:

Structures		
Public nonresidential	$459.8 billion	
Institutional	55.7	
Other nonresidential	288.7	
Residential	682.7	
Farm	50.0	
		1,536.9
Producer durables		377.0
Consumer durables		233.8
Inventories		
Farm	29.5	
Nonfarm	172.7	
Public	14.0	
		216.2
Land		
Farm	152.6	
Nonfarm	418.6	
Public	144.2	
		715.4
	Total	3,079.3 billion

The major part of these assets have ownership paper written against them and serve as an important part of the backing for a credit system.

Unlike financial paper which can be transferred or created and destroyed in a relatively brief time, the real assets give a structure to any economy that can only be modified over time. This is crudely indicated by legal writeoff times:

Land	∞
Buildings	40–60 years
Farm Buildings	25 years
Heavy Machinery	12 years
Office Machinery	10 years

The estimated lives for consumer durables and automobiles vary from around 5 to 11 years.

Obviously in the case of figures on stocks one has to be extremely careful in attaching significance to them, as the fallacy of composition limits their applicability. For example if all, or even a substantial fraction of owners of land tried to sell simultaneously the total worth of land would be far different from the numbers displayed.

Assuming random mass market trading of a sufficiently high level then the stock figures do provide an operationally useful figure to compare with the amount of financial instruments needed for trade when major correlated disturbances are not present.

We need not only production flows and stocks when we consider financing. We also need a measure of trade among existing stock (i.e. ownership changes in the secondhand markets).

Judging from mortgages the turnover in private real estate is about 10–15 percent of the stock. Trade in automobiles is about 50 percent of the stock, of which somewhat more than half is for new products. Trade in corporate plant appears to be considerably smaller as most exchanges are of financial instruments such as stock. Mere replacement of machinery and plant calls for a trade of the order of magnitude of 10 percent.

When we consider consumer goods, to every final sale there are between 4–6 times the volume of derivative sales. In a nonstrategic world in or near equilibrium this would be of little interest. However in a world where ownership change must be primarily achieved by the use of financial instruments, each stage involves different actors in the control of the process.

Goods and paper are used to back paper. Left out of Table 2 was human capital or the worth of all individuals considered as productive plant. By picking a production function such as the Cobb-Douglas or C.E.S. and by fudging the data one might arrive at a number between $30,000 to $60,000 per capita. However the amounts of unsecured loans appear to indicate that human capital is not as directly important as the stock of A.T. and T. or IBM or your house for securing a loan.

Several problems in depreciation and evaluation would have to be solved to get a reasonably accurate estimate of the mortality tables for all goods. Hence instead the legal depreciation lives were used in the estimates above.

Some idea of the faith we can put in legal depreciation rates as a reflection of actual is given by the 1960 housing stock which was approximately 7 million 5 years old or less; 7 million 5–10 years old; 8 million 10–20; 6 million 20–30; and 25 million 30 or older. One then has to adjust these figures for maintenance and repairs.

(3) *Money*: Money is, to the individual in a society with prices, instantaneous liquidity, or a reasonable approximation thereof. In a loosely coupled mass dynamic system with weakly correlated behavior, at any instant of time, money is to the two traders in an exchange approximately a transferable utility in the sense that in a mass market the sale by A to B of 100 shares of A.T. and T., a used car, or 1,000 bushels of wheat should have virtually no effect on reigning prices. The dynamics of adjustment in a mass monetary economy runs on a fallacy of composition which at or near equilibrium is not present or sufficiently weak that it scarcely matters. When, however, behavior is highly correlated such as in panics, or in reactions to major events which bring a mass change in expectations, then the differences between individual behavior and mass behavior are manifested in the dynamics of mass disequilibrium.

Custom and law determine the instruments of individual liquidity. Thus in a society where gold is a legal means of payment it should be counted in the money supply.

Behind custom and law are the enforcement mechanisms; hence what may be liquid for an individual economic unit may or may not be for a country. Thus international and internal means of payment are not necessarily the same. For example, to the Chilean government, gold is a money vis-a-vis the world, but the Escudo may not be.

In the United States in 1970 the monies, near monies and gold were:

Specie and Currency	$57.1 billion
Demand Deposits	165.7
Time Deposits	230.4
Gold	10.6

(4) *Ownership Claims*: In a world that is only partially monetized, where conceptual problems concerning the meaning of complex forms of ownership, depreciation, marketability and hosts of different transactions and information costs exist, where barter transactions take place and many services (such as housewives' services) are rendered without going through a market, it is no surprise that accounting is not an exact science.

In particular not all assets are represented by ownership paper. The more likely it is that an asset can be used to secure a loan, or that the asset cannot be easily transferred physically, but the title can be transferred, the more likely it is that ownership paper will be created. Hence there are deeds, titles and stock certificates.

When one buys a washing machine or an automobile on consumer credit the paperwork concerning ownership documents is considerably more precise than when one buys a washing machine for cash. Many of the lesser consumer durables are less desirable as items for consumer financing and many consumers do not even bother to keep serial numbers or ownership papers for items such as wholly owned television sets.

The stock market is the clearest example of a relatively low transactions cost market designed primarily for the exchange of second hand ownership certificates for money. Investment bankers and brokers also deal with the distribution and exchange of newly created ownership certificates for money.

The other major markets for ownership claims are in real estate. However these markets are far more imperfect. Furthermore the motivations for operating in them are diverse. Leaving aside tax laws which influence motivation considerably, many individuals and firms who buy real estate ownership claims want to actually use the asset represented by the claim. They live in the houses, or operate the manufacturing plant. The separation of ownership and active management is far less than it is in the stock market. The opportunity to be a passive investor trading real estate paper is still less than it is to trade corporate shares. Institutions such as Real Estate Investment Trusts (REITS) have come into being to enlarge the possibilities of passive investment more disassociated from management of real estate than would otherwise be possible. For common stock, the number of ownership claims in 1970 was 21 billion, in the amount of $814 billion. Shares are sufficiently visible and frequently traded (velocity around 1/5 per annum in New York) that a money measure of value can be obtained at any time even though the "value" may fluctuate 100 percent or more in a year.

(5) *Service Contracts*: Few service contracts are negotiable instruments. In the case of salary and wage contracts this is probably related to considerations of bondage as the capital asset providing the backing on one side is a person. In general however service contracts are explicitly bilateral and specialized hence can scarcely be thought of as fungible, in contrast with stock where although the individual trade is bilateral it is essentially anonymous with a mass market mechanism designed to process fungible items.

Out of the four basic futures contracts that can exist it must be concluded that with the exception of leases there is no mass market for trade in service contracts. These contracts, especially wage contracts, however, do provide considerable interlinkage between the present and future. The hiring of labor may vary from hourly payments to life tenure. Thus even though there is little if any negotiability the lock-in to the future may be considerable, as anyone who has tried to fire a secretary in a government bureaucracy can soon find out.

(6) *Futures*: The economic purpose of a futures market is to be able to hedge against changes in price of basic commodities or inputs which may be needed in an industrial process. Thus for example well developed futures markets exist for wheat, eggs, copper and several other commodities.

In general futures cannot be purchased more than one or two years into the future. Longer futures take the form of nonnegotiable contracts between specific and frequently large institutions such as a 30 year contract for a coal supply by a power company. Such contracts will usually have renegotiation or other flexible clauses in them.

A futures contract is essentially an insurance policy based upon different price predictions. As such the actual sale of a contract involves the sale of one piece of paper (the futures contract) for other paper (money). Only a minuscule amount of the money required to purchase the actual commodity may be put up until delivery date.

An estimate of the size of trade in the United States in futures contracts in the year ending June 30, 1973 is 47 million contracts, with a value of $399 billion. Five years previously there were 20 million contracts valued at $81 billion. In recent years it is estimated that around 300,000 outsiders (i.e. others than professional commodity dealers) have traded in this world of the 10 percent margin.

(7) *Promissory Notes*:

> "The Creditors are a superstitious Sect,
> great observers of set days and times."
>
> Ben Franklin

By far the greatest interconnector between the present and the future is the market for loans. It includes much of insurance as well as bonds and other forms of debt (although mutual insurance companies provide equity rather than debt positions).

The major forms of debt which exist in the United States are Federal Government, State and municipal debts and other obligations; consumer credit, mortgages, commercial borrowing and bonds. The amounts outstanding in 1970 are given below.

U.S. Government Debt	$301.3 billion
State and Local	148.1
Corporate	362.9
Mortgages	445.9
Consumer Credit	126.8
Other Business	123.6

Total insurance in force and the assets of the insurance companies were approximately $1,400 billion and 207 billion.

The time structure of all of this paper is key to understanding the constraints of the past and present on the future. For example the V.A. mortgage issues were 5.3 percent under 25 years; 9.8 percent 25 years; 13.4 percent 26–29 years and 71.5 percent for 30 years; but the *de facto* life of a mortgage is between 5–9 years and the turnover rates for real estate around 5–10 years.

(8) *Calls*: In the classification of the basic units of a financially controlled economy perhaps one area where the distinction is not as clear as it might be concerns individually owned assets as contrasted with assets held under joint ownership and limited responsibility. In general in a mass market for corporate stock the exchanges involve a change in paper holdings, not a change in control, although clearly this is not the case when corporate takeovers take place. When an individual buys an option on the purchase of a piece of land it could become a Jesuitical task to decide whether the option should be described as being on the land, or on the ownership papers representing the land.

If we keep in mind that the purpose of defining the various financial and economic units is to be able to provide the basis for constructing a model of the financial control system then as a good first approximation it is reasonable to consider some options on land as goods futures and others as calls. When a corporation buys an option on a land purchase to locate a new plant, essentially a future is being purchased on land; when a land speculator buys an option on land in a suburban area with the intention of selling the land the trade can best be considered as a call.

Calls (and "Puts") have been in existence for many years but do not yet figure as major financial instruments. It is as recently as April 1973 that an organized exchange, the Chicago Board Options Exchange was opened to trade this type of paper. The CBOE thus offers a secondary market for this paper which previously was hardly negotiable. For the first eight months of its operation 1.1 million contracts were written.

4.1. *On Trust and Finance*

The amount of unsecured loans to individuals in our economy appears to be relatively small. It seems as if the financial structure involving the interchange of paper claims backed by physical goods and the law is designed to optimize the availability of credit for a given level of trust and security. If your bank loan is covered several times by the market value of your A.T. and T. stock the bank does not need a large dossier on your personal history.

This paper is devoted to sketching the preliminary factors needed prior to mathematical model building, hence it is several steps from "theorem proving." Nevertheless even at this level of informality it can be seen that:

> For a given level of trust, all other things being equal an economy with more real goods which can be posted as security will have a financial structure as close or closer to optimum than an economy with fewer real assets.

5. The Institutions or Actors Who Create and Destroy the Basic Units

In this section several basic problems in modeling are noted. Few solutions are given beyond suggesting that isolating promising problems is the first step towards obtaining useful solutions.

(1) and (2) *Goods and Services*: Much of microeconomic theory has been devoted to discussing this aspect of the economy. In the United States approximately 1,500,000 corporations, 900,000 partnerships and 9,200,000 proprietorships provide the major sources for the production of goods and services.

Except for the inevitable accounting errors and difficulties in measuring the organizational and informational inputs, on the whole the technocratic-engineering-microeconomic view of goods and services provides a reasonable first approximation, at least in comparison with the microeconomic view (or lack of view) of the financial structure.

(3) *Money*: Among the institutions which participate in the creation of money are the Federal Reserve System, the Treasury and the public in general. A simple picture of causality is not possible. Different groupings may have different potentials for money creation. Active and passive money creation must be distinguished. The system is by no means necessarily linear or reversible. An apt analogy concerning the relationship between active and passive roles in money creation is with an elastic band. If the band has an expansion capacity an outside force can bring about its expansion and maintain it.

The accurate description of the monetary expansion possibilities in an economy cannot be given by merely describing the individual financial units. It has been said that "it takes two to tango," in some instances it may take two or more to create new money[3]. The description of how money can be created or destroyed calls for the specification of the rules of the money game. Morgenstern has made this point elsewhere [2]. The minimum list of institutional actors whose role must be described are:

The Federal Reserve System,
the Treasury,
the tax authorities, and
the banking system.

[3]This extremely important point is discussed further in the Appendix. A *key* distinction must be made between instruments involving only one party and those involving two or more. Concepts such as "neutral banking" depend delicately on this distinction.

(4) *Ownership Claims*: It is frequently forgotten that the stock market is primarily a second-hand market for negotiable paper. The creation and destruction of ownership paper may frequently involve the use of the stock market but the market itself is a brokerage arrangement, not a creator or destroyer of ownership paper *per se*.

The creation of new stock comes through flotations by new or existing firms. In 1970 for example the new issues of common and preferred stock were approximately $9,213 million. Retirements were estimated at $2,411 million. The changes represented around 39,000 new issues and 9,000 retirements. These should be compared with a market value of sales on the exchanges of $131 billion with a turnover of less than 20 percent.

Stock ownership paper is wiped out by failure, merger refinancing or confiscation. It is created by groups who have the power to be believed. To quote from the great example of South Sea Bubble financing:

> A company for carrying out an undertaking of
> great advantage, but nobody to know what it is.[4]

Leaving aside corporations, paper of varying levels of negotiability is constantly being created and destroyed by the changing patterns of real estate holdings and development.

As economies are increasingly monetized and accounting practices and reporting improves we may expect the amount of ownership paper to increase. For example a state or city pawnbroker such as the Dorotheum in Vienna creates tickets which are the owners' ownership claims against property which previously may have even been unrecorded. The ticket is a compound instrument combining an ownership claim with a futures contract.

Among the goods of sufficient importance to merit separate discussion are houses, automobiles and major consumer durables. In economies where decentralized markets are used as the means for the distribution of these items the creation of this paper involves four parties, the producers, consumers, the specialized financial institutions and governmental agencies.

(5) *Service Contracts*: The major parties to the creation of service contracts are individual suppliers of labor, unions, small and large private firms and institutions and governmental bodies. The security aspect of a civil service post or a university tenure does to some extent provide extra borrowing power or creditworthiness, but the labor service contract in general, as has already been noted, is not negotiable paper.

It is my guess that commercial service contracts such as leases, rent rolls, franchises, time sharing and other communications contracts may well be among the most important class for the growth of new negotiable instruments. This is evinced by the sale leaseback activity, the growth of REITS (real estate investment trusts) and the possibility of the computer utility.

(6) *Futures*: The major parties to the creation and destruction of futures are primary producers and industrial users. The trading stories of the few professional

[4]MacKay[1] reports that £500,000 was to be raised by selling 5,000 shares of £100 each. 1,000 subscriptions were sold at £2 each, the £98 to be paid one month later. At this point the promoter disappeared.

brokers and arbitrageurs make good light reading for those who prefer financial detective stories to real detective stories, but that is not really where the action is. A little light comedy is also provided by distinguished mathematical economists or econometricians losing their "grub stakes" or making fortunes.[5]

(7) *Promissory Notes*: The creation of promissory notes, bonds, loans, mortgages, insurance claims and so forth is a major subject by itself. Only one small, but in my opinion extremely important, point is made here concerning loans.

Whenever a loan is created there are at least two identifiable directly active parties (and frequently more). There is almost always an indirectly involved, but important third party—some governmental agency controlling and frequently changing rules. And there are at least two basic types of paper, several subspecies or mixed instruments involved.

When an individual borrows from his bank he usually takes his loan in some mixture of money (or liquidity or "now money") such as cash and a bank deposit. He exchanges this for a piece of paper he, the borrower creates (albeit with the consent of the bank, and in conformity to various regulations and legal requirements). Depending upon the nature of the securing of the loan other paper such as stock or other ownership claims are pledged. But the fundamental feature is that when a loan is made, regardless of the type of money the lender supplies, it is the borrower who is the prime creator of a new type of paper which is the promissory note (the I.O.U. or a future delivery contract for money). If the lender does not control or limit the promises of the borrower, disaster can result. If you as the lender lend me $100, I as the borrower, by several strokes of the pen could write out an I.O.U. note promising to pay you $1,000,000,000 next year in return for your $100. If there were no usury laws or other limitations the contract might be legally correct but the chances of fulfillment would be slight.

When a lender permits a borrower to promise more than he can probably pay he increases the chances for insolvency. This however is directly related to the supply of promissory notes of varying quality and not the supply of money for loans.

Money and promissory notes are different financial instruments and their direct creators are different institutions and individuals. Because a banking system plays an important risk evaluation role in an enterprise economy it is reasonable to expect that it will apply rules not only to its lending but to what it will permit a borrower to offer. Only with relatively shady banking groups in deals of dubious legality will the offer by the borrower to pay 10 times the prime rate improve his chances of getting a loan.

(8) *Calls*: The individual owner of a claim can become the writer of a put or call. The leverage is large and the number of players in this game is few. Puts and calls are instruments for magnifying the gains or losses to be made from fine differences in perception and hence are magnifiers of a dynamic adjustment (or disequilibrium?) process. In a complex world with capital gains taxes and lots of fine print the real and stated purpose of a calls market may be different.

[5]My (possibly incomplete) statistics on this point indicate the former but not the latter.

6. A THEORY OF MONEY AND FINANCIAL INSTITUTIONS

It is my contention that there exists a theoretical basis which can be used to reconcile micro and macroeconomic theory. It involves viewing the processes of financial control of an economy as a mass market noncooperative "money game"[7].

The general equilibrium model of the economy is an extremely limited and not fully defined special case of this particular game. It applies only when the system is in an equilibrium state and the need for and purpose of most financial instruments has disappeared. When one tries to view the general equilibrium theory in a disequilibrium state the difficulties appear. It is not possible to do so without considerable further modeling to describe the adjustment processes.

As a first cut one can try for an arcane noninstitutional array of adjustment processes involving excess supply and demand. This can be tied in with a tatonnement involving "virtual trades" rather than actual trades. However as such a system has so little dynamic structure and totally ignores the intervening financial variables it should come as a surprise to no one if all that is discovered is a set of essentially negative results, i.e. systems behavior which may be globally unstable even if there were a unique state equilibrium[4].

The general equilibrium theory is not congenial to being modified for information conditions. In particular the very concept of the system also does not allow for models where there is an intermixture of mass behavior and oligopolistic or large unit behavior. Yet there is not a country in the world where this is even a good first order approximation. Governmental units, utilities, public services and major investment agencies are by the very nature of things large with respect to trade in vegetables.

It may be that trade in markets with mass suppliers and purchasers can be usefully approximated by a low information content dynamic non-historical process. Trade and major investment by large units involves historical and information rich processes. These processes are highly dependent on detailed institutional and behavioral facts. One must recognize the asset structure of the industry, the level of organization, the quality of management, the age structure of the real assets and the financial paper. They provide the framework for the dynamics.

A little known feature of the formal theory of games is that when one tries to extend the concept of a static noncooperative equilibrium to multiperiod models it becomes necessary to distinguish between equilibria which are obtained from state strategies and those obtained from historical strategies.[6] Attempts to derive an adequate dynamic theory of dynamic noncooperative games lead to the creation of models which melt imperceptibly into behavioral models.[7]

At this point the true believer in general equilibrium might say, "So what, why do we need to bother with a game theoretic formulation if all it is going to do is to lead us to behavior equations? We can easily cut our own out of the whole cloth!"

[6]This point is explained in detail elsewhere[8].
[7]See for example the thesis of A. Sanghvi [3].

The point is that the methodology and the model building power of the game theory provides a natural framework which forces the individual to model sufficient detail to well define a system not merely for its equilibrium states (if they exist), but for every attainable state of the system. In doing so information conditions must be specified in detail and rules of the game such as bankruptcy laws, conditions on how money is issued or financial claims are traded emerge from the need to well-define the model.

The general equilibrium model is not inconsistent with the noncooperative game model; it is merely insufficiently defined to provide an adequate dynamic model of the economy. It can however be regarded as an extremely limited and somewhat misleading special case of a much more general, relevant and interesting class of models.

That it may be regarded as a special case of a more general set of models is in my opinion an indication of both its value and its limitations. It provided an important abstract means to examine relationships between production and consumption over an important range of economic problems. But it provided no insights whatsoever into the problems of economic control.

The general equilibrium theory is simultaneously not mathematically general enough to provide the tools to model information conditions or oligopolistic or strategic behavior. And it is not sufficiently institutionally oriented that it is not possible to extend it in a way to model the mechanisms which govern and transmit strategic behavior.

6.1. *On International Finance*

In this paper no mention has been made previously of international finance. I believe that international finance involves no basically new instruments; however, it does involve new agencies for their creation and destruction.

The essence of a financial system is that it is a control system which has as its neural network "paper" or claims and information flows. In parallel with the real goods and services are the ownership papers which can be alienated from them and hence a continuum of ownership and control relations can be created.

In the resulting loosely coupled system where mass markets exist together with powerful central agencies the individual units which make up the economic mass of a single national society may be controlled or predicted at least to a reasonable first order approximation. When we move to an international context the strategic and control conditions are essentially different. To talk of financial and monetary international control without talking of international law enforcement has about as much operational significance as writing for the Sunday supplements.

6.2. *On Mathematical Institutional Economics*

In a preface to the discussion of information and dynamic economic models I suggested, some time ago, the title "Mathematical Institutional Economics"[6] as the appropriate title for the type of approach needed to specify the economic process. In particular the qualifier "mathematical" refers to the formal specification of the economic models as mathematical structures with the behavior of the individuals described by mathematical operators such as maximization. The qualifier "institutional" refers to the "rules of the game": the laws, the institutions and the details in any society which act as the *carriers of process.* In a truly dynamic economy no magic auctioneer or tatonnement guide exists. Supply and demand are not equalized by a magic noninstitutional matching and rematching of tentative bids and offers. When you blunder, you blunder and you pay for your education in a world where information is expensive, perceptions poor and decisionmaking expensive in many dimensions.

This essay has not been meant as an institutional attack on the true church of microeconomic theory. On the contrary it is an attack from within the church.

Institutional economics, macroeconomics and behavioral models of economic process are *not* alternatives to microeconomic analysis. They are the natural extensions that are implicit in the appropriate formulation of the microeconomic models.

This appropriate formulation requires an abandonment of attempts to patch up the general equilibrium system in a manner not unlike the efforts to patch up the Ptolemeic system in astronomy.

The general equilibrium system was modeled as a tightly coupled essentially static construct with, to say the least, a grotesquely oversimplified view (or nonview) of ownership and control, and perception and information processing.

The model needed calls for a loosely coupled system with money and financial institutions providing the loose coupling and the control mechanism. It has to permit easy modifications to information conditions. It must allow for a mixture of mass markets and large economic units. And lastly, both for scientific and theological purposes it should yield the general equilibrium analysis as a special case. It is my belief that the class of models which satisfies these requirements can be broadly described as dynamic noncooperative "money games." They can be solved in three different ways—for equilibrium state strategies, for historical strategies or for behavioral descriptions of the choices of agents.

The equilibrium state strategies provide the direct linkage to the simplistic assumptions of general equilibrium theory.

6.3. *A Postscript on Economic Theory*

Before the proof comes the theorem; before the theorem comes the conjecture and an insight as to why the theorem is worth proving. Before the conjecture, or at least with it, comes the model. Before the model comes a need to look at the economic world as it is, not merely as it is told to us in the story books. Books, theories and other *dicta* are great helps to enable us to interpret the economic world around us. But they are secondary, not primary, sources. In spite of the elegance of the general equilibrium system, the world does not look like that—even as a crude approximation for many questions we want to answer. If this is the case we need a different world or possibly a different theory.

Appendix

Of the eight instruments suggested, two are real and six are paper. Of the six paper instruments, two are not contracts except possibly in the broadest sense, but four are clearly contracts. Fiat money and ownership paper are basically one party paper. An attempt to "cash" a $1 bill will merely yield another $1 bill.

All four instruments which link two time periods, service contracts, futures contracts, debt instruments, and calls are also between two (or more) parties. For example, if *i* borrows $100 from *j* now on the condition that he will pay back $105 next period, this can be described by the pair $(100_t^{j,i}, 105_{t+1}^{i,j})$ where the first number is *j*'s contractual promise to *i* and the second is *i*'s promise to *j*. Any time a contract is created essentially two commitments are made, *i* to *j* and *j* to *i*. Once this is fully understood then a concept such as neutral banking can be mathematically specified. If *j* is the bank, it may be possible to state a fixed strategy for the bank, much in the same way as the dealer at a Blackjack table must follow a fixed strategy; this means that the power to borrow or not lies with *i*. If he meets the requirements, then *j* must, by the rules of the game, lend to him.

References

[1] MacKay, C., *Memoirs of Extraordinary Popular Delusions*, Vol. 1, R. Bentley, London, 1841, p. 88.

[2] Morgenstern, O., "Thirteen Critical Points in Contemporary Economic Theory," *Journal of Economic Literature*, Vol. 10, No. 4, 1972.

[3] Sanghvi, A., *The Asymptotic Behavior of Some Sequential Games*, Ph.D. Thesis, Yale University, 1973.

[4] Scarf, H., "Some Examples of Global Instability of the Competitive Equilibrium," *International Economic Review*, Vol. 1, No. 3, 1960.

[5] Shapley, L. S. and M. Shubik, "Competition, Welfare and the Theory of Games" (unpublished manuscript), Chapter 2 (available as Rand Report).

[6] Shubik, M., *Strategy and Market Structure*, New York, Wiley, 1959, Ch. 8.

[7] Shubik, M., "A Theory of Money and Financial Institutions. Part IV. Fiat Money and Noncooperative Equilibrium in a Closed Economy," *International Journal of Game Theory*, Vol. 1, No. 4, 1971.

[8] Shubik, M., "The General Equilibrium Model is the Wrong Model and a Noncooperative Strategic Process Model is a Satisfactory Model for the Reconciliation of Micro and Macroeconomic Theory," Cowles Foundation Discussion Paper No. 365, Nov. 1973.

[9] *The Statistical Abstract of the Unitea States*, 1971, p. 311.

[10] *Ibid.*, p. 328.

[12]

Zeitschrift für Nationalökonomie 36 (1976), 49—60

A Noncooperative Model of a Closed Trading Economy with Many Traders and Two Bankers*

By

Martin Shubik, New Haven, Conn., U. S. A.

(Received September 1, 1975)

1. Introduction

In previous papers Shapley and Shubik have formulated a noncooperative game model of a closed trading economy where one commodity is used as a means of payment. The existence of a noncooperative equilibrium was proved; the role of cash flow and the need for credit was examined and the relationship between the noncooperative equilibria and the competitive equilibria was illustrated.

The basic model is as follows:

Consider n traders in a market with $m+1$ goods. We assume that initially each trader i has a bundle of goods $(a^i{}_1, a^i{}_2, \ldots, a^i{}_m, a^i{}_{m+1})$. We assume that the $m+1^{st}$ good is the means of payment and that the convention of this market is that all traders are required to offer all their holdings of the first m goods for sale. This means that individuals must buy back items they own for their own consumption. It is as though the farmer sells all of his milk to the market and buys his consumption milk from the store.

This condition amounts to requiring that all goods are monetized — it is a useful accounting convention and is mathematically easy to analyze. Other conventions can be used which modify this condition.

We may consider the market as consisting of m trading posts where all of each of the m commodities is offered for sale. Each

* The research described in this paper was undertaken by a grant from the Office of Naval Research.

trader i allocates an amount of his means of payment b^i_j to the jth market. If there is no credit mechanism we require that

$$\sum_{j=1}^{m} b^i_j \leqq a^i_{m+1} \quad \text{for } i=1, 2, \ldots, n. \tag{1}$$

In this model price is given by:

$$p_j = b_j/a_j \text{ where } b_j = \sum_{i=1}^{n} b^i_j \tag{2}$$

and the amount that an individual obtains upon bidding b^i_j for good j is:

$$x^i_j = b^i_j/p_j \text{ if } p_j > 0 \quad (\text{where } a_j = \sum_{k=1}^{n} a_j^k) \text{ for } j=1, \ldots, m \tag{3}$$

$$= 0 \quad \text{if } p_j = 0$$

and

$$x^i_{m+1} = a^i_{m+1} - \sum_{k=1}^{m} b_k^i + \sum_{k=1}^{m} a_k^i p_k.$$

Thus the payoff function to the traders can be expressed both in terms of their strategic variables explicitly or the final bundles of goods which implicitly reflect the strategic behavior

$$\Pi^i (b^1, b^2, \ldots, b^n) = u^i (x_1^i, \ldots, x^i_{m+1}). \tag{4}$$

Given the nature of the market mechanism, all traders pay the same price for the same good. Furthermore the markets and price system constitute an externality to all traders in the sense that all trade is interlinked through the markets.

The specification of a mechanism for trade and price formation may limit the attainable set of economic outcomes.

2. On Modeling the Strategic Choices of Economic Actors

In the modeling of a trading economy as a noncooperative game two separate problems are posed. They are (1) the specification of a mechanism for the generation of prices and (2) the examination of any constraints placed on attainable outcomes by the mechanism. In discussion among Scarf, Shapley and Shubik it was observed that if all commodities can be used as a means of payment then it is possible to formulate a noncooperative game model of a closed trading economy with some restrictions on the feasible set of outcomes but with no need for credit. If fewer than all commodities are used as a means of payment then not only are restrictions placed on feasible outcomes but credit may be needed.

The above comments can be made somewhat less cryptic by means of simple examples. In a simple barter model of a trading economy where any two goods can be bartered by any group of individuals owning them, it is *feasible* (though not optimal) for different groups to simultaneously exchange two commodities at different rates of exchange. If all individuals are required to deal through a formal market which aggregates bids and offers there will be restrictions on attainable trades. In particular a common restriction is that all traders may end up paying the same price for the same commodity.

When all goods serve as a means of payment, all of one's assets are immediately available in the markets and no cash flow problem is created, in the sense that an individual might have less means of payment immediately available than he has assets. When fewer, for example, one or two goods serve as the means of payment then a cash constraint (such as condition (1) in Section 1) may appear. This constraint can block optimal trades unless a credit mechanism is introduced which enables the individuals to obtain short term credit to enable them to carry out exchanges which could conceivably involve payments as high as the total market value of any individual's initial assets.

Shapley and Shubik introduced a credit mechanism *ex machina;* one can merely relax condition (1) and assume that short term credit or "trust chips" are available from a mechanism. These chips must be paid back after trade. Otherwise if an individual fails to repay, then the worth of insolvency must be specified and even bankruptcy rules may need to be specified.

There is another way to handle the supply of credit. This involves introducing a new special class of economic agents known as "bankers". These individuals are distinguished from other only by the set of actions they can take in the game model of the economy. Two formulations of the game with credit are given, under one formulation only two bankers are needed for the noncooperative game to yield efficient equilibria; under the other, many bankers are needed to attenuate their oligopolistic influence.

3. Bankers as Special Economic Actors

In order to define the model with bankers we introduce credit formally as the $m+2^{\text{nd}}$ commodity. The means of payment is the $m+1^{\text{st}}$ commodity which may or may not have an intrinsic value. The $m+2^{\text{nd}}$ commodity does not enter into the utility functions. We may imagine the credit lines of the banks to consist of a pile of chips they can lend out, or we can consider them as a set of I. O. U. notes bearing the banker's name. The unit for the $m+2^{\text{nd}}$ commodity is the same as that for the $m+1^{\text{st}}$ commodity. Thus on payback a unit of one is interchangeable with the other.

In this model there is a money (the $m+1^{\text{st}}$ commodity) which is fiat or has consumer worth, which can be regarded as having been created by community custom, law or "rules of the game". The credit (the $m+2^{\text{nd}}$ commodity) has a different issuer, to wit: the banks. If we wish to be careful with the doubleentry book-keeping, we must include an $m+3^{\text{rd}}$ commodity which consists of the I. O. U. notes of the borrowers which the bank holds in return for the I. O. U. notes it issues to serve as currency. Thus in the model we have:

Commodity $m+1$ = Society's accepted means of payment.

Commodity $m+2$ = Banker's I. O. U. notes or guaranteed credits which circulate on par with the means of payment.

Commodity $m+3$ = Individuals' I. O. U.s or promissory notes which are nonnegotiable, held by the banks in return for quantities of $m+2$ and are contracts for payments of $m+1$ or $m+2$ at the end of the period.

In our extended model with $m+3$ commodities, of which only $m+2$ are negotiable and 2 or 3 do not appear in the utility functions we have as the initial holdings for the $n+2$ traders the following bundles:

For Traders:

$$(a_1^i, a_2^i, \ldots, a^i_m, a^i_{m+1}, 0, 0) \text{ for } i=1, \ldots, n. \tag{5}$$

For Bankers:

$$(0, 0, \ldots, 0, a^i_{m+1}, a^i_{m+2}, 0) \text{ for } i=n+1, n+2. \tag{6}$$

The consumption goods ownership and trading aspects of the bankers has been removed by giving them no other consumer good than the means of payment. We need extra rules of the game to separate banking from commercial activities. In particular a bank is not permitted to extend credit to itself for its own consumption purposes. The bank may only spend for consumption purposes up to the amount a^i_{m+1} which can be regarded as representing its capital.

Perhaps a more satisfactory way of representing a bank's initial holdings than (6) above we might consider it to be:

$$(0, 0, \ldots, 0, a^i{}_{m+1}, 0, 0) \tag{7}$$

but that there is a special rule of the game permitting bank i to create or issue an amount $C^i{}_{m+2}$ of the $m+2^{\text{nd}}$ commodity in exchange for individual trader's promissory notes. This convention has the $m+2$ commodity "in being" until it is issued. The convention in (6) gives it a physical existence, but it is "sterilized" as a means of exchange when it is held by a bank.

The interpretation of the limits $C^i{}_{m+2}$ is that by the rules of the game (set, for example by some, as yet, unmodeled central bank) the credit creation powers of the banks are limited.

We note that the making of a loan by the bank involves two creating agencies. The bank creates its I. O. U.s or $m+2$ and the trader creates his I. O. U.s or $m+3$. Three sets of doubleentry bookkeeping statements for a bank and a trader can be made up for this one period game. They are before lending, at the point of loan, before trade and after trade when the loan is due. The books are kept in terms of $m+1$. The books only show the relevant parts of the banker-trader relationship.

	Bank			Trader		
Before Loan	Assets	Liability		Assets	Liability	
	0	0		0	0	
Loan	Assets	Liability		Assets	Liability	
Traders IOU for 120	100	100	Loan of $m+2$	100	100	Present Value of IOU for 120
	Assets	Liability		Assets	Liability	
	120	100 Loan	Value of Assets after Trade	A	120	IOU due in $m+1$, or $m+2$
IOU Due		20 Earned			E	Value of Equity after Trade
				$A = 120 + E$		

We note that the bank issues units of $m+2$ subject to a ceiling, but so far in this model no ceiling on the issue of units of $m+3$

by the borrowing traders has been specified. Thus for example a trader might offer a promissory note of a billion for one unit of $m+2$! At the point of the loan the promissory note is carried on the bank's books at the amount in terms of $m+2$ (or $m+1$ as they are on par) advanced by the bank.

When the trader's note is due, after trade then the "face value" of the promissory note is translated into terms of $m+1$ at par. Thus in the example above an I. O. U. note of 120 was written in return for the bank's note of 100. After trade when the note is due the bank has "earned" the extra 20 and this is reflected in the books.

If a bank announces as its strategy a premium, or rate of interest it wishes to charge for a loan of a unit of $m+2$ it exercises one type of control over the borrowers different from if it lets borrowers bid quantities of $m+3$ for a fixed supply of $m+2$. In either instance if the borrowers create a volume of $m+3$ which is larger than the supply of $m+1$ plus $m+2$ then insolvency must be forced on some borrowers as the supply of "NOW" money or money acceptable to repay the bank loan is insufficient to cancel all trader I. O. U. notes.

3.1. "Price" or Rate of Interest Naming Bankers

First we consider an economy where there are two banks who each name a price for their extension of credit. After they have named their prices, loans are made, at which point all individuals make their bids in the markets for the first m commodities. After this the game is over and individuals automatically are called upon to settle accounts.

The tree diagram in Fig. 1 shows the sequencing of moves and the information conditions. In this instance it is assumed that the banks set their rates each without information on the act of the other. A slight modification could have them move sequentially. It is assumed that all traders are informed of the two bank rates before they simultaneously bid for loans. If the total amount of loans requested is larger than the limits $C^1{}_{m+2}$ and $C^2{}_{m+2}$ then a rationing scheme for loans must be described. In this model we will limit our concern to the case where the $C^i{}_{m+2}$ are sufficiently large that rationing is no problem.

After all loans have been secured it is assumed that all traders are completely informed and all traders (including the bankers) bid in the markets for the first m goods. The bankers however are forbidden from using bank credit.

After the market has functioned there is a "settlement" which is essentially automatic; loans must be repaid or insolvency must be handled.

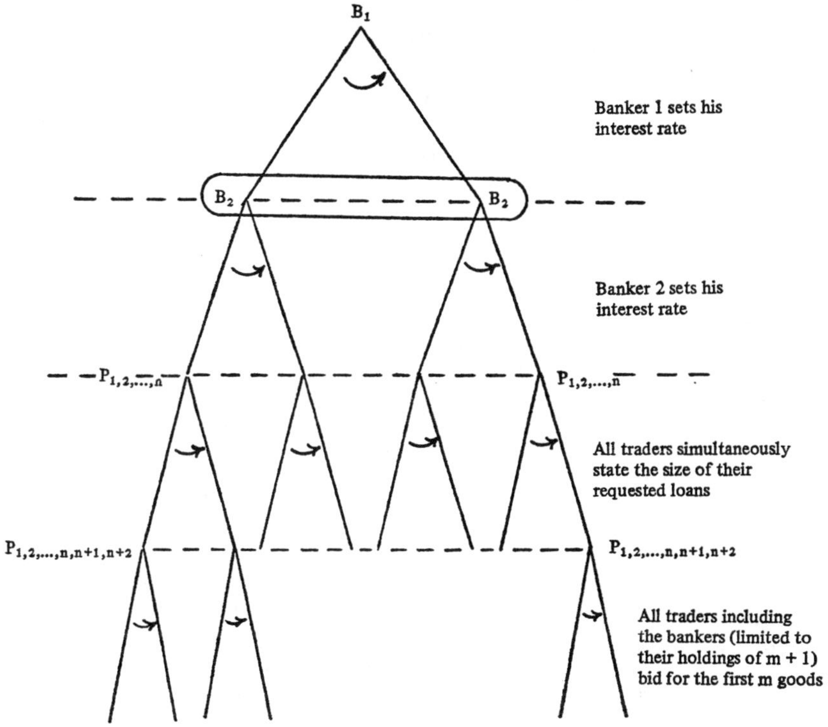

Fig. 1

The existence and nature of the equilibrium point

Shapley and Shubik have shown that a noncooperative equilibrium exists for the last stage. If we solve for a perfect equilibrium point, i. e. one which is not only an equilibrium point in the game as a whole, but also in every subgame, then we may work a backward induction replacing the ultimate branches of the tree by equilibrium payoffs at the penultimate branches.

In the games consisting of selecting the size of a loan the payoffs to each individual are concave and continuous hence these games will have an equilibrium point.

We are now back to the "bankers' game" of setting the interest rate. If we assume that each banker has a credit line so large that he could service the loan requirements of the whole community under any circumstances then our bankers' game is the classical Bertrand duopoly model which has an equilibrium point where the rate fo interest charged on short term credit is zero. This depends on the relatively weak assumption that *ceteris paribus*, borrowers will borrow from the bank charging the lowest rate of interest. When they charge the same rate we may assume that borrowers are indifferent and each obtains half of the loans.

The proof is simple if one bank undercuts the other it obtains all of the market. As each has a costless credit line that is sufficient for all customers (for instance we could merely remove the bounds $C^1{}_{m+2}$ and $C^2{}_{m+2}$) as long as there is a positive rate of interest charged by one bank there is an incentive for the other to just undercut.

In this formulation the oligopoly power of the banks is attenuated as soon as there are two banks.

If we now replicate the number of traders so that we have an economy with $kn+2$ participants this economy will have noncooperative equilibria which approach (as k becomes large) the same relative prices and distribution of the first $m+1$ commodities (assuming that the $m+1^{\text{st}}$ commodity has a worth in consumption) as the competitive equilibria solutions.

3.2. Quantity or Credit Rationing Bankers

A different mechanism would be to have the bankers simultaneously select an amount q^{n+1}_{m+2} and q^{n+2}_{m+2} which are the limits on the credit they are willing to extend. The traders in turn simultaneously and with knowledge offer $s^i{}_{m+3}$ for their loans. Using this mechanism the amount of credit the ith trader obtains is

$$c^i = \frac{s^i{}_{m+3}}{s_{m+3}} \left(q^{n+1}_{m+2} + q^{n+2}_{m+2}\right), \quad s_{m+3} = \sum_{i=1}^{n} s^i{}_{m+3}. \tag{8}$$

The rate of interest or "price of short term credit" which evolves from this model is given by:

$$1 + \text{the rate of interest} = \frac{s_{m+3}}{\left(q^{n+1}_{m+2} + q^{n+2}_{m+2}\right)}. \tag{9}$$

We assume that the rate of interest at equilibrium can never be less than zero because if it were it would pay a bank to reduce the amount of credit offered. This is not quite covered in the model above where credit granting is absolutely costless, but there are several minor modifications which would produce this condition.

This model is the Cournot type model of the market for loans, and like the Cournot oligopoly model may still have a marked oligopolistic force when there are only two firms in control of a market. It appears that for the attenuation of the effect of oligopolistic banking when the banks can control the quantity of credit we need many banks.

4. A Simple Example

A simple example involving $2n$ traders and then 2 or more bankers is computed for illustrative purposes.

Suppose there are n traders with initial endowments $(A, 0, 0, 0)$ where the first commodity is to be put up for auction, the second is the means of payment $(m+1)$, the third bank credit $(m+2)$ and the fourth trader promissory notes $(m+3)$. Let these traders have utility functions $\varphi_i\,(q_1{}^i, q_2{}^i) = q_1{}^i q_2{}^i$.

There are n other traders with the same utility functions and endowments $(0, A, 0, 0)$.

There are two bankers with endowments $(0, 0, \infty, 0)$ and utility functions $\Phi_i\,(q_1{}^i, q_2{}^i) = \Phi\,(q_1{}^i, q_2{}^i)$.

Competitive Equilibrium

Traders of Type 1: 2; $(A/2, A/2, 0, 0)$ = final allocation

$p_1 = p_2 = 1$ = prices.

Noncooperative Equilibrium without Credit

For Type 1

$$\Pi_1{}^i = \left(\frac{n\,x_i\,A}{x+y}\right)\left(\frac{x+y}{n}\right) = x_i A = 0 \text{ as } x_i = 0. \tag{10}$$

For Type 2

$$\Pi_2{}^j = \left(\frac{n\,y_j\,A}{y}\right)(A - y_j). \tag{11}$$

From (11) we obtain:

$$y_j = \left(\frac{n-1}{2n-1}\right) A \tag{12}$$

which for $n = 2$ gives $y_j = A/3$

$n \to \infty$ gives $y_j = A/2$

hence for Traders of Type 1: $(0, A/3, 0, 0)$ = final allocation $n=2$

$(0, A/2, 0, 0)$ = final allocation $n \to \infty$

and for Traders of Type 2: $(A, A/3, 0, 0)$ = final allocation $n=2$

$(A, A/2, 0, 0)$ = final allocation $n \to \infty$

with prices $p_1 = 1/3,\ p_2 = 1$ for $n = 2$,
$p_1 = 1/2,\ p_2 = 1$ for $n \to \infty$.

Noncooperative Equilibrium with Credit Cost of $1+\varrho$

Suppose that an outside agency charges $(1+\varrho)$ for a unit of $m+2$, (or bank credit) then the payoff for traders of Type 1 is:

$$\Pi_1{}^i = \left(\frac{n x_i A}{x+y}\right)\left[\frac{x+y}{n} - (1+\varrho)\, x_i\right] \tag{13}$$

and for traders of Type 2:

$$\Pi_2{}^j = \frac{n y_j A}{x+y}\,(A - y_j). \tag{14}$$

After a certain amount of manipulation from (13) we obtain

$$(x_i + y_j)^2 = \left(2\,(x_i + y_j)\,(x_i) - \frac{x_i{}^2}{n}\right)(1+\varrho) \tag{15}$$

and from (14) we obtain

$$A\left(x_i + y_i - \frac{y_j}{n}\right) = 2 y_j\,(x_i + y_j) - \frac{y_j{}^2}{n}. \tag{16}$$

For a mass market, from (16) with $n \to \infty$ we obtain

$$y_j = A/2. \tag{17}$$

Substituting (17) into (16) and considering $n \to \infty$ we obtain:

$$x_i = \frac{A}{2}\left(\frac{1}{1+2\varrho}\right). \tag{18}$$

Thus we have for traders of Type 1 a final allocation of

$$\left[\frac{A}{2\,(1+\varrho)},\ \frac{A}{2}\left(\frac{1+\varrho}{1+2\varrho}\right),\ 0,\ 0\right]$$

For Type 2

$$\left[\frac{A}{2}\left(\frac{1+2\varrho}{1+\varrho}\right),\ \frac{A}{2},\ 0,\ 0\right]$$

and for the banking system

$$\left[0,\ \frac{A}{2}\left(\frac{\varrho}{1+2\varrho}\right),\ 0,\ 0\right],$$

and

$$p_1 = \frac{1+\varrho}{1+2\varrho},\quad p_2 = 1$$

are the prices.

Noncooperative Equilibrium with "Bertrand" Bankers

Suppose we assume that $\partial \Psi / \partial q^i_2 > 0$, i. e. that the bankers each have a positive marginal utility for the means of payment (this is a statement about $m+1$, *not* $m+2$ or $m+3$). The interest they earn by definition has to be paid in this commodity in the sense that the wiping out of all the credit they advanced retires all of $m+2$ without covering interest, which must be paid in $m+1$ or $m+2$.

If there are two bankers with unlimited credit lines each naming on interest rate, and if borrowers always select the bank with the lowest rate and split on ties then the "duopoly game" of the bankers is as follows

$$\begin{aligned} \Pi_B{}^1 &= 0 && \text{if } \varrho_1 > \varrho_2 \\ &= \Psi\left[0, \frac{A}{4}\left(\frac{\varrho_1}{1+2\varrho_1}\right)\right] && \text{if } \varrho_1 = \varrho_2 \\ &= \Psi\left[0, \frac{A}{2}\left(\frac{\varrho_1}{1+2\varrho_1}\right)\right] && \text{if } 0 \leqq \varrho_1 < \varrho_2. \end{aligned} \tag{19}$$

This is the classical Bertrand duopoly. As long as $\varrho_1, \varrho_2 > 0$ there is a discontinuity in the payoffs associated with just undercutting the competitor. At $\varrho_1 = \varrho_2 = 0$ there is an equilibrium*.

An immediate inspection of final holdings and prices in the noncooperative market with a credit cost of $1+\varrho$ shows that for $\varrho = 0$ we obtain the same distributions and prices as in the competitive market, except that here the bankers are explicit players supplying credit at no gain to themselves.

Noncooperative Equilibrium with "Cournot" Bankers

If we assume that the two bankers face an ocean of traders then at equilibrium if the credit line of banker i is x_i we have the condition derived from (9) and (18)

$$1+\varrho = \frac{\frac{A}{2}\left(\frac{1}{1+2\varrho}\right)}{x_1+x_2} \tag{20}$$

From which we obtain:

$$\varrho = \frac{-3+\sqrt{5+\frac{2A}{x_1+x_2}}}{4}.$$

* There are two fine points not covered. We need to specify a condition that for $\varrho_1 < 0$ the payoff is less than $\Psi(0, 0)$. Furthermore for this and the next model it might be more congenial to imagine the traders to have been fractionated rather than replicated. Otherwise in (19) a factor of n is missing in the interest payments, but for $n \to \infty$ this is unbounded.

The noncooperative game the two banks are in amounts to:

$$\begin{aligned} &\underset{x_1}{\text{maximize}}\ \Pi_{B}{}^{1} = x_1\,\varrho = x_1 \left(\frac{-3+\sqrt{5+\dfrac{2A}{x_1+x_2}}}{4} \right) \\ &\underset{x_2}{\text{maximize}}\ \Pi_{B}{}^{2} = x_2\,\varrho = x_2 \left(\frac{-3+\sqrt{5+\dfrac{2A}{x_1+x_2}}}{4} \right) \end{aligned} \tag{21}$$

Solving these we obtain:

$$x_1 = x_2 \cong 0.134\,A$$

and

$$\varrho \cong 0.131.$$

As in the simple example given by Cournot where the duopolists can offer quantities of a costless product, but perceive their control over supply it pays them to restrict the quantity of credit offered. The final distribution is as already given for the noncooperative equilibrium with a credit cost of $1+\varrho$ and with $\varrho \cong .131$.

It is easy to see that by increasing the number of bankers the oligopoly effect attenuates.

References

L. S. Shapley and M. Shubik: Some Strategic Models Related to General Equilibrium, mimeographed, February 1973.

L. S. Shapley and M. Shubik: Trade Using One Commodity as a Money, mimeographed, March 1974.

Address of author: Prof. Dr. Martin Shubik, Department of Economics, Cowles Foundation for Research in Economics, Box 2125, Yale Station; Yale University, New Haven, CT 06520, U. S. A.

Printed in Austria

[13]

Trade Using One Commodity as a Means of Payment

Lloyd Shapley

Rand Corporation

Martin Shubik

Yale University

A general model of noncooperative trading equilibrium is described in which prices depend in a natural way on the buying and selling decisions of the traders, avoiding the classical assumption that individuals must regard prices as fixed. The key to the approach is the use of a single, specified commodity as "cash," which may or may not have intrinsic value. The model, in several variants, is treated as a noncooperative game, in the spirit of Nash and Cournot. The *rules* of the game, including the price-forming mechanism, are independent of behavioral or equilibrium assumptions, which enter, instead, through the *solutions* of the game.

I. Introduction

In the development of a theory of money and financial institutions, there are three kinds of questions which should be clearly differentiated: (1) Why does money come into use, and how? (2) What keeps it in use? (3) What strategic limitations does its use impose on trade, what additional possibilities does it open up, and what are the institutional implications?

The first would involve both historical research and a study of transactions costs and markets. It would be interesting to explore such questions

We are grateful for suggestions and constructive criticisms from Pradeep Dubey, Robert Slighton, Rodney Smith, Ross Starr, and an anonymous referee as well as many others with whom we have discussed this model during the past 2 or 3 years. We are also grateful to the National Science Foundation (grant GS-31253) and the Office of Naval Research (contract N00014-76-C-0085) for their support.

[*Journal of Political Economy*, 1977, vol. 85, no. 5]

as the smallest size or level of trading activity at which a developing economic community should be expected to adopt a commodity money system, a fiat money system, or a banking system, etc. At least since the time of Cournot (1838), it has been observed that, if m commodities are to be bartered on a one-to-one basis with no use of a monetary medium, as many as $m(m - 1)/2$ pairwise trading markets might be needed, while the use of a generally accepted medium of exchange enables the system to function efficiently with only m or $m - 1$ trading posts. Similarly, a money can enable an economic system to function efficiently through purely bilateral transactions, without requiring tortuous sequences of middlemen who pass the goods along to their ultimate consumers (Ostroy and Starr 1974). In other words, a medium of exchange can "de-couple" cross-interactions both between traders and between commodities; and, if an economic society grows large, the social advantages of this soon outweigh any social disadvantages or costs that may be involved. This is of course not the whole story: information costs, search costs, transportation costs, the desire for safety or anonymity, and myriad other details of transactions technology may be invoked as further explanations of why money and financial institutions come into being (see Jones 1976).

Questions concerning the beginnings of money and finance are basically different, however, from the question of why money stays in use. The fish gets on the hook one way and off it in a radically different manner. Once enough people accept a money and a body of law and custom concerning its use, it is difficult, short of social breakdown, hyperinflation, or revolution, for an individual or small group to "opt out" of the system, even if the conditions attending the system's beginnings have changed. The acceptance of the system has a flavor of a noncooperative equilibrium like the "Prisoner's Dilemma": although a person may view, say, fiat money as being of dubious value as a store of wealth, he knows that most others will continue to use it for trade, and he may be in no position to do otherwise himself.

No modern society is a 100 percent monetized society. Segments with barter trade always exist, and the volume of such trade may vary with many factors such as technology, taxes, faith in financial institutions, constraining laws and regulations, and so forth. Nevertheless, if we accept the facts that money is the major means of exchange in a modern economy and that much of the exchange of money for goods, services, and financial instruments takes place via some form of organized market, be it a stock-market or a supermarket, then we can ask yet another kind of question. What are the restrictions imposed upon economic behavior if we *require* that trade be monetized? How do these restrictions influence the decisions of rational individuals, and how do they limit the range of feasible outcomes? The body of this paper is concerned largely with questions of this type.

Our exposition is based on a multicommodity model of trade as a game in strategic form, to which the noncooperative solution concept of Nash (1951) may be applied. The model is strategically closed, in that, unlike the classical equilibrium models, prices are determined by the actions of the traders, and the system as a whole responds meaningfully to the traders' decisions, even away from equilibrium.[1] The mathematical proof of the existence of a Nash equilibrium and its relationship to the competitive equilibrium of Walras will be presented elsewhere; here we concentrate on an elementary but, we hope, informative exposition, making much use of "Edgeworth" box diagrams. These will enable us to visualize the qualitative effects of monetized trade on such things as price formation, feasibility, Pareto optimality, and equilibrium.

It is worth emphasizing that we are presenting here not a single model with a fixed viewpoint on the nature of money but, rather, a general modeling approach, or framework, within which many different properties and users of money and financial institutions can be analyzed and compared. Our "money" can be a valuable, consumable commodity, or it can be merely paper, of no intrinsic worth. Even in a one-period model, various forms of credit can be introduced, and the trade-inhibiting effects of a limited money supply can be represented. In multiperiod models (not treated here, but see Sec. VIA below), one can go further and introduce lending institutions and derive interest rates. While we make no pretense that we are able to capture all of the complex factors associated with the use of money and financial institutions, it seems reasonable to hope that our approach, extended or elaborated in one direction or another, will prove useful as a basis for more extensive investigations of many facets of this subject.

II. Some Preliminaries on Economics and Modeling

Before proceeding to the description of a specific trading game, it may be useful to discuss some of the highlights of our approach in general terms. In particular, we would like to comment on (*a*) the differences between commodities, commodity money, and fiat money; (*b*) the institutional implications of cash and credit; and (*c*) the distinction between the (legal) rules of the game and the (behavioral) rules of play and the

[1] In his important early paper, Debreu (1952) represents the Walras exchange model as a game in strategic form for the technical purpose of applying a general existence theorem. But as a *descriptive* model his game shares the defect of the Walrasian model of being ill defined, or unrealistically defined, away from equilibrium. Indeed, if only one agent departs from equilibrium, he is presumed to be able to buy and sell at the stated prices, announced by an added fictive player whose objective is to minimize excess demand. But there is no explanation of how the excess demand thereby created is to be satisfied—unless it is out of the bottomless warehouses of the fictive player. See also Arrow and Debreu (1954) and Arrow and Hahn (1971, chap. 5).

reasons for our selection of the Nash noncooperative equilibrium for the analysis of the latter. In Section VI some other general points will be touched upon, in connection with possible variations and extensions of our model.

A. Money and Commodities

In this paper we are frankly concentrating on the "means of payment" role of money. We cannot contend that this is money's only or even its most important reason for being. But we do argue that this is its most conspicuous function in everyday economic life and that the payment *process* has hardly received a fair share of attention in modern mathematical economics. We shall therefore treat money as an element of strategy (in the game-theory sense) and not merely regard it as either an insubstantial price-reporting or book-balancing abstraction, on the one hand, or just another commodity to be traded and consumed, on the other.

Several authors, in the study of *cooperative*-game solutions to economic models, have had recourse to utility functions of the following special form:

$$U^i(x) = u^i(x_1^i, \ldots, x_m^i) + \lambda^i x_{m+1}^i,$$

where the $(m + 1)$th commodity may be regarded as a kind of transferable utility, or "u-money" (see, e.g., Shapley and Shubik 1966, pp. 807–8; 1969*a*, 1972, 1976; Telser 1972, pp. 4–11; and Aumann and Shapley 1974, p. 180). Although the vague phrase "market game with money" would seem to cover both this approach and our present *noncooperative* model, the connection is only superficial. In cooperative solutions such as the core and the value, the underlying strategic form of the game is irrelevant, and all attention is directed to the actions and potential actions of coalitions. The "u-money," if available, serves only as a vehicle for side payments that adjust the final distribution of utility among traders in coalition. There has been little interest in formulating *cooperative* trading games in which money plays a distinguished role in the trading technology; in fact, core theory (with good reason) has traditionally stood aloof from the processes of trade and price formation.[2]

In our present work, we do not assume that the utility of the payment commodity is additively separable, as above. But it is possible that such an assumption might simplify some of our results or proofs or ensure uniqueness or other good behavior on the part of the noncooperative

[2] Telser (1972, chap. 3), however, has analyzed a Cournot-type oligopoly model in terms of the core; see also Shapley and Shubik (1972) for a model whose core has a direct interpretation in terms of prices.

equilibria, just as it does for the classical competitive equilibria.[3] At least the question merits further study.

Viewed as a commodity, real or fictitious, money is distinguished in practice by its near-universal acceptability in exchange for other commodities.[4] The reasons for its acceptance lie as much in the realms of the laws and customs of society as in pure individualistic economic reasoning. In an economic model that is not intended to encompass social and legal developments, it may be taken as axiomatic that the monetary good will be accepted at face value according to the existing conventions of the marketplace, regardless of its intrinsic worth or lack of worth. In our present model, we shall *require* that all exchanges be for money. An interesting consequence of this rule is that the set of attainable redistributions of goods will fail to include many redistributions that would be possible if arbitrary, transaction-cost-free barter were allowed.[5] It should be noted that this curtailment also occurs under the classical rules of Walrasian exchange at stated prices. This is not necessarily unrealistic. A similar curtailment of the feasible set occurs in practice, we believe, in most societies that rely on organized markets and relatively stable price systems for the redistribution of goods.

Because of its general acceptability a means of payment has obvious value to its holder, but this value need not show itself in his utility function.[6] Indeed a *fiat money* would enter the utility function only through the effect of truncation: unspent money at the game's end may be presumed to have buying power in the world to come. (You *can* take it with you!) A *commodity money*, on the other hand, like gold or silver, has utility in its own right, to which we may add any extra buying power it may be considered to possess—an alternative or incremental value conferred by society's acceptance of its special monetary role.[7]

It is obvious that an adequate investigation of the utility of money demands a dynamic treatment in a multiperiod model, with the possibility of durable as well as perishable goods. We forgo any such in-

[3] With an additively separable "u-money" in sufficient supply, the competitive allocations of the other goods are just those that maximize the sum $\sum u^i(x)/\lambda^i$; thus a "fixed-point" situation reduces to a simpler maximization problem.

[4] Clower (1967) goes so far as to assert, as a matter of definition, that *any* commodity universally acceptable in exchange is a "money."

[5] See the end of Sec. IIC.

[6] Many economic investigations include a "money" in the utility functions; see, e.g., Patinkin (1956), Telser (1972 and elsewhere), or the discussion in Samuelson (1947, pp. 119–21).

[7] Such an adjustment for buying power may be called for even in the case of pure barter. A person's utility for a large quantity of wheat, for example, evaluated at a particular point in time, may depend in part on his expectation of being able to trade it later for something more directly useful to him. It is difficult to treat this subject with any precision in a static, one-period model.

vestigation in this paper, adopting, instead, the expedient of reserving a place in the utility functions for the "means of payment" without insisting that it actually be consumable or have utility. Within this framework we can encompass the range between a commodity money of great intrinsic worth, at one extreme, and a perishable fiat money, at the other.

B. Cash and Credit

Our point of departure will be a basic model where "cash" payments are required in advance on all purchases. We then explore some ways of relaxing this condition, via a sort of shopkeepers' credit in the form of deferred payments secured by expected receipts. Since only a one-period model is considered, interest rates and the money market cannot be fairly represented. Nevertheless, some institution-modeling problems do appear, since beyond a narrow zone of "conservative" credit (see Sec. IVB below) the possibility of insolvency and default exists and must be faced if the game is to be well defined. It may seem overelaborate to burden the abstract model with the details of a bankruptcy proceeding in which the assets of the trader who cannot pay his bills are liquidated to satisfy the creditors. Yet important properties of the game and its equilibria may well depend on just how such details are handled, once we relax the "cash on the barrelhead" rule.[8] One simple expedient is to allow negative holdings of "cash" at the game's end while postulating a disutility (suitably concave and continuous) for being "in the red." Implicitly there are loan sharks, swimming outside the model, who pay off the creditors while making life uncomfortable for the debtors, but the burden of providing detailed rules for insolvency is avoided.

Even without default there are institutional overtones in the granting of credit. Suppose a trader pays for some goods with promissory notes rather than cash, expecting to redeem them with cash received from the sale of his own goods. What if some of his customers also use promissory notes? When and how does the redistribution of the actual "means of payment" commodity take place? It appears that a central clearinghouse must be created. This would represent a major change in the nature of the model, which—as will be seen—is otherwise quite clearly decentralized, with respect to both traders and commodities.

While we cannot enter into a detailed discussion here, it is important to note also that credit generally involves a *contract* between two parties, whereas cash generally does not (except perhaps in the weak sense that an

[8] The rule proposed by Postlewaite and Schmeidler (1975), which in effect confiscates and destroys all goods belonging to the bankrupt party, seems both drastic and unrealistic (particularly away from equilibrium), since avoiding insolvency in that model depends not only on personal prudence but on correctly estimating the actions of the other traders.

individual holder of a dollar bill may regard himself as a creditor of the government). We may for convenience describe a person's indebtedness as "negative cash," but it is basically a different instrument. In a society where all pay cash, default and the laws and procedures for dealing with it need not be considered. A society with even the simplest forms of credit is fundamentally more complex than one without credit, and it can adequately be represented only by a fundamentally more complex model.

C. *Rules and Solutions*

The game theorist is usually at great pains to assure himself that the rules of the game are completely defined before he turns to the problem of solving it. The reason for this caution stems from the fact that, while the *descriptive theory*—covering the moves and strategies and information and payoffs, etc.—can be "hard" and mathematically precise, the solution theory is often "soft" and indeterminate, since it expresses the actions of sophisticated, free-willed decision makers. In the multiperson non-zero-sum games common in economics, this indeterminacy shows itself both in the multitude of different solution *concepts* that game theory offers for consideration—like core, value, bargaining set, noncooperative equilibrium, etc.— and in the nonuniqueness of outcome that is so often exhibited by the actual solutions, under any one of these concepts.

We would like to stress, therefore, that we are presenting a well-defined game in the descriptive sense, formulated independently of any assumptions of equilibrium or of what might or might not be "rational" behavior. When the players have made their individual decisions, the market prices and the transfers of goods are *completely determined.* To this game a variety of solution concepts might be applied, but the rules of the game and its solution are two different things.

Having said this, we must admit that we have built our model in a particular way (in a "strategic" form, as opposed, say, to a "coalitional" form) in anticipation of applying a particular solution concept—the "Nash" noncooperative equilibrium. In much of our previous work in this general area we have, instead, used the coalitional form and (with many others) have considered static, essentially noninstitutional games of exchange and/or production, establishing some interesting links between the competitive equilibrium and the core or other *cooperative* game concepts.[9] However, this kind of coalitional approach fails to capture the dynamics of price formation and the essentially individualistic nature of much economic decision making, and so further investigations using *noncooperative* game concepts now seem to be in order.

[9] See, e.g., Shapley and Shubik (1966, 1969*a*, 1969*b*, 1976) or the books of Arrow and Hahn (1971), Scarf (1973), Aumann and Shapley (1974), and Hildenbrand (1974), where many additional references will be found.

The classical model is content to take prices as though given by an "invisible hand," insensitive to the actions of the traders at least in the short run. There is an implicit understanding, not reflected in any mathematical assumption of the classical model, that the traders are so numerous and their individual resources so small that this insensitivity is a good approximation to reality. Our model, on the other hand, has prices that depend in a reasonable way on the individual trading decisions. They are driven upward by increased buying and downward by increased selling. This puts us into a position to examine the validity of the classical assumption of unyielding prices when the traders are *not* individually insignificant and, indeed, to explore the transition zone between perfect competition à la Walras and oligopolistic competition à la Cournot. But to carry out this program, our game-theoretic solution concept must be strategy oriented and collusion free, and the Nash noncooperative equilibrium is ideally suited to this purpose.

It may be of interest to point out in conclusion that if we were to reduce our present model to its characteristic function,[10] we would get *a different cooperative game* from the unrestricted-barter or "Edgeworth" game that is usually considered in connection with the core. The reason is that our rules cause all trade to use a single set of prices. If goods, or goods and money, pass between Traders 1 and 2 in a certain ratio, then they cannot pass between 1 and 3 in a different ratio. From the coalitional standpoint, this restriction on trade can lead to a paradoxical result—a failure of superadditivity. In fact, if we make the natural assumption that subeconomies can form and establish their own price systems, then an economy that is fractured into opposing coalitions may be able to reach allocations on the Pareto surface that cannot be achieved by the same economy united.[11]

III. The Basic Model

In order to describe a well-defined game of exchange in which a specific commodity is used as a means of payment, we must spell out how the prices are formed. The classic general equilibrium model is content to establish the existence of prices (often not unique) *at equilibrium*. For a proper game model, however, we need rules that determine the prices for

[10] The "characteristic function" of a game characterizes the outcomes that can be achieved by any subset of players. See von Neumann (1928), Aumann and Peleg (1960), Scarf (1967), Shapley and Shubik (1973), etc.

[11] It is not generally realized that there is a distinctive "Walras" cooperative game, with a more restrictive characteristic function than the Edgeworth game. Despite the possible failure of superadditivity, and hence of balancedness, the existence of the core in the Walras game is not threatened, since the competitive equilibrium must still have the core property. But the Walras core and Walras-Pareto set will in general be different from the Edgeworth core and Edgeworth-Pareto set. Superadditivity can be restored to the Walras game by the device of taking the "superadditive cover" (see Shapley and Shubik 1969*a*), but the result is still not the Edgeworth game.

positions of disequilibrium as well. With every player free to make an independent decision, the model must yield a well-defined output for every set of inputs.

Several types of price-forming mechanisms might be considered, each placing different restrictions on the strategic possibilities. In particular the traders might control only the quantities offered or demanded, or they might name reservation prices or price ranges or even complete demand curves for their individual transactions. Multistage bargaining might be introduced, or a centralized procedure that converts a set of unilateral price declarations into a unique, market-wide price for each good.[12] Here, as we are interested in general, anonymous exchange, with market-wide prices but with a minimum of ad hoc institutional detail, we adopt what is essentially a generalization of Cournot's original approach. The strategic variables will be *quantities*, not prices, but they will include quantities of the special good that serves as "cash" or "trading money." Indeed, in our simplest version, *only* that good will be subject to strategic choice.

Formally, in our prototype model we shall assume that there are n traders trading in $m + 1$ goods, where the $(m + 1)$th good has a special operational role in addition to its possible utility in consumption. We attribute to each trader an initial bundle of goods,[13]

$$a^i = (a_1^i, \ldots, a_m^i, a_{m+1}^i),$$

and a concave utility function,

$$u^i(x_1^i, \ldots, x_m^i, x_{m+1}^i).$$

We emphasize that u^i need not actually depend on x_{m+1}^i; the possibility of a fiat money is not excluded.

The general procedure will be for the traders to put up quantities of the first m goods to be sold and simultaneously to put up quantities of the $(m + 1)$th good to buy them, all at prices determined by the market-wide supply and demand for each good. For expository purposes, our prototype model will require the traders to offer for sale *all* of their holdings of the first m goods, though they need not spend all of their $(m + 1)$th good. A trader may well buy some of his original goods back, but they must go through the market. In other words, in this version of the model the trader does not own his initial bundle outright; he merely owns a *claim on the proceeds* when the bundle is sold.[14]

[12] Levitan and Shubik (1971, 1972) have given examples of different strategy spaces and Pareto surfaces that are possible in oligopolistic models.

[13] Superscripts will be used consistently to denote traders. Summation over traders will be denoted by a horizontal bar; thus $\bar{a}_j$ means

$$\sum_{i=1}^{n} a_j^i.$$

[14] This simplifying condition is not as unreasonable as it might appear at first glance. In a multiperiod context, it would amount to requiring that all "paper" profits or losses be realized every trading period. Some other modeling possibilities will be discussed in Sec. VI below.

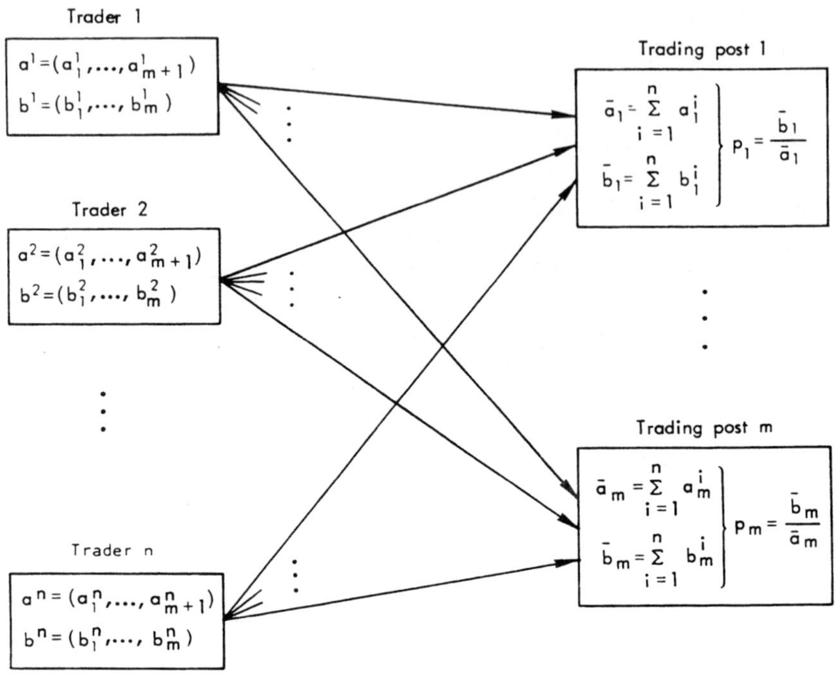

FIG. 1.—Decentralized trade

Let us imagine m separate *trading posts*, one for each of the first m commodities, where the total supplies $(\bar{a}_1, \ldots, \bar{a}_m)$, assumed to be positive, have been deposited for sale "on consignment." Figure 1 illustrates this. Each trader i makes bids by allocating amounts b^i_j of his $(m + 1)$th commodity among the m trading posts, $j = 1, \ldots, m$. We shall denote his *strategy*, in the game-theoretic sense, by the vector $b^i = (b^i_1, \ldots, b^i_m)$. There are a number of possible rules governing the permitted range of bids. In the simplest case, with no credit of any kind, the limits on b^i are given by

$$\sum_{j=1}^{m} b^i_j \leq a^i_{m+1},$$

and

$$b^i_j \geq 0, \qquad j = 1, \ldots, m.$$

The interpretation of this spending limit is that the traders are required to pay *cash in advance*. More generally, we might allow them to defer payment, either in anticipation of receipts or under some other credit arrangement that would have to be made explicit in the model.[15]

[15] See Sec. IVB below. The simplest way to model credit is to remove the spending limit entirely, while extending the domain of the utility function to include negative

A. Price Formation

The prices now emerge in a natural way as a result of the simultaneous bids of all buyers; we define

$$p_j = \bar{b}_j/\bar{a}_j, \qquad j = 1, \ldots, m.$$

Thus bids precede prices. Traders allocate their budgets *fiscally*, committing quantities of their means of payment to the purchase of each good without definite knowledge of what the per-unit price will be. At an equilibrium this will not matter, as prices will be what the traders expect them to be. In a multiperiod multitrader context, moreover, the traders will know the previous prices and may expect that fluctuations in individual behavior in a mass market will not change prices by much. But any deviation from expectations will result in changing the quantity of goods received rather than the quantity of cash spent. In practice, if one allocates a portion of one's budget for purchase of a certain good in a mass market, this will be different—but not too different—from a decision to buy a specific amount at an unspecified price. It is a matter of letting one's stomach rather than one's purse absorb the fluctuations.

The prices in our model are so determined that they will exactly balance the books at each trading post. Note that there is no need for a central clearinghouse and there are no interpost or intertrader transactions. The amount of the jth good that the ith trader receives in return for his bid b_j^i is

$$x_j^i = \begin{cases} b_j^i/p_j & \text{if } p_j > 0, \\ 0 & \text{if}^{16}\ p_j = 0. \end{cases}$$

His final amount of the $(m + 1)$th good, taking account of his sales as well as his purchases, is

$$x_{m+1}^i = a_{m+1}^i - \sum_{j=1}^{m} b_j^i + \sum_{j=1}^{m} a_j^i p_j.$$

His *payoff*, in the game-theory sense, must be expressed as a function of all the traders' strategies; accordingly, we write

$$\Pi^i(b^1, \ldots, b^i, \ldots, b^n) = u^i(x_1^i, \ldots, x_{m+1}^i).$$

amounts of the $(m + 1)$th commodity so as to provide a suitable *disutility* for being caught short of cash when the sales and purchases are all added up. To close the model, an outside source of cash would have to be postulated, to cover the payments due the other traders when one trader defaults. In a dynamic, multiperiod context, we can conceive of durable goods, other than the $(m + 1)$th "payment" commodity, that are carried forward and used as security for the granting of credit for commercial loans (Shubik 1977).

[16] Note that $p_j = 0$ implies $b_j^i = 0$. Thus a trader receives nothing if and only if he bids nothing.

Here the x's depend on the b's according to the three preceding displayed equations. It is noteworthy that the function Π^i is *concave* in b^i for each i; this is important for the existence proof.

Because of the mechanism for price formation and the anonymous allocation of sales, all traders pay the same price for the same good. However, the operation of the market system gives rise to what have been called *pecuniary externalities*, in the sense that the prices paid by one trader are dependent on the monetary actions of the others (Viner 1931; Shubik 1971). In the classic barter market this is not so; trading pairs or groups can form and exchange goods unaffected by the actions of others.

B. The Noncooperative Solution

A noncooperative or "best-response" equilibrium has been defined in game theory as a set of strategy choices by the players with the property that no player, given the choices of the others as fixed, can gain by changing his own choice.[17] Specialized to the present application, this solution concept will be recognized by economists as a close relative of the classic Cournot oligopoly solution (cf. Mañas 1972). It will consist of an n-tuple of strategies:

$$\hat{b} = (\hat{b}^1, \hat{b}^2, \ldots, \hat{b}^n)$$

such that for each i the function $\Pi^i(\hat{b}^1, \ldots, b^i, \ldots, \hat{b}^n)$, considered as a function of b^i alone, is maximized at $b^i = \hat{b}^i$.

The following existence theorem for the noncooperative equilibrium (NE) may be proved by an application of the Kakutani fixed-point theorem to the "best-response" adjustment process or transformation: $b \to b'$, with each b'^i being a best response by player i to the set of choices $\{b^j : j \neq i\}$. The details of the proof, which are rather lengthy, will be given elsewhere.

> *Theorem 1.* For each trader, $i = 1, \ldots, n$, let u^i be continuous, concave, and nondecreasing. For each good, $j = 1, \ldots, m$, let there be at least two traders with positive initial endowments of good $m + 1$ whose utility for good j is strictly increasing. Then a noncooperative equilibrium exists.

Note that there is no assumption that good $m + 1$ has intrinsic value to anyone. It must merely be available to enough people so that nontrivial markets for the other goods can be formed.

[17] The definition of Nash (1951), who, however, includes the possibility of mixed strategies. These have no plausible interpretation in our present model, so we shall be searching only for "Nash equilibria in pure strategies." The term "best-response equilibrium" is due to Robert Wilson.

Our reasons for concentrating on the noncooperative equilibrium rather than on one of the cooperative solutions of game theory have already been discussed in Section IIC. We would like merely to reiterate here that at least one purpose of our work is to formulate and study the *descriptive* game model, without reference to any particular solution concept. Indeed much of the discussion that follows concerns what are sometimes called "pre-solution" concepts like feasibility, Pareto optimality, or individual rationality, which do not depend on equilibrium notions at all.

IV. The Edgeworth Box

The case $m = 1$, $n = 2$ lends itself to simple two-dimensional descriptive analysis based on the familiar Edgeworth box. To avoid the confusion of too many lines and curves in one place, we shall make a sequence of diagrams, using the same labels as far as possible. As we shall see, much of this geometry will apply also to the general case, with many goods and traders, because of the way in which the operation of the system "de-couples" both the traders and the trading posts.

In figure 2, the first trader's holdings are measured up and to the right of the point O^1, while the second's are measured down and to the left of O^2. The dimensions of the box are $\bar{a}_2$ high by $\bar{a}_1$ wide. The point R represents a typical *initial allocation*; thus we have $a_1^1 = |M^1R|$, $a_2^1 = |G^1R|$, etc.[18] The points S^1 and S^2 represent typical *strategy choices*; thus we have $b_1^1 = |M^1S^1|$ and $b_1^2 = |M^2S^2|$. They are restricted to lie along the edges M^1O^1 and M^2O^2, respectively, since we are requiring all of good 1 to be sent to market. (If we did not make this restriction, the strategies would lie arbitrarily in the rectangles $RM^1O^1G^1$ and $RM^2O^2G^2$, respectively; see Sec. VI below.)

Now consider the line joining S^1 and S^2. Its slope is $(b_1^1 + b_1^2)/(a_1^1 + a_1^2)$, which is just the price p_1. Moreover, it divides the line M^1M^2 into two segments which are equal in length to the amounts x_1^1 and x_1^2 of good 1 purchased by Traders 1 and 2, respectively. Similarly, it divides the line G^1G^2 into segments equal to their final holdings x_2^1, x_2^2 of the monetary commodity, or "cash." We see therefore that the *final allocation* is represented by the point F. The vector RF represents the actual transaction that takes place; its slope, naturally, is equal (in absolute value) to the price p_1.

It is essential that the reader understand the nomogram demonstrated in figure 2, as it is the key to all the diagrams that follow.

Figure 3 shows the effect of holding S^2 fixed while varying S^1. As S^1 moves over the interval from M^1 to O^1, the point F traces out the curve A^1RB^1. This curve is a portion of a hyperbola whose asymptotes are the

[18] The notation $|PQ|$ indicates the length of segment PQ.

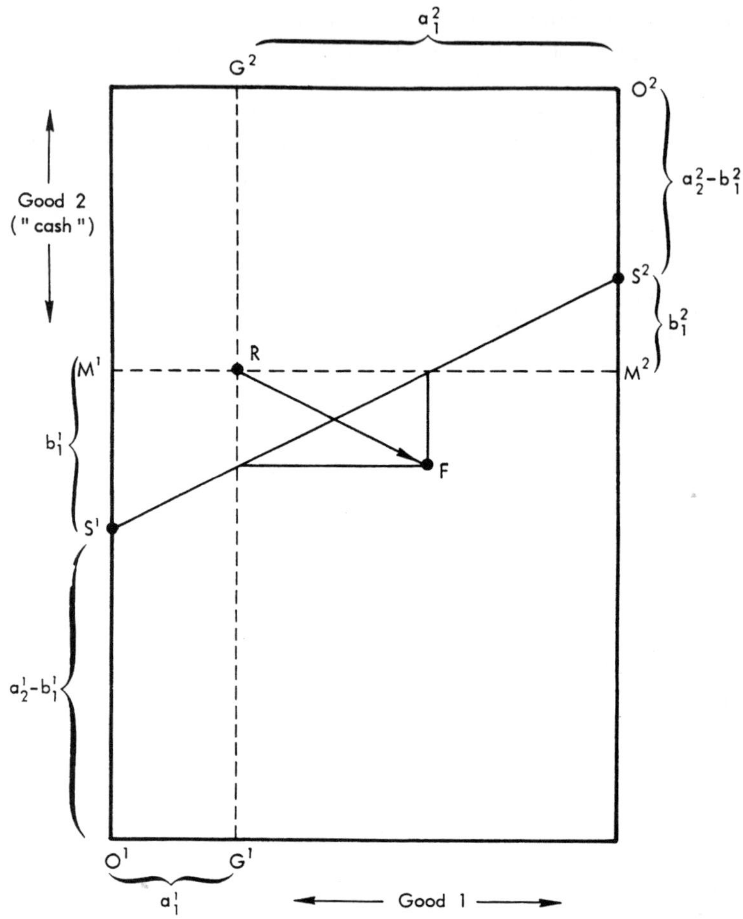

FIG. 2.—Price formation

horizontal and vertical lines through S^2. If we reverse the process and move S^2, holding S^1 fixed, we trace out the curve A^2B^2, which is part of a similar rectangular hyperbola centered at S^1. Endpoints A^1, A^2 correspond to zero bids, while endpoints B^1, B^2 reflect the upper limit on the amount a trader can bid. It happens in this case that we did not allow Trader 2 enough cash to be able to buy back his original holding when Trader 1 plays S^1, so the curve A^2B^2 stops short of R.

These traces are comparable to the price rays or "budget sets" that confront a trader in the classical Walrasian model with its fixed prices. The difference is that in the present case the price is not constant but reacts to variations in a trader's own decisions, so we get a curve instead of a line. The curve is concave, as one would expect, so that if a trader increases his purchase he drives the price up, while if he bids less the price

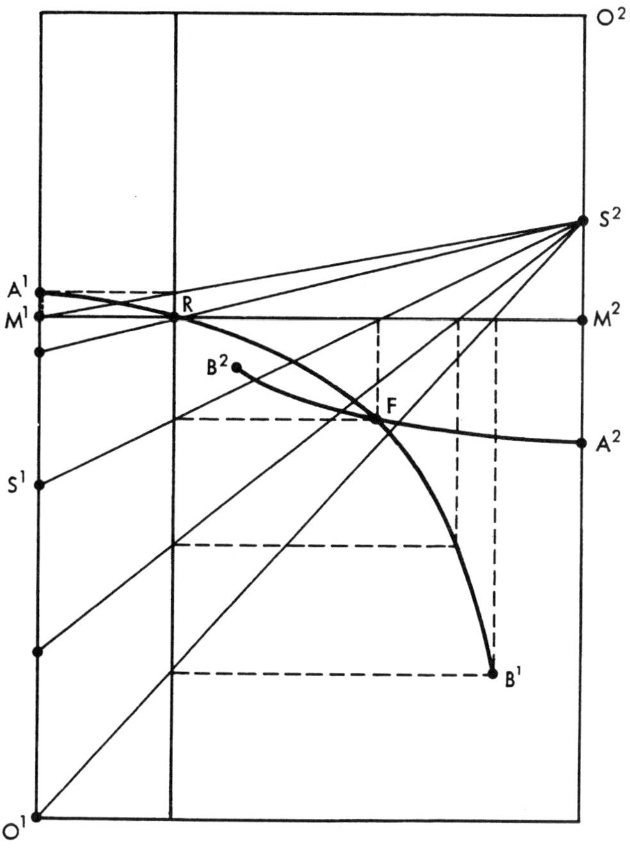

FIG. 3.—"Budget sets"

falls. The connection between the two approaches may be illustrated in figure 3 by supposing the second trader's holdings and bid to be very large, pushing points O^2 and S^2 far off the page. The first trader's hyperbolic "budget set" would then approximate a straight line through R and F, reflecting the fact that he now has little influence on the price.

A. Two Kinds of Equilibrium

So far we have discussed only the mechanics of the rules of exchange. We are now ready to add the traders' preferences to the picture, by superimposing the contours of the utility functions u^1 and u^2. As shown in figure 4, S^1 happens to be the *best response* to S^2, since A^1B^1 is tangent at F to the contour of u^1. (Note that the curvature of A^1B^1 is such that there is always a unique point of tangency.) Similarly, S^2 is the best response to S^1, since A^2B^2 is tangent at F to one of the contours of u^2. Thus figure 4

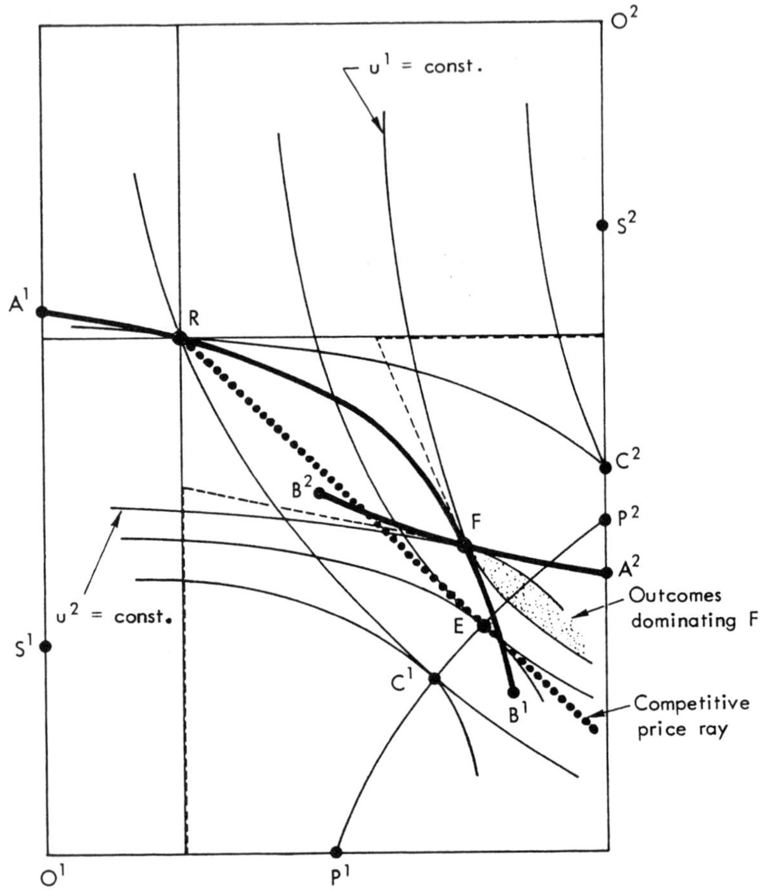

FIG. 4.—Noncooperative (F) and competitive (E) equilibrium

illustrates a noncooperative or "Nash" equilibrium (NE) for the market. Neither trader, knowing the strategy of the other, would wish to change.

A striking feature of this kind of equilibrium is its nonoptimality. Since curves A^1B^1 and A^2B^2 are not generally tangent to each other, the point F cannot be expected to be Pareto optimal or "efficient."[19] In effect, the traders are working with unequal *marginal* prices, represented by the unequal slopes of A^1B^1 and A^2B^2 at F. Any outcome in the shaded region would be preferred by both traders to the NE allocation at F. In particular, they would both profit from increased trade at the *average* price p_1, represented by the slope of RF.

The reader familiar with the Edgeworth diagram will recognize the

[19] Indeed Pareto optimality in a NE can only occur at corners of the indifference curves or in special cases like $F = R$ or $F = B^2$.

contract curve $C^1EP^2C^2$ in figure 4, which is a subset of the more extensive *Pareto set* $O^1P^1C^1EP^2C^2O^2$. The competitive equilibrium (CE) is represented by the competitive price ray RE, which is tangent to both indifference curves at the competitive allocation E. The situation illustrated seems to be typical: there is less volume of trade at the NE than at the CE. But the reverse is also possible. In fact, by a somewhat contorted but perfectly legitimate arrangement of the indifference contours, we could make any given point outside the dotted lines in figure 4 the location of a unique CE, while keeping a unique NE at F.

Although we have not illustrated it, it is also not difficult to construct utility functions for which several distinct strategy pairs (S_1, S_2) are in noncooperative equilibrium. This may or may not be accompanied by a corresponding multiplicity of CE; the two kinds of solution are not directly interlinked. But their general mathematical properties are quite similar. For example, we would expect that even for general m and n, if the utility functions are smooth and the initial allocation is chosen according to a nonatomic probability distribution, then with probability 1 there will be a finite, odd number of NE (cf. Debreu 1970).

B. *The Feasible Set and Credit*

The set of feasible outcomes can be determined by holding one trader's bid at its upper extreme O^i or lower extreme M^i and sweeping through the strategies of the other trader. The feasible set is shown in figure 5 (solid shading); for example, outcome L results if both traders bid their upper limit.[20] This feasible set has nothing to do with rationality or motivation; it merely describes those outcomes that the mechanism *can* be made to produce. It necessarily contains all NE allocations, but, as for the CE allocations, we know a priori only that they lie in the quadrants of the Edgeworth box "northwest" or "southeast" of the point R. It may well be that some or all of the CE allocations fall outside of the feasible set, under our present trading rules.

By introducing *credit* we can enlarge the feasible set, permitting the traders to bid more of the "means of payment" commodity than they actually possess as cash on hand. In diagrammatic terms, the first trader, say, could be allowed the segment C^1M^1 instead of O^1M^1 (fig. 5). The

[20] Most boundary points are attainable, but those on the line M^1M^2 are not, except for R itself. There is, however, an exceptional "null" outcome, not in the Edgeworth box, in which Trader 1 bids and gets M^1 and Trader 2 bids and gets M^2. Our rule is that if a trader bids nothing he gets nothing, even if the price is zero. So if both traders bid nothing, the entire stock of good 1 is lost. (We may imagine that it goes to an otherwise unnoticed "scavenger," who sends infinitesimal bids to each trading post in the hope of making a killing.) This exceptional outcome is manifestly undesirable and unstable, and it has no real effect on the solutions of the game, though it causes technical difficulties in the existence proof.

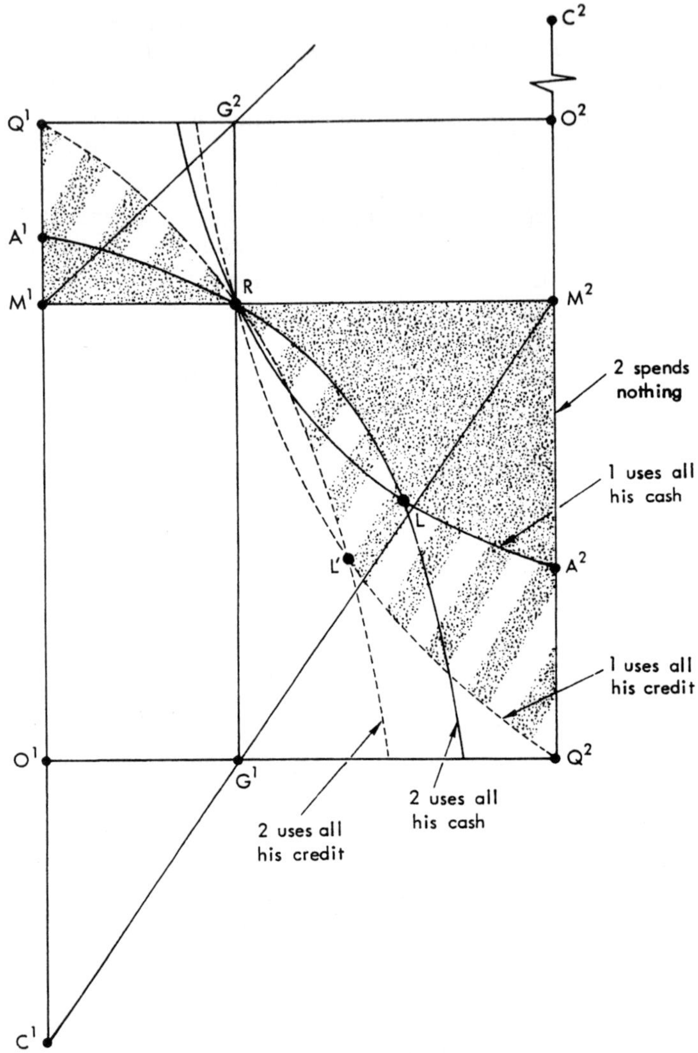

FIG. 5.—The feasible set with and without credit

length $|O^1C^1|$ has a simple interpretation in terms of banking: it is the largest amount of good 2 that could be loaned to Trader 1 with the certainty that he will be able to repay. To see this, consider that the worst case is $S^1 = C^1, S^2 = M^2$, with Trader 1 bidding the limit and Trader 2 bidding nothing. This leads to outcome Q^2, at which Trader 1 is just barely solvent. A similar "conservative" credit limit C^2 for Trader 2 can be determined in the same way.[21]

[21] It will become clear (Secs. IVD and VA) that when there are many traders the amount of "conservative" credit is likely to be very small, since it depends on the amount

If we alter the rules of the game to enable both traders to use this conservative banking credit, the feasible set is extended as indicated by the stripes in figure 5. For example, outcome L' results if both traders bid up to their new limits. The new feasible set still does not fill up the two basic quadrants, however, and so we still cannot be sure that any competitive outcomes will be attainable.

With more liberal credit, the feasible set would cover a larger part of the two basic rectangles and eventually all of them, but it would also include areas above and below the Edgeworth box, representing situations in which one or the other trader is "caught short." If we regard credit as the issue of a financial instrument, the interpretation of a point outside the box is simple and familiar—it amounts to stating that after trade an individual ends up with none of the monetary commodity in hand and with outstanding claims that he cannot meet. In order to complete the model we must either describe the utility to any player at such an outcome or add further moves to the game, corresponding to bankruptcy proceedings (see Sec. IIB).

C. *Individual Rationality*

In the present version of our model we are requiring that all of the initial endowments, except for the "payment" commodity, be put up for sale.[22] Consequently the concept of ownership is somewhat different from that usually assumed in microeconomics and is closer to that in law. All goods are monetized, and trade is virtually anonymous. The economy is an accountant's dream, but by forcing goods to pass through the market ownership rights are weakened. In particular, it may be impossible for a person to recover the bundle he originally started with or obtain an equivalent bundle. On the other hand, he may be able unilaterally to guarantee an outcome superior to his initial holding. (This contrasts with the classical Edgeworth and modern "core" models, where an individual trader can always defend his initial holdings if he so desires, but no more.)

The "defensive" possibilities are easily illustrated. In figure 6 the first trader is fortunate, since he has relatively little of good 1 that must go to market and relatively much of good 2 that he can use at his discretion. He cannot, of course, protect his exact initial holding, since this would mean being able to enforce outcome R regardless of the other's choice of S^2. But he can force the outcome to *have as much utility* for him as R. In fact, by playing S^1 as shown, he restricts the possible outcomes to the set A^2B^2, a set which lies entirely on the "high" side of the indifference

---

of a trader's own money that will surely come back to him from his own bids. In this respect our two-trader exposition may be misleading. We may also remark that "conservative" credit does not require the aid of a central clearinghouse (see Sec. IIB), as any promissory notes can be redeemed at each trading post independently.

[22] But see Sec. VI, where this requirement is removed.

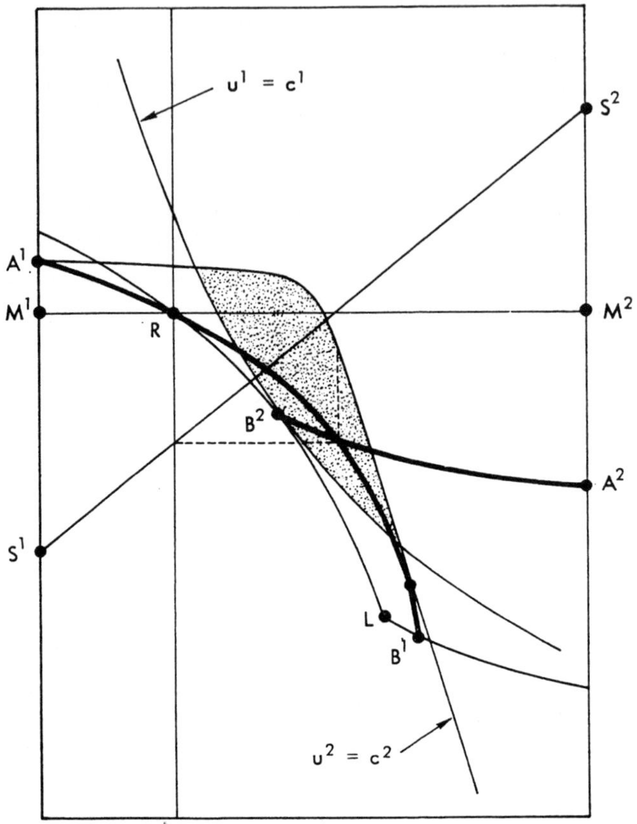

FIG. 6.—"Max-min" strategies and the individual rationality zone

curve $u^1 = c^1$, as drawn. This is in fact the best he can do defensively: S^1 is what would be called his "max-min" strategy, and c^1 is his "max-min" payoff. We see that moving S^1 either up or down would move B^2 into a region of lower payoff, either toward R or toward L.

The second trader, on the other hand, has a relatively poor defensive position. The highest u^2 indifference curve that completely contains one of the A^1B^1 sets is the curve $u^2 = c^2$, which is distinctly inferior to R for him. There are two critical points: the endpoint A^1 and the tangency point near B^1. As S^2 moves, the A^1B^1 curve rotates on R. Raising S^2 would make A^1 worse, while lowering S^2 would make the critical point near B^1 worse, so S^2 is the "max-min" strategy.

The shaded area in figure 6 represents the so-called individually rational allocations, bounded by the curves $u^1 = c^1$ and $u^2 = c^2$. The *feasible* individually rational allocations are just those that lie beneath the horizontal line RM^2, and it is a theorem that every NE solution will lie in that region. In contrast, the Walras-Edgeworth individually rational zone, which contains every CE solution, would be defined by the two indifference curves through R (not shown). It is easy to see that this region could be entirely disjoint from the other. This is not too surprising, as the two regions arise from games with the same economic data but different rules of trade.

D. *Many Traders*

The same diagrams can be used, one-sidedly, when there are more than two traders. This is because our model has the *aggregation* property, with both strategies and outcomes being additive over traders. Viewed by the other traders, a set or coalition of traders acting together is hardly distinguishable from a single, larger trader. In the diagrams, we may regard the point S^2 not as a single bid but as the vector sum of bids by Traders 2, 3, . . . , n, using R as the "origin" from which the vectors are defined. The first trader may notice a quantitative but not a qualitative difference from the two-person case: since the other traders combined may have far more resources than he does alone, the distance of S^2 from R and S^1 may make the slope of S^1S^2 (i.e., the price) almost insensitive to his own choice.

It might appear at first glance that the *feasible set*, when there are many traders, would continue to be a sizable fraction of the set of outcomes that are feasible under unrestricted, Edgeworthian barter. But this is not the case. Let us count dimensions. The unrestricted allocation space has dimension $(m + 1)(n - 1)$.[23] Even in the case $m = 1$, there are $2n - 2$ dimensions of possible outcomes if trade is not required to pass through the "trading post" mechanism. This compares with the n dimensions at most that can arise when n traders bid by each selecting a point on a line as his strategy. Thus, starting at $m = 1$, $n = 3$, the feasible set is only a lower-dimensional surface or manifold in the set of outcomes that would be possible under unrestricted trading.

E. *Many Commodities*

Since the trading posts operate essentially independently of each other, it is possible to continue to use these diagrams when there are more than two commodities (i.e., more than one trading post). But there are two ways in which the markets remain intercoupled: (1) through the spending limits, which apply to all bids combined, so that a trader's upper bound at each trading post will depend on what he spends at the others, and (2) through the utility functions, which in general reflect complementaries,

[23] The "+ 1" here is for the monetary good, while the "− 1" reflects the fact that the sum of the holdings is constant.

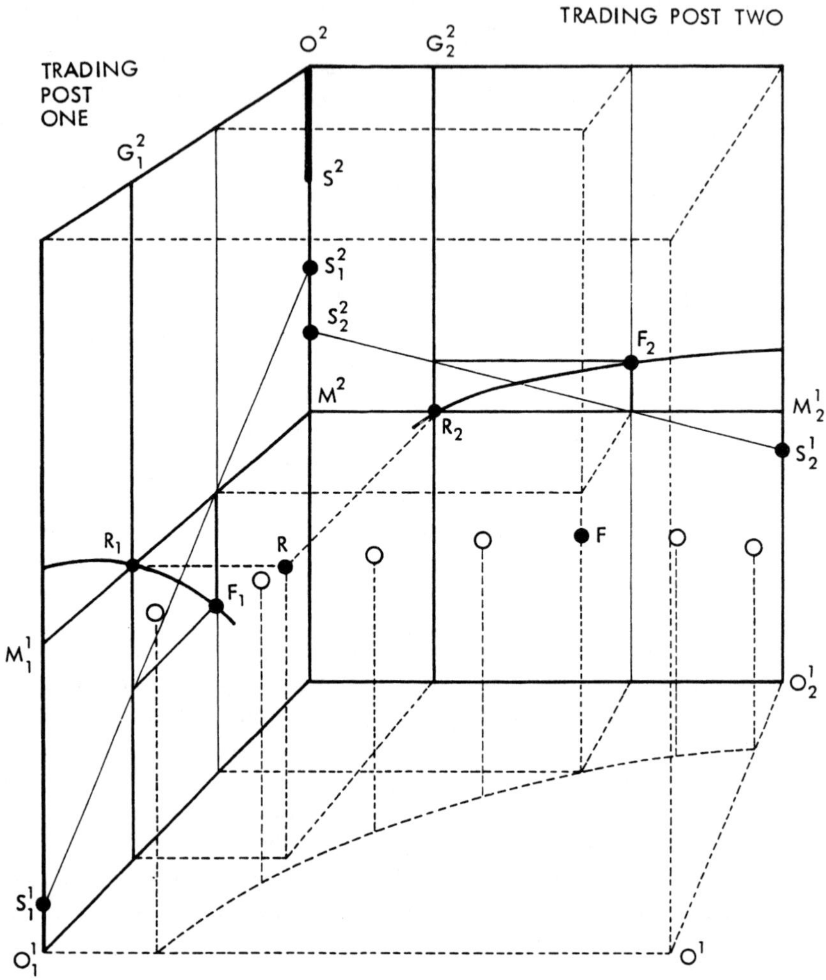

Fig. 7.—Two trading posts

substitution effects, etc., among the different goods, so that the utility maps we superimpose on each Edgeworth box will depend on what is happening in the others. The second effect (as in the classical model) can only be visualized with hyperdimensional eyesight, but the first is still within reach of our diagrams.

Figure 7 illustrates a three-commodity situation. The two "boxes" are erected at right angles to each other, in the back of the three-dimensional figure (solid lines). The initial point, R, projects to R_1 in the first box and R_2 in the second.[24] We may conveniently regard R as the

[24] The labels are generally consistent with the previous diagrams, but with subscripts added to distinguish the two trading posts.

vector sum of M^2R_1 and M^2R_2. Trader 2's pair of bids b_1^2 and b_2^2 are shown at S_1^2 and S_2^2 and their sum at S^2. Thus $|M^2S^2| = |M^2S_1^2| + |M^2S_2^2|$. The unspent balance is represented by the vertical bar O^2S^2.

Trader 1's bids are shown at S_1^1 and S_2^1. As S_1^1 varies over the permitted range $O_1^1M_1^1$, the outcome in the first marketplace sweeps out the curve through R_1 and F_1 in the usual way. Similarly, the curve through R_2 and F_2 describes his possibilities in the second marketplace. But his joint choice of S_1^1, S_2^1 is restricted by the spending limit

$$|M_1^1S_1^1| + |M_2^1S_2^1| \leq |M_1^1O_1^1| = |M_2^1O_2^1|.$$

The case actually illustrated has him spending most of his cash on good 1 and the rest on good 2. (Had he held some back, we would have shown it by a vertical bar at O^1.) The resulting final allocation is found by taking the vector sum of M^2F_1 and M^2F_2; this yields the point F.

Now suppose that Trader 1 changes his allocation of cash between goods 1 and 2, causing points S_1^1 and S_2^1 to move in opposite directions. The point F will then trace out a curve in space, which we have tried to suggest by the open dots perched above the curve's projection on the base. This projection on the base shows the tradeoff between goods 1 and 2 when a fixed amount of cash is bid by Trader 1; note that it is concave to his origin O^1. If Trader 1 bids less than the maximum allowed (strict inequality above), a curved triangular surface is generated, extending out from F and the open dots toward the viewer and rising to an apex directly above O^1. This locus (not shown) indicates ways in which Trader 1 can obtain more of good 3 at the expense of goods 1 and/or 2; it is concave to the origin O^1, and its projection is the large, roughly triangular region in the base of the diagram. The first trader's best response to the other trader's given strategy (S_1^2, S_2^2) is determined by the relationship of this surface to the indifference surfaces of u^1, in the three-dimensional commodity space with origin at O^1.

V. Replication

Inquiries into the behavior of economic models with large numbers of participants often make sweeping assumptions of symmetry, in the hope of keeping the models mathematically tractable and easy to visualize, while capturing at least some of the characteristic effects of large numbers. A favorite technique, involving a high but not total degree of symmetry, is called "replication." Imagine a basic economic system juxtaposed to a large number of identical replicas of itself; then take away all barriers, and form a "common market." Equivalently, assume that all traders in the full model are drawn from a small number of *types*, with an equal number of individuals of each type. Traders of the same type have identical endowments and identical tastes, but they are not constrained to act alike. That is, they are not programmed robots or members of a bloc or cartel but independent decision makers.[25]

The number of members of a type—the "replication number"—provides the modeler with a simple size parameter that he can vary without calling for additional data. Of course we do not intend to treat replication as though it were some kind of actual economic event, like a homogeneous population increase or an accretion of similar countries to a common market. Replication should be considered only as a technical device of comparative statics. As such, however, it has repeatedly proved its worth in explorations of the size effect (see Edgeworth 1881; Shubik 1959, 1968; Debreu and Scarf 1963; Shapley and Shubik 1966, 1969*b*; Debreu 1975; Owen 1975; Shapley 1975; etc.).

A. A Simple Case

The effect of replication on our model is illustrated in figure 8. We have $m = 1$ and $n = 2k$, where k is the replication number. The point S^1 represents the *average* of the bids of the k traders of type 1; similarly S^2 for type 2.[26] The point F therefore indicates the average final bundles for the two types.

Let us focus on a typical member of type 1. His own bid might be, say, at $\tilde{S}^1$. To discover his personal final bundle, $\tilde{F}$, we must draw the line through $\tilde{S}^1$ that is *parallel* to S^1S^2; we then locate $\tilde{F}$ in the usual way. Of course $\tilde{F}$ necessarily falls on the straight line through R and F, since all transactions take place at the same price.

Now suppose our trader changes his bid from $\tilde{S}^1$ to M^1. That is, he decides to buy nothing. This depresses the average bid for type 1, moving it from S^1 to T^1, where $|S^1T^1|/|\tilde{S}^1M^1| = 1/k$. The new price is the slope of T^1S^2, so our trader's new final bundle is $\tilde{A}^1$, determined by the line through M^1 parallel to T^1S^2. Similarly, at the other extreme, if he decides to bid the limit, the result is $\tilde{B}^1$, determined by the line through O^1 parallel to U^1S^2, where $|U^1S^1|/|\tilde{S}^1O^1| = 1/k$. We should not be surprised to see $\tilde{B}^1$ fall outside the box. This means merely that our trader happens to have enough "cash" to buy more than the combined initial bundles of one trader of each type.

The locus of possible final bundles for our variable trader is the curve $\tilde{A}^1R\tilde{F}\tilde{B}^1$, shown in figure 8 for the case $k = 4$. The flattening of the curve, as compared with the analogous curve A^1RFB^1 in figure 3 (which is the case $k = 1$), clearly shows the diminished influence of a single individual on price after replication. Indeed the price variation in the present case

[25] This distinction can be very important in the game-theory approach but is less important in a behavioristic theory. As we shall see, replication changes the NE but not the CE.

[26] More generally, if there were many types, then the analogue of S^2 would be not the average but $1/k$ times the sum of all bids from traders not of type 1.

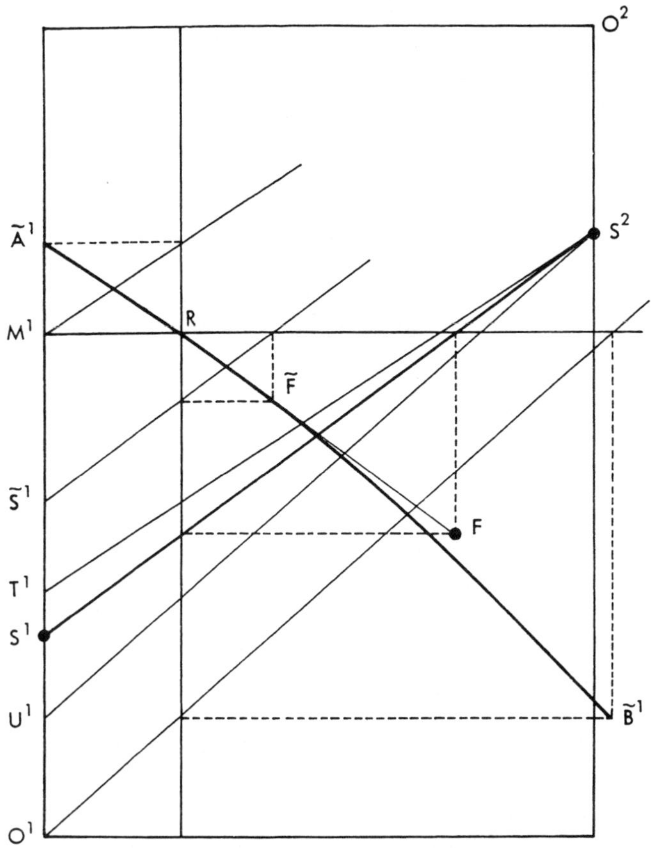

FIG. 8.—Price formation in a replicated market

is encompassed by the thin pencil of sloping lines within the angle $U^1S^2T^1$. In the limit, as $k \to \infty$, the curve becomes the fixed-price "budget set" of the classical CE model—but with one important difference: it is truncated at the $\tilde{B}^1$ endpoint instead of extending to the horizontal axis through O^1, in recognition of the limited quantity of the payment commodity available.[27]

B. *The General Case*

For the replicated form of the general model presented in Section II, we consider a market with T types of traders and k of each type; thus $n = kT$. The initial holding of good j by trader s of type t will be denoted

[27] It may be observed that the "conservative" line of credit for an individual shrinks to zero as $k \to \infty$. A simple calculation shows in fact that, in the present case, the amount of such credit is equal to $a_2^1 a_2^2/[k(a_2^1 + a_2^2) - a_2^1]$.

a_j^{ts}, which we can shorten to a_j^t since it does not depend on s. The total amount of good j (summed over *both* t and s) will be denoted $\bar{a}_j$, as before. A typical strategy will be denoted $b^{ts} = (b_1^{ts}, \ldots, b_m^{ts})$. The defining equations of Section II then become

$$p_j = \bar{b}_j/\bar{a}_j, \qquad j = 1, \ldots, m;$$

$$x_j^{ts} = b_j^{ts}/p_j, \qquad j = 1, \ldots, m \text{ and } p_j \neq 0;$$

$$x_j^{ts} = 0 \qquad j = 1, \ldots, m \text{ and } p_j = 0;$$

$$x_{m+1}^{ts} = a_{m+1}^{ts} - \sum_{j=1}^{m} b_j^{ts} + \sum_{j=1}^{m} a_j^{ts} p_j;$$

and

$$\Pi^{ts}(b^{11}, \ldots, b^{1k}, b^{21}, \ldots, \ldots, b^{nk}) = u^t(x_1^{ts}, \ldots, x_{m+1}^{ts}),$$

where $u^t(x) \equiv u^{ts}(x)$ is the common utility function for members of type t.

Suppose that a NE has been found in which all traders of the same type make the same bids:[28] $\hat{b}^{ts} = \hat{b}^t$, $s = 1, \ldots, k$. It is instructive to look at the equilibrium conditions for the "interior" case, that is, the case where all $\hat{b}_j^t$ are positive and $\sum_{j=1}^m \hat{b}_j^t < a_{m+1}^t$, $t = 1, \ldots, T$. Assuming that the concave functions $u^t(x)$ are differentiable, we let $u_j^t(x)$ denote their partial derivatives with respect to x_j. Then setting $du^t/db_j^{ts} = 0$ gives us

$$u_j^t(x^{ts}) \left\{ \frac{1}{p_j} - \frac{b_j^{ts}}{(p_j)^2 \bar{a}_j} \right\} = u_{m+1}^t(x^{ts}) \left\{ 1 - \frac{a_j^{ts}}{\bar{a}_j} \right\},$$

which must hold at $b = \hat{b}$. Since prices are in terms of good $m + 1$ as numeraire, we can introduce $p_{m+1} \equiv 1$ and rewrite the condition for a symmetric, interior NE as follows:

$$\frac{u_j^t(\hat{x}^t)}{\hat{p}_j} \left\{ 1 - \frac{\hat{b}_j^t}{\bar{\hat{b}}_j} \right\} = \frac{u_{m+1}^t(\hat{x}^t)}{\hat{p}_{m+1}} \left\{ 1 - \frac{a_j^t}{\bar{a}_j} \right\}, \qquad j = 1, \ldots, m,$$

where $\hat{p}$ and $\hat{x}$ are the prices and allocation corresponding to $\hat{b}$.

The expressions in curly brackets reveal the effect of oligopoly, but note that they lie between 1 and $1 - 1/k$. In the limit, therefore, these conditions for NE reduce to the conditions for CE, with the prices of the goods proportional to the marginal utilities of every trader. Thus we may state a convergence theorem, as follows:[29]

> *Theorem 2.* Assume that for infinitely many values of k the market has a symmetric, interior NE, and let $\hat{p}^{(k)}$ be the corresponding m-vector of prices. Let $\tilde{p}$ be any limit point of the $\hat{p}^{(k)}$,

[28] The conditions given in Theorem 1 in Sec. IIIB are sufficient to guarantee the existence of such a symmetric NE. It is quite possible also to have nonsymmetric NEs.

[29] A somewhat more general form of this theorem will be found in Shapley (1976).

and define $\tilde{p}_{m+1} = 1$. Then the $m + 1$ prices $\tilde{p}_1, \ldots, \tilde{p}_m, \tilde{p}_{m+1}$ will be competitive for the market (for any value of k); that is, an allocation x will exist for which $\bar{\tilde{x}} = \bar{a}$ and, for each s and t, $\tilde{x}^{ts}$ maximizes $u^t(x^{ts})$ subject to $x^{ts} \geq 0$ and

$$\sum_{j=1}^{m+1} \tilde{p}_j(x_j^{ts} - a_j^{ts}) = 0.$$

It should be noted that the NE approaches the CE "from below," that is, through outcomes that are not in general Pareto optimal. This contrasts with the convergence of cooperative solutions like the core and the value, which are by definition Pareto optimal all the way (see, e.g., Debreu and Scarf 1963; Shapley and Shubik 1969*b*; Debreu 1975; Owen 1975; and Shapley 1975).

The type of convergence revealed in Theorem 2 depends crucially on having interior solutions, with no trader up against his spending limit. All must have enough cash (or credit, in an extended model), where the meaning of "enough" depends on the payment commodity's marginal utility relative to that of the other goods. Of course there can never be "enough" fiat money in this sense (in a one-period model), for the NE will always have everyone spending the limit and wishing he could spend more. If the CE is considered socially desirable (as one road to the Pareto optimum, if for no other reason), then a society whose trading system resembles our replicated model should make sure that the means of payment is something that is both widely available and generally desirable, either as a consumption item or as a means of payment in future periods.

VI. Variants and Extension

It is not difficult to modify the basic game so that the goods do not necessarily all pass through the market before consumption. A number of different considerations and problems arise, however, depending on just how this is done. We shall review some of them briefly here; for further details see Shapley (1976).

Suppose, first, that a trader can hold back any part of his initial endowment. A strategy for i is now a *pair* of m-dimensional vectors (b^i, q^i), with the b_j^i representing cash bids, as before, and the q_j^i representing quantities of goods sent to the respective trading posts. Thus the amounts $(a_j^i - q_j^i)$ are held back. Of course we require that

$$0 \leq q_j^i \leq a_j^i, \qquad j = 1, \ldots, m.$$

The prices are given by

$$p_j = \bar{b}_j / \bar{q}_j \qquad (\text{if } \bar{q}_j > 0),$$

and we add the convention that if $\bar{q}_j = 0$ any cash sent to the jth trading post is lost. Sales and purchases are calculated as before, but we must remember to add back the amounts $(a_j^i - q_j^i)$ before entering the utility function.

The nomogram in the Edgeworth box works essentially as before, except that the domain of strategies S^i is now the full rectangle $M^iRG^iO^i$ instead of the line segment M^iO^i (see fig. 2). It is easy to see that all outcomes in the northwest and southeast quadrants can now be reached without the aid of credit. But when there are more than two traders, we find the feasible set still sharply restricted by the requirement of trading at fixed prices, as discussed previously in Sections IIC and IVD.

Despite the naturalness of the "hold-back" option, it has one major disadvantage. It turns out that the enlarged strategy spaces cause the conditions that define the noncooperative equilibrium to be *under*determined, making the set of solutions usually infinite.[30] This circumstance leads us to search for ways to sharpen either the solution concept or the rules of the game, without entirely giving up the hold-back feature. The simplest expedient is to prohibit a trader from simultaneously buying *and* selling at the same trading post, by imposing the condition

$$b_j^i q_j^i = 0, \qquad j = 1, \ldots, m.$$

The strategy domains at each trading post are now L-shaped, with a corner at the initial point R. Despite this corner, a trader in an active trading post should not experience any discontinuity in changing roles from buyer to seller, as his budget set is still the familiar, smooth hyperbola through R, as long as there are other traders on both sides of the market. The two-person case is rather trivial, however, there being no meaningful competition to motivate price formation.

This modification does restore some semblance of uniqueness to the NE, but it has its own drawback: the game is *not additive* over players. A coalition has options not available to an individual. Thus, if a trader could split himself into two legal persons, he could buy and sell simultaneously at the same trading post (e.g., in an attempt to stabilize the price). This suggests that the restriction $b_j^i q_j^i = 0$ might be unrealistic in some applications.

Any version of the rules that permits traders to stay out of a market has another peculiarity, namely, the possibility of trading posts that are

[30] This multiplicity is not merely a matter of simultaneous buying and selling by the same trader that could simply be canceled out. Indeed, if at equilibrium Trader i is sending both goods and cash to the same trading post and if the price there is p_j, then he might consider decreasing both q_j^i and b_j^i, in the ratio of 1 to p_j. This would not change his final outcome (assuming that he does not spend the extra cash elsewhere) or the price p_j, but it would change the marginal cost of good j to the other traders and so destroy the equilibrium.

completely inactive. Such a situation is very stable, as no trader would want to enter a market where $\bar{b}_j = \bar{q}_j = 0$. He would either lose his goods or money or, at best, have them returned to him to no advantage. Thus the "null" strategy for all traders is always an equilibrium point, in both of the models above. Moreover, we can arbitrarily declare any subset of trading posts to be inactive and solve the remaining subeconomy —any NE of that subeconomy will be also a NE of the economy as a whole.

We should not completely rule out such inactive or partially inactive solutions as unrealistic on their face. It can plausibly be argued that the real world is full of "latent" markets awaiting discovery which can only become active by an act of faith on the part of the first entrants. But we are interested in active NE as well, and the possibility of inactive ones makes the main existence theorem, comparable to Theorem 1 above, considerably more delicate both to state and to prove. Sometimes markets are "legitimately" inactive, as when the traders start with a Pareto-optimal distribution. The resolution of the problem seems to lie in demanding that inactive trading posts have "virtual prices" at which all traders find it in their best interest neither to buy nor to sell (Shapley 1976).

A. Multiperiod Extensions

Many applications of our general approach can benefit from—and some will require—the adoption of a multiperiod framework. Specifically, we may mention the study of (*a*) credit and bankruptcy, (*b*) nonsymmetric information conditions, (*c*) uncertainty and insurance, (*d*) cyclical variation of endowments and the money market, (*e*) the derived utility of fiat money, (*f*) interest rates and inflation, and (*g*) the role of capital goods and ownership shares. Shubik and others have devised exploratory game models for most of these situations and have worked out a number of tutorial examples (see Shubik 1972, 1973, 1977; Shubik and Whitt 1973; and Dubey and Shubik 1976). There remains, however, much more that can and should be done.

There remain also some serious conceptual problems with the general mathematical theory for multistage models of this type. One kind of difficulty is already apparent if we try merely to extend our basic prototype, without added features, to two or more periods. We should first recall that a vital ingredient in the proof of Theorem 1 (the existence theorem) is the *concavity* of the payoff functions $\Pi^i(b^1, \ldots, b^n)$, with respect to their respective b^i. The natural way to attack the two-period case is to solve the second stage parametrically, as a function of the cash distribution at the end of the first stage. But then the traders' total payoffs will include a term that depends on this cash distribution, and there seems no way to assure that this term will be concave.

VII. Conclusion

In this paper, in an admittedly simplified and abstract setting, we have explored some elementary implications of an explicit market mechanism for the formation of price. We believe that this species of game-theoretic model, by being well defined independently of equilibrium conditions or behavioral assumptions (though capable of accommodating such conditions and assumptions) and by reflecting the decentralized decision making attainable through the use of a tangible money, is far more flexible as an investigatory and explanatory tool than the usual Walrasian model, with its ill-defined causal linkage between individual actions and the action of the system as a whole.

References

Arrow, K. J., and Debreu, G. "Existence of an Equilibrium for a Competitive Economy." *Econometrica* 22 (July 1954): 265–90.

Arrow, K. J., and Hahn, F. H. *General Competitive Analysis*. San Francisco: Holden Day, 1971.

Aumann, R. J., and Peleg, B. "Von Neumann–Morgenstern Solutions to Cooperative Games without Side Payments." *Bull. American Math. Soc.* 66 (May 1960): 173–79.

Aumann, R. J., and Shapley, L. S. *Values of Non-atomic Games*. Princeton, N.J.: Princeton Univ. Press, 1974.

Clower, R. "A Reconsideration of the Microfoundations of Monetary Theory." *Western Econ. J.* 5 (March 1967): 1–9.

Cournot, A. A. "Recherches sur les principes mathématiques de la théorie des richesses." Paris: Hachette, 1838. Translated by N. T. Bacon. New York: Macmillan, 1897.

Debreu, G. "A Social Equilibrium Existence Theorem." *Proc. Nat. Acad. Sci.* 38 (October 1952): 886–93.

———. "Economies with a Finite Set of Equilibria." *Econometrica* 38 (May 1970): 387–92.

———. "The Rate of Convergence of an Economy." *J. Math. Econ.* 2 (March 1975): 1–7.

Debreu, G., and Scarf, H. E. "A Limit Theorem on the Core of an Economy." *Internat. Econ. Rev.* 4 (September 1963): 253–46.

Dubey, P., and Shubik, M. "Trade and Prices in a Closed Economy with Exogenous Uncertainty, Different Levels of Information, Money and Special Futures Markets." Cowles Foundation Discussion Paper no. 410, Yale Univ., January 1976.

Edgeworth, F. Y. *Mathematical Psychics: An Essay on the Application of Mathematics to the Moral Sciences*. London: Kegan Paul, 1881.

Hildenbrand, W. *Core and Equilibria of a Large Economy*. Princeton, N.J.: Princeton Univ. Press, 1974.

Jones, R. A. "The Origin and Development of Media of Exchange." *J.P.E.* 84, no. 4, pt. 1 (August 1976): 757–75.

Levitan, R. E., and Shubik, M. "Noncooperative Equilibria and Strategy Spaces in an Oligopolistic Market." In *Differential Games and Related Topics*, edited by H. Kuhn and G. Szegö. Amsterdam: North-Holland, 1971.

———. "Price Duopoly and Capacity Constraints." *Internat. Econ. Rev.* 13 (February 1972): 111–22.

Mañas, M. "A Linear Oligopoly Game." *Econometrica* 40 (September 1972): 917–22.

Nash, J. F., Jr. "Noncooperative Games." *Ann. Math.* 54 (September 1951): 286–95.

Ostroy, J. M., and Starr, R. M. "Money and the Decentralization of Exchange." *Econometrica* 42 (November 1974): 1093–1113.

Owen, G. "On the Core of Linear Production Games." *Math. Programming* 9 (December 1975): 358–70.

Patinkin, D. *Money, Interest, and Prices: An Integration of Monetary and Value Theory.* Evanston, Ill.: Row, Peterson, 1956.

Postlewaite, A., and Schmeidler, D. "Approximate Efficiency of Non-Walrasian Nash Equilibria." Discussion Paper, Univ. Illinois at Urbana-Champaign, October 1975.

Samuelson, P. A. *Foundations of Economic Analysis.* Cambridge, Mass.: Harvard Univ. Press, 1947.

Scarf, H. E. "The Core of an *n* Person Game." *Econometrica* 35 (January 1967): 50–69.

———. *The Computation of Economic Equilibria.* New Haven, Conn.: Yale Univ. Press, 1973.

Shapley, L. S. "An Example of a Slow Converging Core." *Internat. Econ. Rev.* 16 (June 1975): 345–51.

———. "Noncooperative General Exchange." In *Theory and Measurement of Economic Externalities,* edited by S. A. Y. Lin. New York: Academic Press, 1976.

Shapley, L. S., and Shubik, M. "Quasi-Cores in a Monetary Economy with Nonconvex Preferences." *Econometrica* 34 (October 1966): 805–27.

———. "On Market Games." *J. Econ. Theory* 1 (June 1969): 9–25. (*a*)

———. "Pure Competition, Coalitional Power, and Fair Division." *Internat. Econ. Rev.* 10 (October 1969): 337–62. (*b*)

———. "The Assignment Game. I. The Core." *Internat. J. Game Theory* 1, no. 2 (1972): 111–30.

———. "Characteristic Function, Core, and Stable Sets." Rand Corporation Report R-904/6, Santa Monica, Calif., July 1973.

———. "Competitive Outcomes in the Cores of Market Games." *Internat. J. Game Theory* 4, no. 4 (1976): 229–37.

Shubik, M. "Edgeworth Market Games." In *Contributions to the Theory of Games,* edited by R. D. Luce and A. W. Tucker. Vol. 4. Annals of Mathematics Study no. 40. Princeton, N.J.: Princeton Univ. Press, 1959.

———. "Extended Edgeworth Bargaining Games and Competitive Equilibrium." *Metroeconomica* 20 (September–December 1968): 299–312.

———. "Pecuniary Externalities: A Game Theoretic Analysis." *A.E.R.* 61 (September 1971): 713–18.

———. "A Theory of Money and Financial Institutions: Fiat Money and Noncooperative Equilibrium in a Closed Economy." *Internat. J. Game Theory* 1, no. 4 (1972): 243–68.

———. "Commodity Money, Oligopoly, Credit and Bankruptcy in a General Equilibrium Model." *Western Econ. J.* 11 (March 1973): 24–38.

———. "A Dynamic Economy with Fiat Money, without Banking, but with Ownership Claims to Production Goods." In *Mélanges F. Perroux,* edited by H. Bartoli. Paris, 1977, in press.

Shubik, M., and Whitt, W. "Fiat Money in an Economy with One Nondurable Good and No Credit." In *Topics in Differential Games,* edited by A. Blaquière. Amsterdam: North-Holland, 1973.

Telser, L. G. *Competition, Collusion, and Game Theory*. Chicago: Aldine-Atherton, 1972.

Viner, J. "Cost Curves and Supply Curves." *Zeitschrift Nationalökonomie* 3 (1931): 23–46.

von Neumann, J. "Zur Theorie der Gesellschaftsspiele." *Math. Ann.* 100 (1928): 295–320. (Or see the English translation by Sonya Bargmann in *Contributions to the Theory of Games*, edited by R. D. Luce and A. W. Tucker. Vol. 4. Annals of Mathematics Study no. 40. Princeton, N.J.: Princeton Univ. Press, 1959.)

Vol. 37 (1977), No. 3-4, pp. 337—354

Zeitschrift für
Nationalökonomie
Journal of Economics

The Optimal Bankruptcy Rule in a Trading Economy Using Fiat Money*

By

Martin Shubik** and **Charles Wilson**, New Haven, Conn., U. S. A.

(Received August 18, 1977)

1. Introduction

In several previous papers [1, 2, 3, 4, 5, 6, 7] models of a monetary economy have been solved as a noncooperative game. This problem of granting credit and the possibility of bankruptcy was avoided by the artifact of considering that all traders were supplied with "enough" of a commodity serving as a "money" or means of payment so that there was no need to borrow.

In this paper an outside bank, and borrowing are considered explicitly and the meaning of an optimal bankruptcy rule is considered. We stress that if credit or paper money are introduced into an economy described as a game of strategy, rules describing penalties to be levied against those who cannot pay back what they have borrowed, become a logical necessity in order to fully define all possible outcomes. Our approach is to specify such rules and study them parametrically; i. e. we carry out a sensitivity analysis to see what happens as the severity of the penalties is varied.

* This work relates to Department of the Navy Contract N00014-76-C0085 issued by the Office of Naval Research under Contract Authority NR 047-007. However, the content does not necessarily reflect the position or the policy of the Department of the Navy or the Government, and no official endorsement should be inferred.

** The authors wish to thank Pradeep Dubey and Donald Brown for helpful conversations. This work builds not only on one of the author's previous work but on joint work with Shapley, Dubey, Whitt and Evers.

This paper deals primarily with problems in modelling and interpretation. Thus the argument is carried out in terms of a specific relatively simple example to illustrate the market, banking and bankruptcy mechanism. General proofs pertaining to a broad class of trading models are given in a separate paper [8].

2. The Model

The model is a varient of the model originally suggested by Shubik [1] and investigated by Shapley [3], Shapley and Shubik [9], Shubik [10, 11] and Dubey and Shubik [5, 6]. The paper here however is self contained inasmuch as a complete model is built, although the references noted provide detailed discussion of some aspects of the model and proofs which are not supplied here.

2.1. A Trading Economy Without Uncertainty

The procedure adopted here is to begin by taking a simple model of trade. This is formulated and solved for the standard competitive equilibrium solution. We then take the same economic background and model trade as a noncooperative game with a bank issuing loans to finance trade.

We solve the game for a type symmetric noncooperative equilibrium point (T. S. N. E.). This is an equilibrium point at which traders of the same type obtain equal treatment. Equal treatment is not necessarily a property of a noncooperative equilibrium. We study the conditions under which the T. S. N. E. coincides or fails to coincide with the C. E. in terms of market prices and distribution of resources.

Consider 2 n traders trading in two commodities, n have endowments of $(A, 0)$ and n have endowments of $(0, B)$. Traders of the first type have utility functions of the form

$$u^1 = \log x_1^{\alpha} y_1^{1-\alpha} \tag{1}$$

and the second type

$$u^2 = \log x_2^{\beta} y_2^{1-\beta}. \tag{2}$$

2.2. The Competitive Equilibrium and Pareto Optimal Surface

It is easy to solve for the unique competitive equilibrium and the Pareto optimal surface. We obtain:

$$p_1 = 1,\ p_2 = \left(\frac{1-\alpha}{\beta}\right)\left(\frac{A}{B}\right),\ x = (1-\alpha)\,A,\ y = \beta B, \tag{3}$$

$$\lambda_1 = 1/A \text{ and } \lambda_2 = \beta/A\,(1-\alpha),$$

where $(A-x)$ and y are the consumptions of trader 1; λ_1 and λ_2 are Lagrangian multipliers and p_1 and p_2 are prices.

The Pareto optimal surface is given by

$$\left\{\frac{\alpha-\beta}{\beta(1-\alpha)}\right\} xy = AB - Bx - Ay. \tag{4}$$

2.3. The Money Game

2.3.1. The Trading Mechanism

Let the amount bid by a trader i of type 1 be b^i and by a trader j of type 2 be d^j. Bids are made in quantities of bank money, to be defined later. As traders of type 1 only have the first commodity we may assume that they can offer only it for sale. Let the amount of the first commodity offered for sale by a trader i of the first type be x^i. A trader j of the second type offers an amount y^j of the second commodity.

Let the symbol $b = \sum_{i=1}^{n} b^i$ and similarly for the others.

We may assume in this simple market that traders of type 1 bid only for the second good and traders of type 2 bid only for the first good. It has been shown elsewhere [6] that there is no loss of generality in making this assumption in large markets[1].

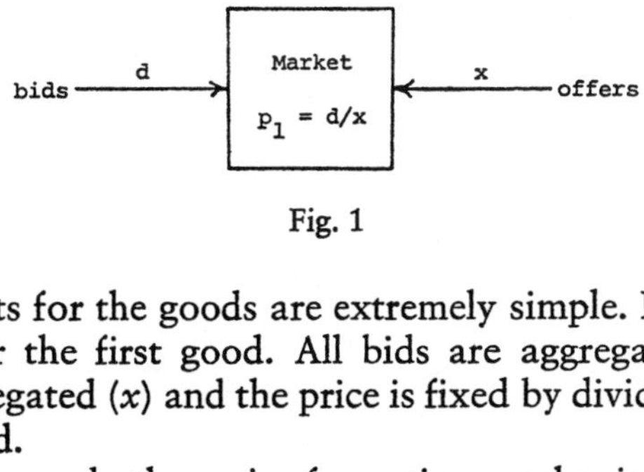

Fig. 1

The markets for the goods are extremely simple. Fig. 1 illustrates the market for the first good. All bids are aggregated (d) and all offers are aggregated (x) and the price is fixed by dividing the amount of good offered.

There are several other price formation mechanisms which could have been selected. These are discussed elsewhere [11]. This was selected because of its basic simplicity.

---

[1] When numbers are few, "wash sales" may be of importance, but we leave this problem aside at this time. A "wash sale" occurs when an individual simultaneously sells and buys back the same item in order to increase the thickness or activity in a market. (See Dubey and Shubik [6] for an example with wash sales.)

When the price has been formed, individual j who has bid d^j for the first good will obtain

$$z^j = \frac{d_j}{p_1} = \frac{d_j}{d} x \tag{5}$$

where

$$p_1 = \frac{d}{x} \tag{6}$$

similarly trader i of the first type obtains

$$w^i = \frac{b^i}{p_2} = \frac{b^i}{b} y \tag{7}$$

where

$$p_2 = \frac{b}{y} \tag{8}$$

2.3.2. Banking: Credit and Promissory Notes

In 2.3.1 we did not specify the currency in which bids are to be made. We now assume that all bids are made using a fiat or bank money which must be obtained from an "outside bank" which is modelled as a mechanism or "dummy" in the sense that it is given a fixed strategy. In particular the bank fixes an amount of money it will issue. In an economy with $2n$ traders let this amount be nK.

The loan mechanism to lend out the nK units will be a simple "money market". Each trader i of type 1 is permitted to create a financial instrument of his own, to wit, a nonnegotiable promissory note of size or denomination u^i. Similarly a trader j of type 2 bids v^j.

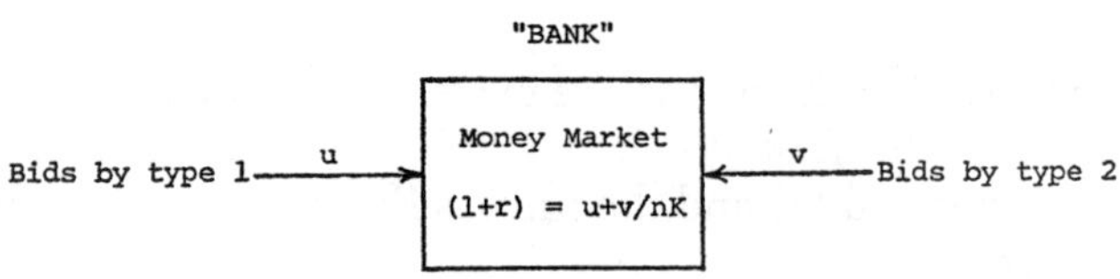

Fig. 2

The traders use the promissory notes to bid for the money supply. The meaning of a promissory note is that it is a contract between an individual i and the bank, which promises to return to the bank an amount of bank money u^i at the end of the period in return for an amount obtained at the start of the period. This amount is determined by the supply of loanable funds nK and the offers of promissory notes, i. e. if s^i and t^j are the loans obtained by i and j they are respectively:

$$s^i = \frac{u^i}{u+v}(nK) \tag{9}$$

and

$$t^j = \frac{v^j}{u+v}(nK). \tag{10}$$

The model is naively simple. Banks in general would be suspicious of borrowers wishing to promise to pay back enormous sums for loans whose size they do not know in advance. However the mechanism is well defined and in a mass market where individuals have estimates of the aggregate money supply and demand this can be interpreted more reasonably, as a single individual will not expect the price of money to move very much in response to his actions.

As can be seen from Fig. 2 the price of money is given by:

$$1+r = \frac{u+v}{nK} \tag{11}$$

where r can be interpreted as a money rate of interest[2].

2.3.3. *Settlement and Bankruptcy*

After the market is over individuals obtain their final allocations of goods, obtain their money incomes and must make their final settlements at the bank.

If an individual ends up with a positive amount of money after having paid the bank, this has no positive value to him. If on the other hand he is unable to honor his debts in full, a penalty is leveled against him. This penalty does not necessarily have to be the same for all individuals. In this model it is proposed to study it parametrically. Two parameters μ_1 and μ_2 are introduced associated with a linear term[3] which measures the "punishment" leveled against anyone who in unable to repay his debts. The specific structure of the payoff functions is shown in 2.3.5 below.

We may consider the bankruptcy act as economic or not directly so. An example of the former might involve a sale of assets of the

---

[2] In a one period model the concept of an interest rate seems somewhat strange. A different way of phrasing the role of r is that it is a loss reserve payment protecting the bank against default.

[3] This term does not need to be linear, as is discussed elsewhere [8]. It is much simpler to make it linear for the example.

debtor, an example of the latter would be a prison sentence or death. In a one period model we may imagine that assets which are confiscated are sold off in a subsequent period.

Regardless of how we regard the punishment, its presence is needed if we wish to influence individuals against making exorbitant repayment commitments to increase the current loans, knowing that they will be unable to meet their obligations.

2.3.4. *The Payoff Functions*

We have seen that the utility functions of the traders [shown in (1) and (2)] involve the only two consumer commodities. We define the payoff functions in terms of strategies and outcomes including the final disposition of money. We may write the payoff to a trader i of type 1 as:

$$\Pi_1{}^i = \log\,(A - x^i)^{\alpha} \left\{\frac{b^i y}{d}\right\}^{1-\alpha} + \mu_1 \min\left[0, \left\{\frac{d}{x} x^i - b^i + \frac{u^i n K}{u+v} - u^i\right\}\right] \tag{12}$$

and the payoff to a trader j of type 2 is:

$$\Pi_2{}^j = \log\left\{\frac{d^j x}{d}\right\}^{\beta} (B - y^j)^{1-\beta} + \mu_2 \min\left[0, \left\{\frac{b}{y} y^j - d^j + \frac{v^j n K}{u+v} - v^j\right\}\right] \tag{13}$$

where the x^i and y^j are considered as functions[4] of the u^i and v^j, and where $0 \leq b^i \leq u^i nK/u+v$ and $0 \leq d^j \leq v^j nK/u+v$.

$$p_1 = d/x \quad \text{and} \quad p_2 = b/y. \tag{14}$$

We note that the bankruptcy conditions enter as linear terms, or as zero.

We assume $0 < \alpha$, $\beta < 1$ to ensure a solution with some trade. The preferences of the individuals are represented by concave utility functions. The introduction of the bankruptcy condition can be considered as though the utility function for an individual had been defined to include money holdings where nothing is added for positive holdings but debt (which may be interpreted as negative hold-

---

[4] If traders have information about prior moves in a multistage market, the selection of subsequent moves may be regarded as a general function of the information state of the trader. If a trader does not know the amount of money he obtained in stage 1 he may still bid, but the bid may be interpreted as a fraction of his wealth.

ings) has a negative worth. Limiting our illustration to one real commodity and money and debt Fig. 3 shows the modified utility function extending into a negative money or debt zone.

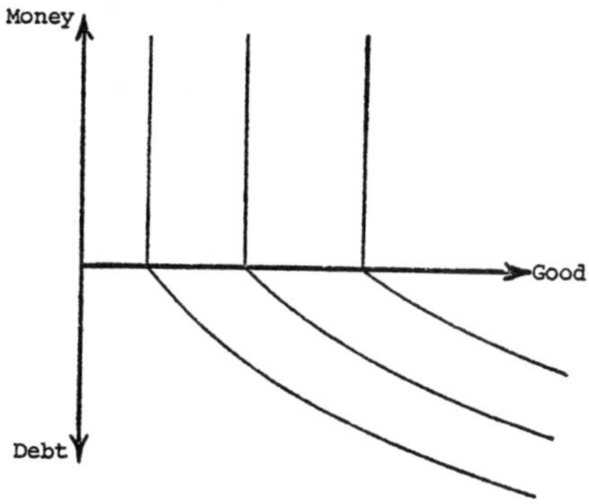

Fig. 3

2.4. The Noncooperative Equilibrium With No Information

Equilibria may be extremely sensitive to information conditions. In multistage games such as this, the less information the traders have the easier it may be to examine the equilibrium point of the games[5].

The simplest situation is where all traders are required to announce simultaneously not only u^i and v^j but[6] also b^i, d^j and x^i, y^j[7].

We may consider that a trader i of the first type attempts to maximize:

$$G_1^i\ (u^i, b^i, x^i) = \alpha \log (A - x^i) + (1-\alpha) \log (yb^i/b) \\ + \lambda_1^i \{dx^i/x - b^i + (M-1)\ u^i\} + \gamma_1^i\ (Mu^i - b^i). \tag{15}$$

---

[5] In the analysis which follows, only the first order conditions for utility maximization are used to describe the equilibrium. The general proof given elsewhere [8] shows that they are noncooperative equilibria.

[6] As an individual may not be sure of his money supply before his bid b^i (d^j) is fixed, we require a feasibility or acceptability condition in the game. For example a nonfeasible bid is replaced by a zero bid or some convention is given which transforms the bid into a feasible act. The simplest convention is to interpret b^i (d^j) as percentages of money available.

[7] It is straightforward to check that degenerate equilibria always exist — that is, equilibria with x^i, $y^i = 0$ and no bankruptcy. In what follows, we will be restricting our attention to non-degenerate equilibria.

There will be a similar expression for a trader j of the second type. We define:

$$M=\frac{nK}{u+v}\geq 0,\ u=\Sigma u^i \quad \text{(and similarly for the others)} \tag{16}$$

where $\lambda_1{}^i$ and $\gamma_1{}^i$ can be regarded as Lagrangian multipliers with the following conditions:

$$\text{if} \quad dx^i/x-b^i+(M-1)\,u^i \begin{cases} >0 & \text{then } \lambda_1{}^i=0 \\ =0 & \text{then } 0\leq\lambda_1{}^i\leq\mu_1 \\ <0 & \text{then } \lambda_1{}^i=\mu_1 \end{cases} \tag{17}$$

$$\text{if} \quad (Mu^i-b^i) \begin{cases} >0 & \text{then } \gamma_1{}^i=0 \\ =0 & \text{then } \gamma_1{}^i\geq 0. \end{cases} \tag{18}$$

First order maximization conditions give:

$$\frac{\partial G_1{}^i}{\partial x^i}=\frac{-\alpha}{A-x^i}+\lambda_1{}^i\,d\left\{\frac{1}{x}-\frac{x^i}{x^2}\right\}=0 \tag{19}$$

and from symmetry:

$$\frac{\alpha}{A-x^i}=\lambda_1{}^i\left(\frac{n-1}{n}\right)\frac{d}{x} \tag{20}$$

Similarly from $\partial G_1{}^i/\partial b^i=0$ and $\partial G_1{}^i/\partial u^i=0$ we obtain

$$(1-\alpha)\left(\frac{n-1}{n}\right)=b^i\,(\lambda_1{}^i+\gamma_1{}^i) \tag{21}$$

and

$$\lambda_1{}^i=M\,(\lambda_1{}^i+\gamma_1{}^i)\left\{\frac{v+\left(\frac{n-1}{n}\right)u}{u+v}\right\} \tag{22}$$

From (20) and $x=nx^i$

$$x\alpha=\lambda_1{}^i\left(\frac{n-1}{n}\right)d\,(A-x^i)$$

gives

$$x^i=\frac{\lambda_1{}^i\,d\,A\left(\frac{n-1}{n}\right)}{n\alpha+\lambda_1{}^i\,d\left(\frac{n-1}{n}\right)} \tag{23}$$

If $\lambda_1{}^i=0$, the only solution consistent with Eq. (23) is $x^i=0$. But this means that each type i trader receives 0 rather than d^i units of money from type j traders. As a consequence, it is easy to verify that no solution exists for $\lambda_1^i = 0$. Therefore, (17) implies

$$d - b + (M-1)\,u \le 0 \quad \text{and} \quad b - d + (M-1)\,v \le 0 \tag{24}$$

or

$$(M-1)\,(u+v) \le 0.$$

Hence if $u+v > 0$ then

$$M \le 1. \tag{25}$$

From (22) and (25) and symmetry[8] then $\gamma_1 > 0$, and similarly $\gamma_2 > 0$; hence

$$Mu = b \quad \text{and} \quad Mv = d. \tag{26}$$

From (21), (22) and (26) and symmetry:

$$u^i = \frac{1-\alpha}{\lambda_1}\left(\frac{n-1}{n}\right)\left\{\frac{v + \left(\frac{n-1}{n}\right) u}{u+v}\right\}$$
$$v^i = \frac{\beta}{\lambda_2}\left(\frac{n-1}{n}\right)\left\{\frac{u + \left(\frac{n-1}{n}\right) v}{u+v}\right\} \tag{27}$$

From (17) and (26):

$$\lambda_1 < \mu_1 \Rightarrow Mv = u \quad \text{and} \quad \lambda_2 < \mu_2 \Rightarrow Mu = v \tag{28}$$

and

$$Mv < u \Rightarrow \lambda_1 = \mu_1, \; Mu < v \Rightarrow \lambda_2 = \mu_2. \tag{29}$$

We now confine our attention to the limiting behavior in markets as $n \to \infty$. From (27) we obtain:

$$u^i \to \frac{1-\alpha}{\lambda_1} \quad \text{and} \quad v^j \to \frac{\beta}{\lambda_2} \tag{30}$$

We wish to consider all λ_1 and λ_2 consistent with (28) or (29).

From (16) and (30)

$$M = \frac{\lambda_1 \lambda_2 K}{\lambda_2 (1-\alpha) + \lambda_1 \beta}$$

hence

$$Mu^i = \frac{K \frac{1-\alpha}{\lambda_1}}{\frac{1-\alpha}{\lambda_1} + \frac{\beta}{\lambda_2}} \quad \text{and} \quad Mv^j = \frac{K \frac{\beta}{\lambda_2}}{\frac{1-\alpha}{\lambda_1} + \frac{\beta}{\lambda_2}} \tag{31}$$

---

[8] Assuming the existence of a symmetric solution we may drop superscripts from the λ and γ.

There are four cases arising from (28) and (29). They are as follows:

(A) $u^i = Mv^j$ and $v^j = Mu^i$

(B) $u^i = Mv^j$ and $v^j < Mu^i$

(C) $u^i > Mv^j$ and $v^j = Mu^i$ and

(D) $u^i > Mv^j$ and $v^j > Mu^i$.

Fig. 4 shows the four cases as μ_1 and μ_2 are varied. The calculations of the values are tedious but straightforward and are given elsewhere [12].

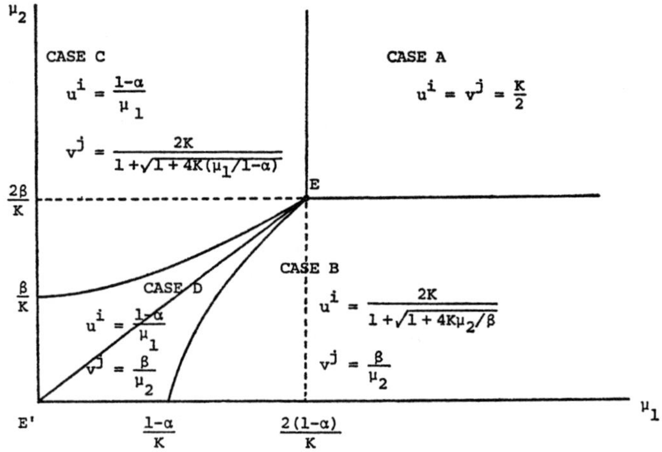

Fig. 4

In zones A and B traders of type 1 stay solvent; in zones A and C traders of type 2 stay solvent.

We observe that at the point E, $\mu_1 = 2(1-\alpha)/K$ and $\mu_2 = 2\beta/K$ where $K = 2A(1-\alpha)$. Hence $\mu_1 = 1/A = \lambda_1$ and $\mu_2 = \beta/A(1-\alpha) = \lambda_2$ which are the same as the Lagrangian multipliers obtained from solving the competitive equilibrium. Thus for these bankruptcy penalties the limit noncooperative equilibrium coincides with the competitive equilibrium.

For other bankruptcy penalties we see that if the μ_1 and/or the μ_2 are set to be less than λ_1 or λ_2 that one or both types will elect bankruptcy. If both go bankrupt and the penalties are selected such that $\mu_1 = g\lambda_1$ and $\mu_2 = g\lambda_2$ where $0 < g < 1$ then trade and prices are the same at the competitive equilibrium along the line EE' in Fig. 4

In Case A there is no bankruptcy and competitive equilibrium trade takes place.

When bankruptcy occurs off the ray EE', the final distribution of trade is not the same as at the competitive equilibrium. It lies on the Pareto optimal surface of the original trading model [see Eq. (4)]. This can be proved generally [8].

2.4.1. *A Finite, Infinite Game Solution Distinction*

We can see from (22) and (25) that for a finite n hoarding cannot take place. This is not so if there were a continuum of nonatomic traders. If each trader felt that his influence on the aggregates were negligible a new solution appears.

Suppose each trader takes x, y, b, d, and M as given, and chooses his optimal x^i, b^i and u^i accordingly. Then the first order conditions for a maximum [given by Eqs. (19), (21), and (22)] simplify to:

$$x^i = \frac{A\lambda_1{}^i d - \alpha x}{\lambda_1{}^i d} \tag{32}$$

$$b^i = \frac{1-\alpha}{\lambda_1{}^i + \gamma_1{}^i} \tag{33}$$

$$\gamma_1{}^i = \frac{1-M}{M}\lambda_1{}^i \tag{34}$$

Using (34), (33) becomes:

$$b^i = M\frac{1-\alpha}{\lambda_1{}^i} \tag{35}$$

Adding (32) n times and simplifying yields:

$$x^i = \frac{A\lambda_1{}^i d}{n\alpha + \lambda_1{}^i d}$$

From Eq. (15), $b^i > 0$ implies $\lambda_1{}^i > 0$, and therefore (34) requires $M \leqq 1$ to keep $\gamma_1{}^i \geqq 0$. Note that $\gamma_1{}^i = 0$ if and only if $M = 1$.

Suppose $M = 1$, then $u^i + v^j = K$. From (35) and an equivalent expression for type 2 traders, $b^i \leqq Mu^i$ and $\lambda_1{}^i \leqq \mu_1$ then imply

$$\frac{1-\alpha}{\mu_1} + \frac{\beta}{\mu_2} = K \tag{36}$$

Furthermore, (24) implies that $b^i = d^j$. Therefore,

$$\frac{1-\alpha}{\lambda_1} = \frac{\beta}{\lambda_2} \tag{37}$$

which does not violate (36) only if:

$$\frac{1-\alpha}{\mu_1}, \frac{\beta}{\mu_2} \leq \frac{K}{2} \tag{38}$$

One can verify that $M=1$ if and only if (38) holds. If (38) does not hold, then the limiting results of the previous section hold. However if (38) is satisfied, not only do both types stay solvent, but the possibility of hoarding occurs. In this case the following values x^i, y^j, b^i, d^j, u^i, v^j satisfy the conditions for a noncooperative equilibrium.

$$x^i = A\,(1-\alpha);\ \ y^j = \beta B \tag{39}$$

$$b^i = d^j \geqq \max\left(\frac{1-\alpha}{\mu_1}, \frac{\beta}{\mu_2}\right) \tag{40}$$

$$u^i + v^j = K;\ \ u^i, v^j \geqq b^i = d^j. \tag{41}$$

Not only may there be an indeterminate solution for u and v, but even when u and v is specified any values of b and d satisfying (40) and (41) will generate a noncooperative equilibrium. The resultant relative prices and distribution coincide with the competitive equilibrium, but the price level may be low enough that not all of M is used.

3. Two Stage Noncooperative Equilibria

3.1. The Extensive Form, Information and Strategies

When we assume that individuals are informed about some aspects of what happened at the first stage of the game prior to selecting their moves in the second stage we may set up many different games which differ only in the shading of information. Two of these are illustrated.

Case 1: The game is played as follows: First, all individuals simultaneously bid for loans. They obtain no information beyond the size of the loan they have secured. After they have obtained their loans they then all bid simultaneously in the markets for goods.

If the individuals know the size of their own loans, as they know their own bids they can calculate the price of money. And if they all know the size of the total money supply they can calculate the aggregate amount of loans obtained by their competitors.

Let $\bar{u}^i$ signify $u+v-u^i$ and $\bar{v}^j$ signify $u+v-v^j$. Then a strategy for a trader i of type 1 is a number u^i and two functions $x^i = \phi_1{}^i(u^i, \bar{u}^i)$, $b^i = \phi_2{}^i\,(u^i, \bar{u}^i)$ and similarly for a trader j of tape 2.

Case 2: If we wished we might assume that trader i is given complete information as to who has borrowed how much. In this case a strategy would be a number u^i and two functions

$$x^i = \phi_1{}^i\,(u^1, u^2, \ldots, u^n;\, v^1, v^2, \ldots, v^n)$$

and

$$b^i = \phi_2{}^i\,(u^1, u^2, \ldots, u^n;\, v^1, v^2, \ldots, v^n).$$

In the first instance the trader bases his plan on *macroeconomic statistics,* in the second, on *microeconomic* detail. The refinement of information in a general noncooperative game could easily create many new equilibrium points associated with a noncooperative solution.

Consider Case 2, do we have guidance as to how to select strategies? Fortunately there is a simple way to pick these functions. Assume that u^i and v^j, $i=1,\ldots, n$, $j=1,\ldots, n$ are given. Any individual i will select the pair (x^i, b^i) by maximizing his payoffs in the one stage game. Any individual j will select the pair (y^j, d^j) in the same manner. We may solve for x^i, y^j, b^i, d^j as functions of the u^i and v^j then solve the bidding-for-loans stage of the market by maximizing with respect to (*wrt*) u^i and v^j.

3.2. Threats and Two Stage Equilibria

In 3.1 we have noted that a strategy in a two stage game can consist of a number in the first period and a function depending upon the information concerning the moves of all others in the first period. The extreme generality of this function enables traders to convey highly implausible threats which may nevertheless give rise to new (and improbable) equilibrium points. Thus in a two stage game, in general, the problem is not with existence of noncooperative equilibria but with a surfeit of them.

Is it possible to distinguish "plausible" or nice equilibria from the others? As yet there does not seem to be a completely general satisfactory way to do so[9]. However, we could use the type of backward solution used in dynamic programming. Unfortunately there are two basic difficulties in doing so. One concerns the information conditions. We need perfect information between the stages so that subgames can be well defined. It is likely that by making use of the special structure of these economic games where it is possible to aggregate moves we could weaken the information requirements, however this is not explored further here. The second difficulty concerns uniqueness. If we wish to replace the second

[9] Although the work of Harsanyi [13] has progressed on this problem.

stage games by the values of their noncooperative equilibria we need them either to be unique, or at least we require some sort of indexing scheme which enables us to associate the equilibrium points we select by some natural property such as continuity. Fortunately in the simple example considered here by appealing to continuity we can choose a unique value of the payoff function so that a backward solution is well defined.

3.3. Symmetric Perfect Equilibria

We have calculated the perfect equilibria for this example. The details of the second stage calculations of the backward solution together with a sketch of the first stage are given elsewhere [12].

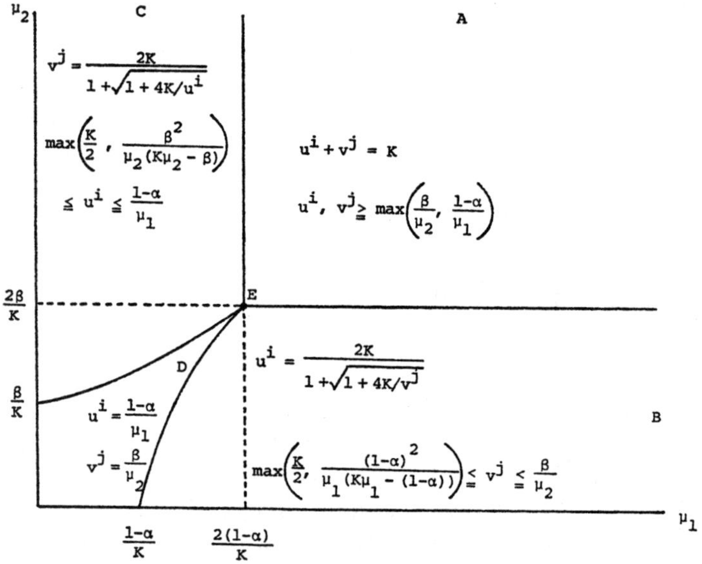

Fig. 5

The results are highly related to the previous analysis, but somewhat different as is shown in Fig. 5 which should be compared with Fig. 4. The qualitative differences of note are that even for the finite model hoarding may take place in regions A, B and C.

4. So What?

Recently there has been considerable interest in the construction of a microeconomic theory with money and with the reconciliation of micro- and macro-economics. There are currently several different approaches as evinced by the work of Clower [14], Hahn [15], Starr [16], Grandmont and Laroque [17] and many others. These approaches differ from each other and from the approach adopted here in both the basic questions asked and models constructed.

The approach adopted here calls for extremely detailed modelling up to the point that the test of completeness and consistency is: is the model sufficiently well specified that it could be played as a game? — i. e. are there rules to describe all moves and cover all contingencies?

In this paper we have labored through a large amount of tedious calculation on a special example in order to at least begin to disclose the fine structure of a general class of economic problems that we believe are of importance.

In particular even though our model may appear to be special certain phenomena have been encountered and problems solved.

(1) Because we wished to construct a model of trade with simultaneous independent bids and offers with a price formation mechanism defined for all positions of equilibrium and disequilibrium we needed to specify bids and offers and market clearance in equilibrium or disequilibrium. When numbers of competitors are large, many mechanisms lead to the same noncooperative outcome [1, 6, 10].

(2) In a previous model [1, 9] it was shown that a mechanism would work if one commodity were used to bid for the others. For the outcome of trade in a mass market to yield results comparable to the competitive equilibrium enough commodity money, appropriately distributed is required.

(3) If there is not enough commodity money present the outcomes will be nonoptimal unless credit is introduced. There are many ways of introducing credit. One is to imagine that each trader has an unbounded open credit line at a bank [9, 18, 19]; another one, adopted here, is to imagine an outside bank which auctions off a fixed amount of its "money" or universally accepted I. O. U. notes in return for individual traders' I. O. U. notes.

(4) No matter which method is adopted for the issue of credit, a rule must be adopted to prevent individuals from issuing unbounded quantities of I. O. U. notes. This rule is embodied here in the bankruptcy penalty.

(5) As there does not appear to be an *a priori* reason for the selection of a specific penalty we have studied the penalty parametrically and have observed that for certain penalty values we

obtain results from the noncooperative game which are comparable with the competitive equilibrium. For other penalties different results are obtained.

(6) We may interpret the ratio of the amount of personal I. O. U.'s offered for bank money as specifying a money interest rate. When the noncooperative and competitive trades coincide without any bankruptcies taking place the money rate of interest is zero, i. e. the credit supplied to finance trade has been provided at no cost beyond the requirement that settlement after trade is on a one for one basis.

(7) The zero rate of interest appears to be most reasonable when we observe that there is no time discount and no production. The only purpose of the outside money is to finance *the float* created by the requirement that individuals bid and offer simultaneously. The creation of the float enables each individual to view the process sequentially, i. e. he borrows, pays for purchases and then obtains income and settles debt. Price formation however appears to take place simultaneously in the "clearing houses" matching bids and offers.

(8) As this is a sequential process the meaning of strategy, the feasibility of moves and the dependence of the process on information conditions becomes crucial. For this reason we have included a brief discussion of the two stage equilibria.

To most of us money and credit are more naturally associated with uncertainty, multiperiod trade and production [20]. Even in this one period model of trade, the float, bankruptcy penalties and the possibility of hoarding appear. A more satisfactory model with many periods and production is being considered — but new phenomena and difficulties appear. In particular (setting aside extra problems due to uncertainty) the model presented here can be immediately generalized for k time periods [8] if we are satisfied with a solution in which intertemporal prices are adjusted by hoarding outside money issued at period 1. This, though logically correct does not appear to be satisfactory. A way of avoiding this is to introduce an inside bank as well as an outside bank. After the traders bid for the outside money supply M using their I. O. U. notes, they then bid for shares in an inside bank using outside money. This model will be presented in a subsequent paper[10].

---

10 There are extra difficulties encountered in defining short term profits, bad loans and roll over conditions on loans, as well as defining the strategies, payoffs and bank failure rules for the inside bank.

References

[1] M. Shubik: Commodity Model, Oligopoly, Credit and Bankruptcy in a General Equilibrium Model, Western Economic Journal *10* (1972), pp. 24—38.

[2] M. Shubik: Fiat Money and Noncooperative Equilibrium in a Closed Economy, International Journal of Game Theory *1* (1971/72), pp. 243—268.

[3] L. S. Shapley: Noncooperative General Exchange, in: S. A. Y. Lin (ed.): Theory and Measurements of Economic Externalities, New York 1976, pp. 155—175.

[4] M. Shubik: Mathematical Models for a Theory of Money and Financial Institutions, in: R. H. Day and Th. Groves (eds.): Adaptive Economic Models, New York 1975, pp. 513—574.

[5] P. Dubey and M. Shubik: Trade and Prices in a Closed Economy with Exogenous Uncertainties, Different Levels of Information, Money and Compound Future Markets, forthcoming, Econometrica *45* (1977).

[6] P. Dubey and M. Shubik: The Noncooperative Equilibria of a Closed Trading Economy with Market Supply and Bidding Strategies, CFDP 422, February 1976, forthcoming, Journal of Economic Theory (1978).

[7] L. S. Shapley and M. Shubik: Models of Noncooperative Exchange, in process, partially available from the authors.

[8] P. Dubey and M. Shubik: Trade Using a Borrowed Means of Payment, with Bankruptcy Conditions, CFDP 448, February 1977.

[9] L. S. Shapley and M. Shubik: Trade Using One Commodity as a Means of Payment, Journal of Political Economy *85* (1977), pp. 937—968.

[10] M. Shubik: A Trading Model to Avoid Tatonnement Metaphysics, in: Y. Amshud (ed.), Studies in Game Theory and Mathematical Economics, New York 1976, pp. 129—142.

[11] M. Shubik: On the Number of Types of Markets with Trade in Money, CFDP 416, January 1976.

[12] M. Shubik and C. Wilson: The Optimal Bankruptcy Rule in a Trading Economy Using Fiat Money, CFDP 424(R), June 7, 1976.

[13] J. Harsanyi: The Tracing Procedure: A Bayesian Approach to Defining a Solution for *n*-Person Noncooperative Games, International Journal of Game Theory *4* (1975), pp. 61—94.

[14] R. W. Clower: A Reconsideration of the Microfoundations of Monetary Theory, Western Economic Journal *6* (1967), pp. 1—9.

[15] F. H. Hahn: On Some Problems of Proving the Existence of an Equilibrium in a Monetary Economy, in F. H. Hahn and F. Brechling (eds.): The Theory of Interest Rates, New York 1965.

[16] R. M. Starr: The Price of Money in a Pure Exchange Monetary Economy with Taxation, Econometrica 42 (1974), pp. 45—54.

[17] J. M. Grandmont and G. Laroque: On Money and Banking, Review of Economic Studies 42 (1975), pp. 207—236.

[18] P. Dubey and L. S. Shapley: Noncooperative Exchange with a Continuum of Traders, CFDP 447, February 1977.

[19] A. Postlethwaite and D. Schmeidler: Approximate Efficiency of Non-Walrasian Nash Equilibria, mimeographed, University of Illinois, September-October 1975.

[20] P. Dubey and M. Shubik: The Money Rate of Interest (A Multiperiod Nonatomic Trading and Production Economy with Outside Money, Inside Money and Optimal Bankruptcy Rules), April 1977.

Address of authors: Prof. Dr. Martin Shubik and Dr. Charles Wilson, Department of Economics, Cowles Foundation for Research in Economics, Box 2125, Yale Station; Yale University, New Haven, CT 06520, U. S. A.

Printed in Austria

Reprinted from JOURNAL OF ECONOMIC THEORY Vol. 17, No. 1, February 1978

A Theory of Money and Financial Institutions. 28. The Noncooperative Equilibria of a Closed Trading Economy with Market Supply and Bidding Strategies*

PRADEEP DUBEY AND MARTIN SHUBIK

Cowles Foundation, Yale University, New Haven, Connecticut 06520

Received April 28, 1976; revised July 18, 1977

1. INTRODUCTION

In several previous papers Shubik [1], Shapley [2], and Shapley and Shubik [3] have proposed and investigated a model of exchange in many markets with the manner of price formation completely formulated and with trade utilizing a money.

The basic model analyzed previously was extremely simple; in some ways not completely satisfactory and certainly not unique in its portrayal of price formation. Shapley and Shubik [4] and Shubik [5] have suggested several alternative models. This paper is devoted to examining a noncooperative equilibrium solution to one of the alternative models and to contrasting this with the noncooperative equilibrium solution to the original model.

2. THE "BID SUPPLY" MODEL

In previous papers Shubik [1] and Shapley [2] have considered trade where all individuals are required to sell all of their endowments of m commodities and bid to make purchases using an $(m + 1)$st commodity.

It may be argued that for many markets and certainiy for markets involving production and the carrying of inventory that the restriction forcing an individual to offer all of his goods for sale is unreasonable. This restriction is relaxed here. Instead we assume that a trader will enter each market as a seller or buyer, as both or neither.

* This work relates to Department of the Navy Contract N00014-76-C-0085 issued by the Office of Naval Research under Contract Authority NR 047-006. However, the content does not necessarily reflect the position or the policy of the Department of the Navy or the Government, and no official endorsement should be inferred.

For a positive integer r we shall denote by I_r the set $\{1,..., r\}$, by E^r the Euclidean space of dimension r, and by Ω^r the nonnegative orthant of E^r.

Let

$$I_n = \text{the set of traders,}$$
$$I_{m+1} = \text{the set of commodities.}$$

The *initial allocation* of trader i is a vector $a^i \in \Omega^{m+1}$, where a_j^i is the amount of commodity j available to i (for $j \in I_m$), and a_{m+1}^i represents the money held by i. The traders' *utility functions* are real-valued:

$$u^i : \Omega^{m+1} \to \Omega^1, \qquad i \in I_n,$$

and are assumed to be concave, continuous and nondecreasing. We shall say that trader i "desires" good j if $u^i(x^i)$ is an increasing function of the variable x_j^i, for any fixed choice of the other variables x_l^i. A trader i for whom $a_{m+1}^i > 0$ will be called "moneyed"; and when $a_j^i > 0$ we will say that he is "j-furnished."

Given $x^i \in \Omega^m$, for $i = 1,..., n$, we will denote $\sum_{i=1}^n x_j^i$ by $\bar{x}_j$. We assume that $\bar{a}_j > 0$ for all $j \in I_m$.

To cast the rules of the market in the form of a game we define the strategy set of trader i to be

$$S^i = \{(q^i, b^i): q^i \in \Omega^m, b^i \in \Omega^m, q_j^i \leqslant a^i, \bar{b}^i \leqslant a_{m+1}^i\}.$$

Here q_j^i is the quantity of commodity j that trader i offers for sale. b_j^i is i's bid on good j; and $\bar{b}^i$ stands for $\sum_{j\in I_m} b_j^i$.

The product $S^1 \times \cdots \times S^n$ will be denoted S; it is a compact convex set of dimension at most $2mn$. $\bar{S}^i$ denotes $S^1 \times \cdots \times S^{i-1} \times S^{i+1} \times \cdots \times S^n$. s (and, s^i, $\bar{s}^i$) will stand for elements of S (and, S^i, $\bar{S}^i$).

The outcome engendered by a particular $s \in S$ is determined in three simple steps. First we calculate a "price vector" $p \in \Omega^m$ by dividing the amount bid for each good by the total supply:

$$p_j = \bar{b}_j/\bar{q}_j, \text{ all } j \in I_m, \qquad \text{where } s^i = (q^i, b^i).$$

(If $\bar{q}_j = 0$, we set $p_j = 0$.) Next we calculate the final allocation that results when the bids are executed:

$$\begin{aligned} \xi_j^i(s) &= b_j^i/p_j + a_j^i - q_j^i, && \text{if } j \in I_m \text{ and } p_j > 0,\\ &= a_j^i - q_j^i, && \text{if } j \in I_m \text{ and } p_j = 0,\\ &= a_j^i - \sum_{j\in I_m} b_j^i + \sum_{j\in I_m} q_j^i p_j, && \text{if } j = m+1. \end{aligned}$$

Finally we calculate the payoff to the traders:

$$p^i(s) = u^i(\xi^i(s)).$$

In this way, we obtain an n-person game Γ in the standard "strategic" (or "normal" form).

A "Nash Equilibrium" (or "N.E.") of this game is defined to be an $\hat{s} \in S$ with the property that, for each $i \in I_n$,

$$p^i(\hat{s}) = \max\{P^i(\hat{s} \mid s^i) : s^i \in S^i\},$$

where $(\hat{s} \mid s^i)$ is $\hat{s}$ but with $\hat{s}^i$ replaced by s^i.

2.1. *The Strategy Set*

In order to give some insight into the nature of the strategy sets of an individual and their relationship to final outcomes we consider a market with one commodity being bought or sold with trade in a commodity money.[1]

In Fig. 1 the point with coordinates (A, M) indicates the initial endowment

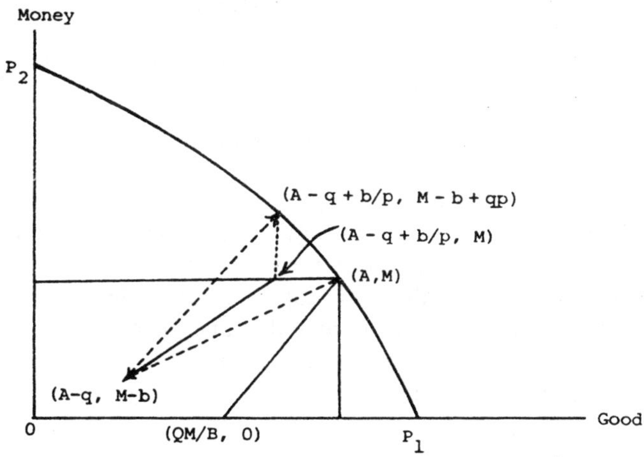

FIGURE 1

of an individual. His strategy set is a rectangle given by $0 \leqslant q \leqslant A$ and $0 \leqslant b \leqslant M$. Any point in this rectangle is a strategy.[2] Consider the strategy (q, b); we wish to study how it transforms the initial point (A, M). Suppose that all other traders have offered Q units of the good and a total of B units

[1] Here $A = a_1^i$ and $M = a_2^i$.

[2] For simplicity in describing the strategies of a single individual the superscript i has been dropped as the meaning should be clear.

of money. Then a bid of (q, b) takes the initial holding of (A, M) to a final holding of

$$(A - q + b/p, M - b + qp),$$

where

$$p = (B + b)/(Q + q) \qquad (Q > 0). \tag{1}$$

We may observe that had the individual bid $(q - b/p, 0)$ this, too, maps (A, M) into

$$(A - q + b/p, M - b + qp).$$

The bids which map (A, M) into itself are given by

$$A = A - q + b/p \qquad \text{and} \qquad M = M - b + qp$$

or

$$b = q((B + b)/(Q + q)) \qquad \text{or} \qquad Qb = Bq. \tag{2}$$

All bids on the line connecting $(QM/B, 0)$ to (A, M) map (A, M) onto itself. Any strategy to the right of this line maps AM onto a point on the curve P_1P_2 to the right.

All points on the curve P_1P_2 can be obtained by bids on the "L-shaped" boundary of the strategy set i.e., bids of the form $(q, 0)$ or $(0, b)$ where $0 \leqslant q \leqslant A$ and $0 \leqslant b \leqslant M$.

Let (x, m) be a point on the curve P_1P_2 then it can be checked that the equation for this surface is given by

$$m = M + (A - x)\, B/(Q + A - x). \tag{3}$$

This is a concave curve.

In an analogous manner we can generalize these observations for m commodities and a money. Suppose the initial holdings of individual i are given by $(a_1{}^i, \ldots, a_m{}^i, a_{m+1}^i)$. All others together have offered $(Q_1, \ldots, Q_m)$ for sale and bid $(B_1, \ldots, B_m)$. A strategy by individual i is a point in a $2m$-dimensional rectangular set of the form $(q_1{}^i, \ldots, q_m{}^i, b_1{}^i, \ldots, b_m{}^i)$.

The initial holdings are transformed to final holdings of

$$\left(a_1{}^i - q_1{}^i + b_1{}^i/p_1, \ldots, a_m{}^i - q_m{}^i + b_m{}^i/p_m\,;\, a_{m+1}^i - \sum_{j=1}^{m} b_j{}^i + \sum_{j=1}^{m} q_j{}^i p_j\right). \tag{4}$$

We may derive the equation for the set of final holdings which can be reached by all bids. Let $(x_1{}^i, x_2{}^i, \ldots, x_m{}^i, x_{m+1}^i)$ be a point on the set of final holdings. Then the equation of this surface is given by

$$x_{m+1}^i = a_{m+1}^i + \sum_{j=1}^{m} \left\{ \frac{B_j(a_j{}^i - x_j{}^i)}{Q_j + a_j{}^i - x_j{}^i} \right\}.$$

This is concave.

In order to fully check that (5) is the surface of final allocations produced by the use of any strategy we can consider an arbitrary strategy (b_j^i, q_j^i) and show that it produces a final allocation satisfying (5). A strategy (b_j^i, q_j^i) produces a final allocation as shown in (4); substituting these values for x_j^i and x_{m+1}^i into (5) we obtain

$$\sum_{j=1}^{m} (q_j^i p_j - b_j^i) = \sum_{j=1}^{m} \left\{ \frac{(B_j/p_j)\{q_j^i p_j - b_j^i\}}{Q_j + q_j^i - b_j^i/p_j} \right\}$$

$$= \sum_{j=1}^{m} \left\{ \frac{(B_j/p_j)\{q_j^i p_j - b_j^i\}}{(Q_j + q_j^i)(b_j^i + B_j) - b_j^i(Q_j + q_j^i)(b_j^i + B_j)^{-1}} \right\}$$

$$= \sum_{j=1}^{m} \left\{ \frac{(B_j/p_j)\{q_j^i p_j - b_j^i\}}{B_j/p_j} \right\} = \sum_{j=1}^{m} (q_j^i p_j - b_j^i). \qquad (6)$$

3. The Existence and Convergence of Noncooperative Equilibria

N.E.'s for Γ exist trivially. For example consider the case where each individual bids and supplies nothing. This choice of strategies clearly constitutes an *N.E.*

However, we wish to establish the existence of nonpathological *N.E.*'s which approach a competitive equilibrium under replication and which can be viewed as being supported in the limit (as the number of competitors increase) by a price system. For this purpose we define an ϵ-modification of the game Γ which we denote by ${}^{\epsilon}\Gamma$ (see below). "Nice" *N.E.*'s are then obtained as a limit of the *N.E.*'s of ${}^{\epsilon}\Gamma$ as $\epsilon \to 0$. We call these *E.P.*'s to distinguish them from the other *N.E.*'s of the game Γ. We demonstrate that these *E.P.*'s (provided they are "interior"—see pg. 14) approach the competitive equilibria (the *C.E.*'s) as the number of traders is increased by replication.

By an "ϵ-modified" game, ${}^{\epsilon}\Gamma$, we mean one in which some outside agency places a fixed bid of $\epsilon > 0$, and a fixed supply of $\epsilon > 0$, in each of the m markets. This does not change the players' strategy spaces, but it does of course change the outcomes and payoffs, which we will denote ξ_ϵ^i and P_ϵ^i.

An *N.E.* ν of Γ will be called an *E.P.* (equilibrium point) if there exist sequences $\{{}^g\nu\}_{g=1}^{\infty}$ $({}^g\nu \in S)$ and $\{{}^g\epsilon\}_{g=1}^{\infty}$ $({}^g\epsilon \in \Omega^1\backslash\{0\})$ such that

(a) ${}^g\nu \to \nu$,

(b) ${}^g\epsilon \to 0$,

(c) ${}^g\nu$ is an *N.E.* of ${}^{g\epsilon}\Gamma$.

THEOREM 1. *Assume*[3] *that* (1) *all traders desire money*; (2) *for any good j there are at least two moneyed traders who desire j, and at least two j-furnished traders. Then an E.P. of Γ exists.*

We begin the proof with some lemmas.

LEMMA 1. *Assume that all traders desire money. Then an N.E. exists for every* ${}^{\epsilon}\Gamma$, $\epsilon > 0$.

For the proof, it will help to build up some notation. Let

$$R_j^i = [0, a_j^i] \times [0, a_{m+1}^i],$$
$$R^i = R_1^i \times \cdots \times R_m^i,$$
$$L^i = \left\{ b \in \Omega^m : \sum_{j \in I_m} b_j \leqslant a_{m+1}^i \right\}.$$

For $Q \in \Omega^m$, $B \in \Omega^m$, let

$$D^i(Q, B, \epsilon) = \left\{ w \in \Omega^m : w_j = a_j^i - q_j^i + \frac{b_j^i(Q_j + \epsilon + q_j^i)}{(B_j + \epsilon + b_j^i)}, \text{ where } s^i \in S^i \right\}.$$

[Here $s^i = (q_1^i, \ldots, q_m^i, b_1^i, \ldots, b_m^i)$.] It is straightforwardly verified that $D^i(B, Q, \epsilon)$ is convex[4]:

$$H^i(Q, B, \epsilon) = \left\{ w \in \Omega^{m+1} : w_{m+1} = a_{m+1}^i + \sum_{j \in I_m} \frac{(a_j^i - w_j)(B_j + \epsilon)}{Q_j + \epsilon + a_j^i - w_j}, \right.$$
$$\left. (w_1, \ldots, w_m) \in D^i(Q, B, \epsilon) \right\}.$$

The analysis of the previous section shows that $H^i(Q, B, \epsilon)$ is the set of holdings that trader i can obtain with his strategies in the ϵ-game if the total supply (bid) to others is Q_j (B_j) in the jth market. Thus, in fact, $H^i(Q, B, \dot{\epsilon})$ is

$$\left\{ w \in \Omega^{m+1} : w_i = a_j^i - q_j^i + \frac{b_j^i(Q_j + \epsilon + q_j^i)}{B_j + \epsilon + b_j^i} \text{ for } j \in I_m \text{ and} \right.$$
$$\left. w_{m+1} = a_{m+1}^i - \bar{b}^i + \sum_{j \in I_m} q_j^i \left(\frac{B_j + \epsilon + b_j^i}{Q_j + \epsilon + q_j^i} \right), \text{ where } s^i \in S^i \right\}.$$

[3] We have elsewhere—see [9]—weakened the hypothesis to: Assume that for each $j \in I_m$ there are at least (1) two moneyed traders who desire j; (2) two j-furnished traders who desire money. The proof then involves the use of the Eilenberg–Montgomery Fixed Point Theorem instead of Kakutani's theorem. For simplicity of exposition we have here dealt with the more restricted theorem.

[4] See Appendix A.

Finally, let

$$\bar{H}^i(Q, B, \epsilon) = \{\hat{w} \in H^i(Q, B, \epsilon): u^i(\hat{w}) = \max_{w \in H^i(Q,B,\epsilon)} u^i(w)\}.$$

We will denote by "Proj" the projection map. Thus, for $(w_1, ..., w_{m+1}) = w \in \Omega^{m+1}$, $\mathrm{Proj}_j(w)$ is w_j, etc.

Proof of Lemma 1. For $\bar{s}^i \in \bar{S}^i$, let $Q_j(\bar{s}^i) = \sum_{k \in I_n/\{i\}} q_j^k$, and $B_j(\bar{s}^i) = \sum_{i \in I_n/\{i\}} b_j^k$, where $s^k = (q_1^k, ..., q_m^k, b_1^k, ..., b_m^k)$ for $k \in I_n \backslash \{i\}$. Then it is straightforward, though laborious, to verify that the correspondence $\phi_1^i : \bar{s}^i \Rightarrow \Omega^{m+1}$, given by

$$\phi_1^i(\bar{s}^i) = H^i(Q(\bar{s}^i), B(\bar{s}^i), \epsilon)$$

is continuous. Since u^i is continuous on Ω^{m+1}, we see by the "Maximum Theorem" [8, p. 116] that the correspondence ϕ_2^i given by $\phi_2^i(\bar{s}^i) = \bar{H}^i(Q(\bar{s}^i), B(\bar{s}^i), \epsilon)$ is u.s.c. We show that, in fact, $\bar{H}^i(\ )$ consists of a single point. For suppose that w and w' both belong to $\bar{H}^i(\ )$. Consider $w'' = \frac{1}{2}w + \frac{1}{2}w'$. Since u^i is concave

$$\begin{aligned} u^i(w'') &\geqslant \tfrac{1}{2}u^i(w) + \tfrac{1}{2}u^i(w') \\ &= u^i(w). \end{aligned}$$

But w_{m+1} is a *strictly* concave function of $w_1, ..., w_m$. Hence there is a $\gamma > 0$ such that $w'' + \gamma e^{m+1} \in \bar{H}^i(\ )$. However, u^i is strictly increasing in the $(m + 1)$st variable, therefore $u^i(w'' + \gamma e^{m+1}) > u^i(w)$, a contradiction.

Put $\tilde{S}_j^i(\bar{s}^i) = \psi_j^i(\mathrm{Proj}_j \bar{H}(Q(\bar{s}^i), B(\bar{s}^i), \epsilon), Q(\bar{s}^i), B(\bar{s}^i))$ and $\tilde{S}^i(\bar{s}^i) = \tilde{S}_1^i(\bar{s}^i) \times \cdots \times \tilde{S}_m^i(\bar{s}^i)$. This is a convex subset of $R_1^i \times \cdots \times R_m^i$ as shown in the preceding section. [Here $\psi_j^i(x_j^i, Q_j, B_j)$ the set of all strategies which give x_j^i of commodity j to i assuming that the total supply (bid) of the others in the jth market is Q_j (B_j).] This is a convex set, and is clearly continuous in its variables. From the fact that ϕ_2^i is u.s.c. and ψ is continuous, we easily deduce that $\phi_3^i : \bar{s}^i \Rightarrow R^i$ given by

$$\phi_3^i(\bar{s}^i) = \tilde{S}^i(\bar{s}^i)$$

is also u.s.c.

We will now define a correspondence ϕ_4^i from $\bar{S}^i$ to S^i as follows:

$$\phi_4^i(\bar{s}^i) = \phi_3^i(\bar{s}^i) \cap L^i.$$

(We interpret $r_j^i \in R_j^i$ as (q_j^i, b_j^i) and see that $\phi_4^i(\bar{s}^i) \subset S^i$.) It is obvious from the definition of H^i that this intersection is nonempty. Consider

$$\phi : S \Rightarrow S,$$

where

$$\phi(s) = \phi_4^1(\bar{s}^1) \times \cdots \times \phi_4^n(\bar{s}^n)$$

ϕ satisfies the conditions of Kakutani's Fixed Point Theorem. Hence there is an $\hat{s} \in S$ such that

$$\hat{s} \in \phi(\hat{s}).$$

This is clearly an N.E. of ${}^\epsilon\Gamma$. Q.E.D.

Remark 1. The subset of $\tilde{S}_j^i(\bar{s}^i)$ which contains only those $s_j^i \in R_j^i$ for which $b_j^i \cdot q_j^i = 0$ is nonempty, and in fact consists of a single point. (see Fig. 1.) We will denote this subset by $\hat{S}_j^i(\bar{s}^i)$. $\hat{\phi}_3^i$ will be the map obtained by replacing $\tilde{S}_j^i$ by $\hat{S}_j^i$. Put $\hat{\phi}_4^i = \hat{\phi}_2^i \cap L^i$. This map is nonempty and satisfies the conditions of Kakutani's theorem (in fact, Brouwer's theorem). Its fixed point will still be an N.E. of ${}^\epsilon\Gamma$, but it will have the nice feature that each trader enters any market either as a buyer or as a seller, but not both. (It may, of course, happen that he neither buys nor sells.)

LEMMA 2. *Assume that for each $j \in I_m$, there exist at least two moneyed traders who desirs j, and at least two j-furnished traders who desire money. Let p_j^ϵ denote the price of the j*th *commodity at an N.E. of ${}^\epsilon\Gamma$. There exist positive constants C_j and D_j, for all $j \in I_m$, such that*

$$C_j < p_j^\epsilon < D_j \qquad \textit{for any } \epsilon > 0.$$

In the proof, we will need the following lemma from [4], *quoted with proof in Appendix B.*

LEMMA C (Uniform monotonicity). *Let $j \in I_m$, let $f(x)$ be a continuous, nondecreasing function from Ω^m to the reals that is actually increasing in the variable x_j, and let H be a positive constant. Then a positive number $h = h(f, j, H)$ exists such that for all x and y in Ω^m, if*

$$\| x \| \leqslant H \qquad \textit{and} \qquad \| y - x \| \leqslant h,$$

then

$$f(y + e^j) > f(x).$$

Proof of Lemma 2.[5] Without loss of generality, let $j = m$; and assume that 1 and 2 are moneyed and desire m, and 3 and 4 are m-furnished and desire money.

[5] This closely follows the proof of Lemma 4 in [4].

Put

$$H = \max\{\bar{a}_j : j \in I_{m+1}\},$$
$$h = \min[h(u^1, m, H), h(u^2, m, H)],$$
$$A = (1/(m+1)) \min[a^1_{m+1}, a^2_{m+1}],$$
$$\hat{h} = \min[h(u^3, m+1, H), h(u^4, m+1, H)],$$
$$\hat{A} = \tfrac{1}{2}\min[a_m{}^3, a_m{}^4].$$

Note that these constants are positive under our assumptions.

First we establish the existence of C_m. Put

$$\delta = p_m{}^\epsilon = (\bar{b}_m + \epsilon)/(\bar{q}_m + \epsilon).$$

Suppose first that the condition

$$b_m{}^i \leqslant \frac{\bar{b}_m}{2} \qquad \text{and} \qquad a^i_{m+1} - \sum_{j \in I_m} b_j{}^i \geqslant A \tag{7}$$

holds for at least one of $i = 1, 2$; say for $i = 1$. Then an increase Δ in 1's bid for m would be feasible if Δ is sufficiently small, say, $0 < \Delta \leqslant \min(\epsilon, A)$, and would have the following incremental effect on his final holding:

$$x_j{}^1(\Delta) - x_j{}^1 = 0 \qquad \text{for } j \in I_{m-1};$$

$$\begin{aligned} x_m{}^1(\Delta) - x_m{}^1 &= \frac{(\bar{q}_m + \epsilon)(\bar{b}_m{}^1 + \epsilon)}{\bar{b}_m + \Delta + \epsilon} - \frac{(\bar{q}_m + \epsilon)\, b_m{}^1}{\bar{b}_m + \epsilon} \\ &= (\bar{q}_m + \epsilon)\,\Delta \left[\frac{\bar{b}_m + \epsilon - b_m{}^1}{(\bar{b}_m + \epsilon)(\bar{b}_m + \epsilon + \Delta)}\right] \\ &\geqslant (\bar{q}_m + \epsilon)\,\Delta \left[\frac{\bar{b}_m/2 + \epsilon/2 + \Delta/2}{(\bar{b}_m + \epsilon)(\bar{b}_m + \epsilon + \Delta)}\right] \\ &= \frac{(\bar{q}_m + \epsilon)\,\Delta}{2(\bar{b}_m + \epsilon)} = \frac{\Delta}{2\delta}. \end{aligned} \tag{8}$$

(The "$\geqslant$" above follows from

$$\bar{b}_m + \epsilon - b_m{}^1 \geqslant \frac{\bar{b}_m}{2} + \epsilon \geqslant \frac{\bar{b}_m}{2} + \frac{\epsilon}{2} + \frac{\Delta}{2});$$

$$x^1_{m+1}(\Delta) - x^1_{m+1} = \left(\frac{q_m{}^1}{\bar{q}_m + \epsilon} - 1\right)\Delta \geqslant -\Delta.$$

Define[5]

$$z = -2\delta e^{m+1}$$

[6] e^j is the vector in Ω^{m+1} whose jth component is 1, and all others are 0.

and note that we have the vector inequality

$$x^1(\Delta) \geqslant x^1 + \frac{\Delta}{2\delta}(z + e^m). \tag{9}$$

We are now in a position to apply Lemma C, taking $f = u^1$, $j = m$, and $y = x^1 + z$. We have $x^1 \in \Omega^m$ and $\| x^1 \| \leqslant H$. So, by Lemma C, if both $x^1 + z \geqslant 0$ and $z \leqslant h$, then

$$u^1(x^1 + z + e^m) > u^1(x^1).$$

Since u^1 is concave, this implies that

$$u^1(x^1 + (\Delta/2\delta)(z + e^m)) > u^1(x^1)$$

holds for sufficiently small Δ, and hence, by (9) and the monotonicity of u^i, that

$$u^1(x^1(\Delta)) > u^1(x^1)$$

for such Δ. But this means that trader 1 could have improved, a contradiction. Hence either $x^1 + z < 0$, or $\| z \| > h$. If the former, we have

$$x^1_{m+1} - 2\delta < 0.$$

But $x_{m+1} \geqslant A$; hence

$$\boxed{2\delta > A}\ . \tag{10}$$

If the latter, we have

$$\boxed{2\delta > h}\ . \tag{11}$$

We now consider the case where (7) fails for both $i = 1, 2$. Without loss of generality we may assume that

$$b_m{}^1 \leqslant \bar{b}_m/2.$$

From the failure of (7) for $i = 1$, we therefore have

$$\sum_{j \in I_m} b_j{}^1 > a^1_{m+1} - A \geqslant mA.$$

Hence $b_j^1 > A$ for at least one $j \in I_m$. If $j = m$, we have $b_m^1 > A$ and so a fortiori

$$\boxed{\delta \geqslant A/\bar{a}_m}\ . \tag{12}$$

If $j \neq m$, then $m \geqslant 2$ and so w.l.o.g. assume $j = 1$. Trader 1 could then decrease b_1^1 by a small $\Delta > 0$ and increase b_m^1 by the same amount, with the incremental effect

$$\begin{aligned} x_1^1(\Delta) - x_1^1 &= \frac{(\bar{q}_1 + \epsilon)(b_1^1 - \Delta)}{\bar{b}_1 - \Delta + \epsilon} - \frac{(\epsilon + \bar{q}_1)\, b_1^1}{\bar{b}_1 + \epsilon} \\ &\geqslant \frac{-(\bar{q}_1 + \epsilon)\,\Delta}{\bar{b}_1 + \epsilon}, \\ x_j^1(\Delta) - x_j^1 &= 0 \qquad \text{for} \quad j = 2,\ldots, m-1; \\ x_m^1(\Delta) - x_m^1 &\geqslant \Delta/2\delta, \end{aligned}$$

by the same calculation as at (8), and

$$x_{m+1}^1(\Delta) - x_{m+1}^1 = \left(\frac{q_m^1}{\bar{q}_m + \epsilon} - \frac{q_1^1}{\bar{q}_1 + \epsilon}\right)\Delta - \left(\frac{q_1^1}{\bar{q}_1 + \epsilon}\right)\Delta.$$

If we define

$$z = \left(-2\delta\,\frac{(\bar{q}_1 + \epsilon)}{(\bar{b}_1 + \epsilon)}\right) e^1 - \left(\frac{2\,\delta q_1^1}{\bar{q}_1 + \epsilon}\right) e^{m+1} \tag{13}$$

then (9) is satisfied as before; so, arguing as before, we find that either $x^1 + z < 0$ or $\| z \| > h$. If the former, then either $| z_1 | > x_1^1$ or $| z_{m+1} | > x_{m+1}^1$. But we have

$$\frac{(\bar{q}_1 + \epsilon)\, b_1^1}{\bar{b}_1 + \epsilon} \geqslant \frac{(\bar{q}_1 + \epsilon)\, A}{\bar{b}_1 + \epsilon}$$

and

$$x_{m+1}^1 \geqslant \frac{b_1^1 q_1^1}{\bar{q}_1 + \epsilon} \geqslant \frac{A q_1^1}{\bar{q}_1 + \epsilon}\,.$$

Thus, referring to (13), we see that in both cases

$$\boxed{2\delta > A}\ . \tag{14}$$

If the latter, then either $|z_1| > h$ or $|z_{m+1}| > h$. In the first case the inequalities $\bar{b}_1 + \epsilon > b_1^1 > A$ and $\bar{a}_1 \leqslant H$ yield

$$\boxed{2\delta > hA/H} \tag{15}$$

in the second, $q_1^1 \leqslant \bar{q}_1$ yields

$$\boxed{2\delta > h}\,. \tag{16}$$

This exhausts all cases. Combining (10)–(12) and (14)–(16), we see that it will suffice to take C_m to be the minimum of $A/2$, $h/2$, $2A$, and $Ah/2H$.

The existence of D_m is established in a similar manner.

Suppose first that the condition

$$q_m^i \leqslant \bar{q}_m/2 \qquad \text{and} \qquad a_m^i - q_m^i \geqslant \hat{A} \tag{17}$$

holds for at least one of $i = 3, 4$; say, for $i = 3$. Then an increase of Δ in 3's supply of m would be feasible if $0 < \Delta \leqslant \min(\epsilon, \hat{A})$, and would have the following incremental effect on his final holding:

$$x_j^3(\Delta) - x_j^3 = 0 \qquad \text{for} \quad j \in I_{m-1};$$

$$x_m^3(\Delta) - x_m^3 \doteq \left(\frac{b_m^3}{\bar{b}_m + \epsilon} - 1\right)\Delta \geqslant -\Delta;$$

$$x_{m+1}^3(\Delta) - x_{m+1}^3 = \frac{(\bar{b}_m + \epsilon)}{\bar{q}_m + \Delta + \epsilon}(q_m^3 + \Delta)$$

$$- \frac{(\bar{b}_m + \epsilon)(q_m^3)}{\bar{q}_m + \epsilon} \geqslant \frac{(\bar{b}_m + \epsilon)\,\Delta}{2(\bar{q}_m + \epsilon)} = \frac{\Delta\delta}{2}\,.$$

Define $z = -(2/\delta)\, e^m$ and note that then

$$x^3(\Delta) \geqslant x^3 + (\Delta\delta/2)[z + e^{m+1}].$$

Arguing as before, we get that either $x^3 + z < 0 \,\|\, z \,\| > h$. If the former,

$$\frac{2}{\delta} > x_m^3 \geqslant \hat{A},$$

hence

$$\boxed{\delta < 2/\hat{A}} \;; \tag{18}$$

if the latter,

$$2/\delta > \hat{h},$$

hence

$$\boxed{\delta < 2/\hat{h}} \;. \tag{19}$$

Now consider the case where (17) fails for both $i = 3$ and $i = 4$. Without loss of generality let

$$q_m{}^3 \leqslant \bar{q}_m/2.$$

From the failure of (17) for $i = 3$ we have

$$q_m{}^3 > a_m{}^3 - \hat{A} \geqslant a_m{}^3/2 \geqslant \hat{A}/2.$$

Hence

$$\delta \leqslant (\bar{b}_m + \epsilon)/(q_m{}^3 + \epsilon),$$

i.e.,

$$\boxed{\delta \leqslant 2\bar{a}_{m+1}/\hat{A}} \;. \tag{20}$$

Pick D_m to be the maximum of $2\bar{a}_{m+1}/\hat{A}$, $2/\hat{h}$, $2/\hat{A}$. Q.E.D.

Proof of Theorem 1. Select a sequence $\{{}^g\epsilon\}_{g=1}^{\infty}$, ${}^g\epsilon \in \Omega^1\backslash\{0\}$, ${}^g\epsilon \to 0$. Let ${}^g\nu \in S$ be an N.E. of ${}^{g\epsilon}\Gamma$, and ${}^gp$ the price at ${}^g\nu$. Since S is compact, and since by the above lemma each ${}^gp$ lies in $T = \mathsf{X}_{j=1}^{m} [C, , D,]$, which is also compact, we can clearly pick a subsequence of $\{{}^g\nu\}_{g=1}^{\infty}$ such that (a) ${}^g\nu$ converges to some $\nu \in S$, and (b) ${}^gp$ converges to some $p \in T$. But since ν is a point of continuity of the payoff functions, it is an N.E. of Γ. Q.E.D.

Remark 2. If at the E.P. ν the aggregate bids and offers in any trading-post are zero, we still associate the *virtual price* given by p with that trading post.

Remark 3. As seen earlier (see Remark 1), we may pick ${}^{g\epsilon}\nu = \{({}^{g\epsilon}b^1, {}^{g\epsilon}q^1),\ldots, ({}^{g\epsilon}b^n, {}^{g\epsilon}q^n)\}$ such that ${}^{g\epsilon}b_j{}^i \cdot {}^{g\epsilon}q_j{}^i = 0$ for every $i \in I_n$ and $j \in I_m$; i.e., a trader enters any market either as a buyer or a seller, but not both. But since ν is a limit of $\{{}^{g}\epsilon\nu\}$, this will hold for ν as well.

Remark 4. If any trader at an E.P. has not bid all his money—which will be the case at interior E.P.'s (see below)—and if he also has a positive amount of $j \in I_m$, then an inactive market j can be made active by his bidding and offering in the ratio of the virtual price associated with that market. This clearly does not disturb the E.P.

3.1. *Limit E.P.'s and the C.E.'s*

Let us consider a "replication sequence" of economies ${}_1\Gamma,\ldots, {}_k\Gamma,\ldots$. There is a fixed number t of types of traders, characterized by their utilities u^i and endowments a^i. The economy ${}_k\Gamma$ has $n = kt$ traders, k of each type. An E.P. in which traders of the same type choose the same strategies is called a *type-symmetric* E.P. (denoted T.S.E.P.)[7] and such a T.S.E.P. ν can be represented as a vector $\bar{\nu}$ in $S = \times_{i=1}^{t} S^i$, where S^i is the strategy set of trader i in ${}_1\Gamma$. Thus, for each k, a T.S.E.P. in ${}_k\Gamma$ gives us a price and an allocation in ${}_1\Gamma$ via ${}_k\bar{\nu} \in S$. We will say that ${}_k\nu$ is an "*interior*" E.P. if, for ${}_k\nu = (s^1,\ldots, s^n)$, where $s^i = (q^i, b^i)$, $\bar{b}^i < a^i_{m+1}$, all $i \in I_n$.

Given a price vector $p \in \Omega^m$, we define the budget set of trader i to be

$$B^i(p) = \{x \in \Omega^{m+1} : \tilde{p} \cdot x \leqslant \tilde{p} \cdot a^i\}$$

with $\tilde{p}_j = p_j$ for $j \in I_m$ and $\tilde{p}_{m+1} = 1$ (i.e., the "price" of money is set equal to 1).

A competitive equilibrium (C.E.) then consists of a price $p \in \Omega^m$ and an allocation $(x^1,\ldots, x^n)$, $x^i \in \Omega^{m+1}$, such that

$$\sum_{i=1}^{n} x^i = \sum_{i=1}^{n} a^i,$$

and

$$u^i(x^i) = \max_{y \in B^i(p)} u^i(y) \qquad \text{for all } i.$$

LEMMA 3. *Let ${}_kp_j$ be the price of $j \in I_m$ at a T.S.E.P. in ${}_k\Gamma$. There exist, for each $j \in I_m$, positive constants C_j and D_j such that*

$$C_j < {}_kp_j < D_j$$

for all $k = 1, 2,\ldots$.

[7] That such a T.S.E.P. always exists is shown in Appendix C.

Proof. First observe that for ${}_k\Gamma$, the inequalities of Lemma 2 hold with the same H (hence h) as for ${}_1\Gamma$. For suppose $x_j{}^i > \bar{a}$, for any trader i at a T.S.E.P. in ${}_k\Gamma$. (Here $\bar{a}, = \sum_{i=1}^{t} a_j{}^i$, i.e., the summation is over the trader types, not over all traders.) Then, since $x_j{}^i = x_j{}^l$ if i and l are of the same type, we have

$$\sum_{\text{of the same type as } i} x_j{}^l > k\bar{a}_j = \text{total of } j \text{ in } {}_k\Gamma,$$

a contradiction. Thus (10), (11), (14), (16), (18), (19) hold. (Note that A and $\hat{A}$ do not vary with k.) To complete the proof, we need to check that (12) and (20) also hold.

In the case of (12),

$$\begin{aligned} & b_m{}^1 > A \\ \Rightarrow\ & kb_m{}^1 > kA \\ \Rightarrow\ & \delta \geqslant kA/k\bar{a}_m = A/\bar{a}_m\,. \end{aligned}$$

In the case of (20),

$$\begin{aligned} & q_m{}^3 > \hat{A}/2 \\ \Rightarrow\ & kq_m{}^3 > k\hat{A}/2 \\ \Rightarrow\ & \delta \leqslant 2k\bar{a}_{m+1}/k\hat{A} = 2\bar{a}_{m+1}/\hat{A}. \end{aligned}$$

Q.E.D.

THEOREM 2. *Suppose we have a symmetric, interior*[8] *sequence ${}_k\nu$ of T.S.E.P.'s with prices ${}_kp$, such that ${}_k\bar{\nu} \to \bar{\nu} \in S$ and ${}_kp \to p$. Let $(x^1,\ldots, x^t)$ be the allocation resulting from $\bar{\nu}$. Then $(p, {}^1x^1,\ldots, {}^kx^1,\ldots, {}^1x^t,\ldots, {}^kx^t)$ is a C.E. for each ${}_k\Gamma$, where ${}^jx^i = {}^lx^i$ for all l and j. (Note that, by Lemma* 3, $p \in \mathsf{X}_{j=1}^{m} [C_j\,, D_j]$.)

Before the proof of the theorem we will need some other resujts. Let ${}^\epsilon_k\Gamma$ stand for the ϵ-modification of ${}_k\Gamma$; and let ${}^\epsilon_k\nu$, ${}^\epsilon_kp$ and ${}^\epsilon_kx$ denote respectively the N.E. of ${}^\epsilon_k\Gamma$, and the prices and allocation at this N.E., *so chosen that* ${}^\epsilon_kp \to {}_kp$, as $\epsilon \to 0$. Further, denote by ${}^\epsilon_kB^i_j$ and ${}^\epsilon_kQ^i_j$ the aggregate bid and supply at ${}^\epsilon_k\nu$ due to all the players except i in the jth trading post. If we drop the superscript i we will mean the aggregate due to all the players.

LEMMA 4. *At ${}^\epsilon_k\nu$ trader i maximizes his utility subject to the constraint that he buy and sell in the m markets at* fixed *prices ${}^\epsilon_kp^i_j$ ($j \in I_m$) given by*

$${}^\epsilon_kp^i_j = ({}^\epsilon_kQ^i_j/{}^\epsilon_kB^i_j) \cdot ({}^\epsilon_kp_j)^2.$$

Proof. Consider the "holdings-surface" H of trader i when all others keep their strategies fixed at ${}^\epsilon_k\nu$. The equation of this surface $w_{m+1} = F$

[8] Conditions for interiority are given in [4]. They require, roughly, that traders have "enough" money and desire money.

$(w_1, \ldots, w_m)$ is of the form given on page 4, from which we observe that it is strictly concave and differentiable. Let H' be the unique hyperplane through ${}^{\epsilon}_{k}x^i \in H$, such that H lies below it. H' is determined by the partial derivatives of F at ${}^{\epsilon}_{k}x^i$, and it can be easily checked that

$$\partial F / \partial w_j = {}^{\epsilon}_{k}p_j^{\,i}.$$

To establish the claim, we must show that ${}^{\epsilon}_{k}x^i$ maximizes u^i on H'. Suppose not. Let $y' \in H'$ (see Fig. 2) be such that y' can be attained by i (if he sells and buys at the fixed prices ${}^{\epsilon}_{k}p_j^i$) and $u^i(y') > u^i({}^{\epsilon}_{k}x^i)$. By the continuity of u^i we can pick a y'' below H' such that $u^i(y'') > u^i({}^{\epsilon}_{k}x^i)$. Consider the straight line joining y'' and ${}^{\epsilon}_{k}x^i$. Since H is strictly concave and H' is the unique hyperplane

Q.E.D.

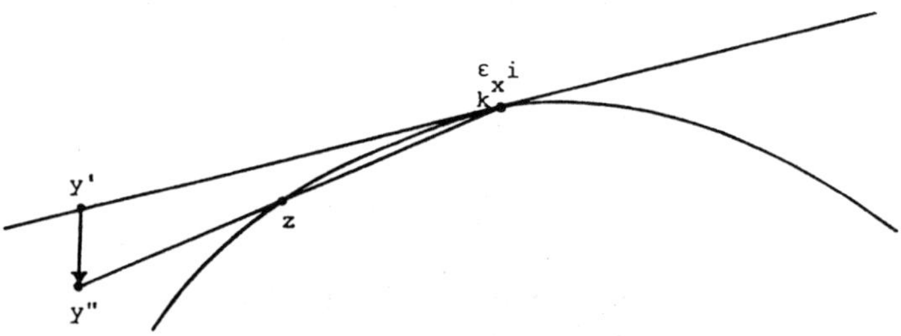

FIGURE 2

that supports it at ${}^{\epsilon}_{k}x^i$, the straight line must intersect H at some point $z \in H$. But then $u^i(z) > u^i({}^{\epsilon}_{k}x^i)$, because u^i is concave. This contradicts the fact that $u^i({}^{\epsilon}_{k}x^i) = \max_{y \in H} u^i(y)$. Q.E.D.

Remark 5. For any $\delta > 0$, there is a k_δ such that $\lim_{\epsilon \to 0} {}^{\epsilon}_{k}p_j^i$ exists for all traders i and all $j \in I_m$, and ${}_{k}p_j - \delta \leqslant \lim_{\epsilon \to 0} {}^{\epsilon}_{k}p_j^i \leqslant {}_{k}p_j + \delta$, if $k \geqslant k_\delta$.

Proof. First observe that

$$1 \leqslant \frac{{}^{\epsilon}_{k}Q_j}{{}^{\epsilon}_{k}Q_j^i} = \frac{Q + (k+1)\,{}^{\epsilon}_{k}q_j^i}{Q + k\,{}^{\epsilon}_{k}q_j^i} \qquad \text{for some } Q > 0$$

$$\leqslant 1 + 1/k.$$

Similarly,

$$1 \leqslant {}^{\epsilon}_{k}B_j / {}^{\epsilon}_{k}B_j^i \leqslant 1 + 1/k.$$

Now, recall that ${}^{\epsilon}_{k}p_j = {}^{\epsilon}_{k}B_j / {}^{\epsilon}_{k}Q_j$, and that ${}^{\epsilon}_{k}p_j \to_{\epsilon \to 0} {}_{k}p_j$. The claim easily follows. Q.E.D.

Proof of Theorem 2. By the continuity of the payoff functions, it follows that ${}_kx^i = \lim_{\epsilon\to 0} {}^{\epsilon}_{k}x^i$ is optimal for trader i if he trades at the fixed prices ${}_kp^i_j = \lim_{\epsilon\to 0} {}^{\epsilon}_{k}p^i_j$. But since he does not bid all his money, ${}_kx^i$ is optimal for i even if we were to relax his money constraint. For ${}_kb^i$ is the solution to the problem

$$\begin{aligned}&\text{Maximize} && u^i(\theta(w))\\ &\text{subject to} && a^i_{m+1} - \sum_{j\in I_m} (w_j) \geqslant 0, \qquad w \geqslant 0,\end{aligned}$$

where $\theta(w)$ is the final bundle that accrues to i if he purchases commodities at fixed prices ${}_kp^i$ with bids w, keeping his sales fixed at ${}_kq^i$. Since u^i is clearly a concave function of w, by the Kuhn–Tucker theorem there exists a nonnegative number λ such that the above problem is equivalent to

$$\text{Maximize } u^i(\theta(w)) + \left[a^i_{m+1} - \sum_{j\in I_m} (w_j)\right]$$

subject to $w \geqslant 0$.
But ${}_kb^i$ is the solution; therefore

$$\lambda\left[a^i_{m+1} - \sum_{j\in I_m} {}_kb^i_j\right] = 0,$$

which implies $\lambda = 0$ since ${}_k\nu$ is an interior T.S.E.P. Thus in fact ${}_kx^i$ is optimal for i on $B^i({}_kp^i)$. Then from the fact that ${}_kx^i \to_{k\to\infty} x^i$, ${}_kp^i \to_{k\to\infty} p$, it is easy to deduce that x^i is optimal for i on $B^i(p)$. Q.E.D.

4. Concluding Remarks

A *Comparison with the All for Sale Model*

In the "all for sale" model [3] all goods in the economy must go through the markets even if many (or all) of them are bought back by their original owners. In the model here, only the amounts intended for sale go through the markets.

We are led immediately to some macroeconomic distinctions. The "all for sale" model measures total *national wealth* "at market" whereas the model here measures *national income*, or *national product*, if there are no wash sales.

$$\text{National wealth } W = \sum_{i=1}\sum_{j=1} p_j a_j{}^i,$$

$$\text{National product} = \text{National income } F = \sum_{i=1}^{n}\sum_{j=1}^{m} p_j q_j{}^i$$

(where $q_j{}^i$ is the amount of good j offered for sale by i).

In particular, we obtain the strange, but accurate, observation that if we started this system at a point of equilibrium the national product and income would be zero! This merely says that as the system begins in equilibrium nothing can be added by trade or production.

Another interpretation which can be given to the amount

$$F = \sum_{i=1}^{n} \sum_{j=1}^{m} p_j q_j^i$$

is that it is the amount of *float* required to run trade in a monetary economy where velocity of money per period is at most one. If all trade were absolutely simultaneous and all individuals had only enough money to cover trade then we observe that the velocity of money would be exactly one and the size of the float would equal the national income.

On Wash Sales

In our original Cowles Foundation Discussion Paper [10] an explicit example is calculated to show the effect of wash sales in creating a thicker market than would otherwise exist when there are few traders. The reader interested in yet another phenomenon which can be illustrated in an Edgeworth box is referred to the original publication!

Appendix A

We verify that $D^i(Q, B, \epsilon)$ is convex. Take w' and w'' in $D^i(\ )$ and consider $\hat{w} = \lambda w' + (1 - \lambda) w''$, where $0 \leqslant \lambda \leqslant 1$. We want to show that $\hat{w} \in D^i(\ )$.

Suppose w' $[w'']$ is obtained through the strategy (q', b') $[(q'', b'')]$. Define $(\hat{q}, \hat{b})$ by

$$\hat{q}_j = \min(q_j', q_j'') \qquad \text{for } j \in I_m,$$
$$\hat{b}_j = \lambda b_j' + (1 - \lambda) b_j'' \qquad \text{for } j \in I_m.$$

Since w_j is a decreasing function of q_j (if we keep b_j fixed),

$$\hat{w}_j' = \xi_j^i(\hat{q}, b') \geqslant \xi_j^i(q', b') = w',$$
$$\hat{w}_j'' = \xi_j^i(\hat{q}, b'') \geqslant \xi_j^i(q'', b'') = w''.$$

But each component of w is concave in b (keeping q fixed), hence

$$\hat{w} = \lambda w' + (1 - \lambda) w'' \leqslant \lambda \hat{w}' + (1 - \lambda) \hat{w}''$$
$$\leqslant \xi^i(\hat{q}, \lambda b' + (1 - \lambda) b'') = w^*.$$

By the intermediate value theorem in calculus, observe that since setting $\hat{q}_i = 0$, $b_i' = 0$, and $b_i'' = 0$ would make $w_i^* = 0$, we may reduce $\hat{q}_i$, b_i', and b_i'' appropriately to get $\hat{w}_i$ for all $i \in I_m$. Q.E.D.

Appendix B

We quote, with proof, the following lemma from [4] that is invoked in the proof of Lemma 3 in Section 3.

Lemma C (Uniform monotonicity). *Let $j \in I_m$, let $f(x)$ be a continuous, nondecreasing function from Ω^m to the reals that is actually increasing in the variable x_j, and let H be a positive constant. Then a positive number $h = h(f, j, H)$ exists such that for all x and y in Ω^m, if*

$$\| x \| \leqslant H \text{ and } \| y - x \| \leqslant h,$$

then

$$f(y + e^j) > f(x).$$

Proof. Let $Q(\alpha)$ denote the cube $\{x \in \Omega^m\colon \| x \| \leqslant \alpha\}$. The function $f(x + e^j) - f(x)$ is positive and continuous, and so has a positive minimum μ in $Q(H + 1)$. The function f itself is uniformly continuous in $Q(H + 1)$, so there exists a $0 < \delta < 1$ such that $x \in Q(H)$, $y \in \Omega^m$, and $\| y - x \| < \delta$ imply that $|f(y) - f(x)| < \mu$, and hence that $f(y + e^j) - f(x) > 0$. So we may take $h = \delta$. Q.E.D.

Appendix C

We will first prove[9] that there always exists a T.S.N.E. of ${}^{\epsilon}\Gamma$. Define S^* to be the subset of S which consists of type-symmetric strategies, i.e.,

$$S^* = \{s \in S : s^i = s^j \text{ if } i \text{ and } j \text{ are of the same type}\}.$$

S^* is clearly convex. Consider the map

$$S^* \xrightarrow{\tilde{\phi}} 2^{S^*}$$

given by

$$\tilde{\phi}(s) = \phi(s) \cap S^*.$$

[9] This proof is the same as the one in [4].

$\tilde{\phi}(s)$ is always nonempty because at any $s \in S^*$ all players of any given type face the same optimization problem, therefore there will be identical strategies that lie in their "best-response sets" $\phi_4^i(\bar{s}^i)$. $\tilde{\phi}$ is, moreover, convex-valued and u.s.c. We appeal to the Kakutani Fixed Point Theorem, and see that the fixed point of $\tilde{\phi}$ is a T.S.N.E. of ${}^{\epsilon}\Gamma$. A T.S.E.P. is then obtained as a limit of T.S.N.E.'s of ${}^{\epsilon_i}\Gamma$ as $\epsilon_i \to 0$.

References

1. M. Shubik, Commodity money, oligipoly, credit and bankruptcy in a general equilibrium Model, *West. Econ. J.* **10** (1972), 24–38.
2. Shapley, L. S., Noncooperative general exchange, *in* "Theory and Measurement of Economic Externalities" (S. A. Y. Lin., Ed.), Academic Press, New York, 1977. (Also P-5286, Santa-Monica, Calif., Rand Corporation, 1974.)
3. Shapley, L. S. and M. Shubik, Trade using one commodity as a means of payment, Rand Corporation, 1975; *J. Polit. Econ.* **85**, 937–968, 1977.
4. Shapley, L. S. and M. Shubik, Models of noncooperative exchange, in preparation.
5. Shubik, M., On the number of types of markets with trade in money, CFDP 416, January 14, 1976.
6. Dubey, P. and M. Shubik, Trade and prices in a closed economy with exogenous uncertainty, different levels of information, money and special futures markets, CFDP 410R, January 1976; *Econometrica*, **45**, 1657–1680, 1977.
7. Shubik, M., "The Optimal Bankruptcy Rule in a Trading Economy Using Fiat Money," CFDP 424; February 9, 1976 (revised with C. Wilson, 1976).
8. Berge, Claude, "Topological Spaces" Macmillan, New York, 1963.
9. Dubey, P. and M. Shubik, A closed economic system with production and exchange modelled as a game of strategy, CFDP 429, April 1976; *J. Math. Econ.* **4**, 1–35, 1977.
10. Dubey, P. and M. Shubik, The noncooperative equilibria of a closed trading economy with market supply and bedding strategies, CFDP 422R, August 1976.

Printed by the St Catherine Press Ltd., Tempelhof 37, Bruges, Belgium.

Journal of Mathematical Economics 6 (1979) 115–134. © North-Holland Publishing Company

BANKRUPTCY AND OPTIMALITY IN A CLOSED TRADING MASS ECONOMY MODELLED AS A NON-COOPERATIVE GAME

Pradeep DUBEY and Martin SHUBIK

Cowles Foundation, Yale University, New Haven, CT 06520, USA

Received December 1977, final version received October 1978

In this paper we establish the existence of (a) an optimal bankruptcy rule which enables us to describe the Walrasian trading economy as a game with trade in fiat money; and (b) non-cooperative equilibrium points of this game which (in terms of prices and the final allocation yielded) include the competitive equilibrium points, and the accompanying money rate of interest (induced by borrowing at a central bank), when the bankruptcy rule is different from optimal.

1. Introduction

In this paper we establish the existence of (a) an optimal bankruptcy rule which enables us to describe the Walrasian trading economy as a game with trade in fiat money; and (b) non-cooperative equilibrium points of this game which (in the sense specified below) include the competitive equilibria of the trading economy.

The construction suggested here raises several questions concerning the uniqueness of competitive equilibria, the role of a money rate of interest in providing reserves against bankruptcy and the information needed by a banking system to influence a market economy. Hopefully an understanding of the answers to these questions may lead to the formulation of some fruitful dynamic models.

In a previous paper [e.g., Shubik and Wilson (1977)] a simple model of a trading economy with two types of traders, two goods, an outside bank, borrowing and bankruptcy rules, has been studied exhaustively. In that paper a specific example was used to illustrate the effect of varying the bankruptcy conditions. It was shown that there were essentially four cases to be distinguished. They are when no firm goes bankrupt, some go bankrupt and all go bankrupt, and also when no one goes bankrupt but some or all hoard money.

The details of the case distinctions can best be appreciated by working through a specific example in full. However, this is both time consuming and lengthy, hence rather than provide a new example here, the reader who wants an example is referred to the other paper [e.g., Shubik and Wilson (1977)].

In this paper we establish in general for a trading economy a series of results concerning games with penalties against those who fail to repay loans. In particular:

(1) We can define a class of games related to the original trading economy for which we can prove the existence of non-trivial noncooperative equilibria.

(2) These equilibria can be compared with and related to the competitive equilibria of the original trading economy.

(3) The severity of the penalty for each individual can be characterized by a single parameter λ^i for each trader i. If the penalties are sufficiently high then the game we can define has among its equilibria all of the competitive equilibria of the original trading economy, in the sense that the distribution of goods and prices associated with any competitive equilibrium (C.E.) are also associated with a non-cooperative equilibrium (N.E.).

(4) With high penalties, the money rate of interest will be zero at equilibrium and hoarding may take place.[1] The money rate of interest in the context of these models is a loss reserve premium to recoup bankruptcy losses rather than a way for the financing of intertemporal trade and terminal capital stock.

(5) If the penalties λ^i are picked to be sufficiently small and if they are not in the ratio of some set of Lagrangian multipliers at a C.E. then there will be a positive rate of interest and the distribution of goods at an N.E. will not coincide with that at a C.E.

(6) If the original trading economy has K competitive equilibria, associated with each C.E. for a given utility function representation of preferences will be a set of Lagrangian multipliers. For the kth C.E. we may have $({}^k\lambda^1, \dots, {}^k\lambda^n)$. These multipliers will vary along with any order preserving transformation of the utility functions. If it is possible to find some utility function representation of the preferences of the traders such that there is no pair of C.E.'s for which ${}^j\lambda^i \geqq {}^k\lambda^i$, $i = 1, \dots, n$, then we may construct K trading games such that each of these games has a specific C.E. as its unique N.E. with a zero interest rate.

In the previous paper of Shubik and Wilson a market with a finite number of traders was considered. It appears to be considerably easier to work with a non-atomic trading model to establish the general results noted here. The recent work of Dubey and Shapley (1978) enables us to adopt this approach in section 3 and subsequently.

[1]This possibility is illustrated by Shubik and Wilson (1977) but is ruled out in our analysis here by a simplification of the strategy spaces.

These rather informally stated assertions are made precise and the appropriate theorems are proved in the subsequent sections.

2. Models and modelling problems

In several previous papers [e.g., Shubik (1972), Shapley and Shubik (1977), and Dubey and Shubik (1976)] trade in an economy using a commodity money has been studied. The conditions under which trade is hampered by a shortage of money have been noted. A shortage of commodity money can be overcome by the introduction of credit. However if the economy is regarded as even implicitly dynamic we must take into account the possibility that individuals may not be able to pay back that which they have borrowed. When this happens, rules must be given concerning the treatment of default.

We may regard an economy which uses fiat money as one in which a form of credit system is used. An outside bank issues a supply of money to all traders in return for their promissory notes. After trade has taken place the traders are required to redeem their promissory notes. Thus we model the economy as a two-stage game where stage one is used to get the fiat money into the system and stage two is used for trade.

2.1. The extensive form

In figs. 1a and 1b two alternative forms of the extensive form of the two-stage game are suggested. In fig. 1a all traders bid simultaneously for fiat money, then all are completely informed. In the second stage they then bid in the market for goods. The trees have been drawn as though the set of moves is finite.

In fig. 1b individuals are not informed of the outcome of the first stage prior to having to make their plans for the second stage. These represent two extremes in information conditions and imply extremely different strategy sets. The strategy sets are discussed in detail in section 2.4 below.

2.2. Endowments and utility functions

Although in section 3 we restrict our interest to the non-atomic game; for convenience in discussion modelling problems prior to setting up this game we consider an economy with a finite number of traders.

Let there be n traders trading m commodities. Trader i has an initial endowment of $(a_1^i, \ldots, a_m^i)$ where $a_j^i \geqq 0$ and $\bar{a}_j (= \sum_i a_j^i) > 0$.

The preferences of a trader i for any vector of resources $(x_1^i, x_2^i, \ldots, x_m^i)$ can be represented by a utility function $u^i(x_1^i, \ldots, x_m^i)$.

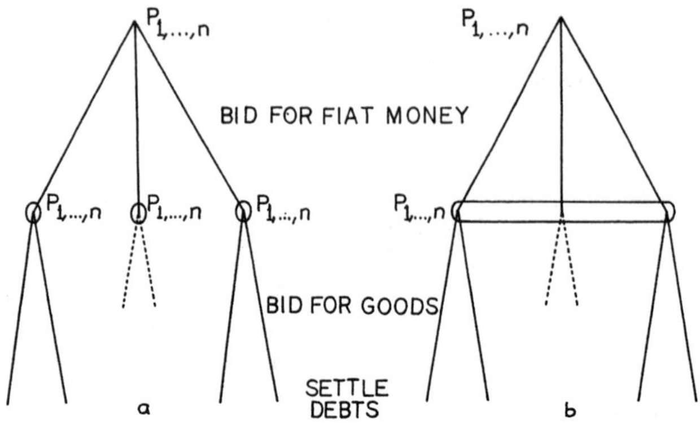

Fig. 1

2.3. *Moves, outcomes and modified utility functions*

Let a move by trader i in the fiat money market be a bid (of a personal I.O.U. note) of size w^i.

Suppose[2] that the total fiat supply is M. He will obtain

$$(w^i/\bar{w})M.$$

A move by trader i in the second stage consists of a set of numbers $b_1^i,\ldots,b_m^i$ which in one model described in section 2.4 can be interpreted as actual expenditures subject to the condition

$$\sum_{j=1}^{m} b_j^i \leqq (w^i/\bar{w})M,$$

in another model a move will be a set of numbers $\eta_1^i,\ldots,\eta_m^i$, which will be interpreted as a percentage allocation of funds[3] thus actual expenditures will be a set of numbers $b_1^i,\ldots,b_m^i$ where $b_j^i=\eta_j^i(w^i/\bar{w})M,\ j=1,\ldots,m$.

We define a set of penalties $\lambda^i>0,\ i=1,\ldots,n$, and a payoff function for each[4]

$$\begin{aligned} \Pi^i(x_1^i,\ldots,x_m^i,x_{m+1}^i) &= u^i(x_1^i,\ldots,x_m^i) && \text{for}\quad x_{m+1}^i \geqq 0,\\ &= \hat{\Pi}^i(x_1^i,\ldots,x_m^i,x_{m+1}^i) && \text{for}\quad x_{m+1}^i<0, \end{aligned}$$

[2]The case where there is no bound on the total fiat supply is analyzed in Dubey and Shapley (1978).

[3]In this formulation hoarding is ruled out.

[4]For the second model the b_j^i are replaced by $^*b_j^i$.

where

$$\hat{\Pi}^i(x_1^i,\dots,x_m^i,x_{m+1}^i)\leqq u^i(x_1^i,\dots,x_m^i)+\lambda^i x_{m+1}^i;$$

where

$$x_{m+1}^i=w^iM/\bar{w}-\sum_{j=1}^m b_j^i+\sum_{j=1}^m p_j a_j^i-w^i;$$

where

$$p_j=\bar{b}_j/\bar{a}_j.$$

2.4. *Strategies, rationing mechanisms and payoffs*

In a game where all players move without information, moves and strategies coincide. When information is available to the players in the process of making a sequence of moves then strategies can be clearly differentiated from moves.

Figs. 1a and 1b show two extreme cases of full information or no information after the bids for money. In the Model 1 a strategy is described by a number and m functions of n variables $(w^i, f_1^i(w^1,w^2,\dots,w^n),\dots,f_m^i(w^1,w^2,\dots,w^n))$. This means that because each individual knows precisely how much money all others have obtained he may base his selection of the b_j^i's upon every w^i. In the Model 2 a strategy and a move coincide. A move or strategy consists of $m+1$ numbers $(w^i,b_1^i,\dots,b_m^i)$.

The most realistic model of the information structure is one in which each individual knows the amount of money he has obtained from the bids in the first stage and the *aggregate* amount for all others. In this game a strategy is described by a number and m functions of two variables $(w^i,f_1^i(w^1,\bar{w}^1),\dots,f_m^i(w^m,\bar{w}^m)$, where $\bar{w}^i=$ where $\bar{w}=\sum_{j\neq 1}^n w^j$.

Model 2 has the least information, and hence the simplest strategies. The price paid for the simplicity of the strategies and the lack of information is the necessity to introduce an extra rationing device, described in section 2.3 which guarantees that all bids are feasible. The device suggested here calls for a simple scaling up or down of bids. One could argue that this *ad hoc* rule is not 'realistic', it is however simple and well defined and is probably one of a large class of rules which lead to models with the same limit non-cooperative equilibria.

In attempting to define trade as a process, apparently arbitrary allocation rules are called for. They may be regarded as the elemental description of the way trading and financial institutions ration resources when systems are in disequilibrium.

2.5. Non-cooperative equilibria

The solution concept employed in this paper is that of the Nash non-cooperative equilibrium (N.E.). Although this solution concept can be formally defined with ease for a multi-stage game, frequently even for games as specially structured as game models of economic exchange, the extreme multiplicity of equilibrium points makes the solution concept less attractive.

In spite of the aesthetic appeal of the competitive equilibrium theory and the formal extension of the equilibrium concept to a multi-period economy, it provides neither an adequate description of the economy in all states of disequilibrium nor a sufficiently comprehensive description of strategy in a multi-period model. In particular because the Walrasian analysis has concentrated primarily upon equilibrium it has been vague about the information conditions, the specifics of price generation and the details of individual selection of strategy. The reformulation of multistage economic trade as a non-cooperative game forces a specificity in describing information conditions and the details of process. Although it may be possible to identify classes of non-cooperative equilibria which appear to be naturally related to the competitive equilibria, it is also possible that other classes of non-cooperative equilibria exist which reflect more complex strategies and communication than are implicitly or explicitly assumed in the study of competitive equilibria or related non-cooperative equilibria.

Even though we regard the general equilibrium analysis to be vague in its treatment of information we may give an interpretation of the C.E.'s in a class of multi-stage games in terms of N.E.'s. Consider the class of multi-stage economies where individuals are equally informed. This class of games contains at one extreme the game where all know everything (as in fig. 1a) and at the other extreme the game where no one obtains further information during the play of the game (fig. 1b). As the amount of information is increased so is the size and complexity of the sets of strategies available to each trader. However the games with more information contain the strategy sets of the games with less information as subsets. In particular for large trading games which have the C.E.'s as N.E.'s (or limit N.E.'s) given complete ignorance during trade, there will be related games with more information which also have the C.E.'s as N.E.'s. More specifically the C.E.'s will be the only N.E.'s belonging to all games which can be constructed from an original game with trade under no information, by the symmetric increase of information.

The replacement of the Walrasian model by the game formulation provides a far more 'open or loosely coupled' model than previously. The use of non-cooperative equilibrium solution provides a more general class of solutions which contains the competitive equilibria as a special case. It must be stressed that we regard the emphasis laid here upon the non-cooperative equilibrium as merely a preliminary to the construction of a truly dynamic theory.

3. The non-cooperative market games

3.1. Notation and definitions

For a positive integer k, Ω^k denotes the non-negative orthant of Euclidean space of dimension k. For any $x \in \Omega^k$, x_j is the jth component of x, and $\bar{x}$ is $\sum_{j=1}^k x_j \cdot \| \ \|$ stands for the maximum norm, i.e., $\|x\| = \max\{|x_j| : j=1,\dots,k\}$. For any x and y in Ω^k, $x > y$ [$\geqq$, $=$] means $x_j > y_j$ [$\geqq$, $=$] for $j=1,\dots,k$. 0 denotes both the number zero and the origin of Ω^k, but the meaning will be clear from the context. e_j is the vector whose jth component is 1 and all other components are 0. H^k is the unit simplex in Ω^k, i.e., $H^k = \{x \in \Omega^k : \bar{x} = 1\}$.

Let $\{I, \mathscr{B}, \beta\}$ be a measure space. If f and g are measurable functions from I to Ω^k, we write $f > g$[$\geqq$, $=$] if $f(x) > g(x)$ [$\geqq$, $=$] for all $x \in I$ except perhaps those x that belong to a null set, i.e., a set $S \in \mathscr{B}$ for which $\beta(S) = 0$. Also we denote $f(x)$ often by f^x.

3.2. The non-atomic market

Let $\{T, \mathscr{C}, \mu\}$ be the measure space of traders, where T is the closed unit interval $[0,1]$ and corresponds to the set of traders, $\mathscr{C}$ is the Borel σ-algebra of coalitions, and $\mathscr{C}$ is a finite, positive, non-atomic, complete measure on $\{T, \mathscr{C}\}$. W.l.o.g. we set $\mu(T) = 1$. The phrase 'all traders' (or 'each trader') will mean all traders (or each trader) except perhaps those that belong to a null set. For a measurable function $f : T \to R^k$; we abbreviate $\int_T f(t)\,\mathrm{d}\mu(t)$ to $\int f$, and $\int_S f(t)\,\mathrm{d}\mu(t)$ to $\int_S f$.

Trade occurs in m commodities. A commodity bundle is represented by a vector in Ω^m. The initial endowment of the traders is given by an integrable function $a : T \to \Omega^m$ where a^t is the endowment of trader t. $\int a_j$ is denoted by $\bar{a}_j$. $\hat{a}$ is the vector $(\bar{a}_1, \dots, \bar{a}_m)$. We assume that $\hat{a} > 0$, i.e., there is a positive net amount of each commodity in the market.

To complete the data of the market we must also specify the preferences of the traders on commodity bundles. This is given (not uniquely) by a function $u : T \times \Omega^m \to \Omega^1$, where $u(t, \cdot)$ – also written $u^t(\cdot)$ – is the utility function of t. Our assumptions on u are as follows [compare with the assumptions in Aumann (1966)]:

(A.1) u is measurable in the product space $T \times \Omega^m$ (where Ω^m is the standard Borel space).

(A.2) u^t is continuous, concave, and non-decreasing for each t.

(A.3) There is an integrable function $v: T \to \Omega^m$, $v > a$, such that:
(i) if $x_j \geqq v_j^t$, then $u^t(x + \Delta e_j) = u^t(x)$ for all $\Delta > 0$;
(ii) if $x_j < v_j^t$, then u^t is strictly increasing in the jth variable at x.

An allocation is a measurable function $x: T \to \Omega^m$ such that $\int x \, \mathrm{d}\mu = \int a \, \mathrm{d}\mu$. It describes a redistribution of the commodities among the traders. A competitive equilibrium (C.E.) of the market is a pair (p, x) where $p \in \Omega^m \setminus \{0\}$ is a price vector, and x is an allocation such that, for all t, x^t is optimal in t's budget set,

$$B^t(p) = \{y \in \Omega^m : p \cdot y \leqq p \cdot a^t\},$$

i.e.,

$$u^t(x^t) = \max \{u^t(y) : y \in B^t(p)\} \quad \text{and} \quad x^t \in B^t(p).$$

An allocation x [a price $p' \in \Omega^m \setminus \{0\}$] is called competitive if there is a $p \in \Omega^m \setminus \{0\}$ [an allocation x'] such that (p, x) [(p', x')] is a C.E.

It is easy to check that, under our assumptions, if p is competitive then $p > 0$. For suppose $p_j = 0$ for some j. Let x^t be any optimal vector in $B^p(t)$. We must have $x_j^t = v_j^t$. So if x is any allocation such that (p, x) is a C.E., $\int x_j = \int v_j > \int a_j$, a contradiction.

Let us also recall the notion of 'shadow prices of income'[5] at a C.E. Suppose (p, x) is a C.E. Then each trader t maximizes $u^t(y)$ subject to $y \in \Omega^m$, $p \cdot y - p \cdot a^t \leqq 0$, and the maximum is obtained at $x^t \in B^t(p)$. By the Kuhn–Tucker theorem there exists a number $\lambda^t \geqq 0$ such that x^t is also a solution of the following unconstrained problem:

$$\max[u^t(y) + \lambda^t(p \cdot a^t - p \cdot y)],$$

subject to

$$y \in \Omega^m.$$

The function $\lambda: T \to \Omega^1$ constitutes a choice of shadow prices at the C.E. (p, x). We show in appendix A that for any C.E. a measurable choice of shadow prices exists, i.e., λ can be picked to be measurable. Note that if (p, x) is a C.E. with shadow prices λ, then for any $K > 0$, (Kp, x) is also a C.E. with shadow prices $(1/K)\lambda$. For any $B > 0$ we will call a C.E. (p, x) B-normalized if $p \cdot \hat{a} = B$.

[5]Henceforth abbreviated 'shadow prices'.

3.3. *Trade using fiat money*

To recast the market in the form of a non-cooperative game we must describe the strategy sets and payoff functions of the traders. We will consider the continuum version of the finite game described earlier. Thus a strategy of trader t is to announce a bid ω^t of promissory notes, and a vector $\eta^t \in H^m$ which constitutes a decision as to how he will divide the fiat money he obtains from the bank into bids on the m trading-posts. Denoting t's strategy set by S^t, we then have

$$S^t = \{\{\omega^t, \eta^t\} : \omega^t \in \Omega^1, \eta^t \in H^m\}.$$

Given a choice of strategies by the traders, how are the trading-posts and the outside bank cleared? We are beset by a fundamental difficulty when we consider the mechanism of section 2. This is because the mechanism calls for aggregating the bids in the bank and in each trading-post. These would be $\int \omega$ and $\int b_j$. But the integrals make sense only if the functions ω and b are assumed to be measurable. It is not clear how we would justify this assumption heuristically. Why should independent decision-makers behave in a jointly measurable way? We refer to section 5 of Dubey and Shapley (1978), where a model of non-cooperative behavior is suggested which leads to measurable strategies.

Assuming then that ω and η are measurable, and letting M stand for the fiat money in the bank, we define

$$1+\sigma = (\textstyle\int \omega)/M.$$

The amount of fiat money obtained by t is[6] $\omega^t/(1+\sigma)$, and thus his bid b_j^t on the jth trading-post is $\eta_j^t \omega^t/(1+\sigma)$. The price p_j is $\int b_j/\bar{a}_j$, and the final holding (of commodities) of t is $x^t \in \Omega^m$, where $x_j^t = b_j^t/p_j$. His credit at the bank is given by

$$x_{m+1}^t = -\omega^t + \sum_{j=1}^{m} p_j a_j^t = -(1+\sigma)\bar{b}^t + \sum_{j=1}^{m} p_j a_j^t.$$

Hence, given a preassigned choice of bankruptcy penalties $\lambda : T \to \Omega^1$, the payoff to t is: $u^t(x^t) + \lambda^t \min[0, x_{m+1}^t]$.

We now have a game in strategic (or normal) form. With the rest of the data fixed, it depends on the choice of $\lambda : T \to \Omega^1$, and hence we will denote it by Γ_λ. A non-cooperative equilibrium (N.E.) of this game is a measurable

[6] Division by 0 is defined to yield 0 throughout this paper.

$s_*: T \to \Omega^1 \times H^m$, $s_*^t \in S^t$, such that, for all t,

$$\Pi^t(s_*) = \max_{s^t \in S^t} \Pi(s_* | s^t),$$

where $(s_* | s^t)$ is the same as s_* except that s_*^t is replaced by s^t.

There exists a trivial N.E. of Γ_λ, namely the collection of strategies in which each trader bids nothing, i.e., $\omega = 0$. We will focus our attention on *active* N.E.'s, namely those which produce positive prices in each trading post. This implies that $1+\sigma>0$, and in fact (as is easily checked) $\sigma \geqq 0$.

We wish to investigate the N.E.'s of Γ_λ as λ varies. For this purpose it will be useful to demarcate certain regions in which λ may lie. Let

$$\Lambda = \{\lambda : T \to \Omega^1 : \lambda \text{ is measurable}\},$$

$$\Lambda_c = \{\lambda \in \Lambda : \lambda \text{ is a choice of shadow prices at some } M\text{-normalized C.E. of the market}\}.$$

For any $^*\lambda \in \Lambda$,

$$\Lambda_{*\lambda} = \{\lambda \in \Lambda : \lambda \geqq {}^*\lambda\},$$

$$\Lambda_{(\lambda,\sigma)} = \{{}^*\lambda \in \Lambda : (1+\sigma)^*\lambda = \lambda\},$$

$$\Lambda' = \Lambda / [\Lambda_c \cup \{\Lambda_{*\lambda} : {}^*\lambda \in \Lambda_c\} \cup \{\Lambda_{(\lambda,\sigma)} : \lambda \in \Lambda_c, \sigma > 0\}.$$

In the following when we say that an N.E. coincides with a C.E., we mean coincidence in prices and allocation.

Proposition 1. If $\lambda \in \Lambda_c$, there is an (active) N.E. of Γ_λ with $\sigma = 0$, which coincides with the C.E. associated with λ.

Proof. Let (p, x) be the C.E. associated with λ. Define $s: T \to \Omega^1 \times H^m$, $s^t \in S^t$, as follows:

$$\omega^t = p \cdot x^t,$$

$$\eta_j^t = p_j x_j^t / \omega^t.$$

It is clear that, at s, $\sigma = 0$ and the prices and allocation produced are p and x. The proof will be finished by showing that s is an N.E. Denote, for any $b \in \Omega^m$, the vector $(b_1/p_1, \ldots, b_m/p_m)$ by $x(b)$. We must show that, for all t, $\max\{u^t(x(b)) + \lambda^t \min\{0, -\bar{b} + p \cdot a^t\}\}$ is achieved at $b = \omega^t \eta^t$. This follows by noting that, since λ is a choice of shadow prices at (p,x),

$$\max\{u^t(x(b))+\lambda^t(-\bar{b}+p\cdot a^t)\}$$

is achieved at

$$b=\omega^t\eta^t=(p_1x_1^t,\dots,p_mx_m^t).$$

and that

$$u^t(x(b))+\lambda^t(-\bar{b}+p\cdot a^t)\geqq u^t(x(b))+\lambda^t\min[0,\ -\bar{b}+p\cdot a^t],$$

for all $b\in\Omega^m$, with '=' when $b=\omega^t\eta^t$. Q.E.D.

Proposition 2. If $^*\lambda\in\Lambda_\lambda$ for some $\lambda\in\Lambda_c$, there exists an N.E. of $\Gamma_{*\lambda}$ with $\sigma=0$, which coincides with the C.E. associated with λ.

Proof. Let s be as in the previous proof, and observe that since $^*\lambda\geqq\lambda$,

$$u^t(x(b))+{}^*\lambda^t\min\{0,\ -\bar{b}+p\cdot a^t\}\leqq u^t(x(b))+\lambda^t\min\{0,\ -\bar{b}+p\cdot a^t\}$$

when '=' when $b=\omega^t\eta^t$. Q.E.D.

Proposition 3. Suppose there is an (active) N.E. of $\Gamma_{*\lambda}$ with $\sigma=0$. Then this N.E. coincides with some M-normalized C.E. (p,x). Moreover, there exists a set of shadow prices λ for (p,x) such that $^*\lambda\in\Lambda_\lambda$.

Proof. Let s be the N.E., (p,x) the prices and allocation produced at s, and x_{m+1} the credit. It is directly verified that $\int x=\int a$. Next note that $x_{m+1}^t\leqq 0$ for all t. For suppose that $x_{m+1}^t>0$ for t in a non-null set S. Then each trader $t\in S$ could improve his payoff by increasing his bids, contradicting that s is an N.E. But since $\sigma=0$, $\int x_{m+1}=0$. Hence $x_{m+1}^t=0$ for all t. It follows immediately that (p,x) is a C.E. Let $\hat{\lambda}\in\Lambda$ be a choice of shadow prices associated with (p,x), i.e.,

$$\max_{b\in\Omega^m}\{u^t(x(b))+\hat{\lambda}^t[-\bar{b}+p\cdot a^t]\}$$

occurs at $\hat{b}^t=(p_1x_1^t,\dots,p_mx_m^t)$. If $^*\lambda^t<\hat{\lambda}^t$, and $^*b^t$ solves

$$\max_{b\in\Omega^m}\{u^t(x(b))+{}^*\lambda^t[-\bar{b}+p\cdot a^t]\},$$

it can be verified that we must have $-{}^*\bar{b}^t \leqq -\bar{\bar{b}}^t$. Suppose $-{}^*\bar{b}^t < -\bar{b}^t$ for t in a non-null set L. Then $-{}^*\bar{b}^t$ is also a solution of

$$\max\{u^t(x(b)) + {}^*\lambda^t \min[0, -\bar{b} + p \cdot a^t] \quad \text{for} \quad t \in L,$$

because

$$u^t(x(b)) + {}^*\lambda^t[-\bar{b}^t + p \cdot a^t] \geqq u^t(x(b)) + {}^*\lambda^t \min[0, -\bar{b} + p \cdot a^t],$$

with '=' when $b = {}^*b^t$ (since $-{}^*\bar{b}^t < \bar{b}^t = p \cdot a^t$). But then $x^t_{m+1} < 0$ for $t \in L$, a contradiction. Hence $-{}^*\bar{b}^t = -\bar{\bar{b}}^t$ for all t such that ${}^*\lambda^t < \hat{\lambda}^t$. Let S' be the set $\{t \in T : {}^*\lambda^t < \lambda^t\}$. Define λ as follows:

$$\begin{aligned} \lambda^t &= \lambda^t \quad \text{if} \quad t \notin S', \\ &= {}^*\lambda^t \quad \text{if} \quad t \in S'. \end{aligned}$$

Then λ is a choice of shadow prices for (p, x), and it is clear that ${}^*\lambda \geqq \lambda$. Q.E.D.

Proposition 4. Suppose there is an N.E. of $\Gamma_{*\lambda}$ with $\sigma > 0$ which coincides with a C.E. Then ${}^*\lambda \in \Lambda_{(\lambda,\sigma)}$ where λ is a choice of shadow prices for the C.E.

Proof. Let (p, x) be the C.E. with shadow prices $\hat{\lambda}$, ${}^*s$ the N.E. of $\Gamma_{*\lambda}$, ${}_*b$ the bids made in the second stage at ${}^*s$. Then

$$\max_{b \in \Omega^m} \{u^t(x(b)) + \hat{\lambda}^t[-\bar{b} + p \cdot a^t]\} \quad \text{is attained at } {}_*b^t, \tag{1}$$

$$\max_{b \in \Omega^m} \{u^t(x(b)) + {}^*\lambda^t \min[0, -(1+\sigma)\bar{b} + p \cdot a^t]\} \quad \text{is attained at } {}_*b^t. \tag{2}$$

Denote the functions in (1) and (2) by $\hat{\xi}^t(b)$ and ${}^*\xi^t(b)$. We will find it useful to define, for $y \in \Omega^1$,

$$\begin{aligned} \hat{\phi}^t(y) &= \max\{\hat{\xi}^t(b) : b \in \Omega^m, \bar{b} = y\}, \\ {}^*\phi^t(y) &= \max\{{}^*\xi^t(b) : b \in \Omega^m, \bar{b} = y\}, \\ \phi^t(y) &= {}^*\phi^t(y) + {}^*\lambda\sigma {}_*\bar{b}^t, \\ \bar{\phi}^t(y) &= \max\{u^t(x(b)) + (1+\sigma){}^*\lambda^t[-\bar{b} + p \cdot a^t] : b \in \Omega^m, \bar{b} = y\}, \\ \bar{u}^t(y) &= \max\{u^t(x(b)) : b \in \Omega^m, \bar{b} = y\}. \end{aligned}$$

Note that, by (A.3), $\bar{u}^t$ becomes constant (say c) for large enough y; also note the relations (where γ^t, $^*\gamma^t$, γ^t, $\tilde{\gamma}^t$ are as in fig. 2)

$$\hat{\phi}^t = \bar{u}^t + \hat{\gamma}^t,$$

$$^*\phi^t = \bar{u}^t + {}^*\gamma^t,$$

$$\phi^t = \bar{u}^t + \gamma^t,$$

$$\tilde{\phi}^t - \bar{u}^t + \tilde{\gamma}^t.$$

Let $\hat{G}^t$, $^*G^t$, $\tilde{G}^t$ and G^t denote the set of $y \in \Omega^1$ on which the max of $\hat{\phi}^t$, $^*\phi^t$, $\tilde{\phi}^t$ and ϕ^t are obtained. Note that these are closed and convex sets since the functions are continuous and concave. It is also easy to see that these sets must be compact as a consequence of the saturation assumption (A.3) (see fig. 2). Hence we may represent them by closed intervals $[\hat{A}^t, \hat{B}^t]$, $[{}^*A^t, {}^*B^t]$,

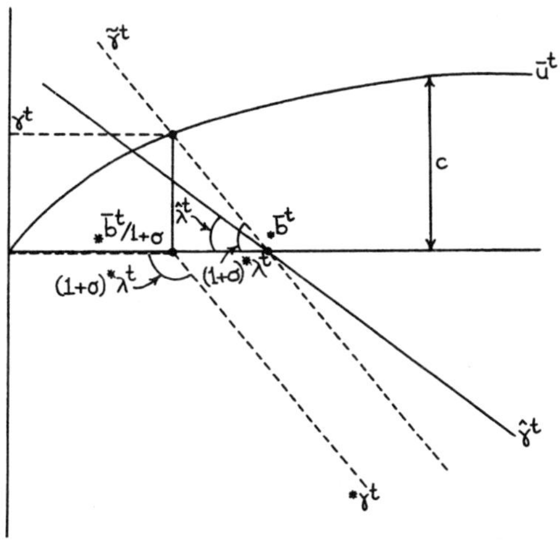

Fig. 2

$[\tilde{A}^t, \tilde{B}^t]$, $[A^t, B^t]$. Moreover $[{}^*A^t, {}^*B^t] = [A^t, B^t]$ because $^*\phi^t$ differs from ϕ^t by a constant. Let $S_1 = \{t \in T : (1+\sigma)^*\lambda^t > \tilde{\lambda}^t\}$ and $S_2 = \{t \in S_1 : \tilde{B}^t < {}_*\bar{b}^t\}$. We will show that $\mu(S_2) = 0$. In the following let t be an element of S_2.

If $\tilde{B}^t \geqq {}_*\bar{b}^t/(1+\sigma)$, then $B^t = \tilde{B}^t$ because $\tilde{\phi}^t(y) \geqq \phi^t(y)$ for all y, and $\tilde{\phi}^t(y) = \phi^t(y)$ if $y \geqq {}_*\bar{b}^t/(1+\sigma)$. But, by (2), $B^t = {}^*B^t \geqq {}_*\bar{b}^t$ for all t. This contradicts that $B^t = \tilde{B}^t < {}_*b^t$. Thus we see that for all $t \in S_2$, $\tilde{B}^t < {}_*\bar{b}^t/(1+\sigma)$. Take a $\delta > 0$ such that $\delta B^t + (1-\sigma){}_*\bar{b}^t = y' \in [{}_*\bar{b}^t/(1+\sigma), {}_*\bar{b}^t)$. Since $\tilde{\phi}^t$ is concave $\tilde{\phi}^t(y') \geqq \delta\tilde{\phi}^t(\tilde{B}^t) + (1-\sigma)\tilde{\phi}^t({}_*\bar{b}^t)$. But $\tilde{B}^t < {}_*b^t$ implies $\tilde{\phi}(\tilde{B}^t) > \tilde{\phi}^t({}_*\bar{b}^t)$. Hence $\tilde{\phi}^t(y') > \tilde{\phi}^t({}_*\bar{b}^t)$. This shows that $\phi^t(y') > \phi^t({}_*\bar{b}^t)$ because $\tilde{\phi}^t(y) = \phi^t(y)$ for $y \geqq {}_*\bar{b}^t/(1+\sigma)$. Consequently $B^t < {}_*\bar{b}^t$. By (2) however $B^t = {}^*B^t \geqq {}_*\bar{b}^t$. This proves that $\mu(S_2) = 0$.

Next let $S_3=\{t\in T:(1+\sigma)^*\lambda^t<\hat{\lambda}^t$, and $S_4=\{t\in S_3:\tilde{A}^t>{}_*\bar{b}^t\}$. We show that $\mu(S_4)=0$. Let t now refer to an element of S_4. Clearly $[\tilde{A}^t,\tilde{B}^t]=[A^t,B^t]$ because $\tilde{\phi}^t(y)\geqq\phi^t(y)$ for all y, and $\tilde{\phi}^t(y)=\phi^t(y)$ if $y\geqq{}_*\bar{b}^t/(1+\sigma)$. On the other hand, for all t, $A^t={}^*A^t\leqq{}_*\bar{b}^t$. This proves that $\mu(S_4)=0$.

Define λ as follows:

$$\begin{aligned}\lambda^t&=\hat{\lambda}^t && \text{if}\quad t\in T\backslash(S_1\cup S_3),\\ &=(1+\sigma)^*\lambda^t && \text{if}\quad t\in S_1\cup S_3.\end{aligned}$$

Recalling that $\hat{\lambda}$ is a shadow price choice for (p,x) and that $\mu(S_2)=\mu(S_4)=0$, we see that λ is also a shadow price choice for (p,x). But clearly ${}^*\lambda=(1+\sigma)\lambda$. Q.E.D.

Proposition 5. Suppose $\lambda\in\Lambda'$. *Then* $\inf\{\sigma:\sigma$ *occurs at an N.E. of* $\Gamma_\lambda\}>0$.

Proof. In the light of the previous propositions, it suffices to show that if $\inf\{\ldots\}=0$, then there exists an N.E. of Γ_λ for which $\sigma=0$.

Suppose ${}^ks$ is a sequence of N.E.'s of Γ_λ with rates of interest ${}^k\sigma$ and prices ${}^kp$, and suppose $\lim_{k\to\infty}{}^k\sigma=0$.

Select a subsequence of ${}^ks$ such that $\lim_{k\to\infty}{}^kp$ exists and is, say, p. By Lemma 1 below $p>0$. Also note that $\|{}^kp\|\leqq F$ for all k, and hence $\|p\|\leqq F$, where $F=\{\max M/\bar{a}_j:j=1,\ldots,m\}$.

Let ${}^kb$ be the bids in fiat money made in stage two at ${}^ks$. Clearly, ${}^kb_j^t<Fv_j^t$. Therefore the functions ${}^kb$ are bounded from above (component-wise) by the integrable function Fv. For each t let $B(t)$ be the set of limit points of ${}^kb^t$ as $k\to\infty$. Let ${}^kB(t)$ be the set consisting of the single point ${}^kb^t$. Then $B(t)=\limsup{}^kB(t)$. Since all the ${}^kB$ are bounded above by the integrable function Fv, by Lemma 5.3 in Aumann (1966), we have

$$\int p\cdot a=\lim\int{}^kp\cdot a=\lim\int{}^kb\in\limsup\int{}^kB\subset\int\limsup{}^kB=\int B.$$

Let ${}^*b$ be such that ${}^*b^t\in B(t)$ for all t and $\int{}^*b=\int p\cdot a$.

First we show that ${}^*\bar{b}^t=p\cdot a^t$ for all t. Let ${}^*b^t=\lim_{r\to\infty}{}^{k_r}b^t$. Now ${}^{k_r}\bar{b}^t\geqq{}^{k_r}p\cdot a^t/(1+\sigma^{k_r})$, hence by letting r tend to ∞, we get ${}^*\bar{b}^t\geqq p\cdot a^t$. But $\int{}^*\bar{b}=\int p\cdot a$, hence ${}^*\bar{b}^t=p\cdot a^t$ for all t.

Let ${}^rx(b)$ be the vector $(b_1/{}^{k_r}p_1,\ldots,b_m/{}^{k_r}p_m)$. Now ${}^{k_r}b^t$ maximizes $u^t(x(b))+\lambda^t\min[0,(1+{}^{k_r}\sigma)\bar{b}-{}^{k_r}p\cdot a^t]$ at ${}^{k_r}b^t$. Since by Lemma 1 ${}^{k_r}p\xrightarrow[r\to\infty]{}p>0$, ${}^rx(b)\to x(b)=(b_1/p_1,\ldots,b_m/p_m)$ for all b. Then recalling that ${}^{k_r}\sigma\xrightarrow[r\to\infty]{}0$, and ${}^{k_r}b^t\to{}^*b^t$, we see that ${}^*b^t$ maximizes $u^t(x(b))+\lambda^t\min[0,b-p\cdot a^t]$. Put $s^t=\{\omega^t,\eta^t\}$ where $\omega^t={}^*\bar{b}^t$, and $\eta_j^t={}^*b_j^t/\omega^t$. It is easy to verify that s is an N.E. of Γ_λ with $\sigma=0$. Q.E.D.

Lemma 1. Let ${}^k p$ *be as in Proposition 5. Assume* $\lim {}^k p \xrightarrow[k\to\infty]{} p$. *Then* $p > 0$.

Proof. Let α be the measure generated by the indefinite integral of v_1, i.e., $\alpha(S) = \int_S v_1$. α is absolutely continuous w.r.t. μ. Therefore for any $\varepsilon > 0$, there is a $\delta(\varepsilon) > 0$, such that $\mu(S) < \delta(\varepsilon)$ implies $\alpha(S) < \varepsilon$.

Let $A = \int v_1 - \int a_1 > 0$. Pick a K and an N^* small enough to ensure that $K + N^* < A$. For any allocation x, define

$$S_x = \{t \in T : x_1^t < v_1^t - N^*\}.$$

We assert that $\mu(S_x) \geqq \delta(K)$, *for any allocation* x. Suppose not. Then $\int x_1 \geqq \int_{T \setminus S_x} v_1 - N^* \mu(T \setminus S_x) \geqq \alpha(T) - K - N^*$, therefore.

$$\begin{aligned} K + N^* &\geqq \int v_1 - \int x_1 \\ &= \int v_1 - \int a_1 > A, \end{aligned}$$

a contradiction.

From now on denote $\delta(K)$ by K^*.

Define $h : T \to R$ as follows:

$$h^t = \min\{u^t(x + e_1) - u^t(x) : x_1 < v_1^t - N^*\}.$$

By our assumptions on u^t, $h^t > 0$ for all t. Furthermore[7] h is measurable by Proposition 3 in Hildenbrand (1974, p. 60). Let $Z_1 \subset Z_2 \subset \dots$ be defined by

$$Z_n = \{t \in T : h^t > 1/n\}.$$

Since $h > 0$, $\bigcup_{n=1}^{\infty} Z_n = T$. Hence $\mu(Z_n) \xrightarrow[n\to\infty]{} 1$. Pick n_1 such that $\mu(Z_{n_1}) > 1 - K^*/2$. Then, *for any allocation* x,

$$\mu(S_x \cap Z_{n_1}) > K^*/4.$$

Finally define $V_1 \subset V_2 \subset \dots$ by

$$V_k = \{t \in T : 2\lambda^t < k\}.$$

Again $\bigcup_{k=1}^{\infty} V_k = T$, so $\mu(V_k) \xrightarrow[k\to\infty]{} \mu(T)$, and there exists a k_1 large enough to guarantee that $\mu(V_{k_1}) > 1 - K^*/8$. Put $W_x = S_x \cap Z_{n_1} \cap V_{k_1}$. Observe that, for any allocation x, $\mu(W_x) > K^*/8$.

[7]We are grateful to Gerard Debreu for this reference.

We are prepared to prove that $p>0$. Suppose not. W.l.o.g. let $p_1=0$ and assume[8] that ${}^l\sigma<1$ for all l. For large enough l, ${}^lp_1<1/n_1k_1$. Consider the set W_{l_x}. It is non-null (in fact its measure is at least $K^*/8$). If any $t\in W_x$ increases his bid on commodity 1 by the amount $1/n_1k_1$ the increase in his final holding is given by

$$\Delta^l x_j^t=0 \quad \text{for} \quad j=2,\dots,m,$$

$$\Delta^l x_1^t \geqq 1/n_1k_1{}^lp_1>1.$$

Hence the increase in the utility of the final bundle of t is greater than $1/n_1$. On the other hand the disutility for going bankrupt is at most $(1+{}^l\sigma)\lambda^t/n_1k_1 < k_1/n_1k_1=1/n_1$. Hence t can improve his payoff by increasing his bid on 1 in the amount $1/n_1k_1$. Since each $t\in W_{l_x}$ can do this, ${}^ls$ cannot be an N.E. of Γ_λ, a contradiction. Q.E.D.

Proposition 6. Let $\lambda\in\Lambda'$. Suppose (p,x) are the prices and allocation at an active N.E. of Γ_λ. Then (p,x) is not a C.E.

Proof. Obvious from the previous propositions.

Proposition 7. Let ${}^*\lambda=(1+\sigma)\lambda$, where $\lambda\in\Lambda_c$. Let (p,x) be the C.E. corresponding to λ. Then there exists an N.E. of $\Gamma_{*\lambda}$ coinciding with the C.E. and with rate of interest equal to σ.

Proof. The following s is easily verified to be the N.E.:

$$s^t=(\omega^t,\eta^t),$$

where

$$\omega^t=(1+\sigma)\sum_{j=1}^{m} p_jx_j^t,$$

$$\eta_j^t=p_jx_j^t \Big/ \sum_{j=1}^{m} p_jx_j^t. \qquad \text{Q.E.D.}$$

4. Concluding remarks

4.1. On uniqueness

Suppose that the penalties have been set at some function λ which corresponds specifically to the choice of shadow prices at some C.E. Can we deduce that the N.E. for this game is unique?

[8]Recall that we are assuming here that $\lim_{l\to\infty}{}^l\sigma=0$.

There appear to be two different questions which must be answered before such a conclusion could be drawn.

It is clear that for an economy with K C.E.'s if there is a utility function representation of preferences such that the shadow prices at any one C.E. are never $\geqq$ (for each trader) the shadow prices at another C.E. then there will only be one C.E. with $\sigma=0$ for any game with penalties set this way.

Question 1. We have not been able to establish that this will always be the case. Lloyd Shapley has provided an example for specific utility functions where the shadow prices at one C.E. strictly dominate those at another. But there are concavity preserving transformations of the utility functions for this case which get rid of the domination. We do not know, however, if this can be done in general. (fig. 3 provided by Shapley shows the example.)

Even supposing that we had an economy where the conditions for λ_c with a unique equilibrium with $\sigma=0$ were met, another question remains.

Question 2. Given a game with a unique N.E. with $\sigma=0$ is it possible that another N.E. with $\sigma>0$ exists? We have not been able to prove this but we suspect that the answer is no without too many qualifications beyond differentiability and perhaps strict concavity.

Fig. 4 is drawn for a non-atomic economy with two types of traders. We consider only those λ which are type-symmetric, hence we may represent them by vectors in Ω^2. Fig. 4 shows that λ-space. We assume that there are a finite number of C.E.'s, and for any C.E. there is a unique choice of (type-symmetric) shadow prices, represented by • in Fig. 4.

From our propositions an intimate relationship emerges between the bankruptcy penalties and optimality. Ignoring those λ that lie in $U\{\Lambda_{(\lambda,\sigma)}:\lambda\in\Lambda_c,\sigma>0\}$, which is in any case a 'small' set (see fig. 4), the active N.E.'s of Γ_λ contain at least one C.E. if and only if $\lambda\in\Lambda_c\cup\{\Lambda_\lambda:\lambda\in\Lambda_c\}$. For such λ $\inf\{\sigma:\sigma$ occurs at an active N.E. of $\Gamma_\lambda\}=0$; indeed, those N.E.'s with $\sigma=0$ are precisely the ones that coincide with C.E.'s. For other λ, this $\inf>0$, and the N.E.'s do not contain any C.E. We can think of this inf as a measure of the non-optimality (or deviation from a C.E.).

Another way of considering this is to conceive of all games defined by taking λ's associated with every point on the Pareto optimal surface of the trading economy. To each game there will be a smallest σ defined, where $\sigma=0$ at the choice of λ's associated with C.E.'s thus we have a function of σ defined on the Pareto optimal surface.

4.2. *On production*

When production is considered a new set of possibilities appear. In particular if the prices for leftover goods at the settlement period are positive

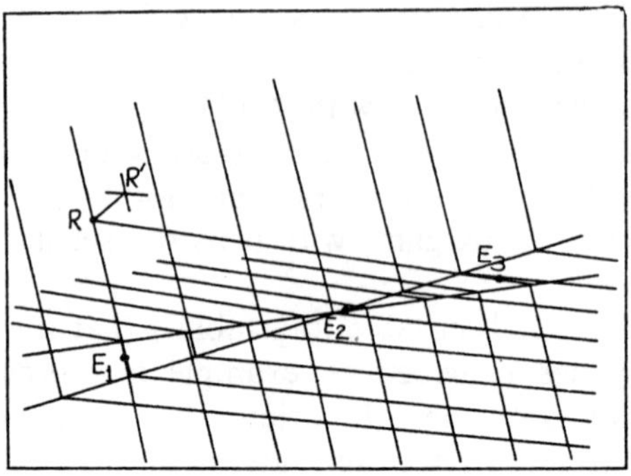

Fig. 3. Piecewise linear utility functions – can be made strictly concave and differentiable. Prices at E_1: (0.8, 0.2), and at E_3: (0.1, 0.9).

$$\frac{\text{I's marginal utility of income at } E_3}{\text{I's marginal utility of income at } E_1} = 3:1.$$

$$\frac{\text{II's marginal utility of income at } E_3}{\text{II's marginal utility of income at } E_1} = 3:1.$$

II is *worse* off at E_3, but has *greater m.u.i.* (I is better off at E_3, and has greater m.u.i.)

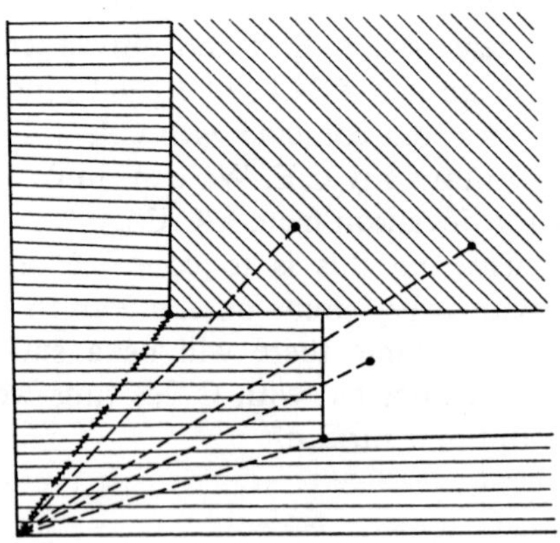

Fig. 4. is Λ', deleting the broken lines, and is Λ_λ. is $U\{\Lambda_{\lambda,\sigma}:\sigma>0\}$, σ increases as we move towards the origin.

then a positive money rate of interest appears even though the bankruptcy penalties are set appropriately (as defined earlier, i.e., from Λ_c). For an analysis of this, see Dubey and Shubik (1977).

Appendix A[9]

Consider the map $T \times \Omega^m \times \Omega^1 \xrightarrow{F} \Omega^1$, where

$$F(t, y, \mu) = u^t(y) + \mu(p \cdot y - p \cdot a^t).$$

ϕ is measurable, and so is $T \times \Omega^1 \xrightarrow{G} \Omega^1$ where

$$G(t, \mu) = \max_{y \in \Omega^m} F(t, y, \mu).$$

Denoting the C.E. under consideration by (p, x), now define subsets L and R in $T \times \Omega^1 \times \Omega^1$ as follows:

$$L = \{[t, \mu, G(t, \mu)] : (t, \mu) \in T \times \Omega^1\},$$
$$R = \{[t, \mu, u^t(x^t)] : t \in T, \mu \in \Omega^1\}.$$

Clearly L and R are measurable, and so is $S = L \cap R$. Let P be the projection of S on $T \times \Omega^1$. By Proposition 3 in Hildenbrand (1974, p. 60), P is measurable. Also for each $t \in T$, $(\{t\} \times \Omega^1) \cap P \neq \phi$ (by the Kuhn–Tucker theorem). Now we invoke the measurable choice theorem [e.g., Aumann and Shapley (1974, prop. 36.5)] to claim that there is a measurable function $\lambda : T \to \Omega^1$ such that $(t, \lambda^1) \in P$ for all t. This λ clearly constitutes a set of shadow prices for (p, x).

[9]We are grateful to Gerard Debreu for suggesting the outline of this argument.

References

Aumann, R.J., 1966, Existence of competitive equilibria in markets with a continuum of traders, Econometrica 34, 1.

Aumann, R.J. and L.S. Shapley, 1974, Values of non-atomic games (Princeton University Press, Princeton, NJ).

Dubey, P. and L.S. Shapley, 1978, Noncooperative exchange with a continuum of traders, Econometrica, forthcoming.

Dubey, P. and M. Shubik, 1976, The noncooperative equilibria of a closed trading economy with market supply and bidding strategies, Cowles Foundation Discussion Paper no. 422R (Yale University, New Haven, CT); curtailed version in the Journal of Economic Theory.

Dubey, P. and M. Shubik, 1977, A theory of money and financial institutions, Part 36: The money rate of interest (A multiperiod nonatomic trading and production economy with outside money, inside money and optimal bankruptcy rules).

Hildenbrand, W., 1974, Core and equilibrium of a large economy (Princeton University Press, Princeton, NJ).

Shapley, L.S. and M. Shubik, 1977, Trade using one commodity as a means of payment, Journal of Political Economy 85, 5.

Shubik, M., 1972, Commodity money, oligopoly, credit and bankruptcy in a general equilibrium model, Western Economic Journal X, 4.

Shubik, M. and C. Wilson, 1977, The optimal bankruptcy rule in a trading economy using fiat money, Zeitschrift für Nationalökonomie 37, 3–4.

Reprinted from JOURNAL OF ECONOMIC THEORY
Vol. 22, No. 2, April 1980

Efficiency Properties of Strategic Market Games: An Axiomatic Approach*

PRADEEP DUBEY

Yale University, New Haven, Connecticut 06520

ANDREU MAS-COLELL

University of California, Berkeley, California 94720

AND

MARTIN SHUBIK

Yale University, New Haven, Connecticut 06520

Received July 23, 1979

The paper investigates the conditions under which an abstractly given market game will have the property that if there is a continuum of traders then every noncooperative equilibrium is Walrasian. In other words, we look for a general axiomatization of Cournot's well-known result. Besides some convexity, continuity, and nondegeneracy hypotheses, the crucial axioms are: anonymity (i.e., the names of traders are irrelevant to the market) and aggregation (i.e., the net trade received by a trader depends only on his own action and the mean action of all traders). It is also shown that the same axioms do not guarantee efficiency if there is only a finite number of traders. Some examples are discussed and a notion of strict noncooperative equilibrium for anonymous games is introduced.

1. INTRODUCTION

In this paper we shall attempt an axiomatization of a central theme of economic theory, namely, the idea that the noncooperative equilibria of

* This paper brings together and supersedes Dubey and Shubik (1978) and Mas-Colell (1978). We are indebted for discussion, suggestions, and amendations to K. Arrow, W. Heller, L. Hurwicz, Ch. Kahn, E. Maskin, A. Postlewaite, and W. Thomson. The usual caveat applies. A. Mas-Colell's research was supported by NSF Grant SOC76-19700A01 and SOC77-06000. The latter grant was at the Institute for Mathematical Studies in the Social Sciences, Stanford University. The work of Dubey and Shubik relates to Department of the Navy Contract N00014-77-C-0518, issued by the Office of Naval Research under contract authority NR047-006.

market-like economies with many relatively small traders are Walrasian (hence efficient) or, in other words, that price-taking behavior is, in a mass market, the natural consequence of "message taking" behavior. The classical partial equilibrium reference is, of course, Cournot's (1838). Two modern general-equilibrium version of Cournot's model are represented by Shubik (1973) [see also Shapley (1976), Shapley and Shubik (1977), Jaynes, Okuno, and Schmeidler (1978)] and by Gabszewicz–Vial (1972) [see also Hart (1979) and Novshek–Sonnenschein (1978)]. The work reported here develops from the Shubik line.

We take as our conceptual starting point the notion of Strategic Market Game, to be understood as a complete specification of trading rules and transaction constraints with respect to which a noncooperative equilibria is reached. This is to be contrasted with the theory of the core of a market, [which originated in Edgeworth (1881); see Hildenbrand (1974) and his references] where final outcomes are completely unconstrained by institutional mechanisms of trade.

The game being played in real economic life is quite complex. It is clear that there is not a unique, somehow given, way to formalize it as a strategic market game. So it is sound research strategy to build a gallery of models emphasizing different important features, their fine structure reflecting the host of institutional features of the economy not usually taken into account by the economic theorist, but of importance in the understanding of marketing and transaction technologies. Nevertheless, experience with different models indicates that a number of facts are quite robust to changes of specification. Among them: the noncooperative equilibria of markets with a finite number of traders tend (this is a suitably vague verb) to be inefficient, and the noncooperative equilibria of markets with many relatively small participants tend to be (almost) efficient. Thus, it seems justified in order to bring out the robustness of these phenomena to engage in axiomatics.

We proceed by sidestepping the modeling problem and assuming (admittedly, this is quite a lot to assume) that a strategic market game is given to us in an abstract manner. We then try to identify general principles yielding the efficiency results. Our axiomatization will be fairly transparent and, from the mathematical point of view, trivial. The basic principles turn out to be:

(i) convexity: traders have available a convex set of strategies.

(ii) anonymity: from the point of view of the market, only the message sent by the trader matters.

(iii) continuity: of outcomes with respect to strategies.

(iv) aggregation: the trading possibilities of any player are influenced by the messages of the other players only through the *mean* of those messages (and not, say, through the variance).

(v) nondegeneracy: it must be possible for individual players to influence to a substantial extent their trading possibilities in the market.

With the given axioms we first prove some inefficiency results for the finite number of traders case and then establish efficiency for the large number of small traders case. As in Aumann (1964) for "large number of small traders," we shall adopt the idealization of the continuum. We do this merely for convenience and in order to sharpen the essential facts. We will, however, devote one section to the asymptotic, in contrast to limit, theory. It is fortunate that the consideration of a large number of small traders (in the limit a continuum) has simplifying implications for noncooperative analysis. Indeed, under general conditions, if individual players are, in whichever is the relevant sense, very small relative to the size of the game, the only sensible expectations (at "equilibrium" or at "disequilibrium" plays) of individual players is that the rest of the players will not react (or react very little) to changes of their actions. Thus, in contrast to games among few, simple Cournot–Nash equilibrium emerges as the natural noncooperative solution. Our treatment of Cournot–Nash equilibrium in the continuum builds on a paper of Schmeidler (1973) which deserves to be better known.

There are clear similarities and connections between our formal structure and the extensive literature growing out of Hurwicz's (1960) seminal contribution on the designing of resource allocation mechanism. But it is important to keep in mind that the spirit of our work is very different. In particular, we do not attach normative significance to our results. For us, the market mechanism is given, and we merely analyze its properties.

The paper is organized as follows: Sections 2–5 present the basic model. Section 6 enumerates the axioms. Section 7 looks at the continuum as a limit of the finite but large situation. Section 8 discusses a nondegeneracy issue for equilibria. Sections 9 and 10 present efficiency results for the finite and continuum cases, respectively. Section 11 looks at examples; and, finally, Section 12 defines and discusses a type of noncooperative equilibria of special relevance to market situations.

2. Commodities, Agents, Assignments

The commodity space is R^l.

We let I, the set of agents names, be either a finite set equipped with the counting measure or the interval [0, 1] with Lebesgue measure (denoted λ). This will allow us to treat simultaneously the finite and the continuum cases.

An *assignment* is an integrable map $\mathbf{x}\colon I \to R^l$ such that $\int \mathbf{x}(t)\, dt = 0$. Note that we are concerned only with net trades.

3. Economics

We now look at the agents in I as the traders of an economy.

A *trader* is characterized by:

(i) a closed, convex, nonempty set $X \subset R^l$ of net trades, interpreted as the set of contracts the trader can honor, and

(ii) a complete and transitive preferences relation $\gtrsim \subset X \times X$. Every X is assumed to satisfy *free disposability*, i.e., if $x \in X$, then $x + R_+^l \subset X$. Preferences $\gtrsim$ are taken to be continuous, convex, and *monotone*. The set of characteristics is $\mathscr{A} = \{(X, \gtrsim)\}$. As in Hildenbrand (1974), one can view $\mathscr{A}$ as a measurable space by endowing it with the σ-field derived from the closed convergence topology on the closed subsets of R^l.

We define an economy à la Aumann (1963), i.e., as a (measurable) map $\mathscr{E}: I \to \mathscr{A}$. We denote $\mathscr{E}(t)$ by $(X_t, \gtrsim_t)$.

An assignment $\mathbf{x}: I \to R^l$ is *feasible* if $x(t) \in X_t$ for a.e. $t \in I$.

The existence of two types of trade constraints is a basic distinction in our treatment. We have the *public constraints* embodied in the fact that the sum of net trades must be zero (i.e., markets should clear) and the *private constraints* requiring that individual net trades belong to the individual trading sets. As a matter of interpretation we take that the private constraints are not publicly observable.

A feasible assignment $\mathbf{x}$ is efficient if there is no other feasible assignment $\mathbf{x}'$ such that $\mathbf{x}'(t) \gtrsim_t x(t)$ for a.e. $t \in I$ and $x'(t) >_t x(t)$ for a set of t's of positive measure.

A feasible assignment $\mathbf{x}$ is *Walrasian* if there is $p \in R^l$ such that for a.e. $t \in I$, $\mathbf{x}(t)$ is $\gtrsim_t$ maximal on $\{x \in X_t : p \cdot x \leqslant 0\}$. It is, of course, well known that, with the hypothesis made, a Walrasian assignment is efficient.

Remark 1. Note that (i) individual production (i.e., firms with a single owner) is encompassed by the model, (ii) consumption or technological externalities are not permitted.

4. Strategic Markets Games

We now look at the set of agents I as a set of players.

There is given a nonempty set S called the message, action, or strategy space. We assume that S is a *subset of a separable Banach space*. It will be conceptually simpler if the reader thinks of S as lying in a finite-dimensional space, i.e., a subset of some R^m.

A *play* is an integrable function $\mathbf{s}: I \to S$.

A *strategie market game* is specified by the *set of players I*, the *strategy space S* and an *outcome rule* Φ, which associates with every play $\mathbf{s}$ an assignment $\Phi(\mathbf{s})\colon I \to R^l$. If Φ is understood, we write $\mathbf{x_s}$ for $\Phi(\mathbf{s})$.

For the measure theoretic treatment to be sensible, we should require that the outcome rule satisfies the following property: *if* $[\mathbf{s}] = [\mathbf{s}']$, *then* $[\mathbf{x_s}] = [\mathbf{x_{s'}}]$. The symbol [ ] denotes the equivalence class of the function (i.e., all functions which are a.e. equal to the given function). Informally speaking, this property, which is nonvacuous only for the case $I = [0, 1]$, has the implication that (again for $I = [0, 1]$) individual actions do not have a macroscopic influence on market outcomes (i.e., pecuniary externalities are absent), which is a well-known heuristic requirement for a competitive process to yield efficient outcomes.

Let us illustrate with some examples. They will be discussed with more detail in Section 11, from where, especially for Example 1, a clearer understanding of the underlying economics will be gained.

EXAMPLE 1. [Shubik (1973)]. There are two commodities. The strategy space is R_+^2. A message (s_1, s_2) is to be interpreted as follows: s_1 (resp. s_2) is the amount of commodity 1 (resp. 2) a trader proposes to deliver to the market. Let $\mathbf{s}$ be a play. Suppose first that $\int \mathbf{s}_1, \int \mathbf{s}_2 > 0$. Then the outcome is given by

$$\mathbf{x_s}(t) = \left(\frac{\int \mathbf{s}_1}{\int \mathbf{s}_2}\, \mathbf{s}_2(t) - \mathbf{s}_1(t), \frac{\int \mathbf{s}_2}{\int \mathbf{s}_1}\, \mathbf{s}_1(t) - \mathbf{s}_2(t)\right).$$

If either $\int \mathbf{s}_1 = 0$ or $\int \mathbf{s}_2 = 0$, we put $\mathbf{x_s} \equiv 0$.

In the next two examples the strategies are whole demand-supply functions.

EXAMPLE 2. Again $l = 2$. The strategy set is $S = \{f \in C([0, r]) : f(0) \geqslant 0, f(r) \leqslant 0\}$, where r is some large number and for $p \in [0, r]$ one interprets $f(p)$ as a proposal to buy ("sell" if negative) $f(p)$ units of the second commodity in exchange for $pf(p)$ units of the first commodity. Let $\mathbf{s}$ be a play. Define

$$p = \max \left\{p \in [0, r]\colon \left(\int \mathbf{s}\right)(p) = 0\right\}.$$

Then the outcome is given by

$$\mathbf{x_s}(t) = (-p\, \mathbf{s}(t)(p), \mathbf{s}(t)(p)).$$

This example shows that there is substance in the implicit hypothesis that to every play $\mathbf{s}$ there corresponds a unique assignment. The rule for choosing one among the possibly many price equilibria is completely arbitrary. In the next example, the equilibrium is naturally unique.

EXAMPLE 3. Identical to Example 2 except that now $S = \{f \in C([0, r]) : f(0) \geqslant 0, f(r) \leqslant 0$ and f is decreasing$\}$.

Remark 2. In the case where S is infinite dimensional (which shall not be emphasized), "integrable" is taken to mean Pettis integrable [see, for example, Yosida (1971)].

Remark 3. We could in the definition of a play require simply that $\mathbf{s}$ be measurable, thus allowing for nonintegrable plays. But not much would be gained, since in fact we will have a need to integrate.

Remark 4. In the present approach the outcomes of a play are not necessarily feasible assignments. This is in keeping with the modelling of market institutions as blind to private characteristics. Any player is free to send any message to the market although the outcome of a play can be realized only if almost every trader can honor the proposed contract without going bankrupt. Thus, our strategies are more in the nature of proposals than actual actions, and to that extent the use of the term "game" may be a little abusive. A modeling via Debreu's Generalized Games (1952) would have been possible. The present situation can be contrasted with the well-known price tatonnement. Here public constraints are always satisfied, but private ones need not be. There private constraints do hold, but not necessarily the public ones (i.e., at given arbitrary prices, there may be excess demand or supply).

Remark 5. The concept of strategic market games introduced in this section is formally related to the concept of allocation mechanism growing out of Hurwicz's (1960) approach to the *design* of resource allocation methods. [See, for example, Hurwicz (1979) and Schmeidler (1978)].

5. NONCOOPERATIVE EQUILIBRIA

For fixed I let an economy $\mathscr{E} : I \to \mathscr{A}$ and a strategic market game Φ with message space S be given.

DEFINITION. A play $\mathbf{s}$ is a Cournot–Nash (CN) equilibrium if $\mathbf{x_s}$ is feasible and for almost every $t \in I$, $\mathbf{x_s}(t)$ is $\gtrsim_t$ maximal on $X_t \cap \{\mathbf{x_{s'}}(t) : \mathbf{s}'(t') = \mathbf{s}(t')$ for all $t' \neq t\}$.

The assignment $\mathbf{x_s}$ corresponding to a CN play will be called a CN assignment.

Since we will very quickly specialize our model, we shall not dwell on a discussion of this definition.

Remark 6. For the continuum, i.e., $I = [0, 1]$, the definition of CN equilibrium has taken inspiration in Schmeidler's (1973) work on the non-cooperative theory of games with a continuum of players.

6. Axioms

Let the strategic market game Φ on a set of players I be given.

We shall restrict our consideration to strategic market games whose outcomes depend only on the strategies played and not on the names of the players. More precisely: (i) if two traders choose the same action, they get assigned the same net trade, and (ii) for a given fixed action of a trader, the net trade assigned depends only on the distribution of the strategies of all traders. Those are particular conditions, but they embody the anonymity principles of well-functioning markets: the trading possibilities of any particular economic agent are limited *only* by their availability of commodities to trade.

Our wish to impose an anonymity axiom explains two aspects of our treatment of strategic market games: the strategy sets of different players were assumed equal and the outcome of a play was not required to satisfy the private feasibility constraints.

Let $\mathscr{M}$ be the set of probability measures on S with finite mean. To every $\mathbf{s} : I \to S$ there corresponds an $\nu_{\mathbf{s}} = \lambda \circ \mathbf{s}^{-1} \in \mathscr{M}$.

Anonymity Axiom. There is a function $G : S \times \mathscr{M} \to R^l$ such that, for all plays $\mathbf{s}$, $x_{\mathbf{s}}(t) = G(\mathbf{s}(t), \nu_{\mathbf{s}})$ for all $t \in I$.

Clearly, for I finite, the function G is not unique, while for $I = [0, 1]$, it is trivially so.

Under the Anonymity Axiom, a play $\mathbf{s}$: $I \to S$ is a CN equilibrium if $G(\cdot, \nu_{\mathbf{s}}) : I \to R^l$ is a feasible assignment and, for a.e. $t \in I$, $G(\mathbf{s}(t), \nu_{\mathbf{s}})$ is $\succsim_t$ maximal on $X_t \cap \{G(s, \nu_{\mathbf{s}}) : \mathbf{s} \in S\}$.

A second, important axiom is:

Continuity Axiom. The function G given by the Anonymity Axiom is jointly continuous when $\mathscr{M}$ is endowed with the topology of the weak convergence [see (Hildenbrand, 1974, p. 48), for a definition].

Under the continuity axiom the uniqueness of G in the continuum case (i.e., $I = [0, 1]$) can be sharpened to: if for all $\nu \in \mathscr{M}$, G and G' are continuous and coincide on supp(ν), then $G = G'$.

A third axiom will be:

Convexity Axiom. The strategy set S is convex.

Given the previous axioms, our key hypothesis is:

Aggregation Axiom. For any $\nu, \nu' \in \mathcal{M}$ if $\int i d\nu = \int i d\nu'$, then $G(\cdot, \nu) = G(\cdot, \nu')$. The symbol i denotes the identity map.

In words: the net trade proposed to a player by the strategic market game does only depend on the message sent by this player and the mean message of all players. Particular as this condition is, we would argue that it reflects well the aggregate nature of the impact of demand and supply in organized markets. To be sure, it is quite possible that an economic trading situation may be most succinctly and naturally described in a way which does not satisfy the Aggregation Axiom. The point is, however, whether or not it can be described in some way for which it does, even if it perhaps involves a much expanded, but still convex, strategy space. For example, a typical competitive problem among firms may involve as individual strategies, price and quality (suppose there are only m possible qualities). The concept of mean quality is then irrelevant to the equilibrium problem. It may not even be defined. So, the Aggregation Axiom is not satisfied. On the other hand, if we view as the set of messages the much larger set of supply (and demand) functions for the m quality goods (with arguments the m vector of prices), then the Aggregation Axiom will hold.

We will postpone a more detailed discussion of the role of the different axioms until after the statement of the efficiency results in Section 10.

If the four axioms of this section are satisfied, it will be simpler to describe the strategic market game Φ by the continuous function $F : S \times S \to R^l$ such that $G(s, \nu) = F(s, \int i d\nu)$.

7. The Continuum as a Model for a Large Number of Participants

Suppose we have a continuum of traders $I = [0, 1]$, an economy $\mathscr{E} : I \to \mathscr{A}$ and a strategic market game Φ, with player set I, which satisfies the Anonymity, Continuity, Convexity, and Aggregation Axioms of the previous section, i.e., the strategic market game can simply be defined by a continuous function $F : S \times S \to R^l$. In which sense does this amount to a model for an economy and strategic market game with a large number of participants? We address this question now. Given the limited nature of the strategic games we deal with (anonymous, aggregable), this is not a difficult problem. Even so we shall not attempt to obtain the most general possible results.

Suppose we are given a sequence of trader sets I_n, $\#I_n < \infty$, economies $\mathscr{E}_n : I_n \to \mathscr{A}$, and strategic market games Φ_n with player set I_n and a common convex strategy space S. We assume that $\#I_n \to \infty$.

As in Hildenbrand (1974), we say that $\mathscr{E}_n \to \mathscr{E}$ if the distribution of characteristics induced by $\mathscr{E}_n$ converges weakly to the distribution of characteristics of $\mathscr{E}$.

For the sequence of strategic market games we assume: (i) there is a function $H : S \times \hat{S} \to R^l$, where $\hat{S}$ is the cone spanned by S, such that for all n, plays $\mathbf{s} : I_n \to S$ and $t \in I_n$, $\Phi_n(\mathbf{s})(t) = H(\mathbf{s}(t), \sum_{t \in I_n} \mathbf{s}(t))$ and (ii) the function H is homogeneous of degree zero in its second arguments, i.e., $H(s, \lambda b) = H(s, b)$ for all $\lambda > 0$. This could be called the Homogeneity Axiom.

We say that $\mathscr{E}_n$, Φ_n *converges to* $\mathscr{E}$, F if (i) $\mathscr{E}_n \to \mathscr{E}$ and

(ii) $F = H \mid S \times S$. Since H, F remain fixed, we identify $(\mathscr{E}_n, \Phi_n)$ with $\mathscr{E}_n$.

A comment to this definition may be useful. The requirement that H be common to all terms of the sequence is in order since we want to formalize the idea that the number of participants in some *fixed* market institutions increases. For the same reason, i.e., the market institutions do not depend on the number of participants, the function H is given in an unnormalized way. Thus the homogeneity axiom emerges as fundamental since it is this axiom which allows us to replace sums by averages and therefore to define a meaningful continuum model embodying the notion of the negligibility of individual participants. The sequences generated as in Examples 1–3 satisfy the axiom.

If assertions about CN equilibria in the continuum are to be relevant, then it should at least be true that the "limits" of CN equilibria for $\mathscr{E}$ are CN equilibria for the limit $\mathscr{E}$. But what is a limit of CN equilibria? We borrow the notion of equilibrium distribution introduced by Hart, Hildenbrand, and Kohlberg (1974). Given a (Borel) distribution μ on $\mathscr{A} \times S$, let $\mu_{\mathscr{A}}$, μ_S be the respective marginals. We say that μ is a CN distribution for $\mathscr{E}_n$ (resp. $\mathscr{E}$) if (i) $\mu_{\mathscr{A}}$ is the characteristic distribution of $\mathscr{E}_n$ (resp. $\mathscr{E}$) and (ii) for all $(X, \succsim, s) \in \text{supp}(\mu)$, $F(s, \int i\mu_S)$ is $\succsim$-maximal on $X \cap \{F(s', \int id\mu'_S) : \text{all } s' \in S\}$, where i is the identity map on S and μ' is the distribution obtained by transferring a total mass of $1/n$ from $(X, \succsim, s)$ to $(X, \succsim, s')$ (resp. $\mu = \mu'$). Note that the concept of CN distribution captures everything essential in the definition of CN equilibria in the sense that if two CN equilibria have the same distribution, then they are identical up to a relabeling of the traders' names.

Then we have:

Proposition 1. *Assume that for all n and $t \in I_n$, $X_t = R^l$. Let μ_n be a CN distribution for $\mathscr{E}_n$ and $\mu_n \to \mu$ weakly. Then μ is a CN distribution for $\mathscr{E}$.*

Proof. If $\mu_n \to \mu$, then $\mu_{n\mathscr{A}} \to \mu_{\mathscr{A}}$ and $\mu_{nS} \to \mu_S$. So the first condition in the definition of CN distribution is satisfied for μ. Now let $(\succsim, s) \in \text{supp}(\mu)$ (we can forget about X as $X \equiv R^l$). Take $(\succsim_n, s_n) \to (\succsim, s)$, $(\succsim_n, s_n) \in \text{supp.}(\mu_n)$. Call $x = F(s, \int id\mu_S)$ and for arbitrary $s' \in S$ let $y = F(s', \int id\mu_S)$. By the continuity of F, $x_n \to x$ and $y_n \to y$, where $x_n = F(s_n, \int id\mu_{nS})$ and

$y_n = F(s', \int i\mu'_{nS})$. Hence $x \gtrsim y$ as $x_n \gtrsim_n y_n$ for all n. So, the second condition is also satisfied. ∎

Proposition 1 is a closed graph property for the CN equilibrium correspondence (defined from distribution on $\mathscr{A}$ to distribution on $\mathscr{A} \times S$). It is well known that in order to prove results of this sort it is essential that constraint sets correspondences be continuous. To guarantee this we take $X_t = R^l$ in Proposition 1. Of course, more general conditions could be given.

A stronger result than Proposition 1 would be the upper hemicontinuity of the equilibrium correspondence. Only then can we assert that the equilibria of large economies are near the equilibria of the continuum. Upper hemicontinuity obtains under the hypothesis of Proposition 1 and, for example, the very strong assumption that S be compact (this is trivial to prove). If upper hemicontinuity fails (as it may well be the case in the context, for example, of an Example 3), then the behavior of the equilibria in arbitrarily large economies may differ markedly from the continuum situation. See Thompson (1978) for examples where precisely this happens. This is an important point which does surely deserve closer investigation.

8. Proper and Full Cournot–Nash Equilibria

Our aim is to establish efficiency properties for CN equilibria. It is clear that, informally speaking, a necessary condition for efficiency is that individual net trades be responsive enough to individual actions. Indeed, if all markets are closed, every play is a CN equilibrium. It will be useful to distinguish two types of not totally unrelated local rigidities. The first would arise if in some markets there were quantitative limits on trades and some traders had reached them. The second would arise if some markets were effectively closed at the CN equilibrium. This motivates the following definitions.

Let the agents set I, economy $\mathscr{E}$ and strategic market game Φ be given. Suppose that $\mathbf{s}$ is a CN equilibrium. For every t let $B_t = \{\Phi(\mathbf{s}')(t) : \mathbf{s}'(t') = \mathbf{s}(t') \text{ for all } t' \neq t\}$.

We say that the CN equilibrium $\mathbf{s}$ is *proper* if for some $n \leqslant l$ and a.e. $t \in I$, a neighborhood of $\mathbf{x}_{\mathbf{s}}(t)$ on B_t is homeomorphic to R^n. If we can take $n \geqslant l - 1$, we say that the equilibrium is *full*.

Note that the notion of proper equilibrium is compatible with traders being at the boundary of their private trading sets.

We shall see in Section 11 that it is quite possible, and even typical, that for the same economy a strategic market game exhibit both full and nonfull CN equilibria. Improper equilibria would appear to be more pathological n "flexible prices" models, but they would be typical in rationing models.

9. On the Efficiency of the CN Net Trade Allocations in the Finite Number of Traders Case

Let $F : S \times S \to R^l$ represent a strategic market game satisfying the four axioms of Section 7. It is fairly clear from the usual examples that if F acts on an economy with a finite number of traders, there is no reason to expect the CN equilibrium allocations to be efficient. We shall here illustrate this point by proving, under a number of simplifying but reasonable hypotheses, that given F and $n \geqslant 2$, there are economies with n traders exhibiting nonefficient full CN equilibria (which, further, cannot be perturbed away).

Let I be a traders index set with $\#I = n$. Given a play $\mathbf{s} : I \to S$ and $t \in I$, it is convenient to define : $\bar{s}_t = 1/(n-1) \sum_{t' \neq t} \mathbf{s}(t')$. Let also $H : S \times S \to R^l$ be given by $H(s, s') = F(s, (1/n)((n-1)\, s' + s))$. Then $B_{s'} = H(S, s')$ is the public budget of any $t \in I$ at any play $\mathbf{s}$ with $s' = 1/(n-1) \sum_{t' \in I\, t' \neq t} \mathbf{s}(t')$.

We shall make the following hypothesis:

(I) S is open and H is C^1.

(II) For all s, s', rank $D_s H(s, s') = l - 1$ and Range $D_s H(s, s') \cap R_+{}^l = \{0\}$.

(III) For some s and s' rank $D_{s,s'} H(s, s') = l$.

These are very reasonable hypotheses. Assumption (II) asserts that it is not possible by an infinitesimal displacement to get something from nothing, but that, subject to this fact, rank $D_s H(s, s')$ is maximal, i.e., $l - 1$. This is a clear requirement if there is to be any hope that the allocations induced by the CN equilibria of the strategic market game be efficient. Assumption (III) asserts that at some combination of actions the strategic market game allows, if cooperation prevails, for (local) unrestricted variation of an individual allocation (i.e., cooperation makes it possible for some individual to get something from nothing). It is satisfied if, for example, $H(S \times S)$ contains an open set.

Proposition 2. *If H satisfies* (I), (II), *and* (III), *there is an economy $\mathscr{E} : I \to \mathscr{A}$ and a play $\mathbf{s} : I \to S$ such that*:

(i) $\mathbf{s}$ *is a CN equilibrium*,

(ii) *the net trade allocation of* $\mathbf{s}$ *is not efficient.*

Proof. Let s and s' be as in Assumption (III). With $I = \{1,..., n\}$, we consider the play $\mathbf{s}(1) = s$, $\mathbf{s}(j) = s'$ for $2 \leqslant j \leqslant n$. We let $s_i = \mathbf{s}(i)$, $s_i' = \bar{s}_i$. By (II), for every i there is a unique $p_i \geqslant 0$, $\| p \| = 1$, such that $p_i \cdot D_s H(s_i, s_i') = 0$. Let $\mathbf{x} : I \to R^l$ be the net trade induced by $\mathbf{s}$. If $\mathbf{x}$ is to be efficient for any economy which satisfies our hypothesis and makes $\mathbf{s}$ into a CN equilibrium, then p_i must be the same for all i. Indeed, for every i, we can take $X_i = R^l$ and choose monotone preferences $\succsim_i$ representable by a C^1 utility function and such that $\mathbf{x}(i)$ is $\succsim_i$ maximal on $B_{s_i'}$. By Assumption (II) this can be done (it is easy to verify that at any s' and $x \in B_{s'}$, $B_{s'} \cap (\{x\} + R_+{}^l) = \{x\}$. See Fig. 1), and it makes $\mathbf{s}$ into a CN equilibrium. By

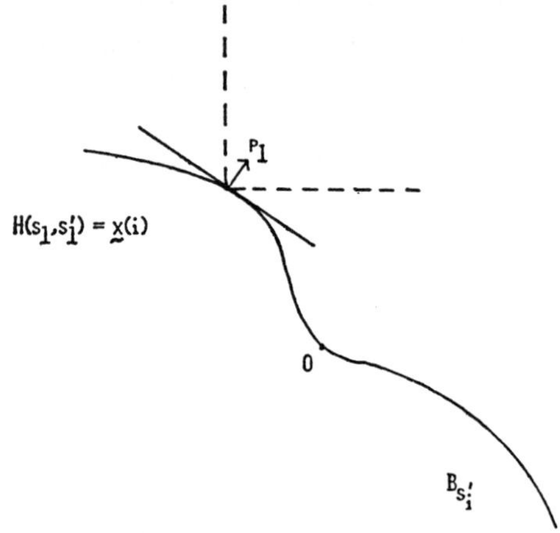

FIGURE 1

the usual efficiency conditions, if $\mathbf{x}$ is efficient, then the normalized gradients of utility functions, which equal the p_i's must be the same.

So, let $p_1 = \cdots = p_n = p$. Differentiating $\sum_{i=1}^n H(s_i, s_i')$ with respect to s_j, we get $D_s H(s_j, s_j') + \sum_{i \neq j} D_{s'} H(s_i, s_i') = 0$ for all j. Therefore, $\sum_{i \neq j} p \cdot D_{s'} H(s_i, s_i') = 0$ for all j. As $p \cdot D_{s'} H(s_j, s_j') - p \cdot D_{s'} H(s_{j'}, s_{j'}') = \sum_{i \neq j} p \cdot D_{s'} H(s_i, s_i') - \sum_{i \neq j'} p \cdot D_{s'} H(s_i, s_i') = 0$ for all j, j', we conclude that $p \cdot D_{s'} H(s_i, s_i') = 0$ for all i. In particular, $p \cdot D_{s'} H(s_1, s_1') = p \cdot D_{s'} H(s, s') = 0$. But then $p \cdot D_{s,s'} H(s, s') = 0$, which implies rank $D_{s,s'} H(s, s') \leqslant l - 1$ and contradicts Assumption (III). Therefore not all p_i's are the same, and we are done. ∎

Remark 7. Proposition 2 is a weak result as it asserts only the existence of an economy exhibiting a nonefficient CN equilibrium. We would surmise that the general picture is as follows:

Postulate Hypothesis (III) for all s, s'. Then restricting ourselves to the class of smooth economies, it would be the case that, but for some exceptional economies (i.e., "generically"), the set of efficient net allocations is a codimension one smooth manifold in the space of net trades allocations and the set of CN allocations induced by the strategic market game is also a smooth manifold. Further both manifolds intersect transversally. This means that if a particular noncooperative theory of market allocation is deterministic in spirit (i.e., generically the set of CN equilibria is finite), then but for coincidental cases, no CN allocation will be efficient. This we have verified for the case dim $S = l - 1$ [see Dubey (1979a), where a particular model is thoroughly analyzed]. More generally, the efficient CN allocations will form a lower-dimensional submanifold of the CN manifold. The investigation of this topic, which does require the methods of differential topology, we have not pursued at much length although we think its clarification is of importance. For a study of the generic properties of CN equilibria in a purely game theoretic context, see Dubey (1978).

Remark 8. Without the Continuity Axiom the conclusions of Proposition 2 may well fail [see, for example, Dubey (1979b) and Schmeidler (1978)] The same is true for the Aggregation Axiom [see Hurwicz (1979) model, where the CN equilibria are efficient, but outcomes depend on the squares of the messages].

10. On the Efficiency of the CN Net Trade Allocations in the Continuum of Traders Case

We now take up an examination of efficiency of CN allocations in the continuum of traders-players situation. We will be led to qualitatively different results from those in the finite case.

From now on $I = [0, 1]$, and we have given a strategic market game satisfying the Anonymity, Continuity, Convexity, and Aggregation Axioms. So, the mechanism is summarized by function $F : S \times S \rightarrow R^l$. After stating our main results, we will discuss the role of each of the axioms.

The first obvious consequence of being in the continuum of agents is that all traders face the same public budget set, as no single trader can affect the average action and anonymity holds. For every $s' \in S$ let $B_{s'} = F(S, s')$. Then given a play $\mathbf{s} : I \rightarrow S$, the common public budget set is $B_{\bar{\mathbf{s}}}$ where $\bar{\mathbf{s}} = \int \mathbf{s}$. Of course, a play $\mathbf{s}$ is a CN equilibrium if for a.e. $t \in I$ $F(\mathbf{s}(t), \bar{\mathbf{s}})$ is $\gtrsim_t$ maximal on $B_{\bar{\mathbf{s}}} \cap X_t$.

The second and fairly obvious consequence of being in the continuum of agents is that for every $s' \in S$, $B_{s'}$ is in fact a convex set. Aggregation is the crucial axiom here.

PROPOSITION 3. *Under the Anonymity, Continuity, Convexity, and Aggregation Axioms, $B_{s'}$ is a convex set for every s'.*

Proof. Let $s' = c$ be given. By the continuity axiom, we can assume without loss of generality that c belongs to the relative interior of S. We proceed to show that $F(\cdot, c) : S \rightarrow R^l$ is then linear. Let $f_c : S \rightarrow R$ be any of the coordinate functions of $F(\cdot, c)$. We know that if $\mathbf{s}$ is a play with $\int \mathbf{s} = c$, then $\int (f_c \circ \mathbf{s})\, dt = 0$. If f_c, which is continuous, is not linear, then an easy argument shows the existence of $s_1, s_2 \in S$ and $0 \leqslant \alpha \leqslant 1$ such that

$$\alpha f_c(s_1) + (1 - \alpha) f_c(s_2) \neq f_c(\alpha s_1 + (1 - \alpha) s_2).$$

Let $c : I \to S$ be a play with $\int \mathbf{c} = c$ and such $\lambda(\mathbf{c}^{-1}(s_1)) > 0$, $\lambda(\mathbf{c}^{-1}(s_2)) > 0$. Such a $\mathbf{c}$ exists because c belongs to the relative interior of S. Now let $\mathbf{c}'$ be as $\mathbf{c}$ except that for a small $\epsilon > 0$ an amount of mass equal to $\alpha\epsilon$ (resp. $(1 - \alpha)\epsilon$) is transferred from $\{s_1\}$ (resp. $\{s_2\}$) to $\alpha s_1 + (1 - \alpha)s_2$. We still have $\int \mathbf{c}' = c$. Therefore, $\int (f_c \circ \mathbf{c})\, dt \neq \int (f_c \circ \mathbf{c}')dt$. But this is impossible because $\int (f_c \circ \mathbf{c})dt = 0 = \int (f_c \circ \mathbf{c})dt$.

Hence, $F(\circ, c)$ is linear and since S is convex, $B_c = F(S, c)$ is convex. ▮

For all s' we also have $0 \in B_{s'}$, since the constant play $\mathbf{s}'(t) = s'$ should yield, by anonymity and aggregation, the null set trade. Let $L_{s'}$ be the linear subspace spanned by $B_{s'}$. Then we have:

PROPOSITION 4. *Under the Anonymity, Continuity, Convexity, and Aggregation Axioms, if* $\mathbf{s} : I \to S$ *is a proper CN play, then for a.e.* t. $F(\mathbf{s}(t), s')$ *is* $\gtrsim_t$ *maximal on* $L_{s'} \cap X_t$, *where* $s' = \int \mathbf{s}$.

Proof. The proof is trivial. Since $\mathbf{s}$ is a CN equilibrium, it is true that for a.e. t $F(\mathbf{s}(t), s')$ is $\gtrsim_t$-maximal on $B_{s'} \cap X_t$. As the CN equilibrium is proper and $B_{s'}$ is a convex set containing 0, we have that for a.e. $t \in I$, $F(\mathbf{s}(t), s')$ belongs to the relative interior of $B_{s'}$. Now suppose that for some of those t we had $x \in L_{s'} \cap X_t$, $x >_t F(\mathbf{s}(t), s')$. As preferences are convex, we could take x in $B_{s'}$, but this is impossible. Hence the proposition is proved. ▮

Proposition 4 could be interpreted as saying that proper CN equilibria yield Walrasian allocations relative to the set of markets which, given the equilibrium strategies, are open. This is particularly clear if $L_{s'}$ belongs to a coordinate subspace. It can well happen (see Section 11) that the set of open markets is different at different CN equilibria.

One could ask about the efficiency properties of proper CN equilibria. Taking a cue from the previous paragraph, one could assert the efficiency of the equilibrium allocations relative to a properly defined subset of net trades allocations. It is not clear, however, if this would amount to much, as the really interesting property would be the efficiency of the CN net trade allocation relative to the net trade allocations which are feasible *within the given strategic market game*. Unfortunately, it is easy to convince oneself that proper CN equilibria are not necessarily efficient in this sense.

So, we are left with the following corollary of Proposition 5, which, being the main result of this paper, we call a theorem:

THEOREM 5. *Under the Anonymity, Continuity, Convexity, and Aggregation Axioms, every full* CN *equilibrium allocation is Walrasian, hence efficient.*

Remark 9. Let's restrict ourselves to strategic market games on $I = [0, 1]$ satisfying the Anonymity and Continuity Axioms. It will be convenient and admissible to identify a game with the continuous function $G : S \times \mathcal{M} \to R^l$. As we are in the continuum case, the public budget set is common to all traders. So, if the game G is understood, we let $B_\nu = G(S, \nu)$ for any $\nu \in \mathcal{M}$.

We show now how given any interval $T \subset R$, continuous function $g : T \to R$, and nondegenerate probability measure $\bar{\nu} \in \mathcal{M}$ satisfying $\int id\bar{\nu} = 0$ and $\int gd\bar{\nu} = 0$, we can construct an anonymous strategic market game G for a two-commodity world with $I = [0, 1]$ and $B_{\bar{\nu}} = \text{Graph } g$.

Indeed, it suffices to define

$$G(s, \nu) = \left(\int | s | \, d\nu \Big/ \int | s | \, d\bar{\nu}\right)$$

$$\times \left(s - \frac{| s |}{\int | s | \, d\nu} \int s \, d\nu, \; g(s) - \frac{| s |}{\int | s | \, d\nu} \int g \, d\nu\right).$$

If $\int | s | \, d\nu = 0$ put $G(s, \nu) = 0$. The Continuity and Convexity Axioms are satisfied, but the Aggregation one is not. Also, if $g(0) = 0$, then $0 \in B_\nu$ for all $\nu \in \mathcal{M}$. Obviously, this G does not have any particular economic interpretation, which is not surprising, given the failure of aggregation.

It is therefore possible to have strategic market games satisfying the Anonymity, Continuity and Convexity Axioms and generating CN equilibria as the ones depicted in Figs. 2a, b, c, d. In every case, the economy is formed by two equal weight types with $X = R^2$ and the indicated preferences. The play distribution is in all cases $\nu(\{s_1\}) = \nu(\{s_2\}) = \frac{1}{2}$.

Figure 2a exhibits an improper equilibria. Figure 2b gives a proper, but not full equilibria. Both equilibria are inefficient. It is obvious that both pictures could be sustained by strategic market games satisfying the Aggregation Axiom. Figures 2c and 2d provide examples of full equilibria failing to be efficient on account of the lack of aggregation.

Remark 10. It is important to notice the interplay between the Aggregation axiom and the convexity hypothesis on the strategy space S. It is the latter which gives force to the first. Indeed, it is always formally possible, by expanding S, to satisfy approximately the aggregation axiom. But the expanded S may fail to be convex. For example, if $S = [0, 1]$ and the outcome net trades depend on plays through the mean and the variance of the message distribution, we could extend the message space to $S' = \{(s, s^2) : s \in [0, 1)\} \subset [0, 1]^2$. Then, the Aggregation Axiom is satisfied, but the Convexity Axiom is not.

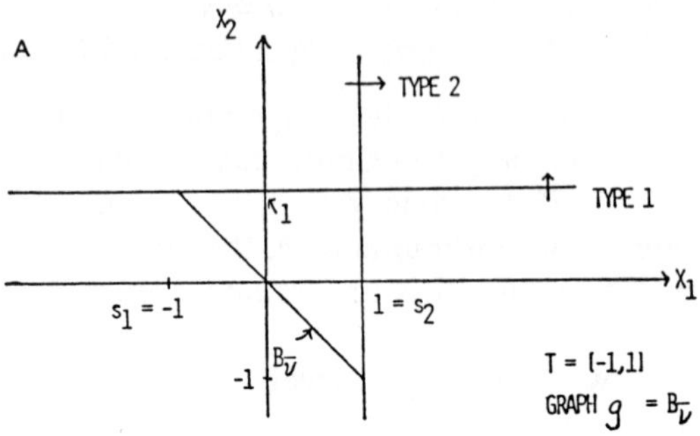

FIGURE 2a

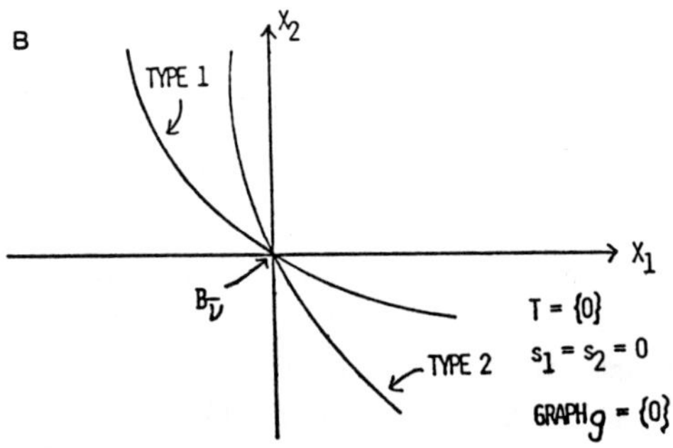

FIGURE 2b

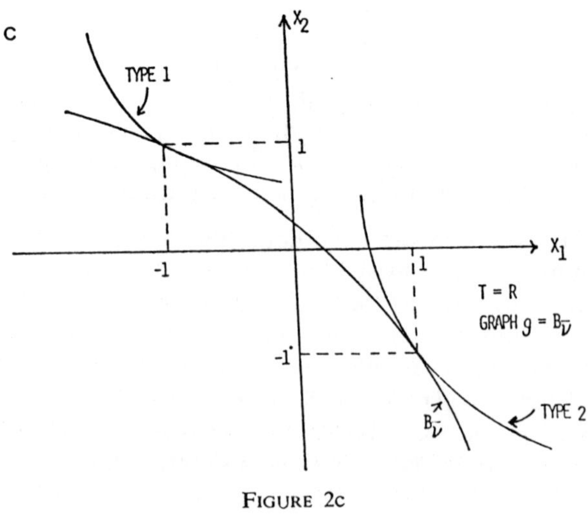

FIGURE 2c

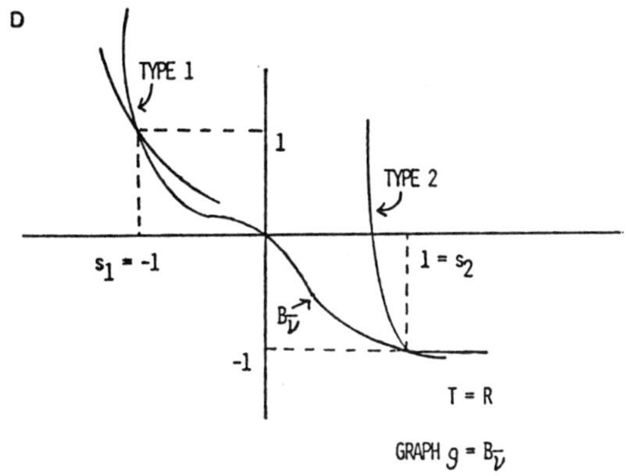

FIGURE 2d

Remark 11. Under the hypothesis made, the Theorem concludes that CN allocations are not just efficient, but in fact Walrasian. Essentially, this is a consequence of the Anonymity Axiom. Under quite general circumstances, it is the case that if the allocation derived from an anonymous continuum strategic market game is efficient then it is Walrasian. See Hammond (1979) for this point. Ch. Kahn (1979) has succeeded in formulating and proving an analog of the Theorem without the Anonymity Axiom. He gets that CN allocations are still efficient but not necessarily Walrasian. A common price system and linear budget sets still obtain, but lump-sum transfers at equilibrium are possible.

Remark 12. The Continuity Axiom is troublesome. It is basic to our approach since, as we saw in Section 7, we need something like it if the continuum model is to be of any use as a representation of a finite but large situation. But there is no denying that there are sensible economic models for which this axiom poses problems. As an instance, the outcome rule may be naturally multivalued (as in Example 2; more generally, this will tend to be the case if the outcome is itself the result of some equilibrium process). Or think of price competition along strict Bertrand lines, or on retaliation strategies, etc. See Green (1979) for a discussion and analysis of the continuity problem in a different (dynamic) context.

At any rate, if continuity fails the properties of equilibria for both the finite and the finite but large models may be very different from the conclusions of this paper. We may have models where equilibria are always efficient, even in the finite case (Bertrand-like models, for example) and models where there is no tendency for the equilibria to become efficient even if the economy is very large [see Green (1979)].

Once we place ourselves in the continuum, the Continuity Axiom plays only a minor role in the proofs of efficiency of a CN play **s**. It is obvious from the proofs that it could be replaced by either of the hypotheses "$\int \mathbf{s} \in \text{Int } S$" or "$F$ is continuous on $S \times \{\int \mathbf{s}\}$".

Remark 13. It is easy to convince oneself that we cannot dispense with the convexity of preferences. It is this hypothesis which, for each trader, makes a local optimum (and this is all we can prove for a CN equilibrium with our axioms) into a global maximizer in the budget set.

11. Some Examples

We shall now examine some specific examples in the direction of the ones introduced by Shubik (1973). See also Shapley (1976) and Shapley and Shubik (1977).

The space of traders is I, finite or infinite.

As a first step, let there be only two commodities, $h = 1, 2$.

The simplest model calls for strategies which are nothing more than the single act of sending messages promising delivery of a certain amount of goods to either side of a single market, i.e., $S = R_+^2$ where by $(s_1, s_2) \in S$ we understand a commitment to supply s_h, $h = 1, 2$, of good h.

Let $\mathbf{s} : I \to S$ be a play. Which rules should we specify for the disbursal of $\bar{s}_1 = \int \mathbf{s}_1$, $\bar{s}_2 = \int \mathbf{s}_2$? Or, in other words, which conditions shall we impose on the outcome $\mathbf{x} : I \to R^2$ of the play? Consider the following three (postulated to be valid for every possible play *and* players space):

(i) Suppose we create a fictitious player space [0, 2]—endowed with Lebesgue measure—and a play $\hat{\mathbf{s}} : [0, 2] \to S$ defined by letting $\hat{\mathbf{s}}(r) = (\mathbf{s}_1(r), 0)$ if $r \leqslant 1$ and $\hat{\mathbf{s}}(r) = (0, \mathbf{s}_2(r - 1))$ if $1 \leqslant r \leqslant 2$. Let $\hat{\mathbf{x}}$ be the outcome of this play. Then $\mathbf{x}(t) = \hat{\mathbf{x}}(t) + \hat{\mathbf{x}}(t + 1)$. In words, there is complete separation of buying and selling and we can, without loss of generality, assume that $I = I_1 \cup I_2$ where

$$I_1 = \{t \in I : \mathbf{s}_2(t) = 0\},\ I_2 = \{t \in I : \mathbf{s}_1(t) = 0\}.$$

(ii) If $t \in I_h$, $h = 1, 2$, then $\mathbf{x}_h(t) = -\mathbf{s}_h(t)$, i.e., the supply of a commodity offered to the market is always taken.

(iii) With the convention that 0/0 qualifies as any number, the ratio $\mathbf{x}_1(t)/\mathbf{x}_2(t)$ (resp. $\mathbf{x}_2(t)/\mathbf{x}_1(t)$) is independent of $t \in I_2$ (resp. $t \in I_1$). This can be interpreted as a kind of anonymity cum arbitrage-freeness.

Then, of course, the outcome is

$$\mathbf{x}(t) = \left(\frac{\bar{s}_1}{\bar{s}_2}\, \mathbf{s}_2(t) - \mathbf{s}_1(t),\ \frac{\bar{s}_2}{\bar{s}_1}\, \mathbf{s}_1(t) - \mathbf{s}_2(t)\right).$$

If $\bar{s}_2 = 0$ (resp. $\bar{s}_1 = 0$) we put the first (resp. the second) entry equal to $-\mathbf{s}_1(t)$ (resp. $-\mathbf{s}_2(t)$). This is nothing but Example 1 in Section 4 (with an inessential difference: to enforce that $\mathbf{x}$ be always an allocation, we required there that $\mathbf{x} \equiv 0$ whenever $\bar{s}_1 = 0$ or $\bar{s}_2 = 0$ while here if, say $\bar{s}_1 \neq 0$ and $\bar{s}_2 = 0$, $\mathbf{x}$ will not be an allocation, since $\bar{s}_1$ is not redistributed to anyone). Note that the Anonymity, Convexity, and Aggregation Axioms are satisfied while Continuity may fail only at the point $\bar{s}_1 = \bar{s}_2 = 0$. It is Condition (iii) which yields anonymity and aggregation. Alternatively, we could have assumed the latter and derive the former. For example, if there was a function $F: S \times S \to R$ such that $\mathbf{x}(t) = F(\mathbf{s}(t), \bar{s})$ for every possible play and players space, then, by Theorem 5, F will have to be linear on the first argument and we recover Condition (iii).

A multicommodity extension of the previous example is readily available. Let there be l commodities. We may define a *simple market* as one in which quantities of a single commodity i are exchanged for another single commodity j. A *market structure* composed of simple markets can be represented by a graph where the points represent commodities and the edges simple markets. See Fig. 3. We could call the edges of the graph *open markets* (and the nonexisting edges *closed markets*). A market structure is *complete* if its graph is connected. A commodity is a *money* if in the market structure there is an edge to every other commodity. So, every market structure with a money is complete. In Fig. 3, Structure 3a is not complete, while the rest are. Structures 3c, 3d, 3e have, respectively, 1, 2, and 4 money commodities.

Given a market structure M we could assume, to focus on the simplest case, that every market functions as the simple market previously discussed. Then the strategy space is $S = R_+^{2M}$ where for every $k \in M$ s_{1k}, s_{2k} represent the quantities offered to the market k of the two commodities transacted in that market. For a given play $\mathbf{s}$ we could let $M_{\mathbf{s}} \subset M$ be the active markets

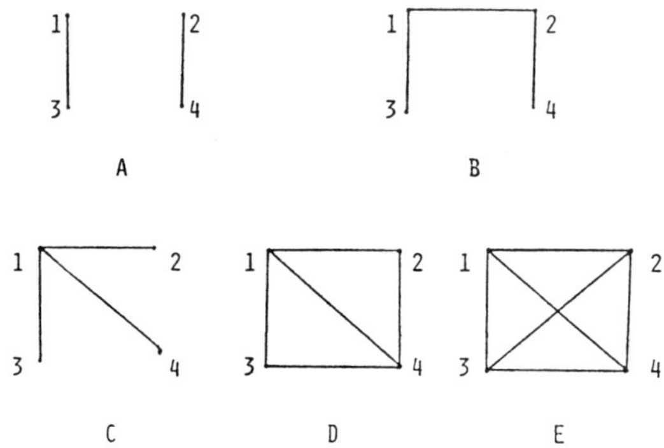

FIGURE 3

at s, i.e., $M_s = \{k \in M: \int s_{1k}, \int s_{2k} > 0\}$ and $S_s = R_+^{2M_s}$ the "active" strategy set.

Suppose s is a CN play. If we forget about nonactive strategies, i.e., imagine that our strategy set is S_s then each one of our axioms is satisfied (see end of Remark 12). So, the efficiency Theorem 5 applies if the equilibrium is full (note that the equilibrium is always proper). Now, the equilibrium will be full if and only if the market structure formed by the *active* markets M_s is complete. For this type of model an inactive market is no better than a closed one.

The reader familiar with Shubik's original model (1973) and with the multicommodity extensions due to Shapley (1976), Jaynes, Okuno and Schmeidler (1976), Postlewaite and Schmeidler (1978), Dubey and Shapley (1977), may wonder why we get efficiency of equilibria in the continuum case merely out of the connectedness of active markets when this was not at all the case in the previous references. The reason lies in the feasibility constraint we are presently using. We completely abstract from transactions problems and for equilibrium we only require the *net* trade to be individually feasible. So, it is possible to use the receipts in one market to buy in another market. If we visualize our simple markets as taking place simultaneously, this amounts to the absence of a liquidity constraint, i.e., to the presence of a perfect credit system.

Another observation is that the previous model will typically have multiplicity of equilibria. In fact if an existence theorem is available (it is, see references of the previous paragraph), we will have an equilibrium for every a priori specification of inactive markets.

The market strategies so far considered are rather extreme. They are commitments to supply a given amount of commodity i to the market in exchange for any quantity of commodity j. More generally, we could envision a supply message (interpretable, perhaps, as a decision rule) to a simple market as being a function $s_i(x_j)$ to be read as a promise to deliver $s_i(x_j)$ units of commodity i in exchange for x_j units of commodity j. In line with Conditions (i), (ii), and (iii) for the simpler case, we could require of market clearing and disbursal rules that at the transactions outcome of a play we had that for all i and j, $s_i(x_j)/x_j$ be independent of the particular supplier. Of course, this is nothing but market clearing at a price system, and if we further constrain every s^i to be nondecreasing and concave, we put ourselves in the anonymous and aggregate situation of Example 3 in Section 4. Indeed, it suffices to redefine a selling strategy as a 0-homogeneous function $\hat{s}_i(p_i, p_j)$ promising to deliver $\hat{s}_i(p_i, p_j)$ units of commodity i in exchange for $1/p_j\ p_i\hat{s}_i(p_i, p_j)$ units of commodity j. Clearly, every s_i strategy induces a $\hat{s}_i$ strategy and equilibrium prices are determined by the condition

$$p_j \int \hat{s}_j(p_i, p_j) = p_i \int \hat{s}_i(p_i, p_j). \qquad \text{See Fig. 4.}$$

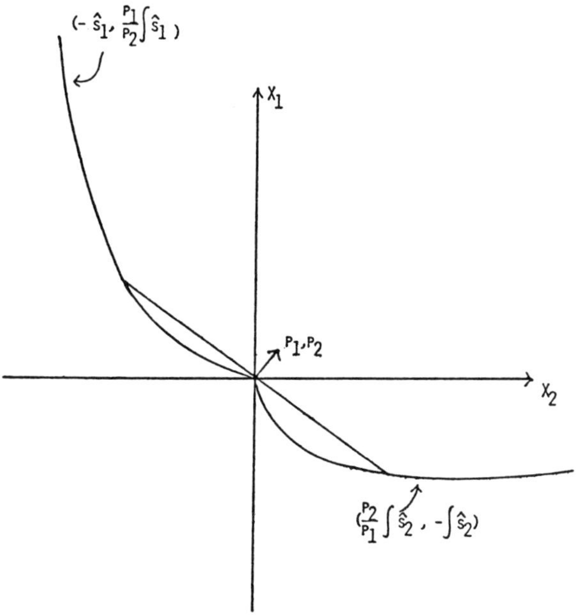

FIGURE 4

The example at the beginning of this section is a limit (the Cournot case) of this latter model. It corresponds to the case $s_i(x^j) = c$. The other limit, $s^i(x_j)$ linear homogeneous, is the Bertrand case. Since outcomes are in this case highly discontinuous on strategies, it is not easily amenable to our analysis.

12. On the Concept of Strict Cournot–Nash Equilibrium in Continuum Economies

Suppose that for $I = [0, 1]$ we have given an economy $\mathscr{E} : I \to \mathscr{A}$ and an anonymous trading game $G : S \times \mathscr{M} \to R^l$. From a noncooperative game theoretic standpoint, a CN play $\mathbf{s} : I \to S$ will be rather fragile if there were traders (strictly speaking, a nonnull set of traders) that could obtain a better net trade by pretending to be n different players (n finite but otherwise arbitrary). Indeed, in an anonymous mass market it may not be possible to prevent an individual trader from entering it several times or simply, using proxies. This observation prompts the following

DEFINITION. A play $\mathbf{s} : I \to S$ is a *strict Cournot–Nash equilibrium* if $x_{\mathbf{s}}$ is feasible and for a.e. $t \in [0, 1]$ $G(\mathbf{s}(t), \nu_{\mathbf{s}})$ is $\gtrsim_t$ maximal on $X_t \cap \bigcup_{n=1}^{\infty} nB_{\nu_{\mathbf{s}}}$.

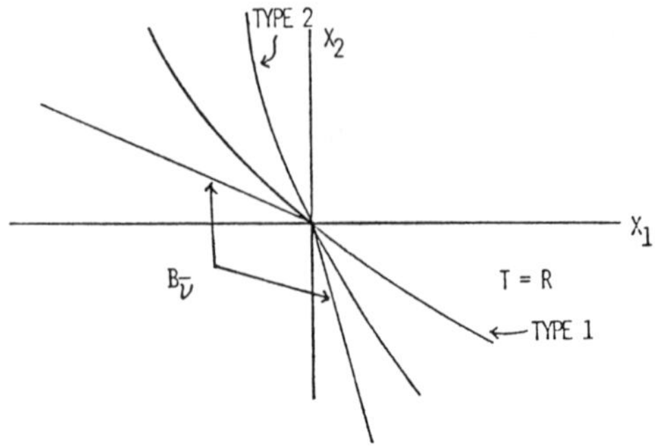

FIGURE 5

From Proposition 4 it is clear that

PROPOSITION 6. *If the Anonymity, Continuity, Convexity, and Aggregation Axioms are satisfied, then every proper CN equilibrium is strict.*

Without the Aggregation Axiom a proper CN equilibrium needs not to be strict. For example if at a CN distribution we have that $B_\nu \cap R^l_{++} \neq \varnothing$ (as in the example of Fig. 2c) then by the monotonicity of preferences, the distribution cannot be a strict CN.

Without the Aggregation Axiom, is every strict Cournot–Nash equilibrium Walrasian? Not necessarily. In Fig. 5 we can see an example where the CN equilibria are proper and full but not Walrasian (the two types indicated have the same weight). By using the technique of Remark 9 we can without difficulty construct economies and strategic market games supporting the figure. In the figure, $B_{\bar{\nu}}$ fails to be smooth. An obvious result is:

PROPOSITION 7. *Suppose that $X_t = R^l$ for a.e. t and that the Anonymity Axiom holds. Let* $\mathbf{s}$ *be a strict CN equilibrium. If $0 \in B_{\nu_\mathbf{s}}$ and $B_{\nu_\mathbf{s}}$ is a $(l-1)$ C^1 manifold, then the equilibrium is Walrasian.*

Proof. Let T be the $(l-1)$ tangent plane to $B_{\bar{\nu}_\mathbf{s}}$ at 0. It is easy to verify that for any $z \in T$, $\bigcup_{n=1}^{\infty} nB_{\nu_\mathbf{s}}$ contains net trades arbitrarily close to z. Henceforth by the continuity of preferences, for a.e. $t \in [0, 1]$, $\mathbf{x}_\mathbf{s}(t) \gtrsim_t z$ for any $z \in T$. Because of monotonicity of preferences, $T \cap R^l_{++} = \varnothing$. Hence we can pick $p \geqslant 0$ such that $T = \{z \in R : p \cdot z = 0\}$. With p as the candidate Walrasian price, the proof proceeds from this point on as in Proposition 4. ∎

As the example of Fig. 2d shows, Proposition 6 holds for strict CN equilibria but not for ordinary ones.

Remark 14. Given an anonymous strategic market game $G : S \times \mathcal{M} \to R^l$ it is possible to build a new one $G^* : S^* \times \mathcal{M}^* \to R^l$ in such a manner that the concept of ordinary and strict CN equilibria coincide in G^* and the ordinary equilibria of G^* correspond to the strict equilibria of G.

One would proceed as follows. Let S be a complete, separable subset of a Banach space. Denote by S^* the set of finite, nonnegative integer valued measures on S. Endow S^* with the weak-star topology and corresponding Borel σ-field. Let $\mathscr{M}^*$ be the (Borel) probability measures on S^* with bounded mean. A play is now an integrable $\mathbf{s}^* : I \to S^*$. The integral is defined by $(\int \mathbf{s}^*)(V) = \int \mathbf{s}^*(t)(V)$ for every Borel set $V \subset S$. It is a nonnegative finite measure on S [see (Mas-Colell, 1975, Sect. 3.1) for the mathematical technicalities of this remark]. Given $\mathbf{s}^*$ we let $\nu^*_{\mathbf{s}^*} \in \mathscr{M}^*$ be the Borel probability measure induced by $\mathbf{s}^*$ on S^*. An element $\nu^* \in \mathscr{M}^*$ induces a distribution ν on A by letting

$$\nu(V) = \frac{}{\int s^*(A)\, d\nu^*} \int s^*(V)\, d\nu^*.$$

Given a play $\mathbf{s}^*$, one let $\nu_{\mathbf{s}^*}$ be the measure in S corresponding to $\nu^*_{\mathbf{s}^*}$. Clearly,

$$\nu_{\mathbf{s}^*} = \frac{1}{(\int \mathbf{s}^*)(A)} \left(\int \mathbf{s}\right).$$

Finally, one defines

$$G^*(s^*, \nu^*) = \int G(s, \nu)\, ds^*.$$

Note that in this new game the Aggregation Axiom is satisfied but S^* is not convex. Also, a proper strict CN equilibrium of G is an ordinary, but not necessarily proper, CN equilibrium of G^*. (See Fig. 5) where B_{ν^*} is the convex hull of B_ν).

REFERENCES

1. R. AUMANN, Markets with a continuum of traders, *Econometrica* **32** (1964), 39–50.
2. A. COURNOT, "Recherches sur les Principes Mathématiques de la Théorie des Richesses," Hachette, Paris, 1838. Translated as "Researches into the Mathematical Principles of the Theory of Wealth," Macmillan, New York, 1929.
3. G. DEBREU, A social equilibrium existence theorem, *Proc. Nat. Acad. Sci. USA* **38** (1952), 886–893.
4. P. DUBEY, "Finiteness and Inefficiency of Nash Equilibria," Cowles Foundation Discussion Paper 508R, 1978.
5. P. DUBEY, Nash equilibria of market games: Finiteness and inefficiency, *J. Econ. Theory* **22** (1980), 363–376.
6. P. DUBEY, "Price-Quantity Strategic Market Games," Cowles Foundation Discussion Paper 520, 1979b.

7. P. Dubey and L. Shapley, Non-cooperative exchange with a continuum of traders," Cowles Foundation Discussion Paper 447 (1977), *Econometrica*, in press.
8. P. Dubey and M. Shubik, "Strategic Market Games and Market Mechanisms," Cowles Foundation Preliminary Paper CF-80915, 1978.
9. F. Y. Edgeworth, "Mathematical Psychics," Kegan Paul, London, 1881.
10. J. Gabszewicz and J. P. Vial, Oligopoly à la Cournot in general equilibrium theory, *J. Econ. Theory* **4** (1972), 381–400.
13. O. Hart, Monopolistic competition with differentiated commodities, *Rev. Econ. Stud.* **46** (1979), 1–30.
14. S. Hart, W. Hildenbrand, and E. Kohlberg, On equilibrium allocation as distributions on the commodity space, *J. Math. Econ.* **1** (1974), 159–167.
15. W. Hildenbrand, "Core and Equilibrium of a Large Economy," Princeton Univ. Press, Princeton, N. J., 1974.
16. L. Hurwicz, Optimality and informational efficiency in resource allocation processes, *in* "Mathematical Methods in the Social Sciences" (Arrow, Karlin, and Suppes, Eds.), pp. 17–47, Stanford University Press, Stanford, Calif., 1960.
17. L. Hurwicz, Outcome functions yielding walrasian and lindhal allocations at Nash equilibrium points," *Rev. Econ. Stud.* **46**, No. 2 (1979), 217–227.
18. J. Jaynes, M. Okuno, and D. Schmeidler, Efficiency in an atomless economy with fiat money, *Internat. Econ. Rev.* **19** (1978), 149–157.
19. Ch. Kahn, "An Extension of the Axiomatic Characterization of Non-cooperative Equilibria in Economies with a Continuum of Traders," manuscript, Harvard University, Cambridge, Mass., 1979.
20. A. Mas-Colell, A model of equilibrium with differentiated commodities, *J. Math. Econ.* **2** (1975), 263–295.
21. A. Mas-Colell, "An Axiomatic Approach to the Efficiency of Non-Cooperative Equilibrium in Economies with a Continuum of Traders," Technical Report 274, Economics, IMSSS (1978), Stanford University, Stanford, Calif.
22. W. Novshek and H. Sonnenschein, Cournot and Walras Equilibrium, *J. Econ. Theory* **14** (1978), 223–267.
23. E. Pazner and D. Schmeidler, "Non-Walrasian Equilibria and Arrow-Debreu Economies," mimeograph, University of Illinois, Urbana, Ill., 1976.
24. A. Postlewaite and D. Schmeidler, Approximate efficiency of non-Walrasian Nash Equilibria," *Econometrica* **46** (1978), 127–137.
25. D. Schmeidler, Equilibrium Points of Non-Atomic Games, *J. Stat. Phys.* **7** (1973), 295–309.
26. D. Schmeidler, "Walrasian Analysis via Strategic Outcome Functions," mimeograph, Tel-Aviv University, Tel-Aviv, Israel, 1978.
29. L. Shapley, "Non-cooperative General Exchange," *in* "Theory and Measurement of Economic Externalities" (S. A. Y. Lin, Ed.), Academic Press, New York, 1976.
30. L. Shapley and M. Shubik, Trade using one commodity as a means of payment, *J. Pol. Econ.* **85** (1977), 937–968.
31. M. Shubik, Commodity money, oligopoly, credit and bankruptcy in a general equilibrium model, *Western Econ. J.* **11** (1973), 24–38.
32. W. Thomson, "Nash Equilibria of Walras and Lindhal mechanism," mimeograph, University of Minnesota, Minneapolis, Minn. 1978.
33. K. Yosida, "Functional Analysis," 3d ed., Springer–Verlag, Berlin/New York, 1971.

Printed by the St. Catherine Press Ltd., Tempelhof 37, Bruges, Belgium

[18]

Vol. 40 (1980), No. 1-2, pp. 25—34

Zeitschrift für
Nationalökonomie
Journal of Economics

A Strategic Market Game with Price and Quantity Strategies*

By

P. Dubey and **M. Shubik**, New Haven, Connecticut, U. S. A.

(Received March 4, 1980)

1. Introduction

The noncooperative game theoretic approach to exchange, in contrast with general equilibrium theory requires the complete specification of the market mechanism through which trade is conducted. Elsewhere several different market mechanisms have been investigated. In particular a "sell-all" model has been considered by Shubik [1] and Shapley and Shubik [2]; Dubey and Shubik [3] have investigated a "bid-offer" model of trade and Shubik has considered a simple market with price-quantity strategies [4] and it has been suggested that there are only a limited number of market mechanisms involving simultaneous bidding in terms of money, goods or prices [5]; and which have intrinsically symmetric roles for all individuals.

A model related to the one suggested here is that of Wilson [8], however he has a special nonsymmetric role for the auctioneer.

2. The Double Auction Market

A mechanism is described for a single market and then it is generalized for m markets. Suppose that n traders have endowments of two commodities, one of which serves as a means of payment and a numeraire. We fix the price of a unit of numeraire at 1.

All traders are required to move simultaneously without knowledge of each other's actions by bidding and offering in a market.

* The research on the original version of this paper was partially supported by the Office of Naval Research. However the content does not necessarily reflect the position or the policy of the Department of the Navy or the Government, and no official endorsement should be inferred.

The endowment of trader i is given by (a_1^i, a_2^i); a move by trader i (which is also his strategy) is described by four numbers $(p^i, q^i; \tilde{p}^i, \tilde{q}^i)$ which are interpreted as follows:

p^i = the maximum price (in terms of the numeraire) that i will pay to buy q^i or fewer units of the first good. We require that $p^i q^i \leq a_2^i$, i. e., he cannot bid more money than he has on hand. We assume $q^i \geq 0$.

$\tilde{p}^i$ = the minimum price that i will accept to sell $\tilde{q}^i$ or fewer units of the first good. We require that $0 \leq \tilde{q}^i \leq a_1^i$, i. e., he cannot offer for sale that which he does not possess.

Although it may be unlikely that a trader will wish both to sell and buy the same item at the same time, there is no *a priori* reason to rule this behavior out, hence the strategies employed here permit an individual to be on both sides of the market if he so chooses.

Figs. 1a, b, and c show possible configurations of supply and demand for the market in aggregate. The supply schedules are obtained by ranking the offer prices in ascending order and cumulating supply; similarly demand prices are ranked, but in descending order.

The market mechanism works as follows. The aggregate supply and demand schedules are calculated and the market price is fixed at the price given by their intersection. All suppliers who require a higher price sell nothing; all buyers who require a lower price buy

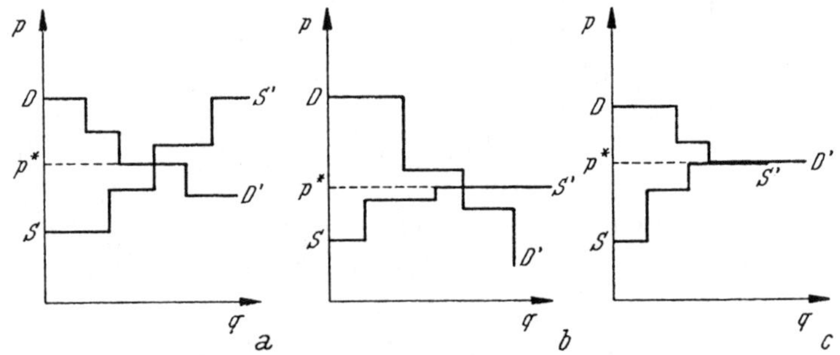

Fig. 1

nothing. At the margin, as is shown in Figs. 1 a and c there may be some buyers willing to pay the market price of p^* for whom there is not enough supply. Similarly there may be sellers willing to supply more at p^* than the buyers wish to buy. This is shown in Fig. 1b.

Fig. 2 illustrates four further possibilities. In Fig. 2a there will be no trade as there is no p^* for which both sides of the market will be active. Fig. 2b illustrates an instance where there is an open

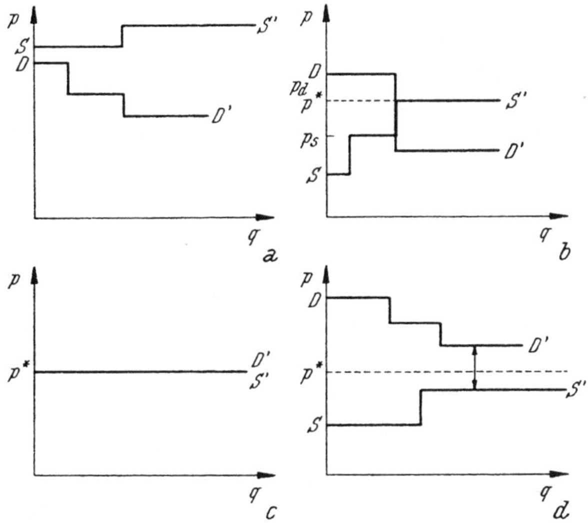

Fig. 2

range of prices at which effective demand will equal supply. Fig. 2c shows an instance where all buyers and sellers have their demands and offers met.

When as in Figs. 1a and 1c there is excess demand at p^* or, as in Fig. 1b there is excess supply we adopt the convention that the marginal buyers or sellers are rationed in proportion to their demands or offers. In a market such as that shown in Fig. 2a no trade will take place. In markets such as shown in 2b and 2d we assume that a price interior to the range $p_s p_d$ is selected by some convention, for example p^* may be selected as the midpoint of the indeterminate range.

2.1 Variations

Before we set up and analyze the m commodity market structure several points are made concerning variants of this model.

In contrast with the mechanisms analyzed in the papers already noted [1, 2, 3] the payoff functions arising from the game are highly discontinuous as functions of the strategies. Intuitively this is the same distinction that can be made between the Bertrand-Edgeworth and the Cournot models of duopolistic competition. Even here, the distinction in results obtained by Cournot and Bertrand are preserved. The Cournot noncooperative equilibria converge slowly towards the competitive equilibria (under the appropriate circumstances) but in general for a finite number of traders the noncooperative equilibria are not efficient [9]. In contrast, under the appropriate circumstances, as was shown by Bertrand full competition may begin with two competitors; i. e. there may be noncooperative equilibria which are also competitive equilibria when there are as few as two competitors on each side of the market.

In actual markets sometimes a trader may buy or sell shares or other items through several accounts, or under several names. This proxy account trading behavior is frequently associated with struggles for control, however even setting aside the corporate control features of stock trading there is no *a priori* reason to rule out multiple account trading. In a separate discussion [10] we have shown that the assumption that there is no gain to be had from multiple account trading is equivalent to showing that the game has strategies which can be aggregated, i. e. individuals are concerned only with their own moves and the sum of the moves of the others.

The model of trade above is only one from a fairly natural class of models which include the conventions of:

(a) maximize trade;

(b) maximize the take of a middleman;

(c) maximize the surplus of one side.

In the first instance in a situation such as that shown in Fig. 1a the highest priced seller who would otherwise be extramarginal could be matched with the highest priced demander.

In the second instance we could introduce a middleman who buys from the sellers and sells to the buyers and keeps the "spread" for himself.

A third convention is to have the sellers sell up to the supply-demand intersection with the buyers all buying at the price they named; i. e. all the surplus goes to the sellers. For example, we may begin by matching the lowest priced seller with the highest priced buyer; if they do not quite match the residual goes to the second seller or buyer and so on.

These somewhat different models are examined elsewhere [11].

3. The Noncooperative Equilibria of a Price-Quantity Strategy Game

Let there be n individuals trading in m commodities using credit as a means of payment. Each trader i has an endowment of $(a_1^i, a_2^i, \ldots, a_m^i)$ where $a_j^i \geq 0$ for all $i=1, \ldots, n$; $j=1, \ldots, m$, and at least one $a_j^i > 0$ for $j=1, \ldots, m$. Also assume that for each j there is some u^i which is strictly increasing in the j-th variable.

A strategy s^i of player i is to announce four vectors in R_+^m: p^i, q_i, $\tilde{p}^i$, $\tilde{q}^i$. How this leads to the disbursement of commodities has already been discussed. In addition we can compute the net credit of i as: money obtained from sales minus spent on purchases. Suppose i ends up with the final bundle $x^i \in R_+^m$, and the credit β^i. Then his payoff $\Pi^i(s^1, \ldots, s^n)$ in the game is $u^i(x^i) + \lambda^i \min[0, \beta^i]$, where λ^i is a preassigned positive constant. This simply says that while having surplus fiat money is useless, being in debt involves a "bankruptcy penalty" which has disutility for the trader (e. g. some of his commodities may be confiscated).

The strategy set S^i of trader i is thus a subset of $R_+^m \times R_{++}^m \times R_+^m \times R_{++}^m$. (Here R_{++}^m is the interior of R_+^m, i. e., traders are not allowed to name zero prices.) Put $S = S^1 \times \ldots \times S^n$. For any $s = (s^1, \ldots, s^n) \in S$, $M \subset N$ and $e = \{e^i : i \in M\} \in \bigtimes_{i \in M} S^i$, let $(s|e)$ denote the element of S obtained from s by replacing s^i by e^i for each $i \in M$. Define s to be *M-efficient* if there does not exist any $e \in \bigtimes_{i M \in} S^i$ such that:

$$\Pi^i(s|e) \geq \Pi^i(s), \quad \text{all } i \in M;$$

$$\Pi^l(s|e) > \Pi^l(s), \quad \text{some } l \in M,$$

If s is $\{i\}$-efficient for each $i \in N$, we call it a *non-cooperative equilibrium* (N. E.); if it is N-efficient, we call it simply *efficient*; if it is M-efficient for all $M \subset N$, we call it a *strong non-cooperative equilibrium.*

This game has certain trivial N. E., e. g. those in which the traders announce that they will buy and sell nothing for some subset of the trading-posts. We focus our attention henceforth on non-trivial N. E. i. e. those in which trade actually occurs in each post.

Finally, recall that $(p; x^1, \ldots, x^n)$ is a *competitive equilibrium* (C. E.) of the market (where p and x^i are in R_+^m) if $\sum_{i=1}^{n} x^i = \sum_{i=1}^{n} a^i$ and x^i maximizes u^i on i's budget set $B^i(p) = \{x \in R_+^m : p \cdot x \leq p \cdot a_j\}$ i. e. $x^i \in B^i(p) = \{x \in \tilde{B}^i(p) : u^i(x) = \max_{y \in B^i(p)} u^i(y)\}$. Also note that with each C. E. we can associate *shadow prices* of income $\lambda^1, \ldots, \lambda^n$ where $\lambda^i \geq 0$ is chosen to ensure that

$$\max_{y \in R_+^m} u^i(y) + \lambda^i (p \cdot a^i - p \cdot y)$$

has x^i as a solution.

We are ready to state our main result:

Theorem. Let $(\hat{p}; \hat{x}^1, \ldots, \hat{x}^n)$ be the prices and allocation produced at an N. E. Then the set of traders who are not optimal on their budget sets i. e. $\{i \in N : \hat{x}^i \notin \tilde{B}^i(p)\}$, has at most $2k$ members. Moreover, each C. E. has corresponding to it a strong N. E. which produces the same prices (up to scaling) and allocation.

Suppose that $\{p^i, q^i, \tilde{p}^i, \tilde{q}^i\}_{i \in N}$ is an active N. E. of $\Gamma(E, \lambda)$ which produces prices $\hat{p}$ and the allocation $\hat{x}^1, \ldots, \hat{x}^n$. First note that if $\beta^1, \ldots, \beta^n$ is the net credit of the traders, then $\sum_{i=1}^{n} \beta^i = 0$. If $\beta^i > 0$ for some i then i could improve his payoff by buying more of a commodity he desires without going bankrupt, a contradiction. Hence $\beta^i = 0$, i. e. $x^i \in B^i(p)$, for each i. Call a trader i "interior" at this N. E. if there is no commodity j at which i is the only active marginal buyer or seller, i. e. at which: i actively buys (sells) j, $p_j^i = \hat{p}_j$, and $p_j^l > \hat{p}_j$ $(p_j^l < \hat{p}_j)$ for all traders l who actively buy (sell) j. We will show that if i is interior then $\hat{x}^i \in \tilde{B}^i(p)$. This is done by a contradiction. W. l. o. g. let 1 be interior and suppose $u^1(\hat{x}^1) < u^1(y)$ for some $y \in \tilde{B}^1(\hat{p})$. Define

$$J = \{j : y_j - \hat{x}_j^1 > 0\},$$

$$J' = \{j : \hat{x}_j^1 - y_j > 0\},$$

T_j = total active sale at trading-post j;

s_j = sale of 1 at trading-post j;

d_j = purchase of 1 at trading-post j;

${}_dp_j$ = $\max\{p_j^i : i \text{ actively buys } j\}$;

${}_sp_j$ = $\min\{\tilde{p}_j^i : i \text{ actively sells } j\}$.

Let $0 < t < 1$ be chosen sufficiently small so as to ensure that:

$$T_j - d_j + t(y_j - \hat{x}_j^1) > 0 \quad \text{for} \quad j \in J;$$

$$T_j - s_j + t(\hat{x}_j^1 - y_j) > 0 \quad \text{for} \quad j \in J'.$$

We now construct a strategy $({}_*p^1, {}_*q^1, {}_*\tilde{p}^1, {}_*\tilde{q}^1)$ for 1 as follows*:

$$ {}_*p_j{}^1 = \begin{cases} ap_j + \varepsilon & \text{if } j \in J \\ p_j{}^1 & \text{otherwise} \end{cases} $$

$$ {}_*\tilde{p}_j{}^1 = \begin{cases} sp_j - \varepsilon & \text{if } j \in J' \\ \tilde{p}_j{}^1 & \text{otherwise} \end{cases} $$

$$ {}_*q_j{}^1 = \begin{cases} q_j{}^1 + t\,(y_j - \hat{x}_j{}^1) & \text{for } j \in J \\ q_j{}^1 & \text{otherwise} \end{cases} $$

$$ {}_*\tilde{q}_j{}^1 = \begin{cases} \tilde{q}_j{}^1 + t\,(\hat{x}_j{}^1 - y_j) & \text{for } j \in J' \\ \tilde{q}_j{}^1 & \text{otherwise} \end{cases} $$

Now suppose 1 deviates to the strategy $({}_*p^1, {}_*q^1, {}_*\tilde{p}^1, {}_*\tilde{q}^1)$ while others hold their strategies fixed. Then the commodity bundle obtained by 1 is $\hat{x}^1 + t\,(y - \hat{x}^1) = z$. At the same time the prices remain unaffected, hence his credit is $\hat{p} \cdot z$. But $\hat{p} \cdot \hat{x}^1 = \hat{p} \cdot a^1$ by Fact 2, and $\hat{p} \cdot y = \hat{p} \cdot a^1$ since $y \in \tilde{B}^1(\hat{p})$. Hence $\hat{p} \cdot z = \hat{p} \cdot a^1$, i. e. the net credit of 1 remains 0 when he deviates. But u^1 is concave, so $u^1(z) > u^1(\hat{x}^1)$. Consequently 1's payoff increases, a contradiction.

Clearly the maximum number of non-interior traders is obtained by having a distinct marginal buyer and seller at each of the k tradingposts. Thus the set $\{i \in N : \hat{x}^i \notin \tilde{B}^i(\hat{p})\}$ has at most $2k$ members.

Next, let $(\hat{p}, \hat{x}^1, \ldots, \hat{x}^n)$ be a C. E. of E with shadow prices $\mu = (\mu^1, \ldots, \mu^n)$. Pick $\alpha > 0$ such that $\alpha\mu^i < \lambda^i$ for each $i \in N$. Consider the n-tuple of strategies $\{p^i, q^i, \tilde{p}^i, \tilde{q}^i\}_{i \in N}$ defined by:

$$ p^i = \tilde{p}^i = \frac{1}{\alpha}\,\hat{p}^i $$

$$ q_j{}^i = a_j{}^i, $$

$$ \tilde{q}_j{}^i = \hat{x}_j{}^i. $$

It is easy to check that these strategies constitute a N. E. of $\Gamma(E, \lambda)$ and yield the prices $\frac{1}{\alpha}\,\hat{p}$ and the allocation $\hat{x}^1, \ldots, \hat{x}^n$.

It remains to show that this N. E. is strong. Suppose some condition $T \subset N$ deviates to new strategies while all the players in $N \backslash T$ hold theirs fixed. Then the members of T can effect two things: (a) trade among themselves, (b) buy from members of $N \backslash T$ at

* For ε small enough ${}_*\tilde{p}^1 \in R^k_{++}$, so the definition is viable.

prices p or more, or sell to them (as before) at prices p. Suppose T ends up with new trades $\{t^i : i \in T\}$. Here $t^i = a^i - y^i$, where y^i is the final bundle of $i \in T$ as a result of the deviation. We can decompose this trade into two parts: the trade $\{\tilde{t}^i : i \in T\}$ which occurs among members of T, and the trade with members of $N \backslash T$. Suppose that the former results in the credit $\{\tilde{\beta}^i : i \in T\}$. If $\tilde{\beta}^i = p \cdot \tilde{t}^i$ for each $i \in T$, then any trader in $i \in T$ can do no better than procure the bundle $\hat{x}^i$ with zero credit. (Recall that $\hat{x}^i$ is optimal for i when he can buy and sell *unrestrictedly* at the prices p, with the rate of bankruptcy penalty equal to μ^i or more, e. g. λ^i). Thus T could not have improved, in this case. So suppose that it is not true that $\tilde{\beta}^i = p \cdot \tilde{t}^i$ for each $i \in T$. Then we claim that for at least one $j \in T$, $\tilde{\beta}^j < p \cdot \tilde{t}^j$. If not, $0 = \sum_{i \in T} \tilde{\beta}^i > p \cdot \sum_{i \in T} \tilde{t}^i = p \cdot 0 = 0$, a contradiction. Consider the trader j. As a result of the deviation, j must be worse off than if he could buy and sell unrestrictedly at prices p, because his credit becomes less favorable. Thus T could not have improved in this other case either.

Q. E. D.

An Example of an N. E. that is not a C. E.

Consider two types of traders with two of each characterized by:

Type 1: $u^1 = x_1^{2/3} y_1^{1/3} + \lambda_1 \min([6p_2 - p_1 x_1 - p_2 y_1], 0)$ with $(0, 6)$

and

Type 2: $u^2 = x_2^{1/3} y_2^{2/3} + \lambda_2 \min([6p_1 - p_1 x_2 - p_2 y_2], 0)$ with $(6, 0)$.

It is easy to see that the unique C. E. is given by $p_1 = p_2 = 1$ (if we choose to normalize) and final holdings for Type 1 of $(4, 2)$ and $(2, 4)$; when $\lambda_1 = \lambda_2 = (4)^{1/3}/3$.

The following set of strategies also form an N. E.

Trader 1 of Type 1: $p_1^1 = 10$, $q_1^1 = 6$; $\tilde{p}_2^1 = 1/2$, $\tilde{q}_2^1 = 4$;

Trader 2 of Type 1: $p_1^2 = 1$, $q_1^2 = 6$; $\tilde{p}_2^2 = 3/2$, $\tilde{q}_2^2 = 4$;

Trader 3 of Type 2: $\tilde{p}_1^3 = 1$, $\tilde{q}_1^3 = 4$; $p_2^3 = 3/2$, $q_2^3 = 8/3$;

Trader4 of Type 2: $\tilde{p}_1^4 = 1$, $\tilde{q}_1^4 = 4$; $p_2^4 = 3/2$, $q_2^4 = 8/3$.

Fig. 3 illustrates this unsatisfactory equilibrium where traders of type 1 are in excess demand in the first market (trader 1) and in excess supply in the second market (trader 1). If trader 1 would lower his bid price sufficiently in market 1 or raise his supply price sufficiently in market 2 the equilibrium would be destroyed.

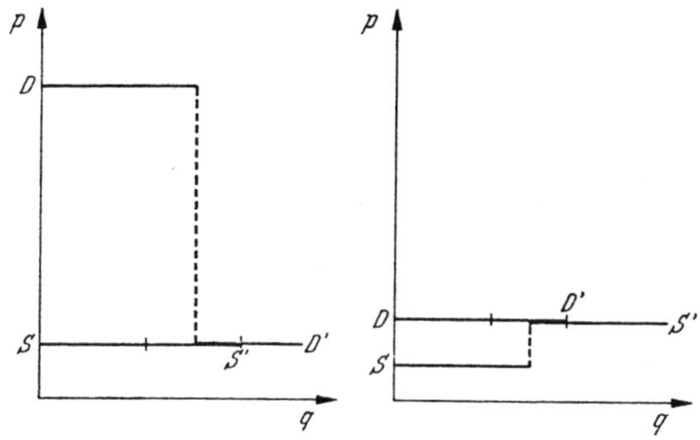

Fig. 3

At equilibrium the payoffs to trader 1 (2) are given by:

$$(6)^{2/3}\,(2)^{1/3} = 4.16 \quad \text{for } 1;$$

$$(2)^{2/3}\,(4-2/3)^{1/3} = 2.65 \quad \text{for } 2.$$

At the C. E. the payoff to all are $(4)^{2/3}\,(2)^{1/3} = 3.175$.

Comment 1. In any economy where there are at least two traders of any type, any type symmetric N. E. will be a C. E.

Comment 2. In an economy with strictly concave utility functions, modelled as a cooperative market game the core will contain only type symmetric imputations, but the example of an N. E. shown above is not type symmetric hence is not in the (Edgeworth) core.

References

[1] M. Shubik: Commodity Money, Oligopoly, Credit and Bankruptcy in a General Equilibrium Model, Western Economic Journal *10* (1972), pp. 24—38.

[2] L. S. Shapley and M. Shubik: Trade Using One Commodity as a Means of Payment, Journal of Political Economy *85* (1977), pp. 937—968.

[3] P. Dubey and M. Shubik: The Noncooperative Equilibria of a Closed Trading Economy with Market Supply and Bidding Strategies, Journal of Economic Theory *17* (1977), pp. 1—20.

[4] M. Shubik: A Trading Model to Avoid Tantonnement Metaphysics, in Y. Amihud (ed.): Studies in Game Theory and Mathematical Economics, New York 1976, pp. 129—142.

[5] M. Shubik: On the Number of Types of Markets with Trade in Money, in V. Smith (ed.): Research in Experimental Economics I, Greenwich, Conn., forthcoming 1979.

[6] P. Dubey and L. S. Shapley: Noncooperative Exchange with a Continuum of Traders, CFDP 447, 1976.

[7] M. Shubik: Strategy and Market Structure, New York, 1959.

[8] R. Wilson: Competitive Exchange Model, Econometrica *46* (1978), pp. 577—586.

[9] P. Dubey: Nash Equilibria of Market Games: Finiteness and Inefficiency, forthcoming CFDP.

[10] P. Dubey, A. Mas-Colell, and M. Shubik: Efficiency of Cournot-Nash Equilibria in Strategic Market Games: An Axiomatic Approach, Journal of Economic Theory, 1980.

[11] P. Dubey: Price-Quantity Strategic Market Games, CFDP 520, March 1979.

Address of authors: Prof. Dr. P. Dubey and Prof. Dr. M. Shubik, Box 2125, Yale Station, New Haven, Connecticut 06520, U. S. A.

Printed in Austria

Journal of Economic Behavior and Organization 2 (1981) 359–385. North-Holland

SOCIETY, LAND, LOVE OR MONEY

A Strategic Model of How to Glue the Generations Together*

Martin SHUBIK

Yale University, New Haven, CT 06520, USA

Received May 1981, final version received January 1982

There are nine and sixty ways
Of constructing tribal lays
And every single one of them is right!
Rudyard Kipling

1. Introduction

In 1958, Paul Samuelson (1958) published a seminal article on the problem of linking the generations together, if the individuals were economic beings with finite lives, concerned only with their own welfare. He suggested that the social contrivance of money as a device for intergenerational transfer would link the generations together.

The purpose of this paper is to show that many different plausible explanations of intergenerational links will all be consistent with an ongoing economy with intergenerational transfers of wealth. Which of the models (or blends of the models) is the best, appears to be both an empirical and a theoretical question. Some suggestions are made.

It would appear that a specific program of interdisciplinary work involving both model building based upon empirical investigation and the development of the mathematical methods of intergenerational or multistage variable person games is called for. In particular, it is argued that biology, anthropology, sociology, demography, political science, law and economics all are concerned with highly different times spans and different degrees of conscious behavior. The emphasis here is not only economic; it is argued

*This work relates to Department of the Navy Contract N00014-77-C-0518 issued by the Office of Naval Research under Contract Authority NR 047–006. However, the content does not necessarily reflect the position or the policy of the Department of the Navy or the Government, and no official endorsement should be inferred. The United States Government has at least a royalty-free, non-exclusive and irrevocable license throughout the world for Government purposes to publish, translate, reproduce, deliver, perform, dispose of, and to authorize others so to do, all or any portion of this work. This supercedes a previous preliminary paper on intergenerational political economy. Many of the ideas in this paper were developed jointly in close collaboration with Brian Arthur over the last few years.

that for the successful development of an economic dynamics the correct links with the other disciplines must be forged.

It is not possible to do justice in a single paper to the many relevant aspects of the demographic, political and social conditions which constrain economic life in an ongoing society. No attempt to do so is made here. Instead the approach is to sketch what we perceive to be the central problem in the study of multigenerational economics. Having stated the central problem we need to specify why it is important.

Given that the problem is stated and deemed to be important what can be done about it? In this paper we proceed by developing (in highly simplified form) the apparatus needed to consider a *multistage variable person game of strategy*. This is applied to several relatively simple examples involving conscious altruism, coded or instinctive behavior, the threat of social custom, the role of ownership, markets and money. We note, but defer to a further projected paper the full development of several other economic models involving taxes and traded capital goods.[1]

In conclusion it is argued that although the examples analyzed are simple they are examples of general phenomena. The level of mathematical analysis itself is kept at a minimum. Further development calls for a considerable enlargement of notation and further discussion of items such as information, games with infinite horizons and the many problems associated with the development of adequate solution theories for games played in extensive form. As the stress is upon the modeling problems of glueing the generations together, the game theory is presented only as a useful method for furthering this purpose. Because it stresses the general concepts of strategic analysis it is our belief that it is particularly suited to the analysis of situations in which economic, social, political and other decisions and motivations must be considered together.

1.1. The problem

If man were individualistic, utilitarian and primarily self concerned, then without an extraeconomic structure to provide the connections there is no guarantee that the generations would hold together in a stable way.

Among animals the problem of glueing the generations together appears to have been taken care of genetically. Care of offspring appears to be programmed in, if for no other reason that a species lacking such a gene would rapidly disappear. The care for offspring needs to be forthcoming until the last of the new generation is selfsufficient and able to reproduce. After the last reproduction and rearing of offspring the parent is of no more 'use' and can die. If we were to accept this argument in full we would see no reason for the care of the aged among animals. Thus, as noted by Brian Arthur, we

[1]For a partial treatment of these in this manner, see Shubik (1980).

would expect to find no old-folks homes or retirement centers for kangaroos, fish or seagulls. But we do expect to find nests, dens and other locals for rearing the young.

It would appear that among the animals, at least some of the human species exhibit some form of built-in perversion which might be most easily explained by a desire for living and the quality of life. This desire may even influence the way the young are looked after or abandoned. It also provides a justification for considering the economics of family planning.

The intergenerational linkage problem, for the reasons given above, seems to be a distinctly human problem, and furthermore it appears to be primarily an old age support problem — although with humans, the genetic programming for parenthood can be overruled by a host of other factors ranging from cultural tradition to individual economic interest.

In the remainder of this paper the society is modeled as though there is a single individual per generation, or it is extended to a mass of homogeneous nuclear families per generation. In the course of human history the structure of the family has varied; depending upon time and culture there have been nuclear families, extended families, tribal groups, monogamy, polygamy, polyandry and so forth. With the variation in family structure comes varying degrees of internalization of insurance, health care and other forms of mutual support. Depending upon this structure the needs for savings banks, loan societies or joint stock ventures will vary as the essential functions of insurance and support are provided by family, private economic, social or political mechanisms. The close relationship between the development of financial institutions and the social structure of the economy merits a detailed discussion which cannot be given here [see Shubik (1970, 1975)].

1.2. Why the problem is important

There are at least four reasons why we believe that the intergenerational connection problem is important. An adequate analysis of the stability of and mechanisms behind generational linkages should be of the following use:

(i) It should help us to better understand the origin and persistence of certain institutions; for example, peasant proprietorships.

(ii) It helps to better understand and more rigorously define welfare questions between generations. The economic growth theorists have frequently used an infinite utility function with a real discount factor in order to provide a bounded optimization model of economic society as a single entity undifferentiated by generations. This appears to be more of a mathematical convenience than an adequate portrayal of reality. We believe that the type of formulation presented here offers several alternatives to this type of model, which are of interest.

(iii) It should help us to better understand the processes of social change and economic development.
(iv) It helps to embed the general equilibrium model of the price system in a richer institutional context yielding an explicit set of mixed socio-economic and politico-economic models which highlight the tradeoffs between strictly economic and other solutions to distribution.[2]

1.3. Solution theory and some remarks on methodology

As the purpose of this paper is to present a sketch of an approach to a broad class of problems deemed to be of importance, it is desirable to avoid technical detail. Yet in doing so the plausibility of the argument should not be damaged. In this section a number of specific points are made concerning the construction and analysis of game theory models.

In spite of having been in existence for around forty years, the basic subtleties of game theoretic modeling and solution methods are not particularly well understood by most social scientists. This lack of understanding has led to several deep misconceptions, where a little knowledge may be worse than none. In particular the usual misconceptions of relevance here are:

(1) The prisoner's dilemma game is somehow more than a trivial example of some didactic value to those being explained some of the paradoxes encountered in game theoretic reasoning.
(2) The game theoretic approach is incompatible with behavioral views of the world.
(3) Game theory in economics is some form of mathematical esoterica designed to push an already unrealistic set of theories of oligopoly even further away from reality.
(4) In particular game theory is even less general and more esoteric than general equilibrium theory.

The reverse of these four points is true. One of the uses of game theoretic techniques here is to enable us to take the first step in putting institutions and behavior back into general equilibrium theory.

The technical essence of the argument is sketched below. The detailed exposition of the game theoretic concepts is given elsewhere [Shubik (1982)].

(1) An attempt to model as a formal game, an economy, a polity, or society with many generations calls for the use of the *extensive* or *strategic* form of a game. In order to do this the rules of the game, in essence, provide

[2]There is now a large literature on economic intergenerational models involving the price system and various extensions of the pure consumption loan model. As this paper does not stress only the price system we defer a discussion of this literature until a subsequent publication.

a rudimentary description of the institutions of a society. The institutions are the carriers of process.

(2) Modeling in the extensive or strategic form is far more flexible than the type of modeling done for the most part in microeconomic theory. In particular the methods of game theory are aimed at the study of competition and cooperation in general, not just economic competition or cooperation.

A key factor in modeling in extensive form is the role of information. Information conditions are manifested in strategies, threats, customs and social conventions.

(3) There are many *solution concepts* which have been proposed to describe the way individuals will behave. No one of them is universally satisfactory. Among the more well known solution concepts suggested in the application of game theory to political economy are the core, the value, the bargaining set, the stable set and the non-cooperative equilibrium [for details see Shubik (1982)].

The non-cooperative equilibrium solution historically goes back to Cournot in its use in economics. It is used as the solution applied to the models constructed here.

(4) It is asserted that the non-cooperative equilibrium solution is far more basic and general than the competitive equilibrium solution. In particular any competitive equilibrium can be regarded as a special case of a non-cooperative equilibrium.

(5) When information conditions are of importance, strategies may become exceedingly complex. In general, when this happens the number of non-cooperative equilibria proliferates. When the full complexity of this proliferation is appreciated it is natural to seek simplifications in terms of limits on behavior, aggregation on information or bounds on calculation. These types of limitation are consistent with behavioral theories.

2. Multistage variable person games

Population. Consider a game with individuals who live at most M time periods. Each individual is strategically active from age m_1 to m_2. From birth to the end of m_1-1 the individual is a child and must be supported if it is to survive. From age m_2+1 to M the individual is deemed to be non-productive in an economic sense and beyond the zone of childbearing and raising.

Suppose that the individual reproduces at m_3 where $m_1 \leqq m_3 \leqq m_2$. The individual has one offspring at the age m_3.

The offspring is a precise replication and lives for at most M periods. Many new more realistic phenomena appear when we introduce uncertainty concerning length of life, birth of children, health and length and quality of productivity. These are suppressed at this time. Furthermore it would be far more realistic to introduce two sexes and the size of family as an endogenous variable. The first look at generational interlinkage can be achieved without these extra realistic but complicating features.

As can be seen from the above, population will at best remain stationary, but there is a chance of early death if not enough sustenance is obtained, and if a sufficient number of individuals die before producing offspring the population will die out.

A digression on 'one or many'. In most of the models here the easiest way of proceeding without considerable elaboration is to consider that there is one representative individual in each generation unless stated otherwise. Thus a game involving a set of N generations is represented by n players (where n is the number of generations). It is asserted but not rigorously formulated here, that most of the pertinent results would equally well apply to a formal model where the single individual per generation is replaced by many identical individuals (i.e., a continuum), although other results might also appear. In some models the existence of many individuals in each generation is assumed explicitly.

Production and consumption. We consider one general aggregate consumption good (food) and one capital good (land). By time t an individual of age m and generation g obtains an amount of food:

$$ {}_gc_1, {}_gc_2, \ldots, {}_gc_m, \tag{1} $$

where ${}_gc_j$ is the amount consumed by an individual of age j and generation g.

At t he also holds ${}_gk_{mt}$ is the amount of land held at time t by an individual of age m and generation g.

We assume that food can be transferred between individuals but cannot be stored beyond a single period.

Survival. We assume that there is a minimum positive subsistence level below which an individual dies within a period. Let this level be D.

Ownership and inheritance. The only durables considered in the various models are 'land' and money or credit. The meaning of ownership of land or money is deeply rooted in the interpretations of the law and the customs of society and the polity. Here we follow a relatively straightforward simplification which is that society is organized so that the young may own but have no strategic freedom to utilize or dispose of land or money until of age. The old have strategic freedom to dispose of their land or money as they see fit.

As soon as a durable is introduced into a model in which individuals die, a strategic game cannot be defined unless *inheritance rules* are specified. The simplest are that the closest surviving offspring inherits all. If an individual dies without heirs in this rudimentary society, we might assume that the state or the king inherits all — but this requires that we either let land or money disappear from the system or model the activities of the state. As a simple way to get around this, we could alternatively assume that any estate without heirs is split up evenly among all.

2.1. Moves and strategies

As our concern is with intergenerational transfers the scope of the strategy of an individual is as follows:

From birth until age m_1 the individual has no strategic freedom.

From m_1 until age m_2 the individual decides upon intergenerational transfers to offspring and parent.

From m_2 until M the individual 'earns' or obtains no food directly but may possibly control land or money and may transfer it.

When there are no markets or money at age m, where $m_1 \leqq m \leqq m_2$, an individual of generation g has a move which consists of at most four non-negative numbers they are ${}_gx^1_m$, ${}_gx^{-1}_m$ and ${}_gy^1_m$, ${}_gy^{-1}_m$ which stand respectively for the amount of food given by an individual of age m and generation g to generations $g+1$ and $g-1$ (denoted by superscripts 1 and -1), and the amount of land given to generations $g+1$ and $g-1$.

At age m where $m_2 < m \leqq M$ an individual has a move which consists of at most two numbers ${}_gy^1_m$, ${}_gy^2_m$. Land can be given directly to children or grandchildren.

Strategy and information. A strategy is a complete policy or plan. It depends upon the amount of information available to the individual, thus it will be a function of the contingencies perceived by the individual. The simplest strategy possible for an individual who lives the full life span of M will contain $4(m_2-m_1+1)+2(M-m_2+1)$ numbers. In general, a strategy will involve functions.

The general strategy set of an individual of generation g is denoted by S_g and a specific strategy by $s_g \in S_g$.

2.2. Outcomes, utility functions and payoffs

The *outcome* from the interaction of the strategies of all individuals will be a set of consumption levels for each period and a terminal amount of assets at death. The *utility* or preferences of an individual will depend upon his own outcomes if we adopt a strictly individualistic view; or it may depend upon outcomes to others as well if we consider a more altruistic concern. The *payoff function* for each individual is a function which relates strategies of all to outcomes and hence to final utility.

2.3. Starting and ending conditions

When trying to model the flow of generations it hardly seems reasonable to have a first and a last generation especially if we are trying to portray conditions for the existence of a stationary state.

One may ask why should one even be bothered with the stationary state. The answer here is that it provides the simplest step towards dynamics and that as such it is a reasonable starting point for analysis.

The method adopted here to approach problems of the infinite horizon is to set up a finite model which can then be examined for a lengthening horizon with the goal being to establish that the solutions to the series of limiting models approach some smooth limit.

In order to set up the multigenerational model described here as a finite n person game we must provide some fairly complex initial and final conditions. In particular the set of individuals who are alive at the start of time may include some middleaged and old who were alive 'before the start of time'. The 'prehistorical' outcomes to them must be specified as parameters of the model. If we decide to 'end the game' after the death of the youngest member of the nth generation; at his time of death there will be many members of the next generation or two alive. Their strategies 'beyond the end of time', as long as they were alive before the end of time might be relevant to the equilibrium conditions. These somewhat cryptic statements will be made precise in the specific games with $M=3$ which are presented below.

3. Multistage games with three period lives

We formulate in full detail several models with three period lives.

The three period life with offspring born at period two enables us to be particularly parsimonious on notation as a new generation appears every period.

Samuelson's model was implicitly a three period life model where he dropped the first period by the assumption that child support was to be purely instinctive and hence not in the analysis.

In Samuelson's model there was only one non-durable consumer good supplied exogenously to individuals aged two periods old.[3] Here we have introduced a durable good as well because we believe that an important essence of socio- and politico-economic relationships is that capital goods plus law provide the complex of escrow, threat and hostage relationships needed to enforce cooperation in an environment that is not intrinsically cooperative. However both for comparison and contrast we consider the cases where the supply of the capital good is 0 or 1.

[3]He actually had two periods of production followed by one without, but this is strategically equivalent here to one and one.

Starting with the case where the amount of capital is 0 we simplify along the lines of Samuelson and assume that 1 unit of the consumer good is earned by any individual who survives to age two. This obliterates the effects of nutrition, education, etc. upon productivity,

$$ {}_g a_{mt} = a = 1. $$

The move of generation g can be described as a distribution x_g^1, x_g, x_g^{-1}, where

$$ x_g^1 + x_g + x_g^{-1} = 1. \tag{2} $$

Subsistence level D is set at some number considerably less than 1.

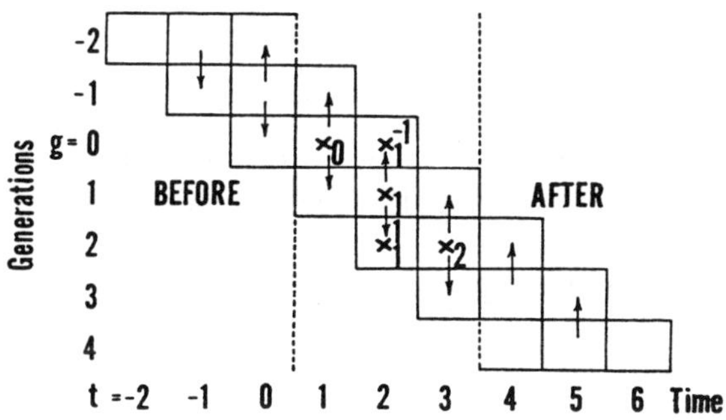

Fig. 1

In fig. 1 we illustrate the game for periods $t=1,2,3$, this encompasses the full life of generation 1 whose parent's (generation 0) strategy may be contingent upon his parent's move (generation -1) in period -1. If we assume that the formulation of an individual's strategy is most dependent upon the actions and income of his parent we need go only two periods back before $t=1$. Initial conditions require that the moves of generations -1 and part of the moves of -2 be specified in advance.

Similarly we will require conditions to be specified for $t=4$ and 5; but the terminal conditions may be somewhat more complicated than the initial conditions because the strategy of generation g at equilibrium may be contingent upon the strategy of generation $g+1$. Thus for terminal conditions we need the strategies of generations 3 and 4 as data.

Given the moves of generations -1 and -2 and the full strategies of generations 3 and 4 the test for a stationary state is, using these as data will there exist a non-cooperative equilibrium in the 3 person game played by generations 0, 1 and 2 which has the same strategies as those assigned to 3 and 4 and produces moves that are the same as those assigned to -1 and -2.

The consumption of generation g of age i is given by ${}_gc_1$, ${}_gc_2$, ${}_gc_3$, where in this simple example

$$ {}_gc_1 = x^1_{g-1}, \qquad {}_gc_2 = x_g \quad \text{and} \quad {}_gc_3 = x^{-1}_{g+1}. \tag{3} $$

3.1. Utility functions

The utility function of an individual of generation g is given by a concave function

$$ U({}_gc_1, {}_gc_2, {}_gc_3) \qquad \text{(individualistic)} \tag{4} $$

if he is modeled as a purely self centered individual.

We may argue that there is some index of love, altruism or concern between the generations. The simplest way to introduce this is to introduce a coefficient of concern or sympathy between the generations. This type of link was suggested by Edgeworth (1881), [see also Collard (1978) and Frisch (1971)].

Several problems in modeling must be faced:

(i) How many generations should be linked?
(ii) Should the linkage go in one direction, down the generational tree, or should it go both ways?
(iii) Should the measure of the linkage be in terms of direct transfer of resources, or in terms of the donor's perception of the satisfaction or welfare of the donee?

Even casual empiricism indicates that personal concern for 'the generations unborn' while a splendid political phrase has little if any operational content beyond possibly a desire for grandchildren. Most of our concern is for those who exist. Humans appear to be the only species[4] that has the concept of grandchild and although there may have been five or even six generations alive at the same time, direct interaction is at most considerable only between parent and child; even in the extended family it is far less between grandparent and child, and beyond that hardly of significance. The approximation suggested here is that only proximate generations need to be linked. For a finer relationship we might wish to include the concern of grandparents for grandchildren.

[4]Possibly baboons or other primates?

The linkage of concern appears to go heavily in one direction, as considerations for the survival of the species would indicate. The support of elderly and weak parents or grandparents calls for claims on resources which could be used on the young and vigorous.

'You will eat your spinach because it is good for you', and 'my happiness is in seeing you happy' present two different views of interpersonal concern, the first involves concern with what the young get, not particularly how they like it. The second is with the preferences of others.

We now sketch some forms for the utility functions reflecting these various conditions.

Proximate generation linked utility functions. The individual's utility function contains the consumption of young offspring and elderly parents as arguments

$$U({}_{g-1}c_3, {}_gc_1, {}_gc_2, {}_gc_3, {}_{g+1}c_1). \tag{5a}$$

We might wish to specialize this structure by introducing two coefficients of concern or sympathy modifying (5a) to

$$U({}_gc_1, {}_gc_2, {}_gc_3) + \eta_{g-1}c_3 + \theta_{g+1}c_1, \tag{5b}$$

where

η = coefficient of concern of offspring for parent,
θ = coefficient of concern of parent for offspring.

Forms (5a) and (5b) portray the 'eat your spinach', or 'make sure they are fed' type of concern. Utility functions expressing preferences for the preferences of others are difficult to operationalize and can give rise to infinite regressions. One way of defining such a function using the separability suggested in (5b) is as follows:

$$\phi_g = U_g({}_gc_1, {}_gc_2, {}_gc_3) + \eta \hat{U}_{g-1}({}_{g-1}\bar{c}_1, {}_{g-1}\bar{c}_2, {}_{g-1}c_3)$$
$$+ \theta \hat{U}_{g+1}({}_{g+1}c_1, {}_{g+1}\hat{c}_2, {}_{g+1}\hat{c}_3). \tag{5c}$$

In (5c) the utility function of g is broken into three separable parts, the first depends only upon his lifetime consumption; the next upon his perceptions of his parent's preference for his lifetime's consumption; here the first two arguments are data. The third component is his perception of his offspring's preference for his lifetime consumption. The last two components have to be estimates as they will be evaluated after the decisions of generation g.

If we assure that all generations have the same preferences and correct perceptions then $\phi_g = \phi$ and if furthermore $\eta = \theta = 1$ then

$$\phi = U({}_g c_1, {}_g c_2, {}_g c_3) + U({}_{g-1}\bar{c}_1, {}_{g-1}\bar{c}_2, {}_{g-1}c_3) + U({}_{g+1}c_1, {}_{g+1}\hat{c}_2, {}_{g+1}\hat{c}_3) \tag{5d}$$

and the successive generations act as if they were a single individual.

3.2. Extensive form and strategies

In order to draw a full game tree for illustration we limit the moves of each generation to three alternatives 1, (0, 1, 0), 2, (1/2, 1/2, 0), and 3, (1/3, 1/3, 1/3) rather than the continuum. We assume $D < 1/3$.

In fig. 2 no information sets are indicated implying that this is viewed as a game with perfect information, i.e., at any stage all previous history is known to all. Thus in this simple example viewed as a finite three stage three player game g_0 has 3 strategies; g_1 has $3^2 = 9$ strategies and g_2 has $3^4 = 81$ strategies. g_3 can recognize 4 positions at which he has 3 choices hence he has $3^4 = 81$ strategies.

In order to capture the effect of continuing generations we need the strategic interlinkage to look the same from any generation, thus we need at least the moves of generations -1 and -2 as data if generation 0 is to have a strategy set of 81 elements.

Similarly in order to be able to calculate the full payoffs to 1 and 2 we must be given the strategies for 3 and 4 as data. The example calculated in section 4.3 will fully specify why this extra information is used and how it is used in calculating the payoffs which are not indicated at the bottom of the game tree in fig. 2.

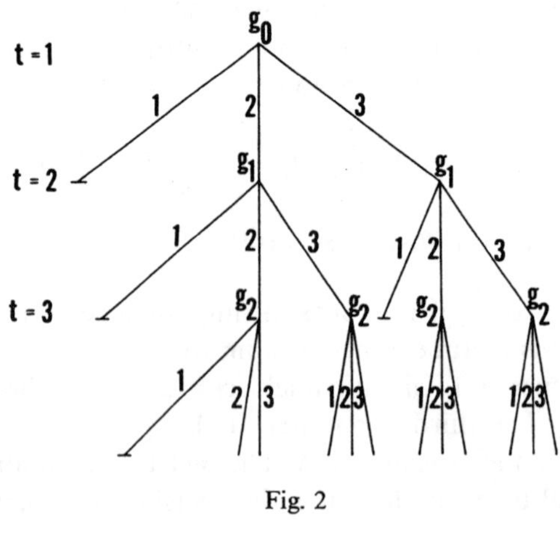

Fig. 2

4. Solutions to multistage games with three period lives

4.1. Historical strategies, threats, aggregation and information

An important contribution of the formal theory of games to an understanding of the logical possibilities of strategic behavior is made in the treatment of information and strategy. Even for a simple matrix game played more than once, as information is made available the domain of the strategy sets expands enormously. A simple example of a 3×3 matrix game played twice serves to illustrate both the expansion of strategies and the concept of threat.

Table 1

	1	2	3
1	5, 5	−5, 10	−10, −5
2	10, −5	0, 0	−10, −5
3	−5, −10	−5, −10	−20, −20

The game shown in table 1 has three strategies for each of two players. A strategy (which here is the same as a move) for each player is to pick a number $i=1,2,3$ or $j=1,2,3$. Suppose that the game is played twice and after the first play both players are informed of each other's choice in the play. Each player can now recognize 9 positions $(1,1)$, $(1,2),\ldots,(3,3)$ at each of these positions he has 3 alternatives thus the number of different strategies available to each is $3^9=19{,}683$. A strategy will contain 9 contingent clauses; more completely it is a number (the first move) followed by a function on the 9 information sets. In total there are $3 \cdot 3^9$ strategies but many are redundant — for example if Player 1 chooses 1 for his first move he really does not need to plan for a position of $(2,2)$ for his second move, as his action has ruled it out.

Two examples illustrate qualitatively different strategies.

The first is the *stationary* or *Markovian* strategy. Whenever you are in the same subgame do the same thing. There are only 3 strategies of this type:

Select i, $i=1,2,3$ then regardless of (i,j) select $k=i$.

The second is a historical strategy which in certain contexts can be reasonably well interpreted as a threat. Consider the following strategy:

Select $i=1$, if he selects $j=1$ select $k=2$,
if he selects $j \neq 1$ select $k=3$.

A glance at table 1 shows that $k=3$ is the punishment for not cooperating by setting $j=1$.

In the example above there are only two players and a threat strategy by one is clearly aimed at the other. In large societies it merits distinguishing between personal and impersonal threats. In a large society, at a personal level involving possibly high information and communication the individual recognizes family and acquaintances and 'the rest of the world' where the rest of the world's behavior and identity is aggregated in some manner. It is possible to distinguish strategies where whole sets of individuals are willing to 'punish' any individual, not necessarily identified by name, but only by action if that individual 'steps out of line' [Shubik (1959)].

4.2. *The non-cooperative equilibrium*

An n-person game has an equilibrium point in pure strategies denoted by a set of n strategies $(s_1^*, s_2^*, \ldots, s_n^*)$ if the following holds true. Let the strategy set for player i be denoted by S_i and an individual strategy by $s_i \in S_i$ and the payoff function for i by f_i; then for all i

$$\max_{s_i \in S_i} f_i(s_1^*, \ldots, s_{i-1}^*, s_i, s_{i+1}^*, \ldots, s_n^*) \Rightarrow s_i = s_i^*. \tag{6}$$

In the game shown in table 1 it should be clear that the strategies for each player 'play $i=2$ each time, regardless' form an equilibrium point, with payoffs of 0 to each. Less obvious is the fact that the threat strategies, if used by both players also form an equilibrium point. That is 'play $i=1$, if competitor replies with 1 then use 2; otherwise 3'. This yields 5 for each as the payoffs.

Although the second equilibrium yields a higher payoff than the first, it is more complicated.

4.3. *Selfish individuals and threat strategies* (*Model 1*)

In this section an example is specified and solved to show the possibility of a stationary state existing via threat. This serves to complete the discussion of fig. 2 in section 3.2.

We consider that an individual of generation g has an individualistic utility function of the type indicated in (4). The conditions for positive marginal utility for each period's consumption are assumed together with decreasing marginal utility for additional consumption. We add the condition that consumption at less than survival level[5] $D < 1/3$ is highly undesirable.[6] In particular the only further conditions imposed are that:

$$U(1,1,0) < \max U(c_1, c_2, c_3), \tag{7}$$

[5]We need the condition $D < 1/3$ for our result as otherwise $3D \geqq 1$ would imply it would not be possible for the society to survive keeping the old.

[6]These conditions amount to stating that a longer life is preferred over a shorter one with higher per period consumption. Furthermore it is no fun to starve at any time.

where $c_1+c_2+c_3=1$, and that the maximum on the right-hand side of (7) is achieved at three numbers c_1^*, c_2^*, c_3^* all greater than or equal to D.

A strategy by an individual of generation g is a function of the information he has on the moves of the two previous generations. In particular we display a specific strategy below, interpret it and show that if all individuals adopt this strategy, the resultant set of strategies form an equilibrium.

A strategy by an individual of generation g is as follows:

$$\text{Set} \quad x_g^1=c_1^*, \quad x_g=c_2^*, \quad x_g^{-1}=c_3^* \quad \text{if information reveals that}$$

$$x_{g-1}^1 \geqq c_1^* \quad \text{and} \quad x_{g-1}^{-1} \geqq c_3^*. \tag{8}$$

$$\text{If however} \quad x_{g-1}^1 < c_1^* \quad \text{or} \quad x_{g-1}^{-1} < c_3^* \quad \text{then set}$$

$$x_g^1=c_1^*, \qquad x_g=c_2^*+c_3^*, \qquad x_g^{-1}=0.$$

In words this states that g will feed his parent and offspring in the manner noted on the right-hand side of eq. (7) if he observes that his parent did that to him and his grandparent. If his parent failed to do so he lets him starve.

We are given the following initial conditions in this game shown in figs. 1 and 2:

$$x_{-2}^1=c_1^*, \qquad x_{-1}^1=c_1^*, \qquad x_{-1}^{-1}=c_3^*.$$

We are also given as data the strategies to be employed by generations $g=3$ and 4. We assume they both employ the strategy set forth in (8). Without this information it would not be possible for generations 1 and 2, who are active players in this game, to calculate their payoffs.

Denote a strategy employed by an individual of generation g by s_g then we may write the payoffs in this particular game with three strategic players as follows:

$$\Pi_0(s_0, s_1)=U({}_0c_1, {}_0c_2, {}_0c_3),$$

$$\Pi_1(s_0, s_1, s_2)=U({}_1c_1, {}_1c_2, {}_1c_3),$$

$$\Pi_2(s_1, s_2)=U({}_2c_1, {}_2c_2, {}_2c_3). \tag{9}$$

We may now check to see that if all use the strategy given by (8) the system is in a stationary equilibrium.

Generation 0 has already been sustained in childhood by -1 at the level ${}_0c_1=c_1^*$. He faces a strategy of the type (8) played by generation 1. If he fails to conform in setting $x \geqq c_1^*$ and $x \geqq c_3^*$ then the most he can obtain is

$(c_1^*, 1, 0)$ but

$$U(c_1^*, 1, 0) < U(c_1^*, c_2^*, c_3^*)$$

hence he is in equilibrium if he sticks to the strategy in (8).

We may now apply the same argument to generation 1 and then generation 2.

It is straightforward to extend this argument for any arbitrary number of generations.

A comment on sensitivity analysis. The threat example given above is relatively simplistic and for those whose game theoretic reasoning tends to be concentrated upon the prisoner's dilemma it looks somewhat like the prisoner's dilemma with a tit-for-tat [see Axelrod (1981)] solution. It is neither. The prisoner's dilemma has strategically symmetric players, here the threats are in sequence.

Not only is the existence of an equilibrium enforceable by threats, robust with respect to variations over period payoffs and classes of threat strategies, it is *too robust.* There is a folk theorem in game theory to the effect that in an infinitely repeated game with disclosure after each move *any* (averaged over time, i.e., Cezaro sum) payoff that is individually rational for all can be achieved as an equilibrium.

Furthermore this example does not depend upon the one person or representative individual feature of the simple model used above. It does not even require a constant population. This assertion can easily be illustrated by a verbal description of strategies that provide such an equilibrium.

Let there be n_g individuals in generation g (a variable birth rate must be exogenously given). Individual i of generation g adopts the following strategy:

> 'I will feed my parents and offspring in the manner noted on the right-hand side of eq. (7) if my parents fed me and my grandparents that way and at least 90 percent of their cohort did the same. Otherwise let them starve.'

It is a tedious but straightforward exercise to develop the notation needed and to specify details such as how resources vary with population in order to mathematize strategies such as the one above and to demonstrate stability more formally. The example above characterized a combination of family or individual threat together with social threat. It must be stressed that as soon as one lets in history and social pressures the combinations leading to stability are enormous. In particular one can introduce a random variable on both features such as the individual's knowledge of population size and resources available and his information on the behavior of the previous generation. This may lead to equilibria or to contingent equilibria, i.e., equilibria which will persist as long as certain factors stay within given bounds. Without building the formal model a verbal example of a strategy which might lead to an equilibrium or contingent equilibrium where there is randomness with respect to numbers, resources and level of information is as follows:

> 'I will feed my parents and offspring in the manner noted on the right-hand side of eq. (7) subject to the availability of food. If there is not sufficient for joint survival I let my parents, then myself starve. This policy is contingent upon my observation that my parents distributed their resources to within around 10 percent of the way I would have distributed them. This policy is further contingent upon my expectation that my resultant behavior was manifested as a norm by at least 80 percent of the population of the previous generation. Otherwise my parents starve.'

4.4. Selfish individuals and simplistic strategies: not enough glue (*Model 2*)

The simplest of strategies are those which ignore all information thus we could restrict the strategies to be equivalent to moves. A strategy by generation g is a triad of numbers

$$(x_g^1, x_g, x_g^{-1}) \quad \text{such that} \quad x_g^1 + x_g + x_g^{-1} = 1.$$

The only information needed from the past to initialize any game in which these types of strategies are employed is that $x_{-1}^1 \geqq D$, i.e., generation 0 survived childhood so that it is alive and strategically active at $t=1$. Suppose that $x_{-1}^1 = k_1 \geqq D$, i.e., some number above the survival level, then it is simple to show that no stationary equilibrium exists. We need only check generation 0. Suppose that there were a stationary equilibrium which would require that $x_1^{-1} = k_2 \geqq D = x_0^{-1}$ if the old were to survive. This cannot be because

$$U(k_1, 1-k_1, k_2) > U(k_1, 1-k_1-k_2, k_2). \tag{10}$$

In words, if generation 0 has been raised by -1 and knows that if it raises generation 1 it will obtain k_2 regardless of its strategy it can gain by failing to support its parent as is shown in (10). Thus if a stationary strategy were to exist it would not include support of the old. Suppose $x_1^{-1} = x_0^{-1} = 0$. If generation 0 is selfish and knows that generation 1 does not intend to provide support in its old age then the optimal strategy will be to set $x_0^1 = 0$ as

$$U(k_1, 1, 0) > U(k_1, 1-k_1, 0). \tag{11}$$

This means that generation 1 will not survive childhood and the species will end if non-contingent strategies are employed.

4.5. Linkage by coding or instinctive behavior (Model 3)

When there is enough food around most animals instinctively feed their young. This can be modeled here by removing the selection of x_g^1 as part of the strategy set and setting it to some number $k_1 \geqq D$. If we do this it is simple to check that there will be a stationary non-cooperative equilibrium with $x_g = 1 - k_1$ and $x_g^{-1} = 0$, each individual will obtain

$$U(k_1, 1-k_1, 0) < U(c_1^*, c_2^*, c_3^*). \tag{12}$$

This is clearly not Pareto optimal, thus instinct to support the young does not necessarily help the survival of the old.

4.6. Linkage by love: the altruism finesse (Model 4)

Suppose that we assume that an individual has a utility function of the shape indicated in (5d), i.e., generation g's concern for its parent and offspring is as for itself. Eq. (9) now become

$$\Pi_0(s_0, s_1) = U({}_{-1}c_1, {}_{-1}c_2, {}_{-1}c_3) + U({}_0c_1, {}_0c_2, {}_0c_3)$$
$$+ U({}_1c_1, {}_1c_2, {}_1c_3),$$
$$\Pi_1(s_0, s_1, s_2) = U({}_0c_1, {}_0c_2, {}_0c_3) + U({}_1c_1, {}_1c_2, {}_1c_3)$$
$$+ U({}_2c_1, {}_2c_2, {}_2c_3),$$
$$\Pi_2(s_1, s_2) = U({}_1c_1, {}_1c_2, {}_1c_3) + U({}_2c_1, {}_2c_2, {}_2c_3)$$
$$+ U({}_3c_1, {}_3c_2, {}_3c_3). \tag{13}$$

Because of the overlap in the arguments of the utility functions it is easy to check that there will be a Pareto optimal stationary non-cooperative equilibrium with simplistic strategies. We check by assuming that all employ $x_g^1 = c_1^*, x_g = c_2^*, x_g^{-1} = c_3^*$.

Consider generation 0 given that $x_{-2}^1 = c_1^*$, $x_{-1} = c_2^*$, $x_{-1}^1 = c_1^*$ then

$$\Pi_0(s_0, s_1^*) = U(c_1^*, c_2^*, x_0^{-1}) + U(c_1^*, x_0, c_3^*) + U(x_0^1, c_2^*, c_3^*). \tag{14}$$

An optimal strategy for generation 0, given information concerning what -1 has done and knowing the strategy s_1^* of generation 1 is one that maximizes

$\Pi_0(s_0, s_1^*)$. A strategy here is to select three numbers

$$(x_g^1, x_g, x_g^{-1}) \quad \text{subject to} \quad x_g^1 + x_g + x_g^{-1} = 1.$$

From the fact that U is concave and has a maximum at (c_1^*, c_2^*, c_3^*) when the consumption is constrained to sum to 1, it follows that $\Pi_0(s_0, s_1^*)$ has a maximum at

$$x_g^1 = c_1^*, \quad x_g = c_2^*, \quad x_g^{-1} = c_3^*.$$

We may now consider $\Pi_1(s_0^*, s_1, s_2^*)$ and show $s_1 = s_1^*$ is in equilibrium, and then consider $\Pi_2(s_1^*, s_2)$.

It should be noted that the assumption of enough love turns a 'game into a team'. Furthermore if love were only of parent for child this would be enough to achieve a stationary state, but not one with the Pareto optimal survival of the old enforced by simplistic strategies. More complex strategies of the variety in Model 1 (section 4.3) will do.

It is suggested here that the coefficients of concern or sympathy are not symmetric. The prime concern of the parent is for the child; the secondary concern is with parents and grandchildren in a Western society with stress on the nuclear family. How much this concern is biologically, sociologically or economically determined appears to be an open question as great variety is seen among different nations, tribes and other social groupings.

Institutions and the game within the game. An important reason for the 'love or altruism' model to be attractive to economists with a non-institutional viewpoint is that enough love minimizes the need for institutions, laws, rules, or any social or political paraphernalia. The need for a systems design of any complexity to produce optimal behavior from individuals with self interest and different goals vanishes if each cares for the other as himself.

Much of the value in considering the non-cooperative equilibrium solution is that it is a solution which requires little explicit cooperation yet can be mutually beneficial (even Pareto optimal) if the rules of the game are appropriately specified. A way of looking at the role of institutions is that they can be regarded as part of the rules of the game called for in the design of self-policing systems.

How institutions change, how fast they change and what factors cause their change, is open to question. A Marxist or a Manchester liberal or Chicago view is that the institutions are heavily formed by economic pressures. A 'laissez faire' view would further have us believe that the appropriate institutions will be moulded quickly and correctly by economic forces. The possibility that institutions and customs take time to form raises the need to try to formalize 'the game within the game' [see Schotter (1981) for instance]. Individuals not only operate in the short term within the rules, but they also devote part of their activity to influencing and changing the rules and institutions in the longer run.

5. Samuelson, money, credit and assets

5.1. Pure consumption loan models

A hundred and fifty years ago, it might be argued that in many parts of Europe and the United States children were economic goods; in other words, parents could have used them on the farm, sold them or hired them out to the mine or mill in such a manner that income from the child, net of cost of raising, was positive. In many societies children have been put to work by parents who took the proceeds. Child labor laws, compulsory schooling, and changes in social custom have wiped out the 'cash crop' view of child maintenance in much of Europe and the more prosperous Western world. Although economic considerations play an important role in education and other aspects of upbringing [see Becker (1975)].

In the remainder of this section we implicitly assume that genetic coding accounts for the nurture of infants and we contrast a threat model of intergenerational linkage with a market model. We need only consider two celled individuals as shown in figs. 3a and b.

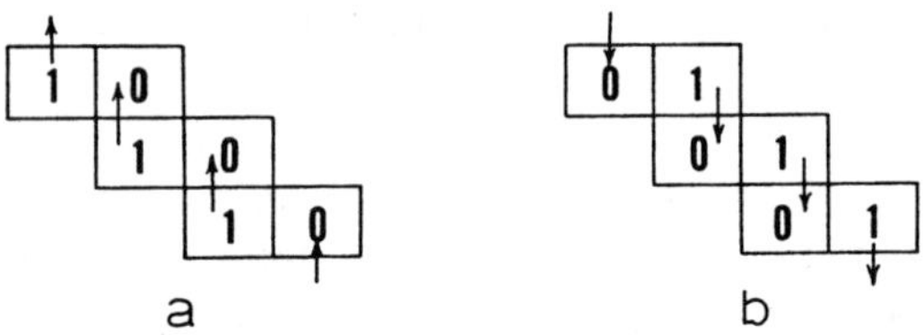

Fig. 3

The analysis in 4.3 fits fig. 3a. The numbers in the cells indicate when an individual obtains his economic resources. A threat by the young to not feed a parent unless the parent fed its parent when young is enough to obtain an equilibrium. But with two celled individuals if it is only the old who have resources (as in fig. 3b) the threat of the young has no influence whatsoever on old. There is a fundamental lack of symmetry in the threats in these models owing to the time sequencing.

The lack of symmetry here is obliterated by the introduction of a market and credit. In order to fully and formally introduce a market and credit more notation is required. We must specify who grants credit, what are the terms, and what happens on default. We consider the model illustrated in fig 3a and fig. 4.

There are a sequence of individuals, all of the same type, where type is defined by endowments and preferences. Any generation has n individuals. Each individual i of generation g has a utility function of the form

$$U({}_g c_1, {}_g c_2) + \mu \min\,[0, (p_g(1 - {}_g c_1) - p_{g+1}\,{}_g c_2)]. \tag{15}$$

The first part of this expression is the standard utility function for consumption. The second part is a penalty term for ending in debt. For simplicity it is given as a separable term of this form; but it can be treated with full generality as has been discussed elsewhere.[7] The important point to stress is that, in essence, the penalty is extraeconomic. It is not enforced by the markets but by government, the law, and society.

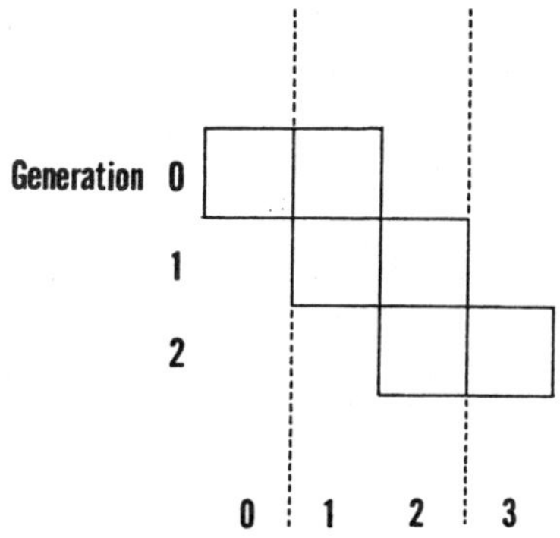

Fig. 4

Each individual i has an endowment of $(1, 0)$ of perishable food or 'manna'; i.e., they each have 1 unit when young and nothing when old.

There is an outside bank which stands to lend individuals government guaranteed credit or money in any amount on the condition that it is paid back or a penalty is exacted. A money rate of interest $\rho \geqq 0$ is set exogenously. The bank accepts deposits and pays ρ on them.

As initial conditions we assume that each individual of generation 0 has 1 unit of money when old. The simplest strategy for an individual ignoring information is three numbers an amount ${}_g x_g^i$ offered for sale by i of generation g at time g; where $0 \leqq {}_g x_g^i \leqq 1$ and an amount of money ${}_g b_g^i$ bid by i to buy food at time g. When i is old, he has no food to sell, hence his only act may be to bid for food ${}_g b_{g+1}^i$.

We assume that banking is instantaneous and automatic. There is no float loss, all income earns interest in the period earned, loans are made by overdraft.

[7]Shubik and Wilson (1977), Dubey and Shubik (1979).

Price is formed in period g as follows:

$$p_g = \left(\sum_{i=1}^{n} {}_g b_g^i + \sum_{j=1}^{n} {}_{g-1} b_g^j \right) \Big/ \sum_{i=1}^{n} {}_g x_g^i. \tag{16}$$

Thus the payoff to an individual i of generation g who does not go bankrupt is given by

$$P_g^i(S_{g-1}, S_g, S_{g+1}) = U(1 - {}_g x_g^i + {}_g b_g^i / p_g; {}_g b_{g+1}^i / p_{g+1}), \tag{17}$$

where S_g is a vector of strategies for members of generation g. In particular, as the numbers per cohort group increase, an individual's effect upon price attenuates thus, for large n we may assume $\partial p_g / \partial_g b_g^i, \partial p_g / \partial_{g-1} b_g^i, \partial p_g / \partial_g x_g^i \simeq 0$.

Suppose that c^*, c_1^* are the unique values for maximizing

$$U({}_g c_g^i, {}_g c_{g+1}^i), \quad \text{subject to} \tag{18}$$

$$p_g(1+\rho)_g x_g^i - {}_g b_{g+1}^i = 0, \quad \text{where} \tag{19}$$

$$1 - {}_g x_g^i = {}_g c_g^i \quad \text{and} \quad {}_g b_{g+1}^i p_{g+1} = {}_g c_{g+1}^i {}_g b_{g+1} / p_{g+1}$$

then it can be seen that c^* and c_1^* can be sustained for all g as an outcome of any of the following class of non-cooperative equilibria.

Starting at $t=1$ with generations 0 and 1 alive with generation 1 holding a unit of money initially a strategy for each individual i of generation g is

$$(c_1^*, 0, (1+\rho)^g), \quad \text{where}$$

$${}_g x_g^i = c_1^*, \qquad {}_g b_g^i = 0, \qquad {}_g b_{g+1}^i = (1+\rho)^g \quad \text{for} \quad g = 1, 2, \ldots.$$

The details of the strategy for $g=0$ are not fully needed as the first part of it is played before $t=1$ and the information is not used by subsequent generations. The only action in the game by generation 0 is to spend to buy food. Here his move will be ${}_0 b_1^i = 1$. *Pro forma* we could conceive of his full strategy as being

$$({}_0 x_0, 0, 1),$$

where the preliminary part of the strategy of generation 0 has been utilized at $t=0$.

Prices will be $p_1 = 1/c_1^* p_t = p_1(1+\rho)^{t-1}$, where $t = 2, 3, \ldots$.

The equilibrium here differs from Sam's[8] original results where he obtained only $\rho=0$ for the stationary state. By slightly changing eq. (19) to the form

$$p_{gg}x_g^i - {}_gb_{g+1}^i = 0 \tag{20}$$

we introduce a float loss, i.e., the money paid to g by $g-1$ during period g is not earning interest for either $g-1$ or g but the outside bank earns the float. Given this condition, we obtain Sam's results that $\rho=0$.

If we consider the model illustrated in fig. 3b, then the initial condition for the existence of an optimal equilibrium is different from the model in fig. 3a. In particular, we postulate that the initial assets of generation $g=0$ at $t=1$ are 1 unit of food and -1 units of money.

Without float loss the cash flow constraint is

$$-(1+\rho)p_{gg}x_g^i + (1+\rho)_g b_{g+1}^i = 0. \tag{21}$$

With float loss the cash flow constraint is

$$-(1+\rho)p_{gg}x_g^i + {}_gb_{g+1}^i = 0. \tag{22}$$

Eq. (21) matches (20) and eq. (22) matches (19).

Using (21), the only equilibrium has $\rho=0$ with $p_1=1/c_1^*$ and $p_g=p_1$ for $g=2,3,\ldots$.

Using (22) then there is an equilibrium for any $\rho \geqq 0$[9] with $p_1=1/c_1^*$ and $p_g=p_1(1+\rho)^{g-1}$.

We may observe that all four solutions for the two models have the same optimal physical outcomes. Two, however, have inflation; the one based on (19) has a constant percentage growth of outside bank debt, the one based on (22) has a growth of individual debt.

In essence the Samuelson model created an outside money. In order to examine conditions for the existence of a stationary state in such a model it is necessary to describe how the money gets into the economy at the start and how it gets out at the end. This calls for the specification of initial and terminal boundary conditions.

5.2. A comment on land, assignats and inheritance

In all of the models noted above, money or credit were only backed by government fiat. There exist credit or money which are 'backed or supported'

[8]Following Professor Samuelson's charming and economical convention abbreviating Bohm–Bawerk to Bohm, Samuelson is abbreviated to Sam.

[9]We do not deal with the cases $\rho<0$ which are logically possible but require 'fussy' conditions concerning hoarding. Various 'funny money' schemes (Major Douglas, for example) have been suggested with negative interest bearing money.

by assets. Gold is the most important of these. But land has been used as is exemplified by the assignats of John Law. It has been suggested that 'possession is nine-tenths of the law'. Property rights and escrow arrangements in some instances are regarded as more enforceable than unsecured trust even given the law.

As soon as we have assets which can outlast their owners, the rules for inheritance must be specified and the strategy sets of the individuals enlarged. Although we do not pursue the formal analysis of the models with land at this time, it is strongly conjectured that the results obtained with fiat money are obtainable with trade in capital goods. Without money, physical assets of productive value extend the strategic possibilities of their owners, if laws are enforced.

Hidden in the pure consumption loan model is a non-cooperative game interpretation where the fiat money is an outside money symbolizing ownership of an unspecified capital stock needed to produce a stationary state economy which must be left at the end if the process is to continue.

6. Concluding remarks

6.1. Money and threats

We can blend the threat model of section 4.3 together with the consumption loan model of section 5.1. It is straightforward to achieve equilibria which are a blend where within a broad range the threat imposed transfers may range from zero to full support of the old by their children. The interpretation of this mixture of market and individual threat can be interpreted as an intermix of social custom and markets.

6.2. Endogenous population and non-stationary solutions

This paper avowedly sketches more models and states more problems than it solves. As the models have been, for the most part, extremely simple one might be concerned that they are artifacts of the stationary state and endogenous population assumptions. Neither appear to be the case. A few comments however, appear to be worth making.

The devising of a formal multigenerational game including endogenous marriage, birth, and up to a point death, together with markets, at least one reproducible capital good, land, labor, one consumer good and even a public good and a rudimentary political structure (see 6.3 below) is a difficult, somewhat tedious but feasible and worthwhile task. Handtailored models with a judicious mixture of demographic, economic and other features should be able to cast further light on topics such as the burden of social security considered by Samuelson (1975) and Arthur and McNicoll (1978) even now.

Specifically however if population is to be made endogenous then although the conceptual task of constructing a variable person multistage game is relatively clear, the empirical problem of deciding upon the causal components of family size is not that clear. Furthermore if population is made endogenous and individuals have a preference for the numbers of offspring some relatively subtle problems in the specification of Pareto optimality appear.

6.3. Politics, voting, economics and demography

Some simple examples of multigenerational models with individual actions aggregated into two or three time periods have been suggested. If we were to consider four periods then an interesting possibility in voting and the modeling of a rudimentary political structure appears. Fig. 5 illustrates the society with four age ranges.

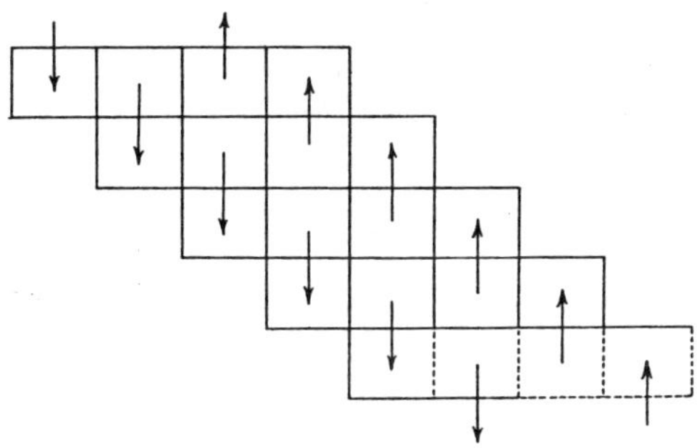

Fig. 5

We may consider the young to be supported by the young middle aged and by education subsidies; and the old to be supported by the older middle aged and social security. The young are neither economic nor political strategic players. The middle aged are both; and the old may not be economic strategic players, but they still may be politically active. The direct burden of the young falls primarily upon the younger middle aged and the direct burden of the old falls upon the older middle aged. But all three, younger, older middle aged and old have voting coalition possibilities with the older middle aged wishing to cut down on the burden of the old, the younger middle aged wishing to cut down on the burden of the young and the old having considerable voting power. As the age profile of the population varies one would expect to see large swings in political programs.

6.4. Economics one or all of the social sciences?

The non-cooperative equilibrium is by no means a completely satisfactory solution concept in the study of economics, but it is useful and also contains the competitive equilibrium as a somewhat restricted special case.

The economising process appears to be common as a means for investigating all of the social sciences. But just because a psychiatrist may wish to explain the actions of an individual as trying to minimize 'tension' does not imply that we can stick tension into a utility function and call psychiatry, economics.

Economic behavior takes place in a societal, legal, political, biological, and psychological context. This is painfully evident in work on development economics. It should be clear in any attempt at theory construction involving the interlinking of generations that there are more forces in the world than economic. The domain of definition for the non-cooperative equilibrium is far broader than that of the general equilibrium model and, thus, we may define and solve mixed economic-social, economic-political or other models and obtain price systems and non-market transfers which are compatible.

Table 2 is given to suggest that there are at least six somewhat different considerations which should be taken into account in the study of intergenerational models. Suggested levels of importance for the United States are given. Even casual empiricism suggests that the elements of importance for the support of the young and old differ considerably.

Table 2

Ways of glueing the generations together.

			Suggested importance	
	Necessary	Sufficient alone	Young	Old
(1) Genetic coding	?	yes	very high	small or 0
(2) Love, altruism	?	yes if enough	moderate	moderate
(3) Economics	no	yes sometimes	very small	moderate
(4) Money, law existing institutions	no	yes if well designed	very small	moderate
(5) Social pressure, threat, tradition	no	yes sometimes	very small	moderate
(6) Politics	no	?	very small	moderate

It may be that some basic elements of genetic coding and love are needed for the survival of the young and even for the old. A totally loveless society might well lead to higher and earlier suicides and even items such as bodily contact between parent and young appear important for survival of the young. The earlier generations may not have survived in high luxury without economics and politics, but they survived. Many mixes of custom, tradition, laws, politics, and markets appear to be able to sustain the generations.

The general equilibrium theory[10] was a masterful first step in seeing the economic problem clearly in a non-institutional or pre-institutional,[11] non-biological, static context. The non-cooperative solution concept provides the means for the analysis of richer and more relevant models for political economy and other social sciences.

[10]Debreu (1959).
[11]Koopmans (1957).

References

Arthur, W.B. and G. McNicoll, 1978, Samuelson, population, and intergenerational transfers, International Economic Review 19, 241–246.

Axelrod, R., 1981, The emergence of cooperation among egoists, American Political Science Review 75, no. 2.

Becker, Gary S., 1975, Human capital (National Bureau of Economic Press, New York, distributed by Columbia University Press, New York).

Collard, David A., 1978, Altruism and economy: A study in non-selfish economies, ch. 2 (Oxford University Press, New York).

Dawkins, R., 1976, The selfish gene (Oxford University Press, New York).

Debreu, G., 1959, Theory of value (Wiley, New York).

Dubey, P. and M. Shubik, 1979, Bankruptcy and optimality in a closed trading mass economy modelled as a noncooperative game, Journal of Mathematical Economics 6, 115–134.

Dubey, P. and M. Shubik, 1981, Information conditions, communication and general equilibrium Mathematics of Operations Research 6, no. 2, 186–189.

Edgeworth, F.Y., 1881, Mathematical psychics (Kegan Paul, London).

Frisch, H., 1971, Die Kontraktkurve bei Interdependenzen im Konsum, Kyklos 24, no. 4.

Hirschleifer, J., 1982, Evolutionary models in economics and law, in: Taylor, ed., Anarchy and cooperation, forthcoming.

Koopmans, T.C., 1957, Three essays on the state of economic science (McGraw-Hill, New York).

Samuelson, P.A., 1958, An exact consumption loan model of interest with or without the social contrivance of money, The Journal of Political Economy 66, no. 6, 467–482.

Samuelson, P.A., 1975, The optimum growth rate for population, International Economic Review 16, no. 3, 531–538.

Schotter, A., 1981, The economic theory of social institutions (Cambridge University Press, Cambridge).

Shubik, M., 1959, Strategy and market structure (Wiley, New York).

Shubik, M., 1970, On different methods for allocating resources Kyklos 23, no. 2, 332–337.

Shubik, M., 1975, On the eight basic units of a dynamic economy with spot and futures markets, Review of Income and Wealth 21, no. 2, 183–201.

Shubik, M., 1980, The capital stock modified competitive equilibrium, in: J.H. Karaken and N. Wallace, eds., Models of monetary economies (Federal Reserve Bank of Minneapolis, Minneapolis, MN).

Shubik, M., 1982, Game theory in the sound sciences (MIT Press, Cambridge, MA) forthcoming.

Shubik, M. and C. Wilson, 1977, The optimal bankruptcy rule in a trading economy using Fiat money, Zeitschrift fur Nationalökonomie 37, 337–354.

Economics Letters 17 (1985) 3–4
North-Holland

PERFECT COMPETITION IN STRATEGIC MARKET GAMES WITH INTERLINKED PREFERENCES

Pradeep DUBEY and Martin SHUBIK

Cowles Foundation for Research in Economics, New Haven, CT 06520, USA

Received 4 June 1984

When the competitive system is viewed as a strategic market game with a continuum of agents, concern for others need not influence outcomes. The generalization of preference conditions is noted and discussed.

This is a remark which, once made, is entirely obvious but merits being put on record. Consider a strategic market game with a continuum of traders T [see, e.g., Dubey, Mas-Colell and Shubik (1980)]. Then the map from strategy-selections to net trades satisfies the condition:

(a) If a single trader unilaterally changes his strategy, this does not affect the trades of any of the others.

We consider two games Γ, $\tilde{\Gamma}$ which differ only in the utilities u^t, $\tilde{u}^t$ ascribed to traders $t \in T$. Let l be the number of commodities in the economy. Then $u^t: R^l_+ \to R$ is the standard utility of personal consumption of the Walrasian model. On the other hand, we incorporate in $\tilde{u}^t$ the possibility that t's utility may depend also on others' consumption. Thus $\tilde{u}^t$ is defined on pairs (α, x), where $\alpha \in R^l_+$ is t's consumption, and $x: T\setminus\{t\} \to R^l_+$ gives the consumption by others. Assume that u^t, $\tilde{u}^t$ are 'compatible' for each t in the following sense:

(b) For any $x: T\setminus\{t\} \to R^l_+$, the function $\tilde{u}^t_x: R^l_+ \to R$ given by $\tilde{u}^t_x(\alpha) = \tilde{u}^t(\alpha, x)$ induces the same preference relation on R^l_+ as does u^t. (This says, in particular, that a trader never considers his own consumption as negligible when compared with that of others.)

Then it is immediate that the Nash Equilibria (N.E.), of Γ and $\tilde{\Gamma}$ coincide. In other words, benevolence or malevolence towards others is washed out by virtue of perfect competition, i.e., by the presence of a continuum of traders. This phenomenon does not appear in a natural way in the Walrasian model, but can be made explicit in the framework of strategic market games.

An asymptotic version of this remark can also be stated, but let us content ourselves with a simple example. Let Γ be a two-person strategic market game, Γ_r its rth replica (thus $\Gamma_1 = \Gamma$) and Γ_* the limit game with a continuum of each type of trader. As is well-known, the N.E. allocations of Γ tend to be inefficient [see Dubey and Rogawski (n.d.)]. One would think that a judicious choice of $\tilde{\Gamma}$, i.e. of $\tilde{u}^t$, might restore efficiency. Let us introduce for each player a coefficient of concern for others, [1] and suppose $\tilde{u}^1$, $\tilde{u}^2$ (the utilities of the two trader-types) to be each a 'social welfare function' of the standard linear separable form. Thus $\tilde{u}^1(\alpha, x, y) = \lambda^1 u^1(\alpha) + \lambda^2 u^1(x) + \lambda^3 u^2(y)$ where $\alpha \equiv$ personal

[1] Called by Edgeworth (1881) a 'coefficient of effective sympathy'.

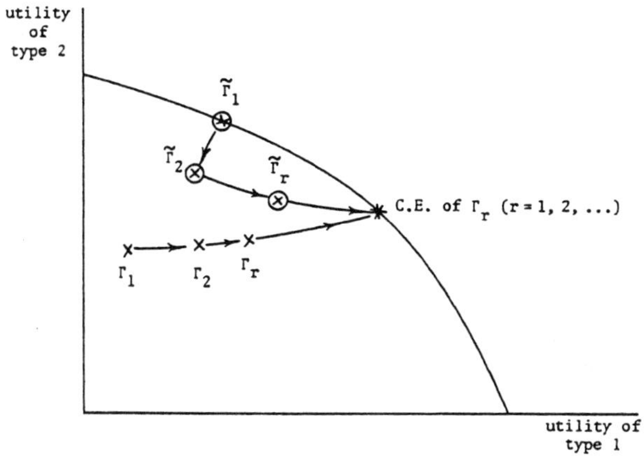

Fig. 1.

consumption of the trader of type 1, $x \equiv$ average consumption by others of type 1, $y \equiv$ average consumption by traders of type 2. (In $\tilde{\Gamma}_1$, since there is only one player of type 1, take the middle term to be zero.) Define $\tilde{u}^2$ similarly. If we choose the coefficients identically (so that $\tilde{u}^1 = \tilde{u}^2$) then $\tilde{\Gamma}_1$ will indeed have an efficient N.E. (simply maximize the common social welfare function on the feasible allocations of the game). But with replication this N.E. will, in general, move off the efficient frontier and converge to it again in the limit. Restricting ourselves to type-symmetric equilibria we have the picture as given in fig. 1.

As indicated by the result in the continuum, both sequences of N.E.'s (of Γ_r $\tilde{\Gamma}_r$) converge to the same limit. Under appropriate conditions [e.g., see Dubey, Mas-Colell and Shubik (1980)] this common limit will be the C.E. of the economy.

References

Dubey, P., A. Mas-Colell and M. Shubik, 1980, Efficiency properties of strategic market games: An axiomatic approach, Journal of Economic Theory.

Dubey, P. and R.J. Rogawski, n.d., Inefficiency of Nash equilibria in a private goods economy, Cowles Foundation discussion paper no. 631.

Edgeworth, F.Y., 1881, Mathematical psychics (Kegan Paul, London) 53.

[21]

The unique minimal cash flow competitive equilibrium*

In a strategic market game with paper money used for exchange the control of the amount of money or credit issued and the selection of the level of the penalty for default make it possible to select a single competitive equilibrium which is feasible as a non-cooperative equilibrium.

1. The exchange economy

An exchange economy $\mathscr{E}$ is characterized by n types of traders and m goods where a trader of type i has a utility[1] function $\varphi_i(x^i_1, ..., x^i_m)$ and an initial endowment $(a^i_1, a^i_1, ..., a^i_m)$.

Suppose that this economy has k competitive equilibrium points (henceforth denoted as CEs). Possible degeneracies which might give rise to a continuum of CEs are set aside. Denote the prices and Lagrangean multipliers associated with the jth CE ($j = 1, ..., k$) by $p^j = (p^j_1, p^j_2, ..., p^j_m)$ and $\lambda^j = (\lambda^j_1, \lambda^j_2, ..., \lambda^j_n)$. As prices are homogeneous of order zero we can multiply p^j by a scalar α without influencing the distribution of goods at the CE. If p^j is multiplied by α (where $\alpha > 0$) then the associated λ^j must be divided by α. Intuitively all that this says is that if prices rise by a factor of α the marginal utility of income is decreased by $1/\alpha$.

It is well known that there are many different ways one can fix a price or normalize prices in the exchange economy. A particular normalization is chosen here. Furthermore it is suggested that it is a natural choice if one wishes to introduce money explicitly into the model.

We want three properties for a good normalization: (a) it should be intrinsically symmetric, i.e., all goods are treated in the same manner; (b) the normalization equation should be dimensionally [303] sound, i.e., the units on both sides of the equation should match; and (c) there should be a good plausible economic interpretation to the normalization chosen.

Three normalizations are noted. They are

$$\sum_{j=1}^{m} p_j = 1 \qquad \text{(prices)}, \qquad (1)$$

$$\sum_{j=1}^{m} p_j a_j = 1 \qquad \text{(total asset value)}, \qquad (2)$$

$$\sum_{i=1}^{n}\sum_{j=1}^{m} \max [(x_j^i - a_j^i), 0] = 1 \qquad \text{(total cash flow)}. \qquad (3)$$

The first norm normalization is one that has been frequently used; it meets the criterion of intrinsic symmetry but fails in dimensional analysis. The dimensions of price p_j are M/Q_1. Thus we have

$$\frac{M}{Q_1} + \frac{M}{Q_2} + \ldots + \frac{M}{Q_m} = ? \qquad (4)$$

The second and third normalizations do not face this problem. Instead of (4) we have

$$\frac{M}{Q_1} Q_1 + \frac{M}{Q_2} Q_2 + \ldots + \frac{M}{Q_m} Q_m = M. \qquad (5)$$

The normalization is in terms of 'money'.

As the choice of units for each good is arbitrary we may normalize the total endowment of each good to be one. Thus as $a_j = 1$ for all j eq. (2) appears to be the same as (1), but the dimensional analysis shows the difference.

Both (2) and (3) have clear economic interpretations. The first (2) is the total value of all wealth in the economy. The second (3) is the value of all trade or total cash flow. The economic interpretation of the normalization suggests that (2) is a better normalization than (3). All CEs are Pareto optimal, but we have no intrinsic way to distinguish among them. This is reflected by using normalization (2) for each. This has the interpretation that total wealth is normalized to be the same at all CEs.

2. The strategic market game with fiat money

The exchange economy $\mathscr{E}$ can be reformulated as a strategic market game. The details of the various formulations are given elsewhere.[2] In particular the point of concern here involves the introduction of a specified amount of credit or fiat money to monetize exchange. Dubey and Shubik (1979) and Shubik and Wilson (1977) have studied the possibility of introducing a fixed amount M of money to finance trade. When one formulates exchange as a game of strategy using any form of credit or fiat money where there is any possibility whatsoever that an individual will be unable to pay back that which he has borrowed, the rules of the game require that the procedure to be followed in case of default must be specified. This is *not* a mere institutional detail but a logical necessity. It is, however, reasonable to expect that one might try to design a default penalty sufficiently harsh to discourage strategic default. [304]

The λ_i^j have an immediate economic interpretation as the marginal utilities to each individual of extra income at CE *j*. In designing a game, if we wanted to discourage strategic bankruptcy we would select a penalty, to be at least as harsh as the largest marginal utility for income.

In a game without a credit restriction any penalty $\mu > 0$ will suffice, as the system is self-normalizing. Prices can always be inflated to the point that the marginal disutility of any penalty, no matter how small, is not worth incurring. The dimensions of μ and λ_i^j are utility/money thus inflating price amounts to increasing the force of the penalty.[3] But if the amount of money is fixed in advance an upper bound is placed on the ability to inflate prices[4] and still obtain a CE, hence a lower bound is set on the value of the λ_t^j.

By utilizing the normalization (2) at each CE *j* we obtain a set of prices $(p_1^j. ..., p_m^j)$ and a set of Lagrangean multipliers $(\lambda_1^j, ..., \lambda_m^j)$. The amount of money required to finance trade at CE *j* is given by

$$\sum_{i=1}^{n} \sum_{g=1}^{m} \max \left[p_g^j (x_g^j - a_g^j), 0 \right] = M_j. \tag{6}$$

If a strategic market game is constructed with a limit on the amount of money of M_j, then in order for the game which we denote by $\Gamma(\mathcal{E}, M_j, \mu^j)$ to be able to achieve the CE *j* as an NE the bankruptcy penalty μ^j must be selected so that $\mu^j \geq \max(\lambda_1^j, \lambda_2^j, ..., \lambda_n^j)$. If μ^j is less than the largest λ_i^j then at least one class of individuals[5] will be motivated to go bankrupt if the prices at CE *j* prevail.

For each CE *j* we can construct a game $\Gamma(\mathcal{E}, M^j, \mu_j)$ where the CE (and possibly other CEs) can be obtained as an NE.

Two games of the *k* games $\Gamma(\mathcal{E}, M_j, \mu^j)$, $j = 1, ..., k$ are distinguished: the game with $M_* = \min[M_1, M_2, ..., M_k]$ and the game with $\mu^* = \min[\mu^1, ..., \mu^k]$. Both of these games may have many NEs but they can each have only one interior NE which coincides with a CE.

The game $\Gamma(\mathcal{E}, M_*, \mu^j)$ is the strategic market game associated with the *minimal cash flow* CE of the exchange economy $\mathcal{E}$. All other CEs of the exchange economy $\mathcal{E}$ require relatively more money (measured as a percentage of total wealth) than M_*. Hence in the game $\Gamma(\mathcal{E}, M_*, \mu^j)$ only one CE is feasible.

The game $\Gamma(\mathcal{E}, M_j, \mu^*)$ is the strategic market game associated with *minimal default penalty* CE of the exchange economy $\mathcal{E}$. If $\mu^j = \max(\lambda_1^j, \lambda_2^j, ..., \lambda_n^j)$ then $\mu^* = \min(\mu^1, \mu^2, ..., \mu^k)$. It is the smallest penalty strong enough for one CE to be attainable as an NE.

An open question is when do these two criteria coincide in their selection of a CE. Even without an answer to this, *minimal cash flow* by itself with its implications of minimal liquidity provides an attractive criterion for selecting among CEs especially if there were a cost associated with liquidity. [305]

Notes

* This work relates to Department of the Navy Contract N00014-77-C-0518 issued by the Office of Naval Research under Contract Authority NR 047-006. However, the content does not necessarily

reflect the position or the policy of the Department of the Navy or the Government, and no official endorsement should be inferred. The United States Government has at least a royalty-free, non-exclusive license throughout the world for Government purposes to publish, translate, reproduce, deliver, perform, dispose of, and to authorize others to do so, all or any portion of this work.

Cowles Foundation Discussion Papers are preliminary materials circulated to stimulate discussion and critical comment. Requests for single copies of a Paper will be filled by the Cowles Foundation within the limits of the supply. References in publications to Discussion Papers (other than acknowledgement that a writer had access to such unpublished material) should be cleared with the author to protect the tentative character of these papers.

1. Enough smoothness, for example C_2, is assumed so that the Lagrangean multipliers will also be a finite set. It is also assumed that the λ^i_j are bounded.
2. See Shubik (1973); Shapley and Shubik (1977); Dubey and Shubik (1978, 1980); Dubey and Shapley (1977); Dubey (1982).
3. The force of inflation causes immediate problems both in bankruptcy and criminal law. In the latter in particular, the theft of money which in one era is clearly a crime, might be still classified as such even though it should be regarded as a misdemeanor.
4. Leaving aside variation of velocity. It is suggested elsewhere that velocity is a red herring [Shubik (1986)].
5. It is argued elsewhere [Dubey and Shubik (1979)] that there is no loss of generality in using a function of the form

$$\varphi_1 (x^j_1, ..., x^j_m) + \mu \min \left[\sum_{g=1}^{m} p_g (x^j_g - d^j_g), 0 \right] \text{ instead of } \varphi_1(x^j_1, ..., x^j_{m+1})$$

to describe the preference conditions involving bankruptcy conditions punative enough to prevent strategic bankruptcy. Furthermore, although a personal bankruptcy penalty for each trader type can be envisioned calling for $(\mu_1, ..., \mu_n)$ rather than a single penalty μ, the single penalty against all types if it is high enough.

References

Dubey, P., 1982, 'Price quality strategic market games', Econometrica, 50, 111-26.

Dubey, P. and L.S. Shapley, 1977, 'Noncooperative exchange with a continuum of traders`, Cowles Foundation discussion paper no. 447 (Yale University, New Haven, CT).

Dubey, P. and M. Shubik, 1978, 'The noncooperative equilibria of a closed trading economy with market supply and bidding strategies', Journal of Economic Theory 17, no. 1, 1-20.

Dubey, P. and M. Shubik, 1979, Bankruptcy and optimality in a closed trading mass economy using fiat money, Journal of Mathematical Economics 6, 115-34.

Dubey, P. and M. Shubik, 1980, 'A strategic market game with price and quantity strategies', Zeitschrift für Nationalökonomie 40, no. 1-2, 25-34.

Shapley, L.S. and M. Shubik, 1977, 'Trade using one commodity as a means of payment', Journal of Political Economy 85, no. 5, 937-68.

Shubik, M., 1973, 'Commodity money, oligopoly, credit and bankruptcy in a general equilibrium model', Western Economic Journal 10, no. 4, 24-38.

Shubik, M., 1986, A game theoretic approach to the theory of money and financial institutions, Cowles Foundation discussion paper no. 805 (Yale University, New Haven, CT).

Shubik, M. and C. Wilson, 1977, 'The optimal bankruptcy rule in a trading mass economy modelled as a noncooperative game', Zeitschrift für Nationalökonomie 37, 337-54 [306]

Economics Letters 27 (1988) 5–6
North-Holland

A NOTE ON AN OPTIMAL GARNISHING RULE *

P. DUBEY and M. SHUBIK

Yale University, New Haven, CT 06520-2125, USA

Received 16 December 1987

A low information, optimal bankruptcy rule is described and analyzed for an exchange economy with credit.

In an exchange economy with many small agents which utilizes credit it is possible that individuals could borrow money which at the end they are unable to pay back. Shubik and Wilson (1977) and Dubey and Shubik (1979) explored models where the government or referee arbitrarily imposed an exogenous and not necessarily economic penalty on any defaulter.

If the original utility function of i is $\phi_i(x_1^i, \ldots, x_m^i)$, we construct a new function:

$$\phi_i = \phi_i\left(x_1^i, \ldots, x_m^i\right) + \mu \min\left[\left\{\sum_{j=1}^{m} p_j\left(a_j^i - x_j^i\right) - \rho s^i\right\}, 0\right], \tag{1}$$

where

μ = the per unit bakruptcy penalty, fixed exogenously,
p_j = the equilibrium prices,
ρ = the endogenous rate of interest,
s^i = the amount of money borrowed by i.

If μ is high enough no one will elect strategic bankruptcy.

Rather than imposing an exogenous penalty to discourage bankruptcy we could consider an endogenous penalty inflicted by garnishing the goods of the bankrupt. There are two simple rules which might be applied, the forgiving rule and the no information rule.

Suppose an individual goes bankrupt for an amount A and obtains a final endowment of $\bar{x}$. The referee removes a vector of resources b such that

$$\max \phi_i\left(\bar{x}_1^i - b_1^i, \bar{x}_2^i - b_2^i, \ldots, \bar{x}_m^i - b^i\right)$$

$$\text{subject to } \bar{p} \cdot b = A,$$

* This work relates to Department of the Navy Contract N00014-86-K0220 issued by the Office of Naval Research under Contract Authority NR 047-006. However, the content does not necessarily reflect the positon or the policy of the Department of the Navy or the Government, and no official endorsement should be inferred.

where $\bar{p}$ are the equilibrium prices. This is the forgiving rule and is exactly the same as though the individual himself were required to adjust to a loss of income A. Thus he is indifferent between not defaulting or defaulting and being subject to this penalty. [1] Elsewhere [see Dubey and Shubik (1979)] we have shown that with such penalties the Nash equilibria of continuum strategic market games will be Walrasian.

The second rule has the important advantage that it requires no knowledge of the individual preferences by the referee. A number k is selected where $0 \le k \le 1$ such that

$$k\bar{p} \cdot \bar{x} = A \quad \text{for} \quad k < 1, \quad \text{otherwise} \quad k = 1.$$

If the referee is able to garnish assets before they have been dissipated this is at least as harsh as the first rule and is an easy economic rule which will discourage strategic bankruptcy. The resources obtained from a bankrupt go to the referee and could be used for redistribution. [2,3,4]

References

Dubey, P. and M. Shubik, 1979, Bankruptcy and optimality in a closed trading economy modelled as a noncooperative game, Journal of Mathematical Economics 6, 115–134.

Shubik, M. and C. Wilson, 1977, The optimal bankruptcy rule in a trading economy using fiat money, Zeitschrift fur Nationalökonomie 7, 334–354.

[1] But, this rule has the considerable inconvenience that it requires, for its implementation, the knowledge of agents' utility functions b.

[2] There is a third rule which we could call the punishing rule. It has the referee act to minimize payoff

$$\min \phi_i\left(\bar{x}_1^i - b_1^i, \ldots, x_m^i - b_m^i\right)$$

$$\text{subject} \quad \text{to } \bar{p} \cdot b = A.$$

This is clearly harsh and could easily destroy the individual. However, as the lenient penalty is sufficient the harshness is unnecessary.

[3] A more subtle rule may be required when there is exogenous uncertainty and incomplete markets. We have not yet resolved this problem.

[4] One problem with our garnishing rule concerns indivisibilities; but fortunately we could modify our rule to 'select any random set of goods such that $\bar{p} \cdot b \geq A$' and this will be sufficiently harsh.

Reprinted from JOURNAL OF ECONOMIC THEORY
Vol. 51, No. 1, June 1990
Printed in Belgium

A Strategic Market Game with Complete Markets*

RABAH AMIR

State University of New York, Stony Brook, New York 11794

SIDDHARTA SAHI

Princeton University, Princeton, New Jersey 08544

MARTIN SHUBIK

Yale University, New Haven, Connecticut 06510

AND

SHUNTIAN YAO†

Yale University, New Haven, Connecticut 06510

Received January 22, 1987; revised October 31, 1989

An exchange economy is modeled as a strategic market game with all pairwise markets available. Existence of noncooperative equilibria is proved. It is shown that if resources are distributed in a skewed manner, in equilibrium prices may not satisfy the no arbitrage condition. One round of trade even with all goods serving as money is not sufficient to provide enough liquidity for efficient trade *Journal of Economic Literature* Classification Numbers: 021, 026, 311.

* This work has partial support from the Department of the Navy Contract N00014-77-C-0518 issued by the Office of Naval Research under Contract Authority NR 047-006. However, the content does not necessarily reflect the position or the policy of the Department of the Navy or the Government, and no official endorsement should be inferred. Partial support from NFS Grant SES-8812051 is also acknowledged. The United States government has at least a royalty-free, nonexclusive and irrevocable license throughout the world for government purposes to publish, translate, reproduce, deliver, perform, dispose of, and to authorize other so to do, all or any portion of this work.

† The authors are indebted to L. S. Shapley and to the anonymous referee for pointing out inadequacies in our original proof of the existence on interior Nash equilibrium. In response to their criticisms, we adopted for our purposes the proof strategy of Peck and Shell [4, Section 2, especially pp. 17–19] and Yao [9, pp. 10–14].

1. The Economic Model and Its Interest

An economic process approach to the study of exchange is by means of a strategic market game (see Shubik [8]; Shapley, [6]; Shapley and Shubik, [7]; Dubey and Shubik, [3]). Price is formed by the simultaneous actions of all agents. One of the simplest models distinguishes one of m commodities as a money. Then $m-1$ markets are considered in which the money can be exchanged directly for one of the other commodities. The direct exchange of other commodities for each other is ruled out. In contrast, when all commodities can be exchanged directly, then for m commodities there are $m(m-1)/2$ markets instead of $m-1$.

A simple geometric representation of trade and markets can be given. Let goods be points and markets be arcs connecting them. Figure 1a shows the market structure for an exchange economy with four goods where the fourth good acts as a money. Figure 1b shows the structure with all markets. Under this description a money is a good which can be exchanged directly for all other goods. In an economy with complete markets all goods are monies.

Suppose that there are m goods and that the mth serves as a money. Dubey and Shubik [3] considered a strategic market game where the strategy of a player α was of the form

$$(q_1^\alpha, \ldots, q_{m-1}^\alpha; b_1^\alpha, \ldots, b_{m-1}^\alpha),$$

where $0 \leqslant q_j^\alpha \leqslant \alpha_j^\alpha$ for $\alpha = 1, \ldots, n$, $j = 1, \ldots, m-1$, where n is the number of traders and a_j^α is the initial endowment of individual α of good j

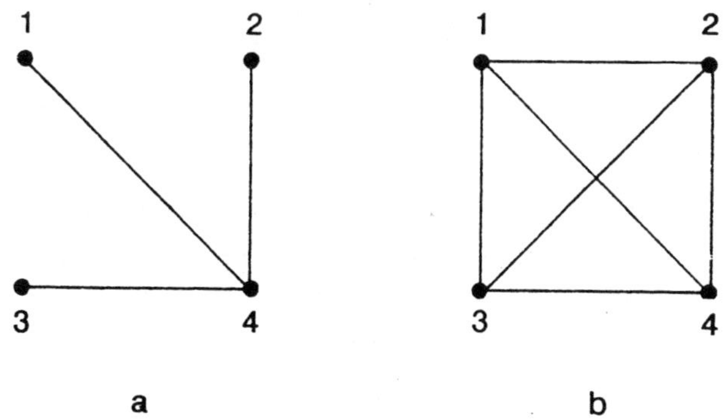

Figure 1

$(j=1, ..., m-1)$; q_j^α is the amount of good j offered for sale by individual α and

$$b_j^\alpha \geqslant 0 \quad \text{and} \quad \sum_{j=1}^{m-1} b_j^\alpha \leqslant a_m^\alpha,$$

where b_j^α is the amount of money committed by trader α to purchase good j and a_m^α is the amount of money in the initial endowment of α. In this game, as can be seen from the bounds on the strategies, there is no trading on credit.

Price (in terms of the money) is formed for good j by dividing the amount of money committed to j by the amount of j placed on sale:

$$p_j = \begin{cases} \sum_{\alpha=1}^{n} b_j^\alpha \Big/ \sum_{\alpha=1}^{n} q_j^\alpha & \text{if } \sum_{\alpha=1}^{n} q_j^\alpha \neq 0 \\ 0 & \text{if } \sum_{\alpha=1}^{n} q_j^\alpha = 0. \end{cases}$$

The second line in the above definition is just one of several possible trading conventions.

The type of price formation mechanism offered by Dubey and Shubik [3] is a symmetric two sided analogue of the Cournot market mechanism. It is possible to define this mechanism for an exchange economy with complete markets. Here a strategy for an individual will have a dimension of $m(m-1)$. There are $m(m-1)/2$ markets and an individual can be on either or both sides of each market. For example in the apples for oranges market an individual could supply both apples and oranges. In the stock market, when an individual simultaneously buys and sells the same shares, this is known as a wash sale.

For the strategic market game with m goods, $m-1$ markets, and one good serving as a money, Dubey and Shubik [3] were able to prove the existence of a pure strategy noncooperative equilibrium (NE)[1] where each player could be on both sides in each market.

In this paper we have extended the model with $(m-1)$ markets to one with $m(m-1)/2$ and have been able to prove the existence of an NE with active trade in all markets when there are at least two j-furnished individuals for every good j (no monopolists) and at least two individuals who have positive marginal utility for j. Our approach remains valid when the strategic exchange economy is endowed with any number of markets between $(m-1)$ and $m(m-1)/2$. In particular, our existence results (Section 2) subsume those in Shapley and Shubik [7] and Dubey and Shubik [3].

[1] We use the abbreviation NE to stand for a pure strategy noncooperative equilibrium.

2. Description of the Model and Existence Results

Let $I_n = \{1, 2, ..., n\}$ and $I_m = \{1, 2, ..., m\}$ be the sets of traders and commodities, respectively. We shall use Greek letters for traders and Roman letters for commodities.

We assume that each trader α has an initial endowment $a_i^\alpha > 0$ of each commodity i. The traders' utility functions are assumed to be strictly concave, increasing, and continuously differentiable functions $u^\alpha : \Omega^m \to \Omega^1$, $\alpha \in I_n$, where Ω^m, Ω^1 are the nonnegative orthants in R^m and R^1, respectively.[2]

A pair $\{i, j\} \subseteq I_m$ is to be thought of as a market between commodities i and j.

A move by a trader α is an $m \times m$ matrix B^α such that:

(i) $b_{ij}^\alpha \geqslant 0$,

(ii) $\sum_j b_{ij}^\alpha \leqslant a_i^\alpha$,

where b_{ij}^α is the amount of commodity i that α sends to trade in the market $\{i, j\}$. We write S^α for the set of possible moves (strategy set) by trader α, and let

$$S = S^1 \times \cdots \times S^n.$$

We shall denote by Γ the game in which the outcome of moves

$$B = (B^1, ..., B^n) \in S$$

is determined as follows:

First, the price matrix P is given by

$$p_{ij} = \begin{cases} \sum_\alpha b_{ij}^\alpha \Big/ \left(\sum_\alpha b_{ji}^\alpha\right) & \text{if } \sum_\alpha b_{ji}^\alpha \neq 0 \\ 0 & \text{if } \sum_\alpha b_{ji}^\alpha = 0. \end{cases} \tag{1}$$

Next, the final holdings are computed by

$$x_i^\alpha = a_i^\alpha - \sum_j b_{ij}^\alpha + \sum_j b_{ji}^\alpha p_{ij}. \tag{2}$$

[2] We may drop the differentiability requirement on the utility functions and only insist that they be continuous. This would entail replacing the infinitesimal argument in the proof of Theorem 1(b) by a more careful "finite" argument as in [1], using the "uniform monotonicity" lemma from [4].

Finally, the payoffs to the traders are

$$\pi^\alpha(B) = u^\alpha(x^\alpha). \tag{3}$$

Given a choice of moves

$$B^{-\alpha} \in S^1 \times \cdots \times S^{\alpha-1} \times S^{\alpha+1} \times \cdots \times S^n$$

by all the traders *except* α, we say that $B^\alpha \in S^\alpha$ is a "best response" by α to $B^{-\alpha}$ if

$$\pi^\alpha(B^{-\alpha}, B^\alpha) = \sup_{T \in S^\alpha} \pi^\alpha(B^{-\alpha}, T).$$

A Nash Equilibrium (NE) of Γ is an n-tuple

$$B = (B^1, ..., B^n) \in S$$

such that each B^α is a best response to $(B^1, ..., B^{\alpha-1}, B^{\alpha+1}, ..., B^n)$.

Nash equilibria of Γ exist for trivial reasons. For instance, the set of moves where $B^\alpha = 0$ for each α clearly constitutes an (autarky) Nash equilibrium; see Peck and Shell [5] for further discussion. To get around this, we consider (as in Dubey and Shubik [3]) a slight modification of Γ:

For each $m \times m$ matrix $\varepsilon = (\varepsilon_{ij})$, $0 < \varepsilon_{ij} \leqslant 1$, we consider the game Γ_ε, where some external agency supplies the fixed amounts ε_{ij} in the various markets, so that the prices are now given by

$$p_{ij} = \left(\sum_\alpha b^\alpha_{ij} + \varepsilon_{ij}\right) \bigg/ \left(\sum_\alpha b^\alpha_{ji} + \varepsilon_{ji}\right)$$

and the other computations are as before.

We define an *equilibrium point* (EP) of Γ to be a pair (B, P) such that

(1) B is an NE of Γ

(2) $p_{ij} > 0$ for all i, j

(3) there is a sequence of matrices $\varepsilon_l \to 0$ and NEs $B(\varepsilon_l)$ of $\Gamma(\varepsilon_l)$ with prices $P(\varepsilon_l)$ such that $B(\varepsilon_l) \to B$ and $P(\varepsilon_l) \to P$.

By ${}^k\Gamma_\varepsilon$ (respectively ${}^k\Gamma$) we shall understand the k-fold replication of Γ_ε (respectively Γ) in which each trader is replaced by k copies of himself. A type-symmetric Nash equilibrium (TSNE) is an NE wherein identical players employ identical strategies. A type-symmetric equilibrium point TSEP is a TSNE of ${}^k\Gamma$ which is the limit in the above sense of a sequence of TSNEs of ${}^k\Gamma_{\varepsilon_l}$.

We say that a good is *active* at an equilibrium if some trader offers a non-zero amount of that good in exchange for some other good.

The first theorem asserts

THEOREM 1. *For each k, $^k\Gamma$ has a TSEP with at least $m-1$ goods active. Furthermore there is a constant D* (independent *of k*) *such that if* (*B, P*) *is* any *TSEP of $^k\Gamma$, then*

$$D \geqslant p_{ij} \geqslant 1/D \qquad \textit{for all} \quad i, j.$$

We start with a lemma which clarifies the underlying geometry of the model.

LEMMA 1. *In the game Γ_ε, fix a choice of strategies by all traders except α; and let H and Π denote, respectively, the set of possible final holdings by α and the prices which form as α varies his strategy in S^α. Then*

(a) *If P and Q are in Π, so is R, where $r_{ij}=(p_{ij}q_{ij})^{1/2}$ for all i, j.*

(b) *H is compact; and if x and y are in H, then there is a z in H such that*

$$z_i \geqslant \tfrac{1}{2}(x_i+y_i) \qquad \textit{for all} \quad i.$$

Proof. Let $c_{ij}=\sum_{\beta\neq\alpha} b_{ij}^\beta+\varepsilon_{ij}$. Then if α makes the move $T=(t_{ij})\in S^\alpha$, the prices will be

$$p_{ij}=(c_{ij}+t_{ij})/(c_{ji}+t_{ji}) \tag{4}$$

and his final holding will be

$$\begin{aligned} x_i &= a_i^\alpha-\sum_j t_{ij}+\sum_j t_{ji}p_{ij} \\ &= a_i^\alpha+\sum_j (t_{ji}p_{ij}-t_{ij}) \\ &= a_i^\alpha+\sum_j (c_{ij}-c_{ji}p_{ij}) \qquad \text{(by (4))} \\ &= a_i^\alpha+\sum_j c_{ij}-\sum_j c_{ji}p_{ij}. \end{aligned} \tag{5}$$

Now, suppose strategies T and U in S^α achieve x and y in H with prices P and Q in Π. Let

$$r_{ij}=(p_{ij}q_{ij})^{1/2} \qquad \text{and} \qquad v_{ij}=[(c_{ij}+t_{ij})(c_{ij}+u_{ij})]^{1/2}-c_{ij}.$$

Then

$$\begin{aligned} (c_{ij}+v_{ij})/(c_{ji}+v_{ji}) &= [(c_{ij}+t_{ij})(c_{ij}+u_{ij})]^{1/2}/[(c_{ji}+t_{ji})(c_{ji}+u_{ji})]^{1/2} \\ &= (p_{ij}q_{ij})^{1/2}=r_{ij}. \end{aligned}$$

So to complete the proof of part(a), it remains only to show that $V \in S^\alpha$; and the only condition to be checked is that

$$\sum_j v_{ij} \leqslant a_i^\alpha.$$

But

$$\begin{aligned}
\sum_j v_{ij} &= \sum_j (c_{ij} + v_{ij}) - \sum_j c_{ij} \\
&= \sum_j [(c_{ij} + t_{ij})(c_{ij} + u_{ij})]^{1/2} - \sum_j c_{ij} \\
&\leqslant \left[\sum_j (c_{ij} + t_{ij})\right]^{1/2} \left[\sum_j (c_{ij} + u_{ij})\right]^{1/2} - \sum_j c_{ij} \\
&\leqslant \max\left\{\sum_j (c_{ij} + t_{ij}), \sum_j (c_{ij} + u_{ij})\right\} - \sum_j c_{ij} \\
&= \max\left\{\sum_j t_{ij}, \sum_j u_{ij}\right\} \\
&\leqslant a_i^\alpha,
\end{aligned}$$

where the first inequality is to Cauchy–Schwartz inequality; and the last inequality holds since T and U lie in S^α.

Part (b) is a simple consequence of part (a) and the arithmetic–geometric mean inequality.

Thus, if V is as before and z is the final holding resulting from V, we have

$$\begin{aligned}
z_i &= a_i^\alpha + \sum_j c_{ij} - \sum_j c_{ji} r_{ij} \qquad \text{(by (5))} \\
&= a_i^\alpha + \sum_j c_{ij} - \sum_j c_{ji} (p_{ij} q_{ij})^{1/2} \\
&\geqslant a_i^\alpha + \sum_j c_{ij} - \sum_j c_{ji} [1/2(p_{ij} + q_{ij})] \\
&= \frac{1}{2}\left[a_i^\alpha + \sum_j c_{ij} - \sum_j c_{ji} p_{ij}\right] + \frac{1}{2}\left[a_i^\alpha + \sum_j c_{ij} - \sum_j c_{ji} q_{ij}\right] \\
&= \frac{1}{2}(x_i + y_i).
\end{aligned}$$

Finally, since S^α is compact and the map from S^α to H is clearly continuous, H must be compact. Q.E.D.

The next step is

LEMMA 2. *For each* $\varepsilon > 0$, Γ_ε *has an* NE.

Proof. Given a fixed set of moves by the other traders, since u^α is strictly concave and continuous, there exists a unique $x \in H$ which maximizes u^α.

It is easily seen that there must be a *unique* price P associated with x. For if Q is also associated with x and $q_{ij} \neq p_{ij}$ for some i, j, then write $r_{ij} = (p_{ij} q_{ij})^{1/2}$ and let z be the final holding corresponding to R. Then by the analysis in the proof of part (b) of Lemma 1 we see that $z \geqslant x$; and by the *strict* artihmetic–geometric mean inequality for (p_{ij}, q_{ij}) in that proof, it follows that

$$z_i > x_i.$$

This contradicts the maximality of x.

Consider all the (best) responses by α which yield the price matrix P. This is clearly a compact, convex, and non-empty set. By the Theorem of the Maximum in Berge [1], it follows that the best response correspondence is upper semi-continuous from

$$S^1 \times \cdots S^{\alpha-1} \times S^{\alpha+1} \times \cdots \times S^n \qquad \text{to subsets of } S^\alpha.$$

As before, let $S = S^1 \times \cdots \times S^n$ and consider Φ: $S \to$ subsets of S given as follows: If $B = (B^1, ..., B^n)$, then $\Phi(B) = \{(T^1, ..., T^n)$: each T^α is a best response by α to $(B^1, ..., B^{\alpha-1}, B^{\alpha+1}, ..., B^n)\}$.

Then Φ is upper semi-continuous, S is compact and convex, and $\Phi(B)$ is compact, convex, and not empty for each $B \in S$. By Kakutani's Theorem, Φ has a fixed point, which is easily seen to be a Nash Equilibrium for Γ_ε.

Q.E.D.

Lemma 2 can be refined to yield

LEMMA 3. *For each* $k = 1, 2, 3, ...,$ *the game* ${}^k\Gamma_\varepsilon$ *has a TSNE.*

Proof. Let ${}^kS$ be the Cartesian product of the strategy sets of all the traders in ${}^k\Gamma_\varepsilon$; and let ${}^kS^*$ ($\approx S$) be the set of type-symmetric strategies in ${}^kS$. Then it is clear that for each $B \in {}^kS^*$, the set $\Phi^*(B) = \Phi(B) \cap {}^kS^*$ is compact, convex and non-empty.

Let Φ^* be the correspondence $B \to \Phi^*(B)$. Then Φ^* is upper semi-continuous; and by Kakutani's theorem, it has a fixed point, which is easily seen to be a TSNE of ${}^k\Gamma_\varepsilon$. Q.E.D.

This can be strengthened to

LEMMA 4. *For each k, ${}^k\Gamma_\varepsilon$ has a TSNE with at least $m-1$ goods active.*

Proof. Given a fixed choice of moves by all traders except one, consider the optimization problem faced by that reader. Let C be the aggregate bid matrix of the other traders (plus ε). Then if the remaining trader makes the move T, the resulting price is P where

$$p_{ij} = (c_{ij} + t_{ij})/(c_{ji} + t_{ji}) \tag{6}$$

Since the best response price is unique, there is a unique T which satisfies (6) for this price, and also the condition $t_{ij} \cdot t_{ji} = 0$ for all i, j. Let us call this *the* "minimal best response." By the theorem of the maximum in Berge [1], the minimal best response map is a continuous function from ${}^kS$ to ${}^kS$.

We can also define the "maximal best response" (with respect to a fixed ordering of the goods) as follows: Let T be a minimal best response by a trader to moves by the others. Suppose that at T, the trader has non-zero amounts of at least two commodities left over after his move. Choose $\{i, j\}$ to be the smallest such pair (with respect to the ordering), and let the trader increase his bids on *both* sides of the (i, j)th market in such a manner that p_{ij} remains unchanged. This can be done until the trader has used up all of at least one of the two commodities i and j. Now let him select the next lowest pair and proceed in this fashion until he reaches a strategy T' in which he has used up all of at least $m-1$ of his goods. Since T' also gives the best response price, it is a best response, which we can call the "maximal best response."

Let Φ': ${}^kS \to {}^kS$ be the maximal best response map; then Φ' is easily seen to be continuous. Furthermore, Φ' maps ${}^kS^*$ (indeed all of ${}^kS$) into ${}^kS^*$. By Brouwer's Theorem, Φ' has a fixed point on ${}^kS^*$.

Since we assumed $a_i^\alpha > 0$ for all α and i, any fixed point of Φ' has the property that *each* trader sends a positive amount (a_i^α) of *at least* $m-1$ goods to the market. So the fixed points of Φ' are TSNEs of the desired kind. Q.E.D.

We proceed next to the proof of boundedness of prices. In the proof of the next lemma and subsequent discussion, we shall abuse the notation slightly and use the index α to represent both a trader and a trader type in ${}^k\Gamma_\varepsilon$.

LEMMA 5. *There is a constant $D \geqslant 1$ (independent of k and ε) such that if P is the price matrix at* any *TSEP of any ${}^k\Gamma_\varepsilon$ (with $k = 1, 2, 3, \ldots$ and $D < \varepsilon_{ij} \leqslant 1$), then*

$$D \geqslant p_{ij} \geqslant 1/D \qquad \textit{for all} \quad i, j.$$

Proof. Let $A = \{x \in \Omega^m : x_i \leqslant L_i \text{ for each } i\}$, *where* $L_i = m + a_i^\alpha$ *for each i.*

It is easy to see that if $\varepsilon \leqslant 1$ and x^α is the final holding by any trader α at a TSNE in the game Γ_ε, then x^α lies in the (compact) set A.

For each α, let ψ_i^α be the partial derivative of u^α in the ith commodity (if $x_i = 0$, ψ_i^α is the right-handed derivative). By virtue of the assumptions on u^α, ψ_i^α is continuous and positive on A. So we may choose a constant $D_1 > 1$ such that

$$D_1 > \psi_j^\alpha(x)/\psi_i^\alpha(x) \qquad \text{for all} \quad x \in A. \tag{7}$$

Now, let $B = (B^1, ..., B^n)$ be a TSNE of ${}^k\Gamma_\varepsilon$ with prices P and final holdings $x^1, ..., x^n$ by the various trader types.

Let α, β be two traders; we may assume that

$$b_{ji}^\alpha \leqslant \frac{1}{2}\left(\sum_\gamma b_{ji}^\gamma + \varepsilon_{ij}\right). \tag{8}$$

For convenience we shall divide the argument into three cases:

Case 1. $b_{ji}^\alpha \geqslant (1/m)a_j^\alpha$

Case 2. $b_{ji}^\alpha < (1/m)a_j^\alpha$ and $\sum_i b_{ji}^\alpha < a_j^\alpha$

Case 3. $b_{ji}^\alpha < (1/m)a_j^\alpha$ and $b_{jh}^\alpha > (1/m)a_j^\alpha$ for some $h \in I_m$.

It is easy to see that these cases exhaust all the possibilities.

In *Case* 1, we have

$$p_{ij} = \left(\sum_\alpha b_{ij}^\alpha + \varepsilon\right) \Big/ \left(\sum_\alpha b_{ji}^\alpha + \varepsilon\right) \leqslant L_i \Big/ \frac{1}{m}\, a_j^\alpha,$$

so if

$$D_2 = \max_{\alpha, i} \max_{j : a_j^\alpha > 0} (m \cdot L_i)/a_j^\alpha, \tag{9}$$

then $p_{ij} < D_2$.

In *Case* 2, we consider the effect on α's utility if he increases b_{ji}^α slightly. Thus, let

$$e = b_{ij}^\alpha, \qquad f = b_{ji}^\alpha, \qquad E = \sum_{\beta \neq \alpha} b_{ij}^\beta + \varepsilon_{ij}, \qquad F = \sum_{\beta \neq \alpha} b_{ji}^\beta + \varepsilon_{ji}.$$

Then

$$p_{ij} = \frac{E + e}{F + f};$$

and if α changes b_{ji}^{α} to $b_{ji}^{\alpha}+\eta$, the price changes to

$$q_{ij}=\frac{E+e}{F+f+\eta},$$

and α's final holding of i increases by

$$\begin{aligned}(f+\eta)\cdot q_{ij}-fp_{ij}&=\frac{(f+\eta)\cdot(E+e)}{(F+f+\eta)}-\frac{f(E+e)}{(F+f)}\\&=\frac{E+e}{F+f}\left(\frac{(f+\eta)(F+f)-f(F+f+\eta)}{(F+f+\eta)}\right)\\&=\eta\cdot\frac{E+e}{F+f}\cdot\frac{F}{F+f+\eta}=\eta\cdot p_{ij}\cdot\frac{F}{F+f+\eta}.\end{aligned}$$

Now by (8), we know that $e<E$; if in addition, $n<E$, then the increase in α's holding of commodity i is greater than $\frac{1}{3}\eta\cdot p_{ij}$. Also the decrease in his holding of j is η.

So if δu^{α} is the increase in α's utility, we must have

$$\begin{aligned}\delta u^{\alpha}&>(\tfrac{1}{3}\eta\cdot p_{ij})\,\psi_i(x^{\alpha})-\eta\psi_j(x^{\alpha})+O(\eta^2)\\&=\tfrac{1}{3}\eta\psi_i(x^{\alpha})\cdot(p_{ij}-3\psi_j(x^{\alpha})/\psi_i(x^{\alpha}))+O(\eta^2).\end{aligned}$$

Since b_{ij}^{α} is a best response, δu^{α} must be negative. Therefore, we must have

$$p_{ij}\leqslant 3\psi_j(x^{\alpha})/\psi_i(x^{\alpha});$$

and if D_1 is as in (7), then

$$p_{ij}\leqslant 3D_1. \tag{10}$$

Finally, in *Case* 3, we consider the change in α's utility if he diverts a small amount η from b_{jh}^{α} to b_{ji}^{α}.

As in Case 2, for sufficiently small η, the increase in his holding of commodity i is greater than $\frac{1}{3}\eta p_{ij}$. It remains to estimate the decrease in his holding of h. Let

$$v=b_{ij}^{\alpha},\qquad V=\sum_{\beta\neq\alpha}b_{hj}^{\beta}+\varepsilon$$

$$w=b_{jh}^{\alpha},\qquad W=\sum_{\beta\neq\alpha}b_{jh}^{\beta}+\varepsilon.$$

Then the decrease in α's holding of h is

$$w \cdot \frac{V+v}{W+w} - (w-\eta)\frac{V+v}{W+(w+\eta)} = \frac{V+v}{W+w} \cdot \frac{W}{W+w-\eta} \cdot \eta.$$

Now in the present situation, we know that $w > (1/m)a_j^\alpha$ and so if $\eta < (1/2m)a_j^\alpha$, the above decrease is less than

$$\frac{L_h \cdot L_j}{((1/m)a_j^\alpha)((1/2m)a_j^\alpha)} \cdot \eta < 2 \cdot D_2^2 \cdot \eta,$$

where D_2 is as in (9).

Therefore if δu^α is the change in α's utility,

$$\delta u^\alpha > (\tfrac{1}{3}p_{ij}\eta)\,\psi_i^\alpha(x^\alpha) - 2D_2^2\eta\psi_h^\alpha(x^\alpha) + O(\eta^2)$$

for sufficiently small η.

Since δu^α must be negative, we must have

$$p_{ij} < 6D_2^2\psi_h^\alpha(x^\alpha)/\psi_i^\alpha(x^\alpha) \leqslant 6D_1D_2^2. \tag{11}$$

So, let $D = D_2 + 3D_1 + 6D_1D_2^2$. Then, by (9), (10), and (11), we see that, in all cases,

$$p_{ij} \leqslant D \qquad \text{for all} \quad i, j.$$

Since $p_{ji} = 1/p_{ij}$, the other half of the inequality follows. Q.E.D.

We are now in a position to prove Theorem 1.

Proof of Theorem 1. Fix k and choose a sequence $\{\varepsilon_l\}_{l=1}^\infty$ decreasing to zero. By Lemma 4, we can find TSNEs $\{(B_{\varepsilon_l}, P_{\varepsilon_l})\}$ of the ${}^k\Gamma_{\varepsilon_l}$ with $m-1$ goods active. Now, by compactness of the strategy spaces and Lemma 5, we may assume (passing to a subsequence, if necessary) that $\{(B_{\varepsilon_l}, P_{\varepsilon_l})\}_{l=1}^\infty$ converges to a limit (B, P) with $m-1$ goods active. Then $D \geqslant p_{ij} \geqslant 1/D$, and it remains only to show that (B, P) is a TSNE of ${}^k\Gamma$.

Let x_{ε_l} (resp. x) be the allocation resulting from B_{ε_l} in ${}^k\Gamma_{\varepsilon_l}$ (resp. B in ${}^k\Gamma$). First of all, we establish that $x_{\varepsilon_l} \to x$, as $l \to \infty$. For this we examine the contribution at B_{ε_l} of each market (i, j) in ${}^k\Gamma_{\varepsilon_l}$ as $l \to \infty$. Clearly, the aggregate (and individual) amounts sent to trade on both sides of the market converge to those at B in ${}^k\Gamma$. So if we aggregate amounts on both sides of the market tend to non-zero limits, then it is clear from (1) and (2) that individual trades converge to those at B in ${}^k\Gamma$. On the other hand, if the aggregate amount on one side tends to zero, so must the other side (by Lemma 5). Consequently the net trade itself tends to zero, and is zero in the limit.

Now, suppose B is not a TSNE of ${}^k\Gamma$. Then there is a trader of type α who can improve on his payoff at B. Let B' be the new set of moves resulting from this trader's change and let

$$\delta = \pi^\alpha(B') - \pi^\alpha(B) > 0.$$

Let $\pi^\alpha_{\varepsilon_l}$ be the payoff function in ${}^k\Gamma_{\varepsilon_l}$; then we can choose l large enough so that

$$|\pi^\alpha_{\varepsilon_l}(B') - \pi^\alpha(B')| < \delta/3.$$

Also, by the continuity of u^α, and the first part of this argument ($x_{\varepsilon_l} \to x$), we may further ensure that

$$|\pi^\alpha_{\varepsilon_l}(B_{\varepsilon_l}) - \pi^\alpha(B)| < \delta/3.$$

So

$$\begin{aligned}\pi^\alpha_{\varepsilon_l}(B') - \pi^\alpha_{\varepsilon_l}(B_{\varepsilon_l}) &> (\pi^\alpha(B') - \delta/3) - (\pi^\alpha(B) + \delta/3)\\ &= \delta - 2\delta/3 = \delta/3 > 0,\end{aligned}$$

which is not possible since B_{ε_l} is an NE for ${}^k\Gamma_{\varepsilon_l}$. So B must be a TSNE of ${}^k\Gamma$.[3]

3. The Relationship between Noncooperative and Competitive Equilibrium

3.1. *The Inequivalence*

We recall the notion of a *competitive equilibrium* (CE) for the *exchange economy* corresponding to Γ. With notation as in the first two paragraphs of Section 2, we proceed as follows:

Given any price *vector* $p \in \Omega^m$, the *budget set* of the trader α is defined (as usual) to be

$$B^\alpha(p) = \{x \in \Omega^m : p \cdot x \leqslant p \cdot a^\alpha\}$$

[3] The case $n = 2$, $m = 2$, $k = 1$ is special; Shapley [6] has provided a construction of an open set of NEs without using a fixed point theorem. It may be possible to give a similar existence argument in general, but this is not clear. In particular, Dubey and Rogawski [2] have shown that for $m > 2$, the set of NE allocations has *strictly* smaller dimension than the space of all allocations.

and a CE for Γ consists of a price vector $p \in \Omega^m$ and allocations $x^\alpha \in \Omega^m$ such that

$$u^\alpha(x^\alpha) = \max\{u^\alpha(y) : y \in B^\alpha(p)\} \qquad \text{for all } \alpha \tag{12}$$

and

$$\sum_{\alpha=1}^{n} x^\alpha = \sum_{\alpha=1}^{n} a^\alpha. \tag{13}$$

It is well known that any exchange economy which satisfies the conditions in the first two paragraphs of Section 2 has a competitive equilibrium.

A strategic market game, in contrast with the CE model of exchange, is "process and transactions"-oriented. Liquidity constraints, which have no role to play in general equilibrium, play a fundamental role in a strategic market game. Thus, in such a game, even with a continuum of traders, there may be CEs which are not NEs and vice versa.

Since, in our model, direct exchanges are permitted between all commodity pairs, it may seem at first that the lack of liquidity will never be a constraint on the actions of the traders. However, this is not so; and we give two examples to show this.

EXAMPLE 1. Consider the case when there are three goods and four trader types. We assume a continuum of each type (the interval $[0, 1]$), characterized by the following utility functions and endowments:

$$\begin{aligned}
\varphi^0(x, y, z) &= (xyz)^{1/3} && \text{and} && (1, 1, 1)\\
\varphi^1(x, y, z) &= (yz^2)^{1/3} && \text{and} && (0, 3, 0)\\
\varphi^2(x, y, z) &= (xy^2)^{1/3} && \text{and} && (3, 0, 0)\\
\varphi^3(x, y, z) &= (x^2 z)^{1/3} && \text{and} && (0, 0, 3)
\end{aligned}$$

There is a TSNE with prices $p_{xy} = p_{yz} = p_{zx} = 2$, where traders 1, 2, and 3 each offer two units of y, x, and z, respectively, in exchange for one unit of goods z, y, and x, respectively, from the traders of type 0. The final endowments of the trader types are $(2, 2, 2)$, $(0, 1, 1)$, $(1, 1, 0)$, and $(1, 0, 1)$, respectively.

On the other hand, the unique CE for the related exchange economy has prices $p_x = p_y = p_z = 1$ and final holdings $(1, 1, 1)$, $(0, 1, 2)$, $(1, 2, 0)$ and $(2, 0, 1)$.

Strictly speaking, this example does not satisfy the positiveness conditions imposed in Section 2. The next example does satisfy these conditions, and is furthermore robust.

EXAMPLE 2. Consider the case of three goods and three players. We assume a continuum of each type (the interval [0, 1]), with the utility functions and endowments

$$\varphi^1(x, y, z) = (x^\lambda yz)^{1/(\lambda+2)} \quad \text{and} \quad (1, 1, \lambda)$$

$$\varphi^2(x, y, z) = (xy^\lambda z)^{1/(\lambda+2)} \quad \text{and} \quad (\lambda, 1, 1)$$

$$\varphi^3(x, y, z) = (xyz^\lambda)^{1/(\lambda+2)} \quad \text{and} \quad (1, \lambda, 1),$$

where λ is some large, positive constant.

By virtue of the symmetry of the model, it is clear that there will be an NE with $p_{xy} = p_{yz} = p_{zx} = p$, say. We would like to deduce a relationship between p and λ. First of all, it is easy to see that if λ is large, then $p > 1$; and also that at an NE strategy, the bids by player 1 must be of the form where $a > 0$, $b > 0$, and $\lambda > a + b$.

If (x, y, z) are the final holdings by player 1, we have

$$x = p + a/p + 1, \qquad y = bp, \qquad z = \lambda - (a + b).$$

Now, by considering the effects of small changes in a and b on the payoff to player 1, we see that

$$p = \lambda z/x = y/z.$$

Finally, since the bids by the other two types are symmetrically determined, we have

$$p = a(b + 1).$$

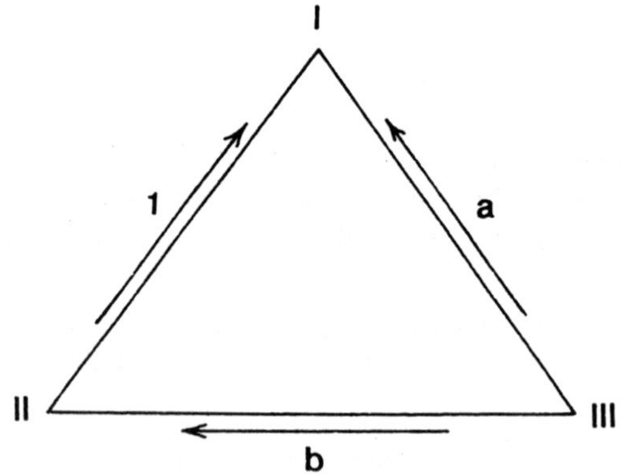

FIGURE 2

Using these six relations, we quickly obtain

$$(\lambda - p)^2 = p(p+2)^2.$$

In particular, if $p=4$ then $\lambda = 16$; and (with a little more calculation) $b=2$, $a=12$, $x=8$, $y=8$, $z=2$. So the final holdings (for $\lambda=16$) are (8, 8, 2), (2, 8, 8), and (8, 2, 8) by the different players.

Also, the unique CE has prices $p_x = p_y = p_z = 1$, and final holdings are (16, 1, 1), (1, 16, 1), and (1, 1, 16). It is of some interest to note that in Example 2 the CE allocations are actually achievable through a particular choice of strategies by the various traders. However, such choices will not constitute an NE.

3.2. *Conditions for Equivalence*

Let (B, P) be a TSEP for ${}^k\Gamma$; we say that the TSEP is *price consistent* if for any three goods $\{h, i, j\}$ we have $p_{hi}\, p_{ij} = p_{hj}$.

If (B, P) is price consistent, then the *matrix* P can be expressed in terms of a price *vector* q such that

$$p_{ij} = q_i / q_j. \tag{14}$$

Price inconsistencies at a TSEP imply the existence of arbitrage possibilities which are not being exploited because of liquidity constraints. In fact, we have

LEMMA 6. *Let (B, P) be a TSEP of ${}^k\Gamma$ such that $p_{i_1 i_2} p_{i_2 i_3} \neq p_{i_1 i_3}$ for a triple $\{i_1, i_2, i_3\}$ of goods. Then for each trader type α, we must have*

$$\sum_j b_{ij}^\alpha = a_i^\alpha$$

for some $i \in \{i_1, i_2, i_3\}$.

Proof. It suffices to prove the lemma for TSNEs of ${}^k\Gamma(\varepsilon)$; but there it must hold, since otherwise such a player would be able to increase his holding of all three goods i_1, i_2, and i_3 by an obvious modification of his bid. Q.E.D.

The examples in Section 3.1 show that it is possible for *all* TSEPs of a game to be price inconsistent.

THEOREM 2. *Let $({}^kB, {}^kP)$ be a sequence of TSEPs of ${}^k\Gamma$; then this sequence has a limit point, say (B, P). Suppose*

(1) *$({}^kB, {}^kP)$ are price consistent;*

(2) *(B, P) has all markets active, i.e., for any two goods i, j, there is $\alpha \in I_n$ with $b^{\alpha}_{ij} \neq 0$.*

Then

(a) *P is consistent and we may choose q as in* (14);

(b) *If ${}^k x^{\alpha}$ are the final holdings at $({}^k B, {}^k P)$, then ${}^k x^{\alpha}$ converges, to x^{α}, say,*

(c) *(q, x^{α}) is a CE for the exchange economies corresponding to Γ^k, $j = 1, 2, 3, \ldots$.*

Proof. We omit the proof since it is completely analogous to Dubey and Shubik [3]. Q.E.D.

We conclude with some remarks and conjectures.

Remark 1. For the game with a continuum of players, fix players' utility functions and the *total* amount of each commodity in the game. Consider the different games that arise as the total commodities are distributed across the players. If a distribution is on the Pareto-optimal surface, then clearly this is the unique CE and TSNE of the corresponding game. Now Lemma 6 and a Cobb–Douglas like condition on the utility functions are sufficient to show that, for distributions in a *neighborhood* of the Pareto-optimal surface, there are active TSNEs which "coincide" with the CEs.

Conjecture 1. For the game with a continuum of players, if all CEs are active TSNEs (with consistent prices) then these are the only TSNEs.

Remark 2. We can show that with three goods, and Cobb–Douglas utility functions, the game ${}^k\Gamma$ always has a TSNE with *all* markets active.

Conjecture 2. With Cobb–Douglas utility functions, ${}^k\Gamma$ has a TSNE with all markets active.

References

1. C. Berge, "Topological Spaces," MacMillan, New York, 1963.
2. P. Dubey and J. Rogawski, Inefficiency of Nash equilibria in strategic market games, Technical Report, University of Illinois, Urbana–Champagne, 1985.
3. P. Dubey and M. Shubik, The noncooperative equilibria of a closed trading economy with market supply and bidding strategies, *J. Econ. Theory* **17** (1978), 1–20.
4. J. Peck and K. Shell, Market uncertainty: Sunspot equilbria in imperfectly competitive economies, University of Pennsylvania, CARESS Working paper #85-21, 1985.
5. J. Peck and K. Shell, Market uncertainty: Correlated equilibrium and sunspot equilibrium in market games, European University Institute, Florence, EUI 86/244, 1986.

6. L. S. SHAPLEY, Noncooperative general exchange, *in* "Theory of Measurement of Economic Externalities" (S. A. Y. Lin, Ed.), pp. 155–175, Academic Press, New York, 1976.
7. L. S. SHAPLEY AND M. SHUBIK, Trade using one commodity as a means of payment," *J. Polit. Econ.* **85** (1977), 937–968.
8. M. SHUBIK, Commodity money, oligopoly, credit and bankruptcy in a general equilibrium model, *Western Econ. J.* **10** (1973), 24–38.
9. S. YAO, "On Strategic Market Games," Ph.D. dissertation, University of California, Los Angeles, 1987.

Printed by Catherine Press, Ltd., Tempelhof 41, B-8000 Brugge, Belgium

The Geneva Papers on Risk and Insurance Theory, Vol. 15, No. 1 (March 1990), 55-71

The Capital Asset Pricing Model as a General Equilibrium With Incomplete Markets*

by John Geanakoplos** and Martin Shubik**

ABSTRACT

We recast the capital asset pricing model (CAPM) in the broader context of general equilibrium with incomplete markets (GEI). In this setting we give proofs of three properties of CAPM equilibria: they are efficient, asset prices lie on a "security market line", and all agents hold the same two mutual funds. The first property requires a riskless asset, the latter two do not. We show that across all GEI only one of these three properties of equilibrium is generally valid: asset prices depend on covariances, not variances. We extend CAPM to many consumption goods in such a way that all three properties hold. But now the definition of a riskless asset depends on preferences and endowments, and so cannot be specified *a priori*.

1. Introduction

This paper is devoted to clarifying the relationship between the CAPM stock market trading model and general equilibrium with incomplete markets. CAPM yields three important views of financial markets. First, that they are efficient. Second that asset prices depend not on the variance, but rather on the covariance of the underlying payoffs with a particular, privileged portfolio. Third, that all portfolio holders may be perfectly happy to hold only a few specially designated mutual funds.

By placing CAPM in the broader context of general equilibrium with incomplete markets (GEI) we find that only the second property, that covariances (and not variances) matter to asset pricing, retains validity. Risk averse agents diversify, to be sure, but in general they will not be satisfied with the same mutual funds. In GEI equilibrium there is no arbitrage, but the final allocations are almost never Pareto optimal. In fact, they are almost never constrained Pareto optimal.

Although there is no variant of either the efficiency principle or the mutual fund principle that is precisely true throughout GEI, one may still wonder if perhaps efficiency is

* We wish to acknowledge assistance of NSF Grant No. 88-12051, and a referee's comments. We are grateful for conversations with H. Polemarchakis, especially concerning Section II.4.

** Cowles Foundation, Yale University.

approximately true if CAPM is a "reasonable" description of reality. CAPM assumes a single consumption good in each state of nature. To regard CAPM as a descriptive model, one must suppose that in reality relative commodity prices are not much affected by redistributions of income. In that case, GEI equilibria do become constrained Pareto efficient, but not Pareto efficient.

When there are many goods, what is the meaning of the riskless asset which is so central to CAPM? Should it promise the same quantity of money in each state, irrespective of the (different) rates of inflation? Should it guarantee equivalent purchasing power measured in terms of some specified basket of goods? How should the basket be chosen? Is there an analogue to the riskless asset in a multi-commodity world?

We shall show that there is a generalization of CAPM to a multi-dimensional world (m-CAPM). To each m-CAPM economy, there corresponds a collection of "riskless" assets. If any *one* of these is marketed, then all the equilibria will be Pareto optimal. But the rub is that this set of riskless assets depends on the underlying m-CAPM economy. It cannot be calculated without knowing the preferences and aggregate endowments of the economy for each possible state of nature. Without a riskless asset the equilibria will be far from Pareto optimal.

In the classical one good CAPM model the presence of a riskless asset is crucial to Pareto optimality. Perhaps the single most obvious policy recommendation that can be derived from the moddel is that the government should always engineer the creation of the riskless asset if it is not there already. Yet (at least in the U.S.) for a very long time there has really been no asset which purported to be riskless, when inflation is taken into account. For the one good CAPM, that is a puzzle, since there is no ambiguity about what the riskless asset should be. In the multi-dimensional CAPM, a riskless asset exists, but knowing how to calculate it might be impossible.

In Part I we place the one good CAPM inside GEI. In the presence of a riskless asset (and quadratic preferences, and other CAPM hypotheses) we derive the Pareto efficiency of equilibrium, the security market line, and the mutual fund theorem. We note that efficiency fails when there is no riskless asset, but that the other two theorems remain valid. In Part II we describe a full-blown multi-commodity GEI model. We note that efficiency and the mutual fund theorem fail, but that an analogue to the security market line does hold. We also describe a special multi-good CAPM in which efficiency holds if there is a riskless asset. But the riskless asset cannot be described *a priori,* independent of the data of the economy.

2. Economic preliminaries

In the past few years there has been a great revival of interest by economists in the theory of general equilibrium with incomplete markets (GEI) anticipated by Arrow [1953] and Diamond [1967], and expressed first in general form by Radner [1972] and Hart [1975]. The GEI theory focuses on the primitive characteristics of the agents and commodities, treating financial assets as limited devices for transferring wealth across states of nature, rather than starting with reduced form preferences defined on the assets themselves. The extra structure preserved by this approach, which is just beginning to be fully exploited, accounts for the renewed interest in the subject. In a series of papers, for example, by Cass [1984], Werner [1985], Duffy [1987], and Geanakoplos-Polemarchakis [1986] existence of equilibrium was proved by using the boundedness from below of the consumption sets for models with assets that all deliver in the same numeraire commodity in each state of nature, such as the quadratic-utility capital asset pricing model (CAPM) we shall consider below. In reduced form representations of CAPM, the existence of equilibrium is not always guaranteed (see Nielson [1985]).

Perhaps the most interesting general question that can be posed in the GEI (but not in any reduced form) theory is: how close will equilibrium allocations, constrained as they are by the limited assets available to the market, come to achieving Pareto optimality? Although no satisfactory definition of distance from optimality has been found for this problem, the work of Hart [1975], Greenwald and Stiglitz [1986], and Geanakoplos-Polemarchakis [1986] shows that except for extremely rare choices of utility-endowment characteristics for the agents, any absence of potential insurance contracts will prevent equilibria from being fully Pareto optimal; in fact the missing asset markets will (with rare exceptions) cause equilibrium allocations to fail to be even "constrained" Pareto optimal.[1] One purpose of this paper is to show that the quadratic utility CAPM model, with a riskless asset, is precisely one of those rare economies for which equilibrium allocations always attain full optimality (if every agent's consumption is strictly positive in every state).

The CAPM name has been used to refer to any model in which it is possible to write for each agent a reduced form utility for asset portfolios depending only on the means and variances of their payoffs. It is a famous property of such models that there is a mutual fund theorem and a security market line theorem. The mutual fund theorem states that all portfolios held in equilibrium can be expressed as a combination of two portfolios, the so-called market portfolio and a "zero-beta" portfolio (see Lintner [1965], Sharpe [1964], and for the case where there is no riskless asset see Black [1972]). When there is a riskless asset, the zero-beta portfolio can be taken to be the riskless asset alone. The security market line theorem asserts that in equilibrium there is a linear relationship between the expected payoffs of assets with price equal 1, and the covariances of their payoffs with the market portfolio payoffs. Of course it is impossible to pose the Pareto optimality question in the reduced form version of the model.

There are two well-known GEI models which give rise to reduced form mean-variance utilities for assets. In one, the state space is taken to be infinite, and all assets are assumed to have normally distributed payoffs, and all commodity endowments are kept at zero. The utilities are arbitrary von Neumann-Morgenstern, with common probabilities. In the second version, the utilities must be quadratic von Neumann-Morgenstern with common probabilities, and the endowments of commodities must be zero, but the asset payoffs can be arbitrary, and the state space finite. It is well-known that the normal distribution is specified by its mean and variance, and that the expectation of a quadratic function depends only on the mean and variance of the underlying distribution, so both of these versions clearly give rise to reduced form preferences on assets of the mean-variance kind. Either version may be considered with or without a riskless asset. The point of Theorem 1 of this paper is that in the quadratic version with a riskless asset, all interior equilibria are fully

---

[1] The reader can consult those papers for definitions of constrained Pareto optimal; Geanakoplos-Polemarchakis [1986] proves (except for rare exceptions) that when some of the asset markets are missing, the equilibrium will not even efficiently use the existing assets, when there are at least two commodities in every state of nature.

Pareto optimal, no matter how few assets there are relative to the number of states of nature. By contrast, in the other three GEI variations, equilibria are almost never Pareto optimal when the asset markets are incomplete. This shows that Pareto optimality is a property distinct from the mutual fund and security market line properties, though closely related. (For prior discussions of Pareto optimality, see Rubinstein [1974], Mossin [1977], Rothschild [1986].)

In Part I we consider in detail a model in which all consumers have quadratic von Neumann-Morgenstern utilities with common expectations. The state space is finite and the asset returns are arbitrary. The consumers are allowed to have endowments, however, provided that these lie in the span of the assets. In Theorem 1 we prove that if there is a riskless asset, all interior equilibria are Pareto optimal.[2] In Theorems 2-3' we derive the mutual fund and security market line theorems with (2, 3) and without (2', 3') a riskless asset from a linear algebraic argument that does not appeal to the mean-variance reduced form representation.

In Part II we give a sufficient condition on preferences for all multi-commodity GEI to be Pareto optimal (Theorem 4). One consequence (see Corollary) is that even in the one commodity model it is possible to obtain Pareto optimality with preferences other than quadratic. (For a discussion of these classes of utilities and their connection to mutual funds, see Cass-Stiglitz [1970]. For an application to efficiency in the one good model, see Rubinstein [1974].) We derive a multiple commodity CAPM model and observe that interior equilibria are Pareto optimal in the presence of a riskless asset, and we note the difficulty of finding a riskless asset in practice (Theorem 6). We note that even without the riskless asset, equilibria in the multiple commodity CAPM are constrained Pareto optimal (Theorem 5). Finally, we show that in all GEI equilibria there is a security market line giving prices in terms of covariances with some privileged asset, but not necessarily with the market (Theorem 7).

I. A model with one good and quadratic utilities

1. The model

Let the set of states of nature be denoted by $S = \{0, 1, \ldots, S\}$. Let there be only one good in each state of nature. The consumption set is taken to be R_+^{S+1}. Each agent $h \in H = \{1, \ldots, H\}$ is characterized by a quasi-concave, monotonic function $V^h : R_+^{S+1} \to R$, and an endowment vector of commodities $e^h \in R_{++}^{S+1}$. In addition agents hold assets, described below.

The set of financial assets is denoted $A = \{1, \ldots, A\}$. Each asset $a \in A$ is represented by a vector $r_a \in R^{S+1}$. Sometimes we shall take the first asset, r_1, to be the so-called riskless asset 1, which pays one unit of the commodity in every state of nature. $r_1 = 1 \equiv \{1, \ldots, 1\}$. Let us denote by the $(S+1) \times A$ matrix R the entirety of assets. A portfolio $\Theta = (\Theta_1, \ldots, \Theta_A)$ is a holding of each asset, and yields a return $R\Theta$ across the $S+1$ states of nature. Each individual h begins with an endowment $\bar{\Theta}^h = (\bar{\Theta}_1^h, \ldots, \bar{\Theta}_A^h) \geq 0$ of assets.

Equilibrium is defined by a price vector $q \in R^A$, and asset holdings Θ_a^h satisfying:

$$\sum_{h \in H} \Theta_a^h = \sum_{h \in H} \bar{\Theta}_a^h$$
$$(1) \qquad \Theta^h \equiv Arg_\Theta Max\{V^h(e^h + R\Theta) \mid q \cdot \Theta \leq q \cdot \bar{\Theta}^h, \text{ and } (e^h + R\Theta) \geq 0\}.$$

[2] It is interesting that in CAPM Pareto optimality holds only for interior equilibria, while in the Arrow-Debreu model interiority has no connection with Pareto optimality.

Notice that first trade takes place in the market for assets, then the state is realized, the assets pay-off, and finally consumption occurs.[3] Since there is only one commodity, there is no need for the markets to re-open once the state of nature is realized. The definition of equilibrium requires that all promises are honored (consumption is $e^h + R\Theta$). In particular, agents are allowed to go arbitrarily short, $\Theta_a < 0$, in any asset, provided they ultimately keep their promise (by reducing their consumption by $-\Theta_a r_a$).

Allowing short sales violates the standard boundedness from below condition used to guarantee the existence of equilibrium in Arrow-Debreu [1954]. Nevertheless, as shown for example in Geanakoplos-Polemarchakis [1986], the possibility of short sales of assets does not interfere with existence of equilibrium, provided that the consumption set is bounded from below. We call an equilibrium $(q, \Theta^h; h \in H)$ interior iff $e^h + R\Theta^h >> 0$ for all $h \in H$.

The assets can be variously interpreted. Some may be thought of as shares in a firm, whose production decision has already been made (perhaps $r_a(0) < 0$ and $r_a(s) \geq 0$ for all $s \in \{1, \ldots, S\}$). On the other hand some may be thought of as permissible contracts. For example, if the initial shares $\bar{\Theta}^h_1$ of the riskless asset $r_1 = \hat{1}$ are zero for every household $h \in H$, and if there is another asset $r_2 = (1, \ldots, 0)$ with $\bar{\Theta}^h_2 = 0$ for all h, then trade between assets 1 and 2 can be considered saving and borrowing between consumption at $s = 0$ and consumption at all other s.

Notice that the distribution of consumption across the states of nature, and across individuals, is constrained by the span of the assets. If $A < S$, there are imaginable insurance contracts, i.e. trades of contingent commodities, that are not feasible with the limited asset markets. It can easily be shown (see for example Geanakoplos-Polemarchakis [1986]) that for "almost all" choices of utilities V^h and endowments e^h, all the resulting equilibria are Pareto *sub*optimal (if $A < S$). The point of our first theorem is to show that for the special case of the quadratic utility capital asset pricing model, full optimality is nevertheless attained even when $A < S$, if there is riskless asset.

2. The assumptions

(A1) There is *one commodity* in every state of nature.

(A2) The *endowments* $e^h \in R^{S+1}_{++}$ satisfy:

$$e^h \in sp[r_1, r_2, \ldots, r_A] \equiv sp[R] \text{ for all } h \in H, \text{ i.e.}$$
$$e^h = R\psi^h \text{ for some } \psi^h \in R^A, \text{ and } e^h + R\bar{\Theta}^h >> 0 \ \forall h \in H.$$

Consumers are able to trade any fraction of their own initial endowments.

(A3) *Quadratic von Neumann-Morgenstern Utilities:* For each $h \in H$ there is some number $\alpha_h > 0$ and probabilities $\pi^h_s > 0$, $\sum_{s \in S} \pi^h_s = 1$, such that

$$V^h(W) \equiv V^h(W_0, W_1, \ldots, W_S) = \sum_{s \in S} \pi^h_s \left[W_s - \tfrac{1}{2}\alpha_h W_s^2\right].$$

Thus the von Neumann-Morgenstern utility of consumption for agent h is

$$u^h(c) = c - \tfrac{1}{2}\alpha_h c^2.$$

---

[3] We have included the state $s = 0$ to make it easy to reinterpret the commodity in state 0 as consumption that occurs simultaneously with the purchase of assets at time 0 provided there is also an asset, say r_2, which satisfies $r_2 = (1, 0 \ldots, 0)$. Under this interpretation the uncertainty $s \in \{1, \ldots, S\}$ only affects consumption at time 1.

(A4) *Common Expectations:* $\pi_s^h = \pi_s^{h'} = \pi_s$ for all $h, h' \in H, s \in S$.

(A5) *Monotonicity:* Let $\bar{\Theta} = \sum_{h \in H} \bar{\Theta}^h$, and let $e = \sum_{h \in H} e^h$. Let $M = R\bar{\Theta} + e$. Then $1 - a_h M >> 0$ for all $h \in H$.

Note that it is infeasible for consumption to exceed M_s in any state s, and $1 - a_h W_s$ is the marginal utility of consumption in state s for agent h, if he is consuming W_s.

We have already noted that under hypotheses A1–A5, existence of equilibrium is guaranteed. We shall shortly show that for any economy satisfying A1–A5, all its interior equilibria are Pareto optimal if there is a riskless asset.

An equilibrium $(q, \Theta^h; h \in H)$ gives rise to a consumption allocation $x^h = e^h + R\Theta^h$, $h \in H$, with $\sum_{h \in H} x^h = M \equiv R\bar{\Theta} + e \equiv R(\sum_{h \in H} \bar{\Theta}^h) + \sum_{h \in H} e^h$. We call the equilibrium Pareto optimal if and only if there is no allocation $(y^h; h \in H)$ with $\sum_{h \in H} y^h \leq M$ and $V^h(y^h) > V^h(x^h)$ for all $h \in H$.

3. CAPM with a riskless asset

In this section we shall assume in addition the presence of a riskless asset:

(A6) *Riskless asset:* $r_1 = \hat{1} \equiv (1, 1, \ldots, 1)$.

At any equilibrium we know from monotonicity that the price of the riskless asset cannot be zero. Moreover it is evident that the equilibrium conditions are homogeneous of degree zero in q, hence without loss of generality we shall suppose that at equilibrium if r_1 is the riskless asset, then $q_1 = 1$.

Pareto Optimality

Theorem 1: Let $(q, \Theta^h; h \in H)$ be an interior equilibrium $(e^h + R\Theta^h >> 0$ for all $h \in H)$ for an economy satisfying assumptions A1–A6. Then the equilibrium allocation $x^h = e^h + R\Theta^h$, $h \in H$, is Pareto optimal.

Proof: Let $\mu^h \in R^{S+1}$ be agent h's marginal utility vector at the equilibrium allocation: $\mu_s^h = 1 - a_h x_s^h$ for all $s \in S$, $h \in H$, i.e. $\mu^h = \hat{1} - a_h x^h$ for all $h \in H$. Since the utilities are concave and differentiable, and since there is agreement on the probabilities π_s, it is enough to show that all the μ^h are colinear.

Observe that on account of monotonicity, w.l.o.g. $q_1 = 1$. Let $\hat{A}$ be the set of assets a for which $q_a \neq 0$. Then $\hat{1} \in \hat{A}$. For each $a \in \hat{A}$, let $\hat{r}_a \equiv (1/q_a) r_a$ be the per dollar return on asset a. Now for any agent $h \in H$, observe that equilibrium requires that $\sum_{s \in S} \pi_s \mu_s^h (\hat{r}_a(s) - 1) = 0$, for otherwise the agent could buy (sell) one unit of the riskless asset $\hat{1}$ and sell (buy) $1/q_a$ units of asset a. Similarly if $a \in A/\hat{A}$, so that $q_a = 0$, then for all h, $\sum_{s \in S} \pi_s \mu_s^h r_a(s) = 0$.

Let us now define an inner product for two vectors x and y in R^{S+1} by $x \,\square\, y \equiv \sum_{s \in S} \pi_s x_s y_s$. We write $x \perp y$ iff $\sum_{s \in S} \pi_s x_s y_s = 0$. Using this notation, we have from above that $\mu^h \perp (\hat{r}_a - 1)$ for all $h \in H$. Let $K = sp[\{(\hat{r}_a - \hat{1}); a \in \hat{A}\} \cup \{r_a; a \in A/\hat{A}\}]$, where for any set T of vectors in R^{S+1}, $sp[T]$ denotes the span of T, i.e. the smallest subset of R^{S+1} containing T. Then $\mu^h \perp K$, for all $h \in H$.

Observe that since $e^h \in sp[R]$ and since consumption $x^h = e^h + \Theta^h$, it follows that $x^h \in sp[R] \equiv sp[1, r_2, \ldots, r_A]$ for all $h \in H$. But recall that $\mu^h = \hat{1} - \alpha^h x^h$, hence $\mu^h \in sp\,[R]$ for all $h \in H$. But clearly $sp[R] = sp\,[K; \hat{1}]$. Hence all the μ^h must lie in the same one-dimensional subspace in $sp[R]$ perpendicular to K; i.e., they are all colinear.■

Q.E.D.

Mutual Fund Theorem 2: Let $(q, \Theta^h; h \in H)$ be an interior equilibrium for a CAPM economy with a riskless asset satisfying assumptions A1–A6. Let $x^h = e^h + R\Theta^h$ be the equilibrium consumption of agent h, and let $M \equiv \sum_{h \in H} e^h + R\bar{\Theta}$ be the market consumption. Then there are scalars s_h and t_h for each agent h such that $x^h = s_h\hat{1} + t_h M$. If R has full column rank and if $e^h = 0$ for all h, then $\Theta^h = s_h\,(1, 0, \ldots, 0) + t_h\bar{\Theta}$, where $\bar{\Theta}$ is the market portfolio of assets. $\bar{\Theta} \equiv \sum_{h \in H} \bar{\Theta}^h = \sum_{h \in H} \Theta^h$.

Proof: We have already seen that if $\mu^h = \hat{1} - a_h x^h$, then the μ^h are all colinear. Hence the consumption vectors $x^h = (1/a_h)\,[\hat{1} - \mu^h]$ all lie in some two-dimensional space, spanned by $\hat{1}$ and some vector μ, colinear with all the μ^h. Since $\sum_{h \in H} x^h = M$, we have that $M \in sp[\hat{1}, \mu]$. Since $M = \sum_{h \in H} x^h = \sum_{h \in H} (1/a_h)\,[\hat{1} - \mu^h] = s\hat{1} - t\mu$ it follows that if M and $\hat{1}$ are colinear, then so are $\hat{1}$ and μ, and hence $x^h = s_h\hat{1}$ for some scalar s_h, for all $h \in H$. If $\hat{1}$ and μ are not colinear, then $sp[\hat{1},\mu] = sp[\hat{1},M]$ and we can write $x^h = s_h\hat{1} + t_h M$, for all $h \in H$.

If $e^h = 0$ for all h, then $x^h = R\Theta^h$ and $M = R\,\bar{\Theta}$. If R has full column rank, then Θ^h is the unique portfolio with $x^h = R\Theta^h$. Hence we must have that in portfolio space, $\Theta^h = s_h\,(1, 0, \ldots, 0) + t_h\bar{\Theta}$. ■

Q.E.D.

Before proving our next theorem, let us recall the definition of expectation and covariance between two vectors x and y in R^{S+1}. Ex denotes the expectation of x with respect to the common priors π, $Ex = x \,\square\, \hat{1} \equiv \sum_{s \in S} \pi_s x_s$. $cov(x,y) \equiv \sum_{s \in S} \pi_s\,(x_s - Ex)(y_s - Ey)$, and $var\, x \equiv cov(x,x)$. Notice that $cov(x,y) = \sum_{s \in S} \pi_s\,[x_s y_s - x_s Ey - y_s Ex + ExEy] = \sum_{s \in S} \pi_s x_s y_s - ExEy = x \,\square\, y - ExEy$.

Security Market Line Theorem 3: Let $(q, \Theta^h; h \in H)$ be an interior equilibrium for a CAPM economy with a riskless asset satisfying assumptions A1-A6, taken with normalization $q_1 = 1$. Let $r : R^{S+1} \to R$ represent the payoffs of any potential asset, marketed ($r = R\psi$ for some ψ) or not ($r \neq R\psi$ for all $\psi \in R^A$). Then there is a unique price $q(r)$ at which every consumer h will be satisfied to continue to hold exactly Θ^h, despite the new opportunity of buying or selling short the new asset at price $q(r)$. (Of course if $r = R\psi$, then $q(r) = q \cdot \psi$.) Moreover, if $q(r) \neq 0$, so that $\hat{r} = r/q(r)$ is well-defined, then again denoting $M = e + R\bar{\Theta}$ and $\hat{M} = M/q(M)$ we must have:

$$(*) \qquad E\hat{r} - 1 = \frac{cov\,(\hat{r}, \hat{M})}{var\,\hat{M}}\,(E\hat{M} - 1)$$

Proof: The existence of the price $q(r)$ follows immediately from the Pareto optimality theorem. Indeed, letting μ^h be the marginal utility vector at equilibrium of any agent h, $q(r) = (\mu^h \,\square\, r)/(\mu^h \,\square\, \hat{1})$. Since in any equilibrium satisfying our hypothesis, $\mu^h \gg 0$, $q(r)$ is well defined and independent of h, since all the μ^h are colinear.

Observe that the security market equation (*) defines a linear relation between $E\hat{r}$ and $cov(\hat{r},\hat{M})$ for all securities $\hat{r}$ with $q(\hat{r}) = 1$. Any line is determined by any two distinct points on the line. It can trivially be checked that $(cov(\hat{1},\hat{M}), E\hat{1}) = (0,1)$ and $(var\ \hat{M}, E\hat{M})$ satisfy (*), for the portfolio payoffs $\hat{r} = \hat{1}$ and $\hat{r} = \hat{M}$, respectively. Thus if there is a linear relationship between $cov(\hat{r},\hat{M})$ and $E\hat{r}$ for all portfolio payoffs $\hat{r}$ with $q(\hat{r}) = 1$, then (*) is the right formula. (If $var\ \hat{M} = 0$, then from Pareto optimality we know that for any h, μ^h_s is a constant independent of s, and so $(\mu^h \Box \hat{r}) / (\mu^h \Box \hat{1}) = E\hat{r}/E\hat{1} = E\hat{r}$, so (*) trivially holds..)

Consider now that $1 = q(\hat{r}) = (\mu^h \Box \hat{r})/(\mu^h \Box \hat{1})$. Recalling that for any two vectors x and y, $x \Box y = cov(x, y) + ExEy$, we have $\mu^h \Box \hat{r} = cov(\hat{r}, \mu^h) + E\mu^h E\hat{r}$. Finally, recall that for any h we could find s_h and t_h with $\mu^h = s^h\hat{1} + t^h\hat{M}$. Hence we have

$$\mu^h \Box \hat{r} = cov(\hat{r}, s^h\hat{1} + t^h\hat{M}) + E\mu^h E\hat{r}, \text{ or}$$
$$\mu^h \Box \hat{r} = t^h\, cov(\hat{r}, \hat{M}) + E\mu^h E\hat{r}.$$

Taking $\mu^h \Box \hat{1}$, t^h and $E\mu^h$ as constants gives us a linear relationship between $cov(\hat{r},\hat{M})$ and $E\hat{r}$. ■

Q.E.D.

Note the necessity of the hypothesis that $q(\hat{r}) = 1$.

Portfolios that are free (for example buying 1 unit of $\hat{M}$ and selling short one unit of the riskless asset 1) have "betas" and returns lying on the line through the origin parallel to the security market line.

4. CAPM without a riskless asset

There are other versions of mean-risk behavior in which the mutual fund theorem and the security market line for marketed securities still holds, but Pareto optimality fails. This shows that Pareto optimality is a separate, stronger property attaching to the quadratic utility, riskless asset version of the mean variance model. To make our point we shall prove the mutual fund theorem and a security market line theorem for marketed assets in a quadratic-utility CAPM model without a riskless asset, in which optimality need not obtain.

Definition: Let $M = e + R\bar{\Theta} = R\Theta_M$ be the market payoff. Let $z = R\Theta_z$. We call Θ_z a zero-beta portfolio iff $cov(z,M) = 0$.

Lemma: Let $r = R\Theta_r$ and $M = r\Theta_M$. Then there is a zero-beta portfolio $\Theta_z = \Theta_r + \lambda\Theta_M$, for some λ, such that $sp[r,M] = sp[z,M]$.

Proof: If $var(M) = 0$, there is nothing to prove. So suppose $var(M) > 0$. Let $\Theta_z = \Theta_r + \lambda\Theta_M$. Then $cov(R\Theta_z, M) = cov(r\Theta_r, M) + \lambda cov(M,M)$ so choose $\lambda = -[cov(R\Theta_r, M)/var(M)]$. Then $cov(z,M) = 0$ and z and M are linearly independent if r and M are noncolinear. Hence $sp[r,M] = sp[z,M]$. ■ Q.E.D.

Mutual Fund Theorem 2': Let $(q, \Theta^h; h \in H)$ be an interior equilibrium for an economy satisfying assumptions A1-A5, that is, possibly without a riskless asset, and let $x^h = e^h + R\Theta^h$ be the final consumption of agent $h \in H$, and let $M = e + R\bar{\Theta}$ be the market portfolio payoff. Then there is a zero beta portfolio payoff z, $z = R\Theta_z$, satisfying $cov(z,M) = 0$, such that for all $h \in H$, there are scalars s_h and t_h for which $x^h = s_h z + t_h M$. Moreover, if R has full column rank and $e^h = 0$ for all $h \in H$, then $\Theta^h = s_h q + t_h\bar{\Theta}$ for all $h \in H$.

Proof: Let $\mu^h = 1 - \alpha_h x^h$ be the marginal utility vector of consumer h. Let $\hat{A}$ be the set of assets a with $q_a \neq 0$, and let $A/\hat{A}$ be the remaining assets a' with $q_{a'} = 0$. Noting that

$M \equiv e + R\hat{\Theta} >> 0$, we must have that $q \cdot \Theta_M \neq 0$ for Θ_M solving $M = R\Theta_M$. Let $\hat{M} = M/(q \cdot \Theta_M)$, and similarly for all $a \in \hat{A}$ let $\hat{r}_a = r_a/q_a$. Then as we saw in the Pareto optimality proof, at an interior equilibrium it must be that $\mu^h \perp (\hat{r}_a - \hat{M})$ for all $a \in \hat{A}$ and $\mu^h \perp r_a$ for all $a \in A/\hat{A}$. Letting $K = sp[\{(\hat{r}_a - \hat{M}) \mid a \in \hat{A}\} \cup \{r_a \mid a \in A/\hat{A}\}]$, $\mu^h \perp K$. But $\mu^h \in sp[K, \hat{1}, M]$ for all $h \in H$. Thus the μ^h all lie in a two-dimensional set. Note the fact that $\hat{1}$ might not be in $sp[R]$ costs a dimension in this argument, and allows for the possibility of Pareto sub-optimality.

If $\hat{1} \in sp[R]$, then we have already proved the result (letting $z = \hat{1}$). If $\hat{1} \notin sp[R]$, then we can still deduce that there is a two-dimensional space V containing all the x^h, $h \in H$. For if there were 3 linearly independent consumption vectors x^h, then the vectors $\mu^h = \hat{1} - \alpha_h x^h$ would vary over at least 3 dimensions, since $x^h \in sp[R]$ for all h.

Finally, let V be the (at most) two-dimensional subspace containing all the x^h. Since $M = \sum_{h \in H} x^h$, $M \in V$. From the lemma, V is spanned by M and a zero-beta portfolio payoff $z = R\Theta_z$. Hence $x^h = s_h z + t_h M$ for all $h \in H$. If $e^h = 0$ and R has full column rank, then $x^h = R\Theta^h$, and $\Theta^h = s_h \Theta_z + t_h \hat{\Theta}$. ■ Q.E.D.

Security Market Line Theorem 3': Let $(q, \Theta^h; h \in H)$ be an interior equilibrium for a CAPM economy satisfying A1-A5, with prices q normalized so that $q \cdot \Theta_M = 1$. Then there is a linear relationship between the covariance of returns with the market return, and expected return, for assets with unit price. Precisely, there are scalars $\lambda_1, \lambda_2, \lambda_3$ such that for any portfolio $\hat{\psi}$ with $q \cdot \hat{\psi} = 1$ and $\hat{r} = R\hat{\psi}$, $\lambda_1 cov(\hat{r},M) + \lambda_2 E\hat{r} + \lambda_3 = 0$. Furthermore, if there exists a zero-beta portfolio $\hat{\Theta}_z$ with $q \cdot \hat{\Theta}_z = 1$, then letting $\hat{z} = R\hat{\Theta}_z$ this linear relationship takes the form:

$$(**) \qquad E\hat{r} - E\hat{z} = \frac{cov(\hat{r},M)}{var(M)}(EM - E\hat{z}).$$

Proof: For any portfolio ψ and any agent $h \in H$, letting $r = R\psi$,

$$q \cdot \psi = \frac{\mu^h \Box r}{\mu^h \Box M} = \frac{cov(\mu^h, r)}{\mu^h \Box M} + \frac{E\mu^h Er}{\mu^h \Box M}$$

Restricting attention to $\hat{\psi}$ with $q \cdot \hat{\psi} = 1$, $\hat{r} = R\hat{\psi}$,

$$\begin{aligned} \mu^h \Box M &= cov(\mu^h, \hat{r}) + E\mu^h E\hat{r} \\ &= cov(\hat{1} - \alpha_h x^h, \hat{r}) + E\mu^h E\hat{r} \\ &= -\alpha_h cov(x^h, \hat{r}) + E\mu^h E\hat{r}. \end{aligned}$$

So $(\mu^h \Box M)/\alpha_h = -cov(x^h, \hat{r}) + (E\mu^h/\alpha_h) E\hat{r}$.

Summing over $h \in H$ gives

$$\sum_{h \in H} \frac{\mu^h \Box M}{\alpha_h} = - \sum_{h \in H} cov(x^h, \hat{r}) + \sum_{h \in H} \frac{E\mu^h}{\alpha_h} E\hat{r}$$

or $\lambda_1 = -cov(M,\hat{r}) + \lambda_2 E\hat{r}$.

Finally, observe that if there is a zero-beta portfolio $\hat{\Theta}_z$ with $q \cdot \hat{\Theta}_z = 1$, then the above linear relationship must take the form (**) since any linear relation is determined by two points, namely (var, M, EM), $(0, E\hat{z})$. ■ Q.E.D.

II. Generalizations to multiple commodities and non-quadratic utilities

In this part we take a broader perspective, allowing for multiple consumer commodities and therefore trade on spot markets and assets, and also for nonquadratic utilities. The essential point that emerges here is that when asset markets are incomplete, there is typically a lack of coordination between the desires of consumers and the desires of shareholders. It is thus not correct to assert that under conditions of information symmetry, perfect competition, etc., asset markets are efficient, or nearly so.

We argue our case in two ways. We quote an argument in Geanakoplos-Polemarchakis [1986], that shows that generically, when the asset market is incomplete, the equilibrium trade in assets is inefficient in the strong sense that all traders could be made better off if they made different asset trades, even if the subsequent spot markets were allowed to clear at competitive prices, on account of the effect on relative commodity prices when the spot markets subsequently clear. When there is only one physical commodity in each state, the desires of shareholders and consumers are necessarily the same, and indeed it is easy to see that no reallocation of *existing* assets can Pareto dominate the competitive allocation.

Second, we consider the case where spot market relative prices are unaffected by asset trades (the case with identical income effects on consumption). This is the case presumably that is represented by the parable of the single commodity, assumed by the previous models. Under this knife-edge hypothesis on income effects, asset reallocations alone cannot Pareto improve on interior equilibria. If, in addition, attitudes toward risk are quadratic, one would expect full Pareto optimality to obtain, as in Theorem 1. But this turns out to be false, unless the assumption of a riskless asset is augmented by a far stronger hypothesis. This abstraction further clarifies the role of the riskless asset in the one-commodity world.

We shall begin by describing an equilibrium with multiple commodities and proving the analogue to Theorem 1 for Pareto optimality in the general case.

From now on the commodity consumption space $R_+^{(S+1)(L+1)}$ consists of $L+1$ physical commodities, $\{0, 1, \ldots, L\}$, in each of $(S+1)$ states of nature. Consumers $(W^h, e^h, \bar{\Theta}^h)$ are characterized by utility functions W^h that are smooth and strictly concave, and by endowments $e^h \in R_{++}^{(S+1)(L+1)}$, and by their holdings $\bar{\Theta}^h$ of initial assets. It is often the case that a utility function W^h can be extended to a convex set X^h containing $R_+^{(S+1)(L+1)}$, while retaining smoothness and strict concavity; the quadratic utility is an obvious example.[4] This is information which can be useful, so we shall take $W^h : X^h \rightarrow R$, where X^h is any closed, convex set, that is bounded from below, containing $R_+^{(S+1)(L+1)}$. Of course we shall always restrict consumption to $R_+^{(S+1)(L+1)}$.

Assets $\bar{r}_a$, $a \in A : \{1, \ldots, A\}$ now yield $(L+1)$ dimensional vectors of commodities in each state, so $\bar{r}_a$ is given by an $(S+1) \times (L+1)$ matrix. The totality of assets is represented by the collection $\bar{R}$. The notation $(\bar{R}\Theta)_s$, for $\Theta \in R^A$, will mean the vector commodity payoff that occurs in state s, given the portfolio $\Theta = (\Theta_1, \ldots, \Theta_A)$. We assume that asset 1 is not

---

[4] We also require that $DW^h(x^h) >> 0$ for all $x^h \in R_{++}^{(S+1)(L+1)}$ that satisfy $x \leqq M$, where M is the aggregate social endowment. It is not necessary to the theory that W^h be monotonic outside the feasible set; indeed, the quadratic utility is not monotonic for large x^h.

identically 0, and that it pays off a nonnegative amount in every commodity, and a positive amount of some commodity in each state. A consequence of this mild assumption is that in equilibrium the price of asset 1 can be taken to be strictly positive.

Asset prices are denoted, as before, by $q \in R^A$. Commodity prices are $p \in R_{++}^{(S+1)(L+1)}$. The product $p_s\,(\bar{R}\Theta)_s$ means the payoff in units of account in state s that is obtained by selling the commodity vector payoff $(\bar{R}\Theta)_s$ at prices $p_s = (p_{s0}, \ldots, p_{sL})$.

As before, we let $\bar{\Theta} \equiv \sum_{h \in H} \bar{\Theta}^h$, and $e \equiv \sum_{h \in H} e^h$ and $M_s \equiv e_s + (R\bar{\Theta})_s$.

An interior equilibrium for the above economy is a tuple $(q, p, (\Theta^h, x^h); h \in H)$ satisfying $(\Theta^h, x^h) \in R^A \times R_{++}^{(S+1)(L+1)}$ for all $h \in H$, $\sum_{h \in H} \Theta^h = \sum_{h \in H} \bar{\Theta}^h \equiv \bar{\Theta}$, $\sum_{h \in H} x^h = \sum_{h \in H} e^h + \bar{R} \sum_{h \in H} \bar{\Theta}^h \equiv M$, and $(\Theta^h, x^h) \in D^h(q,p) \equiv \underset{(\Theta,x)}{ArgMax}\, \{W^h(x) \mid (\Theta,x) \in R^A \times X^h,\ q \cdot \Theta = q \cdot \bar{\Theta}^h$ and $p_s(x_s - e_s^h) \leqq p_s(\bar{R}\Theta)_s$ for all $s \in S\}$, for all $h \in H$.[5]

An extremely useful construction in the following is the asset constrained demand: $D^h(p \mid \Theta^h) = \underset{x}{Arg\,Max}\, \{W^h(x) \mid x \in X^h,\ p_s \cdot (x_s - e_s^h) \leqq p_s(\bar{R}\Theta^h)_s$ for all $s \in S\}$. The choice $x \in D^h(p \mid \Theta^h)$ iff agent h would choose x if he was forced to hold the portfolio Θ^h, but could trade freely on the spot commodity markets at prices $p >> 0$. An asset constrained interior equilibrium is a tuple $(p, (\Theta^h, x^h); h \in H)$ satisfying all the requirements of an equilibrium except that the optimality condition $(\Theta^h, x^h) \in D^h(q,p)$ is replaced by the weaker requirement $x^h \in D^h(p \mid \Theta^h)$. We shall shortly say more about these.

Finally, let $D^h(p) = \cup_{\Theta^h} D^h(p \mid \Theta^h)$ be the set of all conceivable commodity bundles h might demand, if he did not have to worry about his budget constraint for assets, but if he could only spread his limitless wealth across states of nature through assets. Consider for example the case where there is only one commodity per state of nature, and where X^h is all of R^{S+1}. Then the prices p are irrelevant, since there is no spot trade. If the asset span, $sp[R]$, has full rank $S+1$, so that the asset market is complete, then $D^h(p)$ is all of R^{S+1}. If the number of assets $A < S+1$, then $D^h(p)$ is an A-dimensional affine subspace of R^{S+1}(a subspace translated by the endowment e^h).

1. A sufficient condition for Pareto optimality with multiple commodities, non-quadratic utilities, and incomplete markets

For general $X^h \subset R^{(S+1)(L+1)}$, let us give:

Definition: For $x \in X^h$, let $\mu^h(x) \equiv DW^h(x)$ be the marginal utility of agent h.[6] Given commodity prices $p >> 0$, we say that the vector $\mu >> 0$ is *ray p-reachable* for agent h iff there is some $\lambda > 0$ and $x \in (D^h(p) \cap \text{interior } X^h)$ with $\mu^h(x) = \lambda\mu$.

Definition: We say that the map μ^h has the ray property at prices p for the economy $(\bar{R}, (W^h, X^h, e^h, \bar{\Theta}^h); h \in H)$ iff whenever the image $\mu^h(D^h(p))$ of $D^h(p)$ under μ^h contains

---

[5] Notice that we have allowed the choice set for commodities in calculating demand $D^h(q,p)$ to include all of X^h. Since we are restricting attention to demands x^h interior to $R_+^{(S+1)(L+1)}$, this makes no difference on account of the concavity of W^h. We have already mentioned the great importance for Pareto optimality of considering only interior equilibria.

[6] This definition is slightly different from the one given in Section 2, since here it includes the probability weights π_s if they are used in the definition of W^h.

$\mu >> 0$, $\{\lambda > 0 \mid \exists\ x \in D^h(p), \mu^h(x) = \lambda\mu\}$ is an interval and inf $\{W^h(x) \mid \mu^h(x) = \lambda\mu, \lambda > 0, x \in D^h(p)\} < W^h(e^h)$ and sup $\{W^h(x) \mid \mu^h(x) = \lambda\mu, \lambda > 0, x \in D^h(p)\} \geqq W^h(M)$.

Definition (co-reachable hypothesis): We say that the asset structure $\bar{R}$ allows for co-reachable agents at prices p iff (1) every map μ^h has the ray property at p, and (2) a vector μ is ray p-reachable by any agent h iff μ is ray p-reachable by all agents $h' \in H$.

The co-reachable hypothesis means that the images of the $D^h(p)$ under the maps μ^h contain the same rays for all agents $h \in H$, and furthermore that these ray images are large intervals. No matter what the utilities W^h, if $\mu^h(X^h)$ has range including all of $R_{++}^{(S+1)(L+1)}$ (as may be assured by the proper boundary conditions), and if the asset markets are complete, then the co-reachable hypothesis is trivially satisfied. There are cases where the hypothesis is satisfied even when the asset market is incomplete. For example, under the one good, quadratic von Neumann-Morgenstern hypothesis of Section 2, $\mu^h(x) \equiv \hat{1} - \alpha_h x$. If X^h is taken to be R^{S+1} and all the $e^h \in sp[R]$, then the sets $D^h(p)$ are identical vector spaces, and the sets $\mu^h(D^h(p))$ are each identical translations of $D^h(p)$ by the vector $\hat{1}$. If $\hat{1} \in sp[R] = D^h(p)$, then $\mu^h(D^h(p))$ is a vector space, hence it includes all multiples $\lambda\mu$ of any vector μ that it contains. In particular, suppose that $0 << \lambda = \mu^h(x) = \hat{1} - \alpha_h x$. Then letting $x_\lambda = \frac{1}{\alpha_h}-(\hat{1} - \lambda\mu) \in D^h(p) = sp[R]$, we find that for $\lambda \to 0$, $W^h(x_\lambda) \to W^h(\frac{1}{\alpha_h} - \hat{1})$, which is the bliss point or maximum achievable utility, and for $\lambda \to \infty$, $W^h(x_\lambda) \to -\infty < W^h(e^h)$. Hence the ray property holds, and co-reachability is confirmed.

Note that if for all $h \in H$ we choose $X^h = \{x \in R^{S+1} \mid x_i \geq -A\}$, for $A > \max_h 1/\alpha_h$, co-reachability could similarly be confirmed, and at the same time we would have X^h bounded from below. The images $\mu^h(D^h(p))$ would be ray identical (although perhaps not pointwise identical). If $A > 1/\alpha_h$, then it can easily be checked that for any $\mu >> 0$, there is $x \in X^h$, with $\mu^h(x) = \lambda\mu$, for some $\lambda > 0$, and $W^h(x) < W^h(0) < W^h(e^h)$. The significance of the co-reachability hypothesis, when X^h is bounded from below, is illustrated by the following generalization of Theorem 1 to many commodities and possibly nonquadratic utilities.

Theorem 4: Let $(q, p, (\Theta^h, x^h); h \in H)$ be an interior equilibrium for the economy $(\bar{R}, (W^h, X^h, e^h, \bar{\Theta}^h); h \in H)$. Suppose that the co-reachability hypothesis holds at p. Then the allocation $(x^h; h \in H)$ is fully Pareto optimal.

Proof: Let $\mu = \mu^{h'}(x^{h'})$ for some $h' \in H$. We must show that $\mu^h(x^h) = \lambda^h\mu$ for all $h \in H$. For this it suffices to find $(\hat{\Theta}^h, \hat{x}^h)$ with $\hat{x}^h \in D^h(p \mid \hat{\Theta}^h)$ and $\mu^h(\hat{x}^h) = \lambda\mu$, for some $\lambda > 0$, with $W^h(\hat{x}^h) = W^h(x^h)$, since it follows at once that then $\hat{x}^h = x^h$. To see this, note that if h had asset income $q \cdot \hat{\Theta}^h$, his demand at prices (q,p) would be $(\hat{\Theta}^h, \hat{x}^h)$, since there the first order conditions are satisfied. But $q \cdot \hat{\Theta}^h$ is either greater, less, or equal to $q \cdot \bar{\Theta}^h = q \cdot \Theta^h$. If less, then from the fact that W^h is monotonic near x^h, and from the availability of the nonnegative asset $\bar{r}_l$ it follows that it would be possible to make a better choice than $(\Theta^h, x^h$ at prices (q,p) and asset income $q \cdot \bar{\Theta}^h$.[7] A similar argument shows that $q \cdot \hat{\Theta}^h > q \cdot \bar{\Theta}^h$ would contradict the optimality of $(\hat{\Theta}^h, \hat{x}^h)$. Hence the conclusion that $q \cdot \bar{\Theta}^h = q \cdot \hat{\Theta}^h$, and $x^h = \hat{x}^h$, follows from the strict concavity of W^h.

---

[7] The choice $(\alpha\Theta^h + (1-\alpha)\hat{\Theta}^h, \alpha x^h + (1-\alpha)\hat{x}^h)$ ist strictly preferred to (Θ^h, x^h), for $0 < \alpha < 1$, and satisfies the state by state budget sets exactly, while leaving asset income to spare. For α near 1, W_h is monotonic at $\alpha x^h + (1-\alpha)\hat{x}^h$, and so spending the extra asset income on $\bar{r}_l$ makes W^h higher still.

From the ray property there exists (Θ_1^h, x_1^h) and (Θ_2^h, x_2^h) with $W^h(x_1^h) < W^h(e^h) < W^h(x^h) < W^h(x_2^h)$ and $\mu^h(x_1^h) = \lambda_1\mu$ and $\mu^h(x_2^h) = \lambda_2\mu$. Note that $\lambda_2 < \lambda_1$, for by concavity of W^h, we must have that $(\lambda_1\mu) \cdot (x_2^h - x_1^h) > 0$. From the diminishing marginal returns that are a consequence of strict concavity, we must have $0 < (\lambda_2\mu) \cdot (x_2^h - x_1^h) < (\lambda_1\mu) \cdot (x_2^h - x_1^h)$, hence $\lambda_2 < \lambda_1$.

Again from the ray property, to each $\lambda_2 \leq \hat{\lambda} \leq \lambda_1$, there is $(\hat{\Theta},\hat{x})$ with $\mu^h(\hat{x}) = \hat{\lambda}\mu$, and $\hat{x} \in D^h(p \mid \hat{\Theta})$. From the last paragraph, we know that all such $\hat{x}$ satisfy $W^h(\hat{x}) \leqq W^h(x_2^h)$, and hence $q \cdot \hat{\Theta} \leq q \cdot \Theta_2^h$. But from the fact that $p >> 0$, and (q,p) is an equilibrium, we know that q does not permit arbitrage, i. e. there is no Θ with $\bar{R}\Theta > 0$ and $q \cdot \Theta \leqq 0$. Since X^h is closed and bounded from below, the set of x that satisfy the budget constraints with asset income less or equal to $q \cdot \Theta_2$, at asset prices q, is compact. A standard argument now shows that there must be some $(\hat{\Theta},\hat{x})$ as above with $W^h(\hat{x}) = W^h(x^h)$.

Note first that μ^h is a one-to-one function. Take $y \neq z$; by strictly diminishing returns, $\mu(y) \cdot (y\text{-}z) < \mu(z) \cdot (y\text{-}z)$, so $\mu(y) \neq \mu(z)$. Hence we can write $\hat{x}(\hat{\lambda})$, for $\lambda_2 \leqq \hat{\lambda} \leqq \lambda_1$. Moreover, the function $\hat{x}$ is continuous in $\hat{\lambda}$. If $\hat{\lambda}_t \to \lambda^*$, then since the $\hat{x}(\hat{\lambda}_t)$ lie in a compact set, by passing to subsequences we have $\hat{x}(\hat{\lambda}_t) \to x^*$ and $\mu^h(x^*) = \lambda^*\mu$. So the function $\varphi : (\hat{\lambda} \mid \lambda_2 \leqq \hat{\lambda} \leqq \lambda_1 \} \to R$ given by $\varphi(\hat{\lambda}) = W^h(\hat{x}(\hat{\lambda}))$ is continuous. The continuous image of a connected set must be connected, so the image of φ must contain $W^h(x^h)$. ■ Q.E.D.

As a first consequence of Theorem 4, let us consider the models of Section 2, but now extended to allow for a class of nonquadratic utilities:

Corollary: Under assumptions (1), (2), (4), (5), (6), let $W^h(x) = \sum_{s \in S} \pi_s u^h(x_s)$, where all the u^h belong to one of the following exclusive classes: (a) $u^h(z) = \frac{1}{b}(1 + \alpha_h z)^b$, $\alpha_h > 0$, $b < 1$, $b \neq 0$, (b) $u^h(z) = \log(1 + \alpha_h z)$, $\alpha_h > 0$, (c) $u^h(z) = -e^{-(1+\alpha h z)}$, $\alpha_h > 0$. Any interior equilibrium for such an economy must be fully Pareto optimal.

Proof: We need only apply Theorem 4, taking advantage of the fact that each u^h can be extended to $X^h = \{x \in R^{S+1} \mid x_s \geq -1/\alpha_h + \epsilon\}$, where $0 < \epsilon$. Consider case (a), where $\mu_s^h(x) = \alpha_h(1 + \alpha_h x_s)^{b-1}\pi_s$. Note that for any $x \in X^h$, $\mu^h = \mu^h(x)$, $\mu_s^h = \alpha_h \pi_s m_s^{b-1}$, where $m \in sp[R]$. So for any $h' \in H$, take $x_\lambda^{h'} = (1/\alpha_h')\,[\lambda m - 1] \in sp[R]$. For λ bigh enough, $x_\lambda^{h'} \in X^{h'}$ and $\mu^{h'}(x_\lambda^{h'}) = (\alpha_{h'}/\alpha_h)\,\lambda^{b-1}\mu^h$. As $\lambda \to \infty$, $W^{h'}(x_\lambda^{h'}) \to \infty$. As $\lambda \to 0$, eventually $-(1/\alpha_h')1 << x_\lambda^{h'} << 0$, so $W^{h'}(x^{h'}) < W^{h'}(e^{h'})$, and the co-reachability hypothesis is confirmed. A similar argument works for cases (b) and (c). ■ Q.E.D.

2. Constrained optimality

The most important application for us of Theorem 4 is to economies with many goods, but where the relative prices in each state of nature can be taken to be independent of the distribution of income. Consider now utilities of the form: $W^h(x) = \sum_{s \in S} \pi_s^h u^h(v_s^h(x_{s0}, x_{s1}, \ldots, x_{sL}))$, where $v_s^h : R_+^{L+1} \to R$ represents "in state" utility, and $u^h : R \to R$ represents the attitude toward risk of agent h. We shall suppose that the concave and strictly quasi-concave, monotonic v_s^h give rise to the same "income effects". To be precise, let $x_s^h(p_s, I) \equiv \underset{x_s}{Arg\,Max}\{v_s^h(x_s) \mid x_s \in R_+^{L+1}, p_s x_s \leqq I\}$ be the state s demand of agent h.

Assumption 1.1: $x_s^h(p_s, I)$ is a differentiable function at all (p_s, I) for which $x_s^h(p_s, I) >> 0$. Furthermore, if at (p_s, I^h), $x_s^h(p_s, I^h) >> 0$ for all $h \in H$, then $[\partial x_s^h(p_s, I^h)/\partial I] = \bar{z}_s(p_s) \geqq 0$, independent of I^h and h.

Two example of utilities satisfying 1.1 are (1) all individuals have the same homogeneous utility $v_s = v_s^h$ or (2) $v_s^h (x_{s0}, x_{s1}, \ldots, x_{sL}) = x_{s0} + \bar{v}_s^h (x_{sl}, \ldots, x_{sL})$ for all $h \in H$. In the first case, $\bar{z}_s (p_s) = x_s^h (p_s, I)/I$, which is independent of I and h, but not of p_s. In the second case, $\bar{z}_s (p_s) = (1,0, \ldots, 0)$ (if $p_{s0} = 1$), which is independent of h, I, and P_s. It is well-known, after Gorman, that all distributions of income that allow for strictly positive spot market clearing prices p_s give rise to the same price vector $p_s = \bar{p}_s$ (if $\bar{p}_{s0} = 1$).

For our two examples, these spot market clearing prices $\bar{p}_s$ are easy to calculate. In (1), $\bar{p}_s = Dv(M_s)/[\partial v(M_s)/\partial x_{SO}]$. In (2) $\bar{p}_s$ is the marginal utility vector for each consumer at (any) allocation which maximizes the sum of utilities, given the aggregate endowment M_s. Notice that in both cases $\bar{p}_s$ will depend on s, if the aggregate endowment is not constant.

The consequences of assumption 1.1 can be seen immediately for constrained optimality.

Theorem 5: Let $(\bar{R}, (W^h \Leftrightarrow (u^h, v^h, \pi^h), e^h, \bar{\Theta}^h)$ be an economy in which the $(v^h, e^h, \bar{\Theta}^h)$ satisfy assumption 1.1, and suppose that $(q, p, (\Theta^h, x^h); h \in H)$ is an interior equilibrium. Then the allocation $(x^h; h \in H)$ is constrained Pareto optimal.

Proof: Since all interior asset constrained equilibria $(\hat{p}, (\hat{\Theta}^h, \hat{x}^h); h \in H)$ satisfy $p = \hat{p}$, it follows that if $W^h (\hat{x}^h) > W^h (x^h)$, then $q \cdot \hat{\Theta}^h > q \cdot \bar{\Theta}^h$. But then $q \cdot \sum_{h \in H} \hat{\Theta}^h > q \cdot \sum_{h \in H} \bar{\Theta}^h$, contradicting $\sum_{h \in H} \hat{\Theta}^h = \sum_{h \in H} \bar{\Theta}^h$. ■ Q.E.D.

3. CAPM with multiple commodities: the riskless asset revisited

Assumption 1.1 generalizes the requirement that there is only one physical commodity per state of nature, at least for the purposes of obtaining constrained optimality. A natural conjecture is that if in addition we carry over assumptions 2-6, including the hypothesis that the u^h all belong to one of the same utility classes (quadratic, or exponential, etc.) given by the corollary to the last theorem, then we might be able to prove full Pareto optimality for interior equilibria. A moment's reflection in the one commodity world, where 1.1 and 1.2 hold trivially, shows however that even when all the in "state utilities" v^h are identical and homothetic, full Pareto optimality cannot be expected. It is clearly necessary that the "in state" marginal utility of income is constant; otherwise it has implications for risk aversion, and we know that only special risk averse behavior can give rise to full optimality. Note, incidentally, that both of our examples (1) and (2) of acceptable v^h do give rise to constant "in state" marginal utility of income. But there is a more interesting problem, having to do with the meaning of the riskless asset in a many commodity world.

When there is only one commodity, in every state, the riskless asset should provide for delivery of one unit of the good in each state, or 1 unit of account in every state, if we take the price of the good to be one. But when there are multiple commodities, with different relative prices across the states of nature, then what should the riskless asset provide? Should it identify some significant commodity bundle, and pay off in constant purchasing power relative to this bundle? The right answer is that it should distribute purchasing power across the various states so that the increase in "within state" utility is the same across all states.

Definition: Let $W^h (x) = \sum_{s \in S} \pi_s u^h (v_s^h (x_s))$, and suppose that at equilibrium, consumption is $\bar{x}^h = (\bar{x}_s^h; s \in S)$. A riskless asset r for consumer h satisfies $Dv_s^h (\bar{x}_s^h) \cdot r_s = 1$ for each $s \in S$.

Assumption 6.1: The collection of assets includes a riskless asset for at least one agent $h \in H$.

Notice that the riskless asset for consumer h cannot be defined without knowledge of the spot market clearing consumptions $\bar{x}^h$ (or equivalently, without knowing the spot market clearing prices $\bar{p}$), unless there is only one commodity per state. On the other hand, observe that under the hypothesis that all the "in state" utilities v^h are identical and homogeneous, the same asset r is a riskless asset for every consumer. Furthermore, under this homogeneity hypothesis, all initial endowment allocations $(\bar{R}, (e^h, \bar{\Theta}^h); h \in H)$ that give rise to the same aggregate endowment M also give rise to the same riskless asset, at every interior equilibrium. Thus the following theorem is not without content.

Theorem 6: Let $(\bar{R}, (W^h \Leftrightarrow (u^h, v^h, \pi^h), e^h, \bar{\Theta}^h); h \in H)$ be an economy in which all $v^h = v$ are identical and homogeneous of degree one. Furthermore, suppose that assumptions 2 (the span of $\bar{R}$ includes e^h for all $h \in H$), 3.1 (that all u^h are drawn from one of the same special classes designated earlier), 4 (the priors π^h are common), 5 (monotonicity), and 6.1 (there is a riskless asset for at least one individual h) hold. Then any interior equilibrium is fully Pareto optimal.

The proof of the above follows immediately from Theorem 4. Let us repeat that the above theorem shows that the capital asset pricing model can be extended to a model with a genuine multiplicity of commodities, if all consumers have the same homogeneous preferences of degree one within each state. To achieve full Pareto optimality, only one asset is required, beyond the initial endowments of all the consumers. By choosing the spot market price normalizations appropriately, one can of course arrange it so that this asset pays off commodities with value equal to one in every state. We have therefore called this the riskless asset, by analogy to the one commodity model of the last section (to which it clearly reduces when there is indeed only one physical commodity). On the other hand, when there are many commodities the correct price normalization cannot be predicted in advance, i.e. it is impossible to know what asset will serve as the riskless asset without knowing the "in state" preferences and the aggregate endowment.

Under the conditions of Theorem 6, let us normalize prices $\bar{p}_s \equiv DV_s(M_s)$. Let the money payoffs of each asset $r \in R^{(S+1)(L+1)}$ be $\bar{r} \in R^{S+1}$ defined by $\bar{r}_s = \bar{p}_s \cdot r_s$, $s = 0, 1, \ldots, S$. If in addition the u^h are quadratic for all h, then one can easily derive the exact analogues of Theorems 2, 3, 2' and 3' in this multiple commodity world.

4. Security market line

A central relationship in the capital asset pricing model is the security market line, which suggests that the return to an asset is linearly related to the covariance of the asset return and the market return. The fact that it is the covariance, and not the variance, which is important to the pricing of assets is one of the enduring lessons of CAPM.

But this is a lesson which holds much more *generally* than for the CAPM model. We shall derive a security market line for any GEI model, provided that we are permitted to substitute an arbitrary asset $\mu \in R^{(S+1)(L+)}$ for the market asset. In particular, after making this substitution we can always find a "riskless" asset without making any assumptions about the return matrix $\bar{R}$ or the preferences W^h. Let $(\bar{R}, (W^h, X^h e^h, \bar{\Theta}^h); h \in H)$ be any multi-commodity general equilibrium model with incomplete markets. Let $(q, p, (\Theta^h, x^h); h \in H)$ be an interior equilibrium. For each $r \in R^{(S+1)(L+1)}$ let $\bar{r} \in R^{S+1}$ be defined by $\bar{r}_s = p_s \cdot r_s$, $s = 0, 1, \ldots, S$.

Theorem 7. Let $(\bar{R}, (W^h, X^h, e^h, \bar{\Theta}^h)\ h \in H)$ be a multi-commodity GEI. Let $(q, p, (\Theta^h, x^h), h \in H)$ be an interior equilibrium. Let $\pi \in R^{S+1}_{++}$ be arbitrary, and satisfy $\sum_{s \in S} \pi_s = 1$. Then there is always a renormalization of prices p that maintains equilibrium, such that with respect to these prices there is $z \in sp[\bar{R}]$ with $\bar{z}_s = K$ for all $s \in S$, and there is $\bar{\mu} \in R^{(S+1)(L+1)}$ such that for all $r \in sp[\bar{R}]$ with $q(r) = 1$,

$$[E_\pi \bar{r} - E_\pi \bar{z}] = \frac{Cov_\pi(\bar{\mu}, \bar{r})}{Var_\pi(\bar{\mu})} [E_\pi \bar{\mu} - E_\pi \bar{z}].$$

Remark. Note that the security market line holds only for *marketed* secutities (with price equal 1).

Proof. Choose any $h \in H$. From the separating hyperplane theorem there is a vector $\hat{\mu}^h \in R^{(S+1)(L+1)}_{++}$ such that for all $y \in R^{(S+1)(L+1)}_{+}$, if $W^h(y) \geq W^h(x^h)$, then $\hat{\mu}^h \cdot y \geq \hat{\mu}^h \cdot x^h$. Clearly $\hat{\mu}^h_s = \lambda^h_s p_s$ for some $\lambda^h_s > 0$, for each $s = 0, 1, \ldots, S$. Moreover, there is some $\lambda > 0$ such that for each $r \in sp[\bar{R}]$, $\lambda q(r) = \hat{\mu}^h \cdot r$. Let $\mu \in R^{(S+1)(L+1)}_{++}$ be defined by $\mu_s = (1/\lambda\pi_s) \hat{\mu}^h_s$. Then for all $r \in sp[\bar{R}]$, $q(r) = \sum_{s \in S} \pi_s \bar{\mu}_s \bar{r}_s \equiv \bar{\mu} \square_\pi \bar{r}$.

Let us observe that without loss of generality we can always renormalize p so that there is $r \in sp[\bar{R}]$ such that $p_s \cdot r_s = k$ for all $s = 0, 1, \ldots, S$, and $\bar{\mu}$ and $\bar{r}$ are linearly independent. Asset r_1 satisfies $p_s \cdot r_s(1) > 0$ for all $s = 0, 1, \ldots, S$ (recall the definition of GEI in the beginning of Section II) so that r_1 will do unless $\bar{r}_1$ and $\bar{\mu}$ are colinear. If all assets $r \in sp[\bar{R}]$ with $q(r) = 1$ yield money payoffs $\bar{r}$ colinear with $\bar{\mu}$, the theorem is vacuously true. If there is some $r \in sp[\bar{R}]$, with $q(r) > 0$, and $\bar{r}$ not colinear with $\bar{r}_1$ and $\bar{\mu}$, then let $z = ar_1 + (1-a)r$, with $0 < a < 1$. Then $q(z) > 0$, and if a is close enough to 1, $p_s \cdot z_s > 0$ for all $s = 0, 1, \ldots, S$. Moreover, $\bar{z}$ is linearly independent of $\bar{\mu}$ and $\bar{r}_1$. So we can renormalize $\bar{z}$ to be the riskless asset. Thus without loss of generality suppose we have $z \in sp[\bar{R}]$, with $\bar{z}_s = k$ for all s, and $\bar{\mu}$ and $\bar{z}$ linearly independent.

If $q(r) = 1$ for $r \in sp[\bar{R}]$, then $1 = \bar{\mu} \square_\pi \bar{r} = Cov_\pi(\bar{\mu}, \bar{r}) + e_\pi \bar{\mu} E_\pi \bar{r}$. Hence there is a (negative) linear relation between $Cov_\pi(\bar{\mu}, \bar{r})$ and $E_\pi \bar{r}$. As long as $Var_\pi \bar{\mu} > 0$, the claimed equation must describe this relationship. ∎

Q.E.D.

REFERENCES

ARROW, K. J. [1953]: "Le rôle des valeurs boursières pour la répartition la meilleure des risques," *Econométrie, Colloques Internationaux du CNRS* 11, 41-47. English translation in *Review of Economic Studies* 31 [1964], 91-96.

ARROW, K. J. and DEBREU, G. [1954]: "Existence of Equilibrium for a Competitive Economy," *Econometrica* 22, 265-290.

BLACK, F. [1972]: "Capital Market Equilibrium with Restricted Borrowing," *Journal of Business* 45, 444-454.

CASS, D. [1984]: "Competitive Equilibrium with Incomplete Financial Markets," CARESS Working Paper, University of Pennsylvania.

CASS, D. and STIGLITZ, J. [1970]: "The Structure of Investor Preferences and Asset Returns, and Separability in Portfolio Allocation: A Contribution to the Pure Theory of Mutual Funds," *Journal of Economic Theory* 2, 122-160.

DIAMOND, P. [1967]: "The Role of the Stock Market in a General Equilibrium Model with Technological Uncertainty," *American Economic Review* 57, 759-776.

DUFFIE, D. [1987]: "Stochastic Equilibria with Incomplete Financial Markets," *Journal of Economic Theory* 41, 405-416.

GEANAKOPLOS, J. and POLEMARCHAKIS, H. [1986]: "Existence, Regularity, and Constrained Suboptimality of Competitive Allocations when Markets are Incomplete," in W. Heller, R. Starr and D. Starrett (eds), *Essays in Honor of Kenneth Arrow,* Volume 3. Cambridge: Cambridge University Press, 65-95.

GREENWALD, B. and STIGLITZ, J. [1986]: "Pecuniary and Market Mediated Externalities," *Quarterly Journal of Economics* 101, 229-264.

HART, O. [1975]: "On the Optimality of Equilibrium when the Market Structure is Incomplete," *Journal of Economic Theory* 11, 418-443.

LINTNER, J. [1965]: "The Valuation of Risky Assets and the Selection of Risky Investments in Stock Portfolios and Capital Budgets," *Review of Economics and Statistics* 47, 13-37.

MOSSIN, J. [1977]: *The Economic Efficiency of Financial Markets,* Lexington Books.

NIELSON, L. [1985]: "Preference Structure and Equilibrium in the Classical Capital Asset Pricing Model," mimeo.

RADNER, R. [1972]: "Existence of Equilibrium of Plan, Prices and Price Expectations," *Econometrica* 40, 289-303.

ROTHSCHILD, M. [1986]: "Asset Pricing Theories," in W. Heller, R. Starr and D. Starrett (eds), *Essays in Honor of Kenneth Arrow,* Volume 3. Cambridge: Cambridge University Press.

RUBINSTEIN, M. [1974]: "An Aggregation Theorem for Securities Markets," *Journal of Financial Economics* 1, 225-244.

SHARPE, W. [1964]: "Capital Asset Prices: A Theory of Market Equilibrium under Conditions of Risk," *Journal of Finance* 19, 425-442.

WERNER, J. [1985]: "Equilibrium in Economies with Incomplete Financial Markets," *Journal of Economic Theory* 36, 110-119.

MANAGEMENT SCIENCE
Vol. 38, No. 6, June 1992
Printed in U.S.A.

ON MATCHING BOOK: A PROBLEM IN BANKING AND CORPORATE FINANCE*

M. SHUBIK AND M. J. SOBEL
Cowles Foundation, Yale University, 30 *Hillhouse Avenue, New Haven, Connecticut* 06520
312A *Harriman Hall, State University of New York, Stony Brook, New York* 11794-3775

In each of the asset and liability markets in which the banking firm is an intermediary, typically there are instruments with differing maturities. The bank matching book problem is to manage the term structures of assets and liabilities. In our first model, the bank borrows and lends only short run. In our second model, the bank borrows only short run but can lend short run and long run. The criterion in both models is the expected value of the present value of dividends issued. Both models yield dynamic programming problems. Broad aspects of optimal policies are indicated. (BANKING; MATCHING BOOK; DYNAMIC MODEL; INTEREST RATES; INVENTORIES)

1. Introduction

A firm may face a fluctuating demand for product produced from long-lived equipment. The equipment may need long-term financing although the fluctuation in income is short term. Most ordinary individuals can lend at a lower rate of interest than the rate at which they can borrow; this is also true of many firms. There are some large firms whose financial power is sufficiently large that the reverse may be true. They are implicitly, or even openly in part, finance companies. Banks in general are in a position to lend at rates higher than those at which they borrow.

Both banks and firms strive to match their short-term and long-term financial flows, but a bank's problem is somewhat simpler. In general the production processes of the firm are far more complex than those of a bank. In the bank matching book problem, the bank tries to optimize the structure of its long-run and short-run borrowing and lending. Although our primary interest is in the finance of the firm, the bank's structure is a natural preliminary to the more difficult corporate finance problem. The criterion in our model is maximization of the expected value of the sum of the discounted dividends issued by the bank.

The banking firm has a matching problem due to its role as an intermediary between asset and liability markets. Most of the literature on the theory of the banking firm consists of static (one-period) models that optimize the bank's behavior in one of these markets or the other; see Baltensperger (1980) and Santomero (1984) for surveys.

The treatment of asset and liability management as independent problems can be justified in a particular static model [cf. Klein 1971], but the independence is not warranted in other static models or in dynamic models [Baltensperger 1980 and Santomero 1984]. Deshmukh, Greenbaum and Kanatas (1982, 1983) present static and two-period models of coordinated asset and liability management. In their static model they deduce the effects of increased interest rate uncertainty, and in their two-period models they study a deposit-funding environment and a liability management environment.

An early dynamic model in Daellenbach and Archer (1969) concerns optimal short-run cash balance adjustments via borrowing and lending. In the dynamic model in O'Hara (1983), the bank selects short-run levels of consumption, deposit services and loans; the dividend is an exogenous function of each period's state.

* Accepted by William T. Ziemba; received July 1989. This paper has been with the authors 1 month for 1 revision.

"Liquidity risk" is the possibility that depositors may withdraw some or all of their funds, and "default risk" is the possibility that borrowers may not repay all their debts when due. It seems to us that default risk should be imbedded in a model that explicitly describes the effects of bankruptcy and the attitude to risk of the banking firm's manager or owner. Our models include liquidity risk, exclude default risk, and are consistent with risk neutrality. Our results can be viewed as stepping stones to the analysis of more elaborate models with risk sensitivity, default risk, and bankruptcy provisions.

The paper has the following organization. §2 discusses the modeling of a bank's borrowing and lending operations. §3 investigates a model in which the bank matches only its short-term borrowing and lending opportunities in order to maximize the expected present value of the time stream of its dividends. Each period, the bank examines the amount of its cash reserves (a) and observes the current interest rates to borrow (ρ) and lend (η) short-term funds. It decides how much of a dividend (p) to issue and how much to borrow (D); the exogenous demand for short-term loans (Q) is revealed, and the amount of the cash reserves is updated.

§3 shows that the bank's problem corresponds to a dynamic program with the following properties. There is an unconstrained optimization problem with parameters ρ and η and two decision variables; say $[y(\rho, \eta), d(\rho, \eta)]$ is an optimal solution. Then it is optimal for borrowing and the dividend to be nil ($p = D = 0$) if the cash level is too low and a leverage constraint is not binding. If the cash level is high enough and a leverage constraint is binding then the optimal amount to borrow is $D = D(\rho, \eta)$ and the optimal dividend is $a - y(\rho, \eta) - \rho D(\rho, \eta)$. Moreover, if η rises, the dividend does not increase and the amount borrowed does not decrease. §3 ends with a numerical example. §4 extends the model in §3 by permitting the probability distribution of the interest rate at which the bank borrows funds to depend on the amount of the bank's cash reserves.

The model in §5 includes opportunities for the bank to lend funds either short term (x) or long term (z) even though it can secure only short-term funds (D). As in the model in §3, the bank maximizes the expected present value of the time stream of its dividends. Short-term loans are repaid at the end of the same period in which they are borrowed; long-term loans are repaid at the end of the subsequent period. Each period, the bank observes interest rates for lending funds short term (η) and long term (β) and for borrowing funds short term (ρ). Also, it recalls two data from the previous period: the interest rate for lending funds long term (β') and the amount of funds lent long term (Z); let $w = (1 + \beta')Z$. Augmenting the model in §3 and §4 with an additional state variable and an additional decision variable dramatically complicates the analysis.

The dynamic program that corresponds to the bank's problem in §5 is shown to have the following properties. There is an unconstrained optimization problem that depends on parameters w and ρ and two decision variables; say $[y(\rho, w), d(\rho, w)]$ is an optimal solution. If the available deposits are not too high, namely $D < d(\rho, w)$, and the level of liquid assets is high enough, namely $a \geq y(\rho, w)$, then it is optimal to accept all of the available deposits. If the level of liquid assets is too low, namely $a < y(\rho, w)$, then it is optimal neither to issue a dividend nor to accept any deposits. The amounts loaned short term and long term, respectively, are nondecreasing functions of the amounts demanded short term and long term.

§6 extends the results in §3, §4, and §5 to an infinitely long planning horizon. Other considerations, such as bankruptcy, are mentioned in §7.

2. Modeling Considerations

There are no canonical models of the decision problem faced by a bank which borrows from sources of funds (depositors, savings accounts, central bank or elsewhere) and lends to customers. Building even a simple model of such a bank entails numerous choices in the modeling of process details. So the significance of our results depends on the quality of our choices. These choices are discussed in this section to motivate the modeling and to illustrate the distinctions in realism, relevance and tractability of the models.

A. *The Rate of Interest for Borrowing by the Bank*

(i) *The Competitive Assumption.* The bank can obtain money at a given rate ρ_t. Its borrowing at ρ_t could be subject to a bound on size.

(ii) *The Oligopolistic or Imperfect Competition Assumption.* The bank can obtain funds at the rate $\rho_t = f(s_t, D_t)$ where s_t is the level of assets held by the bank and D_t is its level of debts at time t. We make no assumptions about $f(\cdot, \cdot)$ in this paper although one might expect it to have certain properties. These include homogeneity of order one, i.e. doubling assets and debt should not influence the rate at which the bank borrows. If debt is increased for a fixed level of assets, the borrowing interest rate of the bank may increase, thus $\partial f/\partial D_t > 0$ and $\partial^2 f/\partial D_t^2 > 0$ and if assets increase for no extra debt the borrowing position is improved, thus $\partial f/\partial s_t < 0$. We might expect $\partial^2 f/\partial s_t^2 \geq 0$.

B. *The Rate of Interest for Lending by the Bank*

(i) *The Competitive Assumption.* The bank lends at a fixed rate η.

(ii) *The Imperfect Competition Assumption.* The bank can control the rate of interest at which it lends. The relationship between borrower demand and interest rate is given by $Q_t = g(\eta_t, \zeta_t)$ where Q_t is borrower demand and ζ_t is a random variable which represents environmental perturbations.

C. *Default Conditions*

We may model default of loans made with a random variable that represents percentage of failures. As a first approximation we ignore default.

D. *Payoffs and Dividends*

What does the bank maximize? It is easy to answer this question in a competitive market without uncertainty. There is one rate of interest and the bank maximizes the present value of the income stream. But when there is uncertainty and when the borrowing and lending rates are different, what rate is to be used for the bank's profits? Possibly we need a discount rate ψ; the constraint $(1 + \rho)^{-1} \leq \psi \leq (1 + \eta)^{-1}$ can be defended on an ad hoc basis. For any finite horizon we need both the evaluation of an appropriately discounted expected dividend stream and a salvage value to the firm.

E. *Bank Failure*

If the bank cannot pay back its loan either we must specify failure conditions or we must consider the possibility of a rollover of debt. We assume that there is a lender of last resort who will lend at a price α_t adjusted to the circumstances and from which the bank must borrow if it cannot repay depositors from its cash reserves. This structure combined with the absence of default by borrowers avoids the issue of bank insolvency.

F. *Lending and Borrowing: Short and Long*

The minimal complexity required to investigate the problems of matching book is to divide either sources, or uses or both into long term and short term. In modeling long-term borrowing or lending several new features appear which must be made explicit. Is long-term financing done every period? The general and realistic answer is yes. Can a long-term loan be paid back in full before it is due? In practice the answer is often yes, but with a special clause calling for a penalty or even a bonus.

G. *Set Up Costs, Administration and Overheads*

An important feature in financial transactions in general is the cost of transactions. Items such as paperwork may not vary materially with loan size over broad ranges. Frequently refinancing introduces considerable cost. These nonconvex features are not dealt with directly in these models, although part of the reason for the difference between long-term and short-term rates is the difference in transactions costs.

H. *On Leverage*

In modeling the relationship between lending and a bank's asset structure, both law and custom have dictated the ratios between assets and liabilities. Thus there is a leverage ratio (m) between loans and deposits.

3. The Matched Book Bank

In this and the subsequent section we consider two basic models and note some variants. In particular the first model here is that of the individual bank which has a fully matched policy of borrowing and lending short term. In the model in §5 we consider the bank which borrows short term, but can lend long term or short term.

3.1. *The Model*

We assume that the bank pays out dividends which its owners consume and its objective function is to maximize the present value of the expected dividend stream.

Let a_t be the cash level at the start of period t at which time the borrowing and lending interest rates (ρ_t and η_t, respectively) are known. After the bank decides how much to borrow (D_t) and the amount of dividends to issue (p_t), the demand for loans (Q_t) is revealed. The constraints on borrowing and dividends are

$$0 \le D_t \le ma_t, \qquad 0 \le p_t, \qquad p_t + \rho_t D_t \le a_t, \tag{1}$$

where m is the leverage parameter. We assume that $(\rho_1, \eta_1, Q_1), (\rho_2, \eta_2, Q_2), \cdots$ are independent and identically distributed triples; let $(\rho, \eta, \mathbf{Q})$ have the distribution that is common to these triples.

We assume only in §3.3 that

$$P\{(1+\rho)^{-1} < \psi < (1+\eta)^{-1}\} = 1$$

where ψ is the bank's single-period discount factor. This "cheap money" assumption corresponds to

$$P\{\psi(1+\eta) < 1 < \psi(1+\rho)\} = 1. \tag{2a}$$

The event $\{\psi < (1+\eta)^{-1}\}$ has the interpretation that the marginal revenue from loans is higher than the bank's hurdle rate for the profits of new short-term investments. The event $\{(1+\rho)^{-1} < \psi\}$ has the interpretation that the bank's hurdle rate is higher than that of the source of funds from which D_t is borrowed. Corresponding to (2a), we assume that generic values of the parameters satisfy

$$\psi(1+\eta) < 1 < \psi(1+\rho) \tag{2b}$$

which implies $\eta < 1/\psi - 1$. We also assume $\rho < \rho^*$ for some $\rho^* < \infty$.

The dynamics are

$$a_{t+1} = y_t + \eta_t \cdot \min\{Q_t, y_t + D_t\} \qquad \text{where} \tag{3a}$$

$$y_t = a_t - \rho_t D_t - p_t. \tag{3b}$$

In words, cash at the beginning of the next period equals cash at the beginning of this period minus interest paid minus dividends issued plus interest received (on the lesser of cash available and loans demanded). From (3a,b)

$$p_t = a_t - y_t - \rho_t D_t$$
$$= y_{t-1} - y_t - \rho_t D_t + \eta_{t-1} \cdot \min\{Q_{t-1}, y_{t-1} + D_{t-1}\}.$$

Let ψ be the bank's single-period discount factor. Then the present value B of its dividend stream is

$$B = \sum_{t=1}^{\infty} \psi^{t-1} p_t$$
$$= \sum_{t=1}^{\infty} \psi^{t-1}(a_t - y_t - \rho_t D_t)$$
$$= \sum_{t=1}^{\infty} \psi^{t-1}[\psi\eta_t \cdot \min\{Q_t, y_t + D_t\} - \rho_t D_t - (1-\psi)y_t] + a_1. \tag{4a}$$

Let $\psi^{t-1}u_t$ be the summand in (4a). Since a_1 is known, a policy maximizes $E(B)$ if and only if it maximizes

$$E(B) - a_1 = E\left(\sum_{t=1}^{\infty} \psi^{t-1}u_t\right).$$

For the remainder of this section, we analyze the simpler problem of maximizing

$$E\left(\sum_{t=1}^{n} \psi^{t-1}u_t\right) \tag{4b}$$

for $n < \infty$. However, we show in §6 that the qualitative results for criterion (4b) are valid for criterion (4a).

It follows from (1) and

$$u_t = \psi\eta_t \cdot \min\{Q_t, y_t + D_t\} - \rho_t D_t - (1-\psi)y_t \tag{4c}$$

that the maximization of (4b) corresponds to the following dynamic programming recursion:

$$v_n(a, \rho, \eta) = \max\{J_n(y, D; \rho, \eta) : 0 \le y \le a, 0 \le D \le ma\}, \tag{5}$$
$$J_n(y, D; \rho, \eta) = \psi\eta E[\min\{Q, y + D\}] - \rho D - (1-\psi)y$$
$$+ \psi E[v_{n-1}(y + \eta \cdot \min\{Q, y + D\}, \rho, \eta)] \tag{6}$$

for $n = 1, 2, \ldots$, where $v_0(\cdot, \cdot, \cdot) = 0$. We assume that the distribution of (ρ, η, Q) permits the expectations in (6) to exist finitely and that the maximum in (5) is always achieved.

In the following subsection and later in the paper we partially order sets T of pairs of real numbers (u_1, u_2) as follows. If $(u_1, u_2) \in T$ and $(u'_1, u'_2) \in T$ then $(u_1, u_2) \le (u'_1, u'_2)$ if $u_1 \le u'_1$ and $u_2 \le u'_2$. We say that (u_1^*, u_2^*) is *maximal* if $(u_1^*, u_2^*) \in T$ and $(u_1^*, u_2^*) \ge (u_1, u_2)$ for all $(u_1, u_2) \in T$.

3.2. *Solution Properties*

An inductive proof establishes that $v_n(\cdot, \cdot, \cdot)$ and $J_n(\cdot, \cdot; \rho, \eta)$ are continuous concave functions on their respective domains for all n, $\eta \ge 0$, and $\rho \ge 0$.

Let $T_n(\rho, \eta)$ be the set of (y, D) that maximizes $J_n(\cdot, \cdot; \rho, \eta)$ on the nonnegative orthant. It follows from Topkis (1978) that (i) for each ρ and η, $T_n(\rho, \eta)$ has a maximal element $[y_n(\rho, \eta), d_n(\rho, \eta)]$ and (ii) $\rho \geq \rho'$ and $\eta \leq \eta'$ implies $[y_n(\rho, \eta), d_n(\rho, \eta)] \leq [y_n(\rho', \eta'), d_n(\rho', \eta')]$. Concavity implies $(y, D) = [y_n(\rho, \eta), d_n(\rho, \eta)]$ is optimal in (5) if $y_n(\rho, \eta) \leq a$ and $d_n(\rho, \eta) \leq ma$. If $a < y_n(\rho, \eta)$ and $d_n(\rho, \eta) \leq ma$, then $y = a$ is optimal; hence $p = D = 0$. If $a \geq y_n(\rho, \eta)$ and $d_n(\rho, \eta) > ma$ then $D = ma$ is optimal. If $y_n(\rho, \eta) > a$ and $d_n(\rho, \eta) > ma$ then it is optimal either for $y = a$ (hence $D = 0$) or $D = ma$ (or both).

In summary, an optimal policy specifies

$$p = a - y_n(\rho, \eta) - \rho d_n(\rho, \eta) \quad \text{and} \quad D = d_n(\rho, \eta)$$
$$\text{if} \quad y_n(\rho, \eta) \leq a \quad \text{and} \quad d_n(\rho, \eta)/m \leq a,$$
$$p = D = 0 \quad \text{if} \quad d_n(\rho, \eta)/m \leq a < y_n(\rho, \eta),$$
$$D = ma \quad \text{if} \quad y_n(\rho, \eta) \leq a < d_n(\rho, \eta)/m.$$

Suppose that $\mathbf{Q}$ and (ρ, η) are independent. This assumption implies that demand is totally inelastic with respect to "price" ρ. In §5, we discuss a more realistic model with elastic demand. Concavity, (5), and (6) imply that $y_n(\rho, \eta)$ and $d_n(\rho, \eta)$ are nonincreasing in ρ and nondecreasing in η. Therefore, (i) the dividend, p, is nonincreasing as the rate of interest at which the bank lends, η, increases, (ii) the amount it borrows, D, is nondecreasing as the unit revenue it receives, η, increases. The dividend, p, is *not* a monotonic function of ρ as the next section illustrates.

3.3. *Last Period*

When $n = 1$ in (5),

$$J_1(y, D; \rho, \eta) = \psi\eta E[\min\{\mathbf{Q}, y + D\}] - \rho D - (1 - \psi)y.$$

Let $\Phi(\cdot)$ be the (marginal) distribution function of $\mathbf{Q}$ and let $\Phi(0) = 0$.

We now show that cash reserves in the last period are split completely between dividend payment and interest payment for deposits. Also there are two cases which determine the amount of deposits the bank should accept. If $\psi\eta \leq \rho$ then $D = 0$; if $\psi\eta > \rho$ then $D = \Phi^{-1}[1 - \rho/\psi\eta]$ where $\Phi^{-1}(\alpha) = \sup\{x: \Phi(x) \leq \alpha\}$. Let $J_n^{(1)}$ and $J_n^{(2)}$ denote right-hand derivatives $\partial J_n/\partial y$ and $\partial J_n/\partial D$, respectively. Then

$$J_1^{(1)}(y, D; \rho, \eta) = -(1 - \psi) + \psi\eta[1 - \Phi(y + D)], \tag{7}$$
$$J_1^{(2)}(y, D; \rho, \eta) = -\rho + \psi\eta[1 - \Phi(y + D)]. \tag{8}$$

Since $\Phi(\cdot)$ is a distribution function,

$$J_1^{(1)}(y, D; \rho, \eta) \leq (1 - \psi) + \psi\eta[1 - \Phi(d)] \leq -(1 - \psi) + \psi\eta, \tag{9}$$
$$J_1^{(2)}(y, D; \rho, \eta) \leq -\rho + \psi\eta[1 - \Phi(y)] \leq -\rho + \psi\eta. \tag{10}$$

From (1b), $J_1^{(1)} \leq 0$ for all y; so $y = 0$ (hence $a = p + \rho D$) is optimal. In (10), $y = 0$ and $\Phi(0) = 0$ imply $J_1^{(2)} \leq -\rho + \psi\eta$; so $D = 0$ is optimal if $\psi\eta \leq \rho$.

From (8) and $y = 0$, an interior optimum with respect to D satisfies

$$D = \Phi^{-1}[1 - \rho/(\psi\eta)]. \tag{11}$$

So $D = 0$ if $\psi\eta \leq \rho$.

In summary, in the last period, $y = 0$ is optimal, i.e.

$$p = a - \rho D \tag{12}$$

and (11) specifies D. So $\psi\eta \le \rho$ implies $D = 0$ and $p = a$, i.e. all cash is distributed as a dividend, no funds are borrowed, and none are lent. If $\psi\eta > \rho$ then (11) specifies the amount borrowed and (12) specifies the dividend.

EXAMPLE. Suppose that $\mathbf{Q}$ has a marginal distribution which is exponential with $E(\mathbf{Q}) = 1/\mu$:

$$\Phi(x) = 1 - e^{-\mu x} \qquad (x \ge 0, \mu > 0).$$

Then

$$\Phi^{-1}(\alpha) = -[\log(1-\alpha)]/\mu \qquad (0 \le \alpha < 1).$$

So (11) is

$$D = -\{\log[(\rho/\psi\eta)]\}/\mu \qquad (\rho < \psi\eta),$$

and the dividend (12) is

$$p = a + (\rho/\mu)\log[(\rho/\psi\eta)] \qquad (\rho < \psi\eta).$$

Therefore, p increases with ρ if $1 < \psi\eta/\rho < e$ and decreases with ρ if $e \le \psi\eta/\rho < \infty$.

If the bank can obtain funds at 6% and lend at 10%, and if its discount factor corresponds to 8%, then $\rho = 0.06$, $\eta = 0.01$, and $\psi = (1.08)^{-1}$. So (11) is

$$D = -E(\mathbf{Q})\log[(\rho/\psi\eta)] = E(\mathbf{Q})\log(0.648) = 0.188E(\mathbf{Q}).$$

That is, in the last period the bank borrows approximately 19% of the amount which it expects to be demanded. With a rectangular distribution on $\frac{1}{2}$ to $\frac{3}{2}$ and the other parameters as above (11) gives $D = 0.85E(\mathbf{Q})$.

4. Interest as a Function of Liquidity

Recall that liquidity risk is the possibility that depositors may elect to withdraw part or all of their funds. One of the ways to model this risk is to relate the bank's condition to the interest that the bank must pay to obtain funds. In this section, we let ρ_t be a function of the cash level (a_t), the loan size (D_t), or minimum balance (y_t). That is, for a function $g(\cdot)$ let ρ_t be given by $g(s_t)$, $g(D_t)$, or $g(y_t)$. Then in (4) the term $-\rho_t D_t$ is replaced by $-D_t g(a_t)$, $-D_t g(D_t)$, or $-D_t g(y_t)$. Similarly, the term $-\rho D$ in (6) is replaced by $-Dg(s)$, $-Dg(D)$, or $-Dg(y)$.

If $g(\cdot)$ is a nondecreasing convex function on $[0, \infty)$, then the concavity properties previously asserted for v_t and J_t remain valid. It follows that the previously stated monotonicity properties of an optimal selection of y and D are preserved. However, if $g(\cdot)$ is nonincreasing and convex as is more likely, the concavity properties may be in peril.

4.1. *Last Period*

Suppose $\rho_t = g(D_t)$ for each t where $g(\cdot)$ is a nondecreasing, convex, and differentiable function. Then the previous analysis which concludes that $y = 0$ in the last period is unaltered. So $y = 0$ is optimal here. Then (8) becomes

$$J_1^{(2)}(y, D; \eta) = \psi\eta[1 - \Phi(D)] - g(D) - Dg'(D).$$

So $D = 0$ is optimal if $g(0) \ge \psi\eta$. If there is an interior optimum,

$$\psi\eta = g(D) + Dg'(D) + \psi\eta\Phi(D). \qquad (13)$$

Similarly, if $\rho_t = g(a_t)$ for each t and (1b) is valid when ρ is replaced by $g(0)$, then the same analysis is valid as when ρ_t is exogenous. That is, in the last period $y = 0$ is optimal and

$$D = \Phi^{-1}[1 - g(a)/(\psi\eta)].$$

If $\rho_t = g(y_t)$ for each t and (1b) is valid when ρ is replaced by $g(D)$, then an optimal selection is $y = 0$ and

$$D = \Phi^{-1}[1 - g(0)/(\psi\eta)].$$

For example, if $\Phi(x) = 1 - e^{-\mu x}$, then

$$D = -\{\log [g(0)/(\psi\eta)]\}/\mu$$

if $g(0) \leq \psi\eta$. Otherwise, $D = 0$ because the "prime rate," $g(0)$, is too high.

4.2. *Loss Reserves*

Suppose that ξ_t is the fraction of loan dollars on which borrowers in period t default. Then (3a) is altered to

$$a_{t+1} = y_t + (\eta_t - \xi_t) \min \{Q_t, y_t + D_t\}.$$

If $(\rho_1, \eta_1, \xi_1, Q_1), (\rho_2, \eta_2, \xi_2, Q_2), \cdots$ are independent and identically distributed vectors such that

$$P\{\eta_1 \geq \xi_1\} = 1, \tag{14}$$

then the results thus far remain valid when η is replaced by $\eta - \xi$. If (14) is false, then the model is more difficult to analyze.

4.3. *Description of Policy and Sensitivity Analysis*

We have shown above that there is an optimal policy which is more or less consistent with our intuition. The decision variables are dividends and how much to borrow (or levels of deposits to be accepted). These variables will be influenced by changes in the basic parameters of the problem. In particular these are (i) the initial capital, (ii) the cost of money to the bank, (iii) the lending rate, (iv) the bank's profit discount factor and (v) changes in the probability distributions.

5. The Bank With Short-Term Borrowing and Long- and Short-Term Lending

In this model we consider the added complication introduced by a term structure in loans. This complication can enter on both the asset and liability sides. But as a simplification for the sake of more tractability we omit long-term borrowing by the bank. Thus we concentrate on how much capital the bank should have and how much long-term lending it might wish to assume given that it borrows short term.

The bank can obtain short-term money from three sources: (1) its own capital which we denote initially by a_1, (2) short-term deposits $D_t = f(\rho_t, \zeta_t)$ [the level of deposits is a function of the interest offered by the bank and a random component ζ_t], and (3) the lender of last resort or the central bank which lends funds at the rate α_t which may be fixed in advance or be a function of the capital and size of debt of the bank.

The bank can lend money short or long. We shall, as a crude approximation, model this with one period for short and two periods for long. Let the annual rate for short be η_t and for long be β_t. Long-term rates can be larger or smaller than short-term rates depending on transactions costs and expectations.

The needs of bank clients are differentiated by their business requirements; thus there may be some substitutability between long-term and short-term borrowing but it is not necessarily high. We may represent the demand for short-term and long-term borrowing as $S_t = g_1(\eta_t, \beta_t, \zeta_t)$ and $L_t = g_2(\eta_t, \beta_t, \zeta_t)$ where $\zeta_1, \zeta_1, \cdots$ are independent and identically distributed random variables.

The competitive bank, if small, will more or less have to accept the rates at which it lends and borrows, as dictated by outside forces. Thus its strategy is its policy with respect to capital structure, borrowing and lending.

We still need to specify a payoff function for the bank. This, as before, may be the expected present value of dividends (plus the salvage value) where the discount rate ψ has to be justified.

5.1. *Model*

Let S_t and L_t be the exogenous demands for short-term and long-term loans in period t, respectively. Repayments on S_t and L_t are due at the ends of periods t and $t + 1$, respectively. However, the bank may regulate the term structure of its loan portfolio by electing to lend less than S_t or L_t. Let x_t and z_t be the respective amounts of short-term and long-term loans made in period t at interest rates η_t and β_t. Of course,

$$0 \le x_t \le S_t, \qquad 0 \le z_t \le L_t. \tag{15}$$

Credit rationing occurs if $x_t < S_t$ or $z_t < L_t$. Belloc and Freixas (1988) discuss the roles of adverse selection and moral hazard in credit markets as consequences of banks' decisions to grant or withhold credit.

We assume that S_t and L_t are revealed after the dividend p_t is announced. Then x_t and z_t are selected subject to (15) and a liquidity constraint we shall specify below. Suppose that the level of demand deposits and the interest rate on those deposits, D_t and ρ_t, respectively, are known when the dividend p_t is selected. Let d_t be the amount of the deposits which is accepted; so

$$0 \le d_t \le D_t \qquad \text{and} \qquad d_t \le ma_t \tag{16}$$

where m is the leverage parameter. We assume that the assets must be sufficient to cover the dividends and the interest on deposits. Let a_t be the bank's liquid assets at the beginning of period t. So

$$0 \le a_t - p_t - \rho_t d_t, \qquad 0 \le p_t.$$

Let

$$y_t = a_t - p_t - \rho_t d_t, \tag{17}$$

so

$$0 \le y_t \le a_t. \tag{18}$$

The bank has $y_t + d_t$ available to lend, but it may elect to lend a total, $x_t + z_t$, which exceeds the available funds. In this event, the excess must be obtained from a central bank of last resort at a (presumably) high interest rate α_t. We assume that α_t, η_t, and β_t are known when x_t and z_t are chosen.

Under these assumptions

$$a_{t+1} = y_t + \eta_t x_t - \alpha_t(x_t + z_t - d_t - y_t)^+ + (1 + \beta_{t-1})z_{t-1} - (1 - \beta_t)z_t. \tag{19}$$

The last two terms on the right side of (19) refer to repayment of the previous period's long-term loans with interest, and the long-term loans made in the current period.

We assume that the bank operates in a regulatory environment that mandates non-negative cash reserves. So it must select x_t and z_t in order to satisfy $a_{t+1} \ge 0$. From (19), this constraint corresponds to

$$\alpha_t(x_t + z_t - d_t - y_t)^+ - \eta_t x_t - (1 - \beta_t)z_t \le y_t + (1 + \beta_{t-1})z_{t-1}. \tag{20}$$

We shall not analyze the model of a bank in a less restrictive regulatory environment in which cash reserves plus long-term loans must be nonnegative. In that case, the right side of (20) would be increased by z_t.

It follows from (17) and (19) that

$$\begin{aligned} p_t &= a_t - \rho_t d_t - y_t \\ &= -\rho_t d_t - y_t + a_{t-1} + \eta_{t-1}x_{t-1} - \alpha_{t-1}(x_{t-1} + z_{t-1} - d_{t-1} - y_{t-1})^+ \\ &\quad + (1 + \beta_{t-2})z_{t-2} - (1 - \beta_{t-1})z_{t-1}. \end{aligned} \tag{21}$$

The substitution of (21) and rearrangement of terms yields the present value

$$B = \sum_{t=1}^{\infty} \psi^{t-1} p_t = a_1 + \sum_{t=1}^{\infty} \psi^{t-1}\{-y_t - \rho_t d_t + \psi a_t + \psi \eta_t x_t - \psi[1 - \psi - (1 + \psi)\beta_t]z_t - \psi \alpha_t (x_t + z_t - y_t - d_t)^+\}. \tag{22a}$$

As in §3, for the remainder of this section we analyze a finite-horizon version of (22a). Let $\psi^{t-1}u_t$ be the summand in (22a). Since a_1 is known, an optimal policy maximizes

$$E(B) - a_1 = E\left(\sum_{t=1}^{\infty} \psi^{t-1}u_t\right)$$

Thus, we analyze the problem of maximizing

$$E\left(\sum_{t=1}^{n} \psi^{t-1}u_t\right) \tag{22b}$$

and show in §6 that the qualitative results remain valid as $n \to \infty$.

Let $\lambda_t = (\alpha_t, \beta_t, \eta_t, S_t, L_t)$. In the stochastic process $(\rho_1, D_1, \lambda_1), (\rho_2, D_2, \lambda_2), \ldots,$ we assume that the distribution of (ρ_t, D_t) depends only on $(\beta_{t-2}, \lambda_{t-1})$ and the distribution of λ_t depends only on (β_{t-1}, ρ_t). These assumptions encompass a wide variety of autocorrelated interest rate processes.

We use (15), (16), and (18) to formulate the maximization of (22b) as a dynamic program. Let λ be the information which is revealed after the dividend and deposit (p_t and d_t) are selected, namely $\lambda = (\alpha, \beta, \eta, S, L)$ and let β' and Z be the values of β and z in the previous period. It follows from (20) that $w = (1 + \beta')Z$ is an element of the constraint on x and z, given the other information. The state of the dynamic program is $[a, \rho, D, (1 + \beta')Z]$ and the decisions (y, d) are made at the beginning of the period. Then λ is revealed and (x, y) are selected. Let $v_0(\cdot) \equiv 0$ and for $n \geq 1$,

$$v_n(a, w, \rho, D) = \psi a + \max\{-y - \rho d + E[K_n(y, d, w, \lambda)]: 0 \leq y \leq a, d \leq ma, 0 \leq d \leq D\}, \tag{23a}$$

$$K_n(y, d, w, \lambda) = \psi \max\{J_n(x, z, y, d, w, \lambda): 0 \leq x \leq S, 0 \leq z \leq L, \alpha(x + z - d - y)^+ - \eta x - (1 - \beta)z \leq y + w\}, \tag{23b}$$

$$J_n(x, z, y, d, w, \lambda) = \eta x - [1 - \psi - (1 + \psi)\beta]z - \alpha(x + z - y - d)^+ + E\{v_{n-1}[y + \eta x - \alpha(x + z - y - d)^+ + w - (1 - \beta)z, (1 + \beta)z, \rho, \mathbf{D}]\}, \tag{23c}$$

We assume that the maxima in (23a,b) are achieved and that expectations exist and are finite.

5.2. *Solution Properties*

An inductive proof that uses (23a) and $\psi \geq 0$ establishes that $v_n(\cdot, w, \rho, D)$ is a nondecreasing function for each (w, ρ, D) and that $v_n(\cdot)$, $K_n(\cdot, \lambda)$, and $J_n(\cdot, \lambda)$ are continuous concave functions on their respective domains (for each λ with nonnegative components).

Let $\xi = (\alpha, \beta, \eta, y, d, w)$, let $T_{n1}(w, \rho)$ be the set of (y, d) that maximize the objective in (23a) on the nonnegative orthant, and let $T_{n2}(\xi)$ be the set of (x, z) that maximize the objective in (23b) on the nonnegative orthant. It follows from Topkis (1978) that (i) $T_{n1}(w, \rho)$ has a maximal element $[y_n(w, \rho), d_n(w, \rho)]$ for each n, w, and ρ, and (ii) T_{n2} has a maximal element $[x_n(\xi), z_n(\xi)]$ for each n and ξ.

Concavity and (23a,b,c) lead to the following conclusions (in which the arguments of y_n, d_n, x_n, and z_n are suppressed) for optimal values of decision variables:

$$p = a - y_n - \rho d_n \quad \text{and} \quad d = d_n$$

$$\text{if} \quad y_n \le a \quad \text{and} \quad d_n/m + (a - D/m)^+ \le a, \tag{24}$$

$$p = d = 0 \quad \text{if} \quad d_n/m \le a < y_n \quad \text{and} \quad d_n \le D,$$

$$d = \min\{ma, D\} \quad \text{if} \quad y_n \le a < d_n/m + (a - D/m)^+,$$

$$x = x_n \quad \text{and} \quad z = z_n \quad \text{if} \quad x_n \le S, \quad z_n \le L \quad \text{and}$$

$$\alpha(x_n + z_n - d - y)^+ - \eta x_n - (1 - \beta) z_n \le y + w,$$

$$x = S \quad \text{if} \quad x_n > S, \quad z_n \le L, \quad \text{and}$$

$$\alpha(x_n + z_n - d - y)^+ - \eta x_n - (1 - \beta) z_n > y + w, \tag{25}$$

$$z = L \quad \text{if} \quad x_n \le S, \quad z_n > L \quad \text{and} \quad (25),$$

$$(\alpha - \eta)x + (\alpha + \beta - 1)z = y + w + \alpha(d + y) \quad \text{if} \quad (25),$$

$$(y + w)/\alpha \le L, \quad \text{and} \quad (y + w)/(\alpha - \eta) \le S \quad (\alpha > \eta).$$

More generally, there is an optimal solution in (22b), $(x, z) = [X_n(S, L), Z_n(S, L)]$ (we have suppressed parameters except for S and L), such that for all $\gamma > 0$,

$$0 \le X_n(S, L + \gamma) - X_n(S, L) \le \gamma,$$

$$0 \le X_n(S + \gamma, L) - X_n(S, L) \le \gamma,$$

$$0 \le Z_n(S + \gamma, L) - Z_n(S, L) \le \gamma,$$

$$0 \le Z_n(S, L + \gamma) - Z_n(S, L) \le \gamma. \tag{26}$$

That is, the amount of loans made of each type is monotonic in its demand, but does not increase faster than the demand. Similarly, there is an optimal solution $(y, D) = [Y_n(a, D), \delta_n(a, D)]$ in (23a) such that

$$0 \le Y_n(a + \gamma, D) - Y_n(a, D) \le \gamma,$$

$$0 \le Y_n(a, D + \gamma) - Y_n(a, D) \le \gamma,$$

$$0 \le \delta_n(a + \gamma, D) - \delta_n(a, D) \le \gamma,$$

$$0 \le \delta_n(a, D + \gamma) - \delta_n(a, D) \le \gamma \tag{27}$$

for all $\gamma > 0$. The interpretation is similar to that of (26). Also, the accepted deposit δ_n is a nonincreasing function of the deposit interest rate ρ and Y_n is invariant with respect to ρ; so the dividend must be a nondecreasing function of ρ. The short-term and long-term loans (X_n and Z_n) are nonincreasing functions of the interest rate for "last resort" funds (α), X_n is a nondecreasing function of η (the interest rate on short-term loans) and Z_n is a nondecreasing function of β (the interest rate on long-term loans).

6. Matching with an Infinite Horizon

The analyses in §3 and §5 replaced infinite horizon planning horizons in (4a) and (22a) with finite horizon planning horizons ($n < \infty$) in (4b) and (22b). In this section we show that the earlier conclusions regarding the qualitative properties of an optimal policy remain valid for (4a) and (22a). Our method draws on Heyman and Sobel (1984, sections 8-5 and 8-6) and outlines a proof that (a) the value functions of the finite horizon dynamic programs converge as $n \to \infty$, (b) the limit value function satisfies the functional equation of dynamic programming, (c) as $n \to \infty$, the finite horizon optimal policies converge to a policy that is optimal in the functional equation, and (d) the limit policy inherits the qualitative properties of the finite horizon optimal policies.

Model in §3

First, we consider the model in §3 where each period the bank reviews its cash level (a_t) and decides how much to borrow (D_t) and to issue as dividends (p_t). The following argument shows that v_n and J_n, the dynamic programming value functions in (5) and (6), converge as $n \to \infty$.

Let $u = E(\eta Q)\psi/(1 - \psi)$. It follows from (1), (3a,b), and (4c) that v_n, the value function for the maximization of (4b), satisfies

$$v_n \le E \sum_{t=1}^{n} \psi^{t-1}\psi\eta_t Q_t \le u \tag{28}$$

and in (6) $J_n \le \psi E(Q) + \psi u$ under the assumption $P\{\eta \le 1\} = 1$. For lower bounds, we observe in (5) that $v_n(0, \cdot, \cdot) \equiv 0$ and $v_n(a, \rho, \eta) \ge 0$ because $y = D = 0$ is always feasible. Finally, the assumptions $P\{\rho \ge 1 - \psi\} = 1$ and $P\{Q \le K\} = 1$ imply $v_n(a, \rho, \eta) \le v_{n+1}(a, \rho, \eta)$ and $J_n(y, D; \rho, \eta) \le J_{n+1}(y, D; \tilde{\rho}, \eta)$ for all n. Therefore, v_n and J_n are bounded monotone sequences and there exist

$$v(a, \rho, \eta) = \lim_{n\to\infty} v_n(a, \rho, \eta) \qquad \text{and} \qquad J(y, D; \rho, \eta) = \lim_{n\to\infty} J_n(y, D; \rho, \eta). \tag{29}$$

Continuity and concavity of $v_n(\cdot)$ and $J_n(\cdot, \cdot; \rho, \eta)$ on their respective domains for each $\rho \ge 0$, $\eta \ge 0$, and n and monotone convergence imply that $v(\cdot)$ and $J(\cdot, \cdot; \rho, \eta)$ inherit these properties. Since the set of feasible actions is compact at each state, it follows that

$$v(a, \rho, \eta) = \max\{J(y, D; \rho, \eta): 0 \le y \le a, 0 \le D \le ma\}. \tag{30}$$

Let $T(\rho, \eta)$ be the set of (y, D) that maximize the objective in (30) on the nonnegative orthant. It follows from Topkis (1978) that (i) for each ρ and η, $T(\rho, \eta)$ has a maximal element $[y(\rho, \eta), d(\rho, \eta)]$ and (ii) $\rho \ge \rho'$ and $\eta \le \eta'$ implies $[y(\rho, \eta), d(\rho, \eta)] \le [y(\rho', \eta'), d(\rho', \eta')]$. If $y(\rho, \eta) \le a$ and $d(\rho, \eta) \le ma$ then $(y, D) = [y(\rho, \eta), d(\rho, \eta)]$ is optimal in (27). If $a < y(\rho, \eta)$ and $d(\rho, \eta) \le ma$, then $y = a$ and $D = 0$ are optimal. If $a \ge y(\rho, \eta)$ and $d(\rho, \eta) > ma$ then $D = ma$ is optimal. If $y(\rho, \eta) > a$ and $d(\rho, \eta) > ma$ then it is optimal either for $y = a$ (hence $D = 0$) or $D = ma$ (or both).

In summary, an optimal policy specifies

$$p = a - y_n(\rho, \eta) - \rho d_n(\rho, \eta) \quad \text{and} \quad D = d_n(\rho, \eta) \quad \text{if} \quad y_n(\rho, \eta) \le a \quad \text{and} \quad d_n(\rho, \eta)/m \le a,$$

$$p = D = 0 \quad \text{if} \quad d_n(\rho, \eta)/m \le a < y_n(\rho, \eta),$$

$$D = a \quad \text{if} \quad y_n(\rho, \eta) \le a < d_n(\rho, \eta)/m.$$

Model in §5

We outline here how the same argument as in the previous subsection leads to the conclusion that the infinite horizon counterparts of (23a,b,c), (24), (26), and (27) are valid. A bound similar to (28) can be specified for (23a,b,c) which yields versions of (29) for v_n, K_n, and J_n with n suppressed. Hence, the counterpart of (30) is valid, namely (23a,b,c) with n suppressed. Also, $v(\cdot, w, \rho, D)$ is nondecreasing for each (w, ρ, D) and $v(\cdot)$, $K(\cdot, \lambda)$, and $J(\cdot, \lambda)$ are continuous concave functions on their respective domains. Therefore, the infinite horizon counterparts of (24), (26), and (27) are valid.

7. Further Considerations

The analysis presented above hopefully offers some insight into the problems of matching long-term and short-term portfolios. But this is a highly oversimplified view of the many features involved in understanding time structure in financing. Some of these factors are noted below.

7.1. *Partial Equilibrium versus General Equilibrium*

The models we have considered are open or partial equilibrium models. If one tries to close them, several key logical difficulties emerge. How do we explain differences in borrowing and lending rates beyond the existence of transactions costs? Is it possible for individuals to hold different expectations over time without convergence to consistent rational expectations"?

7.2. *On Bankruptcy*

It generally simplifies the mathematics to treat the problems of illiquidity and insolvency continuously by stipulating costlier financing when these events arise. In actuality, banks or firms are liquidated or reorganized. In the reorganization the assets of a manufacturing firm tend to be far less liquid and more difficult to evaluate than those of a financial firm. Furthermore, the partial equilibrium open model of bankruptcy does not adequately illustrate the efficiency or distributional effects in restructuring debt.

7.3. *Time Structure, Technology and Expectations*

In actuality the needs of corporations, other institutions and individuals require a debt structure that can run out for 20 or 30 years. The durability, immovability and special purpose of many buildings, machines and equipment causes the problem of matching book to be far harder for the nonfinancial firm than for the financial institution.

The need to match long and short structure is present even with consistent expectations, as an appropriate defense against random fluctuations. The bank's (or firm's) capital is a money inventory to protect against fluctuations. What constitutes an optimal level of capital can be answered for the bank or firm in the type of models noted above. But although we believe that such a question merits answering it casts only partial light on the questions of the relative merits of the carrot of the availability of a lender of last resort or the whip of bankruptcy if protection provided by one's own capital is inadequate.[1]

[1] This paper was prepared with partial support from NSF grants SES-8408951, ECS-8705698 and SES-8812051, and was written while Dr. Shubik was visiting at SUNY—Stony Brook.

References

BALTENSPERGER, E., "Alternative Approaches to the Theory of the Banking Firm," *J. Monetary Economics*, 6 (1980), 1–37.

BELLOC, B. AND X. FREIXAS, "Equilibrium and Rationing in the Credit Market: A Sequential Equilibrium Approach," D.P. 60, S.E.E.D.S., GREMAQ, Université des Sciences Sociales de Toulouse 1, Toulouse, France, 1988.

DAELLENBACH, H. G. AND S. B. ARCHER, "The Optimal Bank Liquidity: A Multi-period Stochastic Model," *J. Financial and Quantitative Anal.*, 4 (1969), 329–343.

DESHMUKH, S. D., S. I. GREENBAUM AND G. KANATAS, "Bank Forward Lending in Alternative Funding Environments," *J. Finance*, 37 (1982), 925–940.

———, ——— AND ———, "Interest Rate Uncertainty and the Financial Intermediary's Choice of Exposure," *J. Finance*, 38 (1983), 141–147.

HEYMAN, D. P. AND M. J. SOBEL, *Stochastic Models in Operations Research*, *Vol. II*, McGraw-Hill, New York, 1984.

KLEIN, M. A., "A Theory of the Banking Firm," *J. Money, Credit, and Banking*, 3 (1971), 205–218.

O'HARA, M., "A Dynamic Theory of the Banking Firm," *J. Finance*, 38 (1983), 127–140.

SANTOMERO, A. M., "Modeling the Banking Firm," *J. Money, Credit, and Banking*, 16 (1984), 576–602.

TOPKIS, D. T., "Minimizing a Submodular Function on a Lattice," *Oper. Res.*, 26 (1978), 305–321.

[26]

Vol. 55 (1992), No. 2, pp. 123–150

Journal of Economics
Zeitschrift für Nationalökonomie

A Strategic Market Game with a Mutual Bank with Fractional Reserves and Redemption in Gold
(A Continuum of Traders)

By

M. Shubik and **D. P. Tsomocos,** New Haven, CT, USA*

(Received February 18, 1991; revised version received January 30, 1992)

We utilize the strategic market game approach to analyze the role and function of a mutual bank with variable fractional reserves, redemption in gold, and endogenous interest rate formation. We specify the conditions of enough money and its distribution. Using the continuum of traders model, we show existence and optimality for the case of *no* bankruptcy as well as for the case in which there exists the potentiality of bankruptcy. Finally, we analyze the relationship of the gearing ratio and the bankruptcy penalty with respect to the resulting equilibrium allocations.

> On these principles it will be seen that it is not necessary that paper money should be payable in specie to secure its value; it is only necessary that its quantity should be regulated according to the value of the metal which is declared to be standard.
>
> Ricardo (1965, p. 239)

> Experience, however shows that neither a state nor a bank ever have had the unrestricted power of issuing paper money without abusing that power; in all states therefore, the issue of paper money ought to be under some check and control; and none seems so proper for that purpose as that of subjecting the issuers of paper money to the obligation of paying their notes either in gold coin or bullion.
>
> Ricardo (1965, p. 241)

---

* The authors wish to acknowledge the help of discussions held with Pradeep Dubey, John Geanakoplos, Jingang Zhao and the comments of two anonymous referees.

1. The Financing of Trade

In a one period economy the only need for a money or means of exchange is to finance the mechanism of short term trade. In two previous papers, examining exchange with a single commodity money (Shubik and Wilson, 1977; Shubik, 1990) it was observed that three conditions prevail concerning the amount of gold or other commodity money used as a means of exchange. They are: (1) Enough money, well distributed; (2) enough money, poorly distributed; and (3) not enough money.

(1) The strategic market game Γ where individual i has initial endowment $a^i \in \mathbb{R}_+^{m+1}$ and utility function $\varphi^i : \mathbb{R}_+^{m+1} \to \mathbb{R}$, will have enough money that is well distributed with respect to a Walrasian Equilibrium W.E. (x,p)[1] if:

$$\sum_{j=1}^{m} p_j \max[0, (x_j^i - a_j^i)] \leq a_{m+1}^i$$

for all i, where p_j are the prices at the W.E. (x,p) and x_j^i is the final holding of good j by individual i at W.E. (x,p).[2] There are $m+1$ goods and the $m+1^{\text{st}}$ is the money.

(2) The game Γ will have enough money that is badly distributed with respect to a W.E. (x,p) if:

$$\sum_{i=1}^{n} \sum_{j=1}^{m} p_j \max[0, (x_j^i - a_j^i)] \leq \sum_{i=1}^{n} a_{m+1}^i ,$$

and for some i

$$\sum_{j=1}^{m} p_j \max[0, (x_j^i - a_j^i)] > a_{m+1}^i .$$

---

[1] For a definition of W.E., see Sect. 2.1. Moreover, W.E. (x,p) is the associated Walrasian Equilibrium, of a finite number of potential equilibria, of the game Γ. The formal description of Γ is given in Sect. 2.1.

[2] In the game price is given by $\sum b_j^i / \sum q_j^i$. It is also important to note that the inequalities are specified for the instance where the individuals are assumed to either buy or sell; buying *and* selling in the same market would call for more money.

(3) The game Γ will not have enough money with respect to a W.E. (x, p) if

$$\sum_{i=1}^{n}\sum_{j=1}^{m} p_j \max[0, (x_j^i - a_j^i)] > \sum_{i=1}^{n} a_{m+1}^i \, .$$

1.1 From Gold to Fractional Reserves

In the previous papers noted a money market is introduced where individuals can lend gold, but all payments are still made in gold. When there is an absolute shortage of gold needed for transactions a positive rate of interest appears and trade is not Pareto optimal.

In this paper a way to overcome the shortage of gold is considered by introducing a mutual bank with paper currency redeemable in gold. The amount of paper issued is a multiple of the gold held.

The essential observation made here is that exchange with a full bodied commodity money is trustless trade. The noncooperative equilibria are self policing, all exchanges give at least value for value. In contrast exchange with fiat, no matter how it is modelled requires a specification of how or why the fiat is trusted. At one extreme we can adopt the chartalist position of Knapp (1905) which argues that the acceptance of money is a pure example of the force of the state. Another way is to examine only an infinite horizon economy with sufficiently high expectations to support the use of money (Grandmont and Younes, 1972; Grandmont, 1983); another way is to invoke custom and trust in the actions of the state. It is suggested here that all of the above are probably in part true and different economies show different inclinations to depend more or less on one over the other. The basic contrast between fully valued coin and pure paper, from the point of view of game theory is that one requires no enforcement rules and the other requires trust in or enforcement by government and banks.

In this paper we observe that there is a continuous class of models which can be defined connecting trade with enough gold, all the way to trade with paper only via the use of fractional reserves introducing progressively smaller percentages of fractional reserves. Historically, since the introduction of coinage by the Lydians around 600 B.C. (Vicker, 1975) the trend has been from fully valued money to debased or symbolic coinage, then to fractional reserve backed paper and to paper backed by nothing more than the faith in the state and, or compliance with its laws.

An economy in which all pay cash, from individual endowments, does not require failure to pay, default or bankruptcy laws. One with any form of unsecured credit, in general, must have these rules specified if the system can reach a state where it is not possible to honor a commitment. An economy which uses paper money requires that its value be supported in some manner. A way to model this support is to have in the rules that it be redeemed in some manner or the other for an item of value. Furthermore, if money is owed and not paid back, the individual defaulting should suffer a loss. The first is achieved by having a fractional reserve ratio between the paper and the amount of gold in the bank and the second calls for a default penalty [which could be economic such as garnishing or capturing assets (see Dubey and Shubik, 1988) or uneconomic, such as debtor's prison or even death or slavery (see the ancient Roman Laws of the Twelve Tables, 450 B.C., or the criminal code of Draco, 623 B.C.)].

When there is enough gold, the default penalty need only be as harsh as the marginal utility of gold as can be calculated at any associated W.E. When a reserve ratio requiring less than one unit of gold per gold certificate is used, then the default penalty must be higher in order to prevent strategic bankruptcy. In essence the closer the reserve ratio is to one between a gold certificate and the unit of gold, the less there is need for trust in the system. But as long as any one is borrowing *even if* trade were in gold coin the need for a default law would still be required.

2. A Mutual Bank with Redemption in Gold

In constructing a playable exchange game making use of a mutual bank the granting of credit and the accumulation of debt can be centralized.

The phrase "playable game" means a game which fully defines a process, up to the point that one could actually play it with a group of individuals. This apparently naive requirement will convince anyone who tries to construct a playable game with borrowing that some form of default penalty is not merely an institutional curiosity but a logical and technological necessity. Instead of having to be concerned with individual indebtedness between i and j, the bank is a full intermediary. If i has a line of credit from the bank and pays j, i's default is against the bank, not j. The financial damage to j by i's default may come in an aggregated form through j's part ownership of the bank.

A mutual bank exists to facilitate trade and thus is not necessarily a profit maximizer. As it is a mutual bank, if it incurs losses these may be prorated among the depositors. The definition used here of a mutual bank is that all of its customers are simultaneously owners where their ownership claims are reflected in the amount of gold they have deposited. The bank itself is a strategic dummy, i.e., it is required to loan up to the amount permitted by its gearing ratio. The gearing ratio multiplies the amount of gold by some factor permitting the bank to issue more gold certificates than it has gold. The effect of this ratio is that it allows the "paper gold" to expend until there is enough to achieve some W.E. of the underlying exchange economy as a Type Symmetric Nash Equilibrium (T.S.N.E.)[3] of the noncooperative game.[4]

Figure 1 shows the extensive form of the game. At the first move individuals $P_1, \ldots, P_n$ may deposit gold in the mutual bank. The bank ownership is assigned in proportion to the amount deposited. The amount of gold deposited is multiplied by a gearing ratio (given as a parameter of the system) and this determines the total amount of credit offered to all borrowers. At the second move all who wish to borrow bid for bank money. At the third move they exchange goods for bank money. We avoid the complex functional forms of strategies which may arise when traders have perfect information by assuming a continuum of traders, simultaneous moves at all stages, and a minimum of information.[5]

2.1 Description of the Model

There is a continuum of traders $T = [0, n)$ which consists of n types of traders, $T_i = [i-1, i)$ where $a^t = a^i$, $\phi^t = \phi^i$ for $t \in T_i$. The endowments of traders are $a^i = (a^i_1, \ldots, a^i_m, a^i_{m+1})$ where $a^i \in \mathbb{R}^{m+1}_+$. For $1 \leq i \leq n$ the utility functions are, $\phi^i : \mathbb{R}^{m+1}_+ \to \mathbb{R}$. We assume

(i) $\sum_{i=1}^{n} a^i \gg 0$;

(ii) $a^i \neq 0$, for all $i = 1, \ldots, n$;

(iii) ϕ^1 is continuous, concave, and strongly monotonic for all $i = 1, \ldots, n$.

---

[3] The definition of Type Symmetric Nash Equilibrium (T.S.N.E.) is given in Sect. 2.1.

[4] It must be noted that there are some problems in trying to extend this model to a multiperiod game where the money supply must be varied from period to period and hence one cannot merely use the device of letting the bank passively lend all the paper it is permitted to print.

[5] There are also several important and interesting problems arising from nonsymmetry of information which can be considered, as has been noted by Dubey et al. (1987), but these are not dealt with in this discussion.

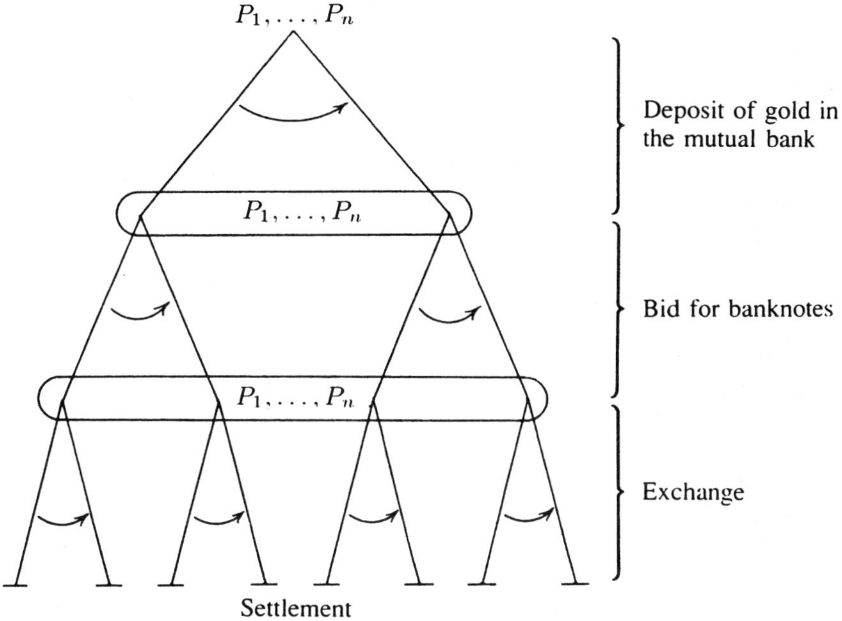

Fig. 1. Trade with a mutual bank
(Note that the labelling $P_1, \ldots, P_n$ indicates that all agents move simultaneously.)

The strategy set of trader i is of the form,

$$S^i = \{u^i, v^i, b_1^i, q_1^i, \ldots, b_m^i, q_m^i, b_{m+1}^i, q_{m+1}^i\} \ , \tag{1}$$

where the arguments appearing in S^i are constrained as follows,

$$\left.\begin{array}{ll} \text{(a)} & 0 \leq u^i \leq a_{m+1}^i \ ; \\ \text{(b)} & 0 \leq v^i \leq k \sum_{i=1}^{n} a_{m+1}^i, \quad k \geq 1 \ ; \\ \text{(c)} & \sum_{j=1}^{m+1} b_j^i \leq 1, \quad \text{for all } j = 1, \ldots, m \ ; \\ \text{(d)} & q_j^i \leq a_j^i \quad \text{for all } j = 1, \ldots, m \ ; \\ \text{(e)} & q_{m+1}^i + u^i \leq a_{m+1}^i \ , \end{array}\right\} \tag{2}$$

with

$u^i \equiv$ amount of gold deposited in the bank;
$k \equiv$ the gearing ratio for bank note issue;
$v^i \equiv$ the amount of I.O.U. notes bid by i for banknotes;
$q^i_j \equiv$ amount of good j offered for sale;
$b^i_j \equiv$ percentage of banknotes bid on good j.

It should be noted that one may naturally expect that traders would not offer to repay more money than the total money supply in the economy. Also the presence of a default or bankruptcy penalty provides a motivation against unreasonably high bidding, but the physical bound is sufficient.

A trader type i's share of ownership of the mutual bank is specified by:

$$\theta^i = \frac{\int_{T_i} u^t \, dt}{\int_T u^t \, dt} = \frac{u^i}{\int_T u^t \, dt} \,. \tag{3}$$

At the second stage of the game the bank, which is a strategic dummy (that is why we suppress its move in the extensive form of Fig. 1) offers to issue banknotes up to the level of $M = k\bar{u}$, where $\bar{u} = \sum_{i=1}^{n} u^i$ and k is exogenously set.

An endogenous interest rate is formed and is given by:

$$1 + \rho = \frac{\int_T v^t \, dt}{k \int_T u^t \, dt} = \frac{\int_T v^t \, dt}{M} \,. \tag{4}$$

The last stage of the game Γ consists of the offer-for-sale model as in Dubey and Shubik (1978). In order to study the simplest set of strategies we may imagine that all agents are completely uninformed at each stage, thus strategies remain a set of numbers rather than complex functions of information. This can be done by having players name ratios for the amount that they intend to spend in different markets as in (2) (c).

The actual amount of banknotes in possession of i after the second move is given by:

$$w^i = \frac{v^i M}{\int_T v^t \, dt} \,. \tag{5}$$

Using the restrictions (2) on S^i, we can specify the strategy set as,

$$\sum^i = \{s \in S^i \text{ subject to (2)}\}. \tag{6}$$

It is easily seen that $\sum^i$ is compact and convex. We will employ the Dubey–Shubik offer-for-sale price formation mechanism. Thus,

$$p_j = \begin{cases} \dfrac{\int_T b_j^t w^t \, dt}{\int_T q_j^t \, dt} & \text{if } \int_T b_j^t w^t \, dt > 0 \text{ and } \int_T q_j^t \, dt > 0 \, , \\ 0 & \text{otherwise} \, . \end{cases} \tag{7}$$

The final allocations of the first m goods is given by:

$$x_j^i = a_j^i - q_j^i + \frac{b_j^i w^i}{p_j} \quad \forall j = 1, \ldots, m. \tag{8}$$

The final amount of gold which also indicates the trader's debt position at the end of exchange is specified by:

$$\begin{aligned} Gi = {} & \{a_{m+1}^i - u^i\} + \left\{\frac{v^i}{1+\rho} - \hat{v}^i - p_{m+1} g_L^i\right\} + \\ & + \left\{\sum_{j=1}^{m+1} p_j q_j^i - \sum_{j=1}^{m+1} b_j^i w^i\right\} + \\ & + \theta^i \left\{M - \frac{\sum_{i=1}^n v^i}{1+\rho^*} + \sum_{i=1}^n \hat{v}^i + p_{m+1} \sum_{i=1}^n g_L^i\right\} , \end{aligned} \tag{9}$$

where

$\hat{v}^i$ ≡ the part of repayment of his loan made with banknotes;
g_L^i ≡ the part of repayment of his loan made with gold;
$p_{m+1} g_L^i$ ≡ par value of gold used for repayment;
ρ ≡ *ex ante* interest rate;
ρ^* ≡ *ex post* interest rate.

The first two terms are the amount of gold retained. The next three are the size of loan and the size of repayment. These terms which will be denoted by D^i specify the trader's debt. The next two terms are the balance from buying and selling commodities. Finally, the last terms represent the liquidation of the mutual bank, the payment of profits, and the return of the original deposit to a depositor in the mutual bank. Note that $\hat{v}^i$ and g_L^i are functions of the remaining terms of expression (9).

Since we allow repayment of loan either using remaining banknotes after exchange or by gold at one-to-one basis (i.e., $p_{m+1} = 1$), we can potentially encounter six different cases depending on whether a trader bankrupts or not. In Appendix 1, we present the cases of the general formulation of (9) and present an explicit calculation for the case where there is no bankruptcy and all loans are paid back only by banknotes. It is important to note that the specification of repayment of loan either by banknotes or by gold at one-to-one basis is what bounds the interest rate from above and allows us to show optimality.

Several items must be noted. If we restrict the use of gold so that it is not a direct means of payment, but can be used only as bank reserves or for consumption purposes then we can have a market for gold where it is bought for banknotes. If gold can be used for payments in trade then the money supply in the hands of the public includes both the banknotes issued against the gold deposited and the gold held by the public.

The interest rate formed in the second stage and denoted by ρ is the contractual (or *ex ante*) rate which is used to determine how much any borrower has agreed to pay. It may be that a borrower is unable to meet his obligation and defaults. In this case the depositors of the mutual will receive less than promised from the mutual. There will be an effective rate ρ^* where $\rho^* < \rho$ which is the *ex post* rate of interest taking into account the default of various traders.

The calculation of ρ^* is complicated by a possible "domino effect" in default among traders who are both depositors and debtors. If individual i goes into default he lowers the dividend to be received by j, but this could cause j to default. In Appendices 2 and 3 we provide flow diagrams for the calculation of the *ex post* interest rate for both the case of finite number of traders and the continuum case we analyze in this paper.

In our model and its main result we consider equilibria where there is in fact no strategic bankruptcy, hence *ex post* and *ex ante* interest rates match, but even so, for the game to be well defined the possibility that failure may occur must be covered. In essence one must describe an iterative process (as is done in the appendices) where all immediate failures are added together. These are removed from the loan payback and result in a lower *ex post* interest rate and hence in a lower payback to all owners of the mutual. If all are still solvent, even given the diminished payback, matters end at that point. If some are converted into debtors a recalculation must take place and a still lower *ex post* rate of interest results. Sooner or later this process converges and the final *ex post* rate and bankruptcies are determined.

The final payoff[6] to trader type i is given by:

$$\Pi^i(s) = \phi^i\big(x^i(s)\big) - \mu D^i_+(s) \ , \tag{10}$$

where $D^i_+(s) = \max[0, D^i(s)]$ and μ is the bankruptcy penalty in case of default.

In the next section we will show existence and optimality for the case where there is no bankruptcy (i.e., $D^i \leq 0$)[7] and for the case where the potentiality of bankruptcy is present. Before proving these results we state some definitions.

An allocation (x, p) of an exchange economy $E(n, \phi^i, a^i)$ where $x^i \in \mathbb{R}^{m+1}_+$, $p \in \mathbb{R}^{m+1}_{++}$ is a *Walras Equilibrium* (W.E.) if

(i) $\displaystyle\sum_{i=1}^n x^i = \sum_{i=1}^n a^i$

and

(ii) $x^i = \arg\max\{\phi^i(y) : y \in \mathbb{R}^{m+1}_+,\ py = pa^i\} \quad \forall i = 1, \ldots, n.$

A *Nash Equilibrium* (N.E.) of $\Gamma(n, \phi^i, a^i, s^i)$ is an integrable s such that $\forall t \in T$ and $\alpha^t \in \sum = \underset{i=1}{\overset{n}{X}} \sum^i$,

$$\Pi^t(s/\alpha^t) \leq \Pi^t(s) \ ,$$

where (s/α^t) is s with s^t replaced by α^t.

If $s^t = s^{t'}$ for every $t, t' \in T_i$ and $\forall i = 1, \ldots, n$, s is a type symmetric play and that *N.E. is Type-Symmetric* (T.S.N.E.).

An *Active T.S.N.E.* of $\Gamma(n, \phi^i, a^i, s^i)$ is a T.S.N.E. with $\rho > -1$ and $p_j > 0 \ \forall j = 1, \ldots, m+1$.

---

[6] When we introduce the potentiality of bankruptcy, the payoff becomes a function of not only utility but also of debt $D^i \in \mathbb{R}$.

[7] Bankruptcy can be avoided by selecting a sufficiently harsh penalty so that nobody will be induced to strategically bankrupt. This can be achieved by setting $\mu = \max(\lambda_1, \ldots, \lambda_n)$ where the λ's reflect the marginal utility of an extra unit of gold at equilibrium. For more on this see Shubik (1986) and Shubik and Wilson (1977).

2.2 Existence and Optimality[8]

Theorem 1: The strategic market game $\Gamma(n, \phi^i, a^i, s^i)$ with a continuum of traders of n types and a mutual bank for any non-zero amount of gold and any gearing ratio $k \geq 1$ and a sufficiently harsh bankruptcy penalty μ which results into no bankruptcy has an active T.S.N.E.

Proof. Step 0:
(i) Let an outside agency place a fixed bid and a fixed offer $\epsilon > 0$ in each trading post as well as in the mutual bank. So, let the modified strategy space be,

$$\left(\sum^i\right)^\epsilon = \{s \in \sum^i : s \geq \epsilon,\ \forall s \in \sum^i\}\ .$$

Note that we do not need to place upper bounds because by construction we have naturally imposed bounds on the arguments of the strategy sets.
(ii) Γ^ϵ is the ϵ-modified game where we restricted the strategies of the traders to $\left(\sum^i\right)^\epsilon$. Then for an n-tuple s, the map from strategies to outcomes, $(s/\alpha^t) \rightarrow (x^t, D^t)$ where $D^t = 0$ since utilities are locally non-satiated, is a linear function where $\alpha^t \in \left(\sum^i\right)^\epsilon$ since α^t cannot affect either p or ρ in the continuum. Thus $\Pi^t(s/\alpha^t) = \phi^t\left(x^t(s)\right)$ is continuous and concave. Also,

$$\left(BR^i\right)^\epsilon : \sum^\epsilon \rightarrow \left(\sum^i\right)^\epsilon \text{ is convex ,}$$

where

$$\left(BR^i\right)^\epsilon(s) = \arg\max\left\{\Pi^t(s/\alpha^t) : \alpha^t \in \sum^\epsilon,\ t \in T_i\right\}$$

and

$$\sum^\epsilon = \underset{i \in T}{X} \left(\sum^i\right)^\epsilon\ .$$

Hence $\left(BR^i\right)^\epsilon$ is compact valued and upper-semicontinuous, by the "Maximum Theorem" (Berge, 1963, p. 116).

Step 1: Γ^ϵ has at least one T.S.N.E.
The correspondence,

$$BR^\epsilon(s) = \underset{t \in T}{X} \left(BR^i\right)^\epsilon : \sum^\epsilon \rightarrow \sum^\epsilon\ ,$$

[8] The arguments used here are as in Dubey and Geanakoplos (1989a, b).

where s is a type symmetric play, satisfies all the conditions of Kakutani fixed point theorem, and therefore admits a fixed point $BR^{\epsilon}(\hat{s}^{\epsilon}) \ni \hat{s}^{\epsilon}$. It is easily verified that $\hat{s}^{\epsilon}$ is a T.S.N.E. Thereafter $\hat{s}^{\epsilon}, \hat{p}^{\epsilon}$ indicates a T.S.N.E.

Step 2: $\rho \geq 0$ at any T.S.N.E.

If $\rho < 0$ then $D^t < 0 \; \forall t \in T_i$ and some trader type i. Then t could bid for more of a commodity without ending insolvent and increasing his utility by monotonicity. Therefore a contradiction.

Step 3: There exists $c > 0$ such that $\forall \epsilon$ sufficiently small $\hat{p}^{\epsilon} > ce$, where $e = (1, \ldots, 1)$ of suitable dimensionality.

Suppose $\lim_{\epsilon \to 0} p_j^{\epsilon} \to 0$ for some j. Recall that at equilibrium,

$$p_j = \frac{\int_T b_j^t w^t \, \mathrm{d}t + \epsilon}{\left(\int_T q_j^{\epsilon} \, \mathrm{d}t + \epsilon\right) p_{m+1}} \, ,$$

where $p_{m+1} = 1$. If for a commodity $\ell \neq j$, $p_{\ell}^{\epsilon} \to 0$ then trader t could have made a bid smaller by Δ or otherwise if all $p_j^{\epsilon} \to 0 \; \forall j = 1, \ldots, m$, then there definitely exists a trader of type i with $\Delta > 0$ of banknotes. The argument also goes through the gold market where $p_{m+1} = 1$. Thus, this type i could bid Δ amount of money to obtain $\Delta / p_j^{\epsilon} \to \infty$ of j since $p_j^{\epsilon} \to 0$. Let

$$\left\{ \frac{\partial \phi^t}{\partial x_j^t}(y) : y \leq \bar{G} \right\} \, ,$$

where $\bar{G} = k \sum_{i=1}^{n} a_{m+1}^i$.

Since nobody can acquire more than $\bar{G}$ at equilibrium we see that because of strict monotonicity of utilities, the net gain in utility for $t \in T_i$ as a result of his action is at least

$$\left\{ \frac{\partial \phi^t / \partial x_j^t(y)}{p_j^{\epsilon}} - 1 \right\} \Delta$$

for a Δ sufficiently small. This quantity in parentheses has to be non-positive since we are at equilibrium. So,

$$\frac{\partial \phi^t}{\partial x_j^t}(y) < p_j^{\epsilon} \, .$$

We will construct the T.S.N.E. $\hat{s}$ and the corresponding vector of prices $\hat{p}$ of Γ as a limit of Γ^ϵ and its equilibrium. So, we take a sequence of ϵ's which defines a limiting $\hat{s}$.

Step 4:

Select a sequence of ϵ and subsequences of subsequences so that:

(i) $\lim_{\epsilon\to 0} \hat{s}^\epsilon = s$. This is possible since strategy spaces are *bounded.*

(ii) Choose a subsequence so that $E^\epsilon = \sum_{j=1}^{m} p_j^\epsilon$ is convergent to either $\Psi < \infty$ or ∞.

(iii) $p_j^\epsilon / \sum_{j=1}^{m} p_j^\epsilon$ converges $\forall j = 1, \ldots, m$.

Also, – if $E^\epsilon < \infty$ define $\lim_{\epsilon\to 0} p_j^\epsilon = p_j$ and $p_{m+1} = 1$,

– if $E^\epsilon = \infty$ define $\lim_{\epsilon\to 0}\left[(p_j/E)^\epsilon\right] = p_j$ and $p_{m+1} = 0$.

Step 5: If $E^\epsilon \to \infty$

(i) $p_j^\epsilon \to \infty$, $\forall j = 1, \ldots, m$.

Clearly $p_j^\epsilon \to \infty$ for some j. Let p_ℓ^ϵ be bounded for some $\ell \neq j$. Take trader $t \in T_i$ who has a positive amount of j. He can sell Δ to obtain an enormous amount of banknotes B so as to bid and acquire $B/p_\ell^\epsilon \gg 0$. So he has a *positive* net gain in utility. So, we get a contradiction that we were at a T.S.N.E. In addition, bankruptcy could be problematically prevented since $B \gg k \sum_{i=1}^{n} u^i$.

(ii) $q_j^\epsilon \to 0$ and $\dfrac{\left(b_j^t\right)^\epsilon w^t}{p_j^\epsilon} \epsilon \to 0 \ \forall i = 1, \ldots, n$

$$p_j^\epsilon \leq \frac{k \sum_{i=1}^{n} a_{m+1}^i + \epsilon}{\sum_{i=1}^{n} \left(q_j^i\right)^\epsilon + \epsilon} \quad \text{and} \quad \sum_{j=1}^{m+1} \left(b_j^i\right)^\epsilon w^i \leq k \sum_{i=1}^{n} a_{m+1}^i \,.$$

Thus, by (i) $\left(q_j^t\right)^\epsilon \to 0$ and $\dfrac{\left(b_j^t\right)^\epsilon w^t}{p_j} \to 0$.

Step 6: If $E^\epsilon < \infty$ and $p_{m+1} = 1$ then by Step 3, $p^\epsilon > 0 \ \forall j = 1, \ldots, m$. Moreover, $\hat{s}$ is a T.S.N.E. It follows from Step 4 that $\lim_{\epsilon\to 0}\left(BR^i\right)^\epsilon = BR^i$. From Step 5 we know that $\hat{s}_i = \lim_{\epsilon\to 0}\left(s^i\right)^\epsilon$ $\forall i$ except when $p_{m+1} = 0$. So, $\hat{s}$ maximizes ϕ^t on Γ and therefore $\hat{s}$ is a T.S.N.E.

Step 7: Any limit s^ϵ of T.S.N.E. for Γ^ϵ is an active T.S.N.E.

By Step 2, $\rho > -1$ and by Steps 5 and 6, $p_j > 0$.

Q.E.D.

Theorem 2: For a sufficiently high gearing ratio k and a sufficiently harsh penalty μ which entails *no* bankruptcy, the strategic market game

$\Gamma(n, \phi^i, a^i, s^i)$ will produce $\rho = 0$ and give the same relative prices and distribution as the Walrasian equilibrium of the associated exchange economy $E(n, \phi^i, a^i)$.

Proof. Step 1: $\rho \to 0$ as $k \to \infty$ at any T.S.N.E.

By Step 2 of Theorem 1, $\rho \geq 0$. Now let $\rho > 0$. From (4),

$$\int_T v^t \, \mathrm{d}t > k \int_T u^t \, \mathrm{d}t > 0 \,. \tag{*}$$

Since $p_{m+1} = 1$ and $a^i_{m+1} < \infty \; \forall i = 1, \ldots, n$, pick $\bar{a}^i = \max[a^1_{m+1}, \ldots, a^n_{m+1}]$ and k such that,

$$\frac{k \int_T u^t \, \mathrm{d}t}{n} > p_{m+1} \bar{a}^i \,. \tag{**}$$

This is possible by the upper bound on u's by (2) since k is an unbounded exogenous parameter. By (*), (**), for trader i $D^i > 0$. Similarly, for the other traders. So, contradiction. Thus, there exists k high enough for which $\int_T u^t = k \int_T v^t$.

Step 2: (x, p) arising from the optimal play $\hat{s}$ is a W.E.

By the no bankruptcy condition and Step 1, $px^i \geq pa^i \; \forall i = 1, \ldots, n$. So, by utility maximization and local non-satiation $\sum_{i=1}^n x^i = \sum_{i=1}^n a^i$ which implies $px^i = pa^i \; \forall i = 1, \ldots, n$.

Q.E.D.

Theorem 3: The strategic market game $\Gamma(n, \phi^i, a^i, s^i)$ with a continuum of traders of n types and a mutual bank for any non-zero amount of gold and any gearing ratio $k \geq 1$ and bankruptcy penalty μ has an active T.S.N.E.

Proof. Step 0:

(i) Now since there exists the potentiality of bankruptcy the effective payoff function is

$$\Pi^i(s) = \phi^i\big(x^i(s)\big) - \mu \max[0, D^i(s)] \,.$$

(ii) From Step 0 of Theorem 1, we know that $\phi^i\big(x^i(s)\big)$ is concave. Also, 0 is trivially concave. It is easily seen that D^i is concave (linear) as well. So, $\Pi^i(s)$ is concave and continuous. Hence

$$\left(BR^i\right)^\epsilon : \Sigma^\epsilon \to \left(\Sigma^i\right)^\epsilon \quad \text{is convex} \,.$$

So, $\left(BR^i\right)^\epsilon$ is compact valued and upper-semicontinuous, by the "Maximum Theorem" (Berge, 1963, p. 116).

The rest of the formulation is as in Step 0 of Theorem 1.

Step 1: Γ^ϵ has at least one T.S.N.E.

As in Step 1 of Theorem 1.

Step 2: $\rho \geq 0$ at any T.S.N.E.

As in Step 2 of Theorem 1.

Step 3: There exists $c > 0$ such that $\forall \epsilon > 0$ sufficiently small $\hat{p}^\epsilon > ce$, where $e = (1, \ldots, 1)$ of suitable dimensionality.

Suppose $\hat{p}_j^\epsilon \to 0$ as $\epsilon \to 0$ for some j. Then choose player t. He could borrow Δ of banknotes to buy $\Delta / p_j^\epsilon \to \infty$ since $p_j^\epsilon \to 0$. This transaction would incur him a bankruptcy penalty $\mu\Delta < \infty$ since both Δ and μ are bounded from above. Now let

$$\left\{ \frac{\partial \phi^t}{\partial x_j^t}(y) : y \leq \bar{G} \right\} .$$

Since nobody can get more than $\bar{G}$ at equilibrium, we see that because of strict monotonicity of preferences, the net gain in utility of trader t because of his action is at least

$$\left\{ \frac{\partial \phi^t / \partial x_j^t(y)}{p_j^\epsilon} - \mu\rho \right\} \Delta ,$$

for a Δ sufficiently small. Note that by Step 2, $\rho \geq 0$. In addition ρ is bounded from above. Since $k \int_T u^t(1+\rho)\,\mathrm{d}t$ has to be repayed at the end of trade to the mutual bank and consequently some trader t has to pay a bankruptcy penalty equal to at least $\mu \left(\frac{k \int_T u^t \rho \,\mathrm{d}t}{n} \right)$. Therefore, if $\rho \to \infty$ then i could have been less worse off than he is now at our assumed equilibrium either by bidding less or if he had deposited less gold at the mutual bank and therefore deriving excess utility from gold directly as a commodity at the end.

The expression in the brackets has to be non-positive,

$$p_j^\epsilon \geq \frac{\partial \phi^t / \partial x_j^t(y)}{\mu\rho} .$$

We will construct again the T.S.N.E. and the corresponding ρ of Γ as a limit of Γ^ϵ and its associated equilibrium. For this purpose take a sequence ϵ which defines a limiting pair $\hat{s}$.

Step 4: (i)–(iii) as of Step 4 of Theorem 1.

Step 5: If $E^\epsilon \to \infty$ then
(i) $p_j^\epsilon \to \infty \ \forall j = 1, \ldots, m$.

Clearly $p_j^\epsilon \to \infty$ for some j. Let p_ℓ^ϵ be bounded for some $\ell \neq j$. Consider a trader who has a positive amount of j. Then this trader is in a position to borrow a small amount Δ of banknotes and offer for sale $(1+\rho)\Delta/p_j$ of commodity j and afterwards buy $\Delta \left(b_j^t\right)^\epsilon w^t/p_\ell^\epsilon$ where in this case $\left(b_\ell^t\right)^\epsilon = 1$. Consequently he can return $(1+\rho)\Delta$ banknotes to the mutual bank. Therefore, the net gain in utility is positive and therefore we get a contradiction.
(ii) $(q_j)^\epsilon \to 0$ and $\left(b_j^t\right)^\epsilon w^t/p_j^\epsilon \to 0 \ \forall t \in T_i$ and $i = 1, \ldots, n$:

$$p_j^\epsilon \leq \frac{k\sum_{i=1}^n a_{m+1}^i + \epsilon}{\sum_{i=1}^n \left(q_j^i\right)^\epsilon + \epsilon} \quad \text{and} \quad \sum_{j=1}^{m+1} \left(b_j^i\right)^\epsilon w^i \leq k \sum_{i=1}^n a_{m+1}^i \,.$$

So by (i), $\left(q_j^t\right)^\epsilon \to 0$ and $\left(b_j^t\right)^\epsilon w^t/p_j^\epsilon \to 0$.

Step 6: If $E^\epsilon < \infty$ and $p_{m+1} = 1$ then by Step 3, $p^\epsilon > 0 \ \forall j = 1, \ldots, m$. In addition $\hat{s}$ is a T.S.N.E. Same argument as in Step 6 of Theorem 1.

Step 7: Any limit s^ϵ of T.S.N.E. for Γ^ϵ is an active T.S.N.E.

As in Step 7 of Theorem 1.

Q.E.D.

Theorem 4: For a sufficiently high gearing ratio k and a sufficiently harsh default penalty μ, the strategic market game $\Gamma(n, \phi^i, a^i, s^i)$ will have T.S.N.E.'s which do not involve bankruptcy, produce an endogenous interest rate $\rho = 0$, and give the same relative prices and distribution as the Walrasian equilibrium of the associated exchange economy $E(n, \phi^i, a^i)$.

Proof: It only suffices that for a high enough k at any T.S.N.E. $\rho \to 0$ and furthermore that no agent bankrupts. Then we are back at the argument of Theorem 2.

Step 1: No agent bankrupts when $k \to \infty$ if $E^\epsilon = \infty$.

Recall that D^t is the debt of a trader t. If $D^t > 0$ then trader t could have always borrowed Δ less amount of banknotes and *not* bid them for a commodity j whose $(p_j^\epsilon) \to \infty$ and finally obtain

$$\left(\hat{x}_j^t\right)^\epsilon = \left(x_j^t\right)^\epsilon - \frac{\Delta}{p_j^\epsilon} .$$

Therefore, he is saving the bankruptcy penalty $(1+\rho)\Delta\mu$ without reducing his utility for a sufficiently small Δ as $p_j^\epsilon \to \infty$.

Step 2: $\rho \to 0$ as $k \to \infty$ at any T.S.N.E. Also, no agent bankrupts in general.

Let $\rho > 0$ along a subsequence of ϵ. Then consider a type-symmetric play. Then for some trader t, $D^t > \rho k \int_T u^t/n$ if bankruptcy has occurred. If not then go back to Step 1 of Theorem 2. Now for the case of bankruptcy, if

$$-\frac{\mu\rho k \int_T u^t}{n} + \phi^t\big(\sum_i a^i\big) < \min_{1 \le t' \le n} \phi^{t'}(0) - \frac{m+1}{M}$$

for sufficiently high k, then each $t \in T_i$ could have improved upon this allocation by borrowing $(m+1)/M$ less banknotes and selecting $s' = \{\ldots;\ 1/M, \ldots, 1/M;\ \ldots\}$. Therefore, we arrived at a contradiction.

Finally, note that easily one can show the converse of the theorem along the same lines of Dubey and Geanakoplos (1989a).

Q.E.D.

2.3 An Example

In a previous paper an example of a loan market for gold was given. We extend this example here to a bank with fractional gold reserves. The initial endowment of traders of type 1 is assumed to be $(2, 0, 0)$ and that of traders of type 2 is $(0, 2, A)$ where the third commodity is gold and can be used as money. A strategy by a trader of type 1 is of the form (v^1, b_j^1, q_j^1) where $j = 1, 2$ and v^1 is the amount of I.O.U. notes bid to obtain banknotes. A strategy by a trader of type 2 is of the form (u, v^2, b_j^2, q_j^2) where $j = 1, 2$ and u is the amount of gold deposited in the mutual bank and v^2 is the amount of I.O.U. notes bid to obtain banknotes.

In this example because of the special structure and the assumption that all traders are of insignificant size we can simplify the notation even further. We assume that the price of money is set at $p_3 = 1$. The other prices are denoted by p_1 and p_2. Traders of the first type only will borrow and those of the second type only will form the mutual bank and then borrow from it. Because of the initial endowments only those of the first type can sell the first good and only of those of the second type can sell the second good.

We assume that there is a type symmetric noncooperative equilibrium and that setting $\mu = 10$ provides a sufficiently harsh penalty to avoid strategic bankruptcy. We furthermore assume that traders borrow only to spend on buying goods (i.e., they do not borrow to hoard). Given these simplifications we can write the payoff function as follows[9]:

$$\Pi_1 = \sqrt{(2-q_1^1)(v^1/(1+\rho)p_2)} + \left(p_1 q_1^1 - v^1\right)^+ + + \mu\left(p_1 q_1^1 - v^1\right)^- \tag{11}$$

$$\Pi_2 = \sqrt{(2-q_2^2)(v^2/(1+\rho)p_1)} + \left(p_2 q_2^2 - v^2 + k\rho A\right)^+ + + \mu\left(p_2 q_2^1 - v^2 + k\rho A\right)^- , \tag{12}$$

where $p_1 = v^2/(1+\rho)q_1^1$ and $p_2 = v^1/(1+\rho)q_2^2$. Also $1+\rho = (v^1+v^2)/ku$ and $v^1 = p_1 q_1^1$ and $v^2 = p_2 q_2^2$.

Treating p_1 and p_2 as parameters and making the observation that if $k > 1$ and the rate of interest is greater than zero all gold will go into the bank, the first order optimization conditions give:

$$\tfrac{1}{2}\sqrt{v^1/(1+\rho)p_2(2-q_1^1)} = p_1 , \quad \text{and} \tag{13}$$

$$\tfrac{1}{2}\sqrt{(2-q_1^1)/(1+\rho)p_2 v^1} = 1 ; \tag{14}$$

$$\tfrac{1}{2}\sqrt{v^2/(1+\rho)p_1(2-q_2^2)} = p_2 , \quad \text{and} \tag{15}$$

$$\tfrac{1}{2}\sqrt{(2-q_2^2)/(1+\rho)p_1 v^2} = 1 . \tag{16}$$

We can check that for $A = .5$ the following values result (shown in Table 1).

---

[9] The notation $()^+$ or $()^-$ stands for the positive or negative part of the item in the brackets, or zero.

Table 1

k	ρ	q_1^1	q_2^2	v^2	v^1	p_1	p_2
4	0	1	1	1	1	1	1
2	0	1	1	.5	.5	.5	.5
1.8	.1096	1	.8967	.4734	.4266	.4733	.4759
1	.9091	1	.4306	.3280	.1719	.3281	.3992

3. Sensitivity Analysis

3.1 Inflation and the Money Supply

In Table 1 the first entry has $k = 4$, thus the supply of banknotes is $M = 2$. At equilibrium the rate of interest cannot be negative if loans are available because an individual could arbitrage by borrowing, thus the bids will be $v^1 = v^2 = 1$. However price will no longer be uniquely determined. Any set of prices $.5 \leq p_1 = p_2 \leq 1$ is in equilibrium. If prices for the two goods are between .5 and 1 individuals hoard the extra banknotes. If we assume that gold although used for reserves is not a means of payment we can have a market for gold in the hands of the public, its price in terms of banknotes is twice that of the two other goods. In essence it has two prices because of the strategic restrictions on it. Its price in trade is determined in the open market and has to bear the appropriate marginal relationship to other goods in competitive trade, but its price in settling loans at the bank is one to one with banknotes.

There is yet another solution here where $u = .25$, thus $M = 1$ and the price level is $p_1 = p_2 = .5$. If the gold remaining in circulation is counted as part of the money supply, then it can be used instead of banknotes in exchange. However it will be either hoarded or traded only in an extra money market between gold and banknotes where its price will be k in terms of banknotes.

Legally the total money supply is 1.25 of which $4/5$ are banknotes and $1/5$ gold. But gold in this model will be no more used as a direct means of payment than $ 20 United States gold coins are used as a means of payment at the local grocery. In the United States, as recently as the 1950s it was not uncommon to be given silver dollars in payment; they are still legal tender and can be used by those who choose to pay with them, but a new market exists between them and paper money and although as a *legal* means of payment they trade at par with a federal reserve note, as a consumer (or potential producer) good they trade at a premium.

3.2 The Role of the Bankruptcy Penalty and the Gearing Ratio

In our model exchange is modelled as a game of strategy in which credit is granted and the medium of exchange is a partially backed paper (i.e., the paper does not enter the utility function).[10] In general it will be feasible for the system to reach a state in which an individual is unable to honor his commitments. It is a logical necessity and not merely an institutional convenience that a default penalty be specified in order to complete the rules of the game so that payoffs can be computed in the case of default.

Penalties can be highly economic (such as the confiscation and sale of goods) or societal such as exile, prison, or loss of civil rights. The details are highly institutional. If, for example we wish to consider garnishing or confiscation of property we must specify what can be removed and how it is to be used to extinguish the loan. For our purposes here there are two items that matter in the one period model. They are that the penalty serves as a disincentive to the borrower to default and that in the case of default the penalty is such that the loss to society as a whole is in some sense minimized. Moreover, the bankruptcy penalty sets a lower bound on prices.

In an economy without exogenous uncertainty and enough money[11] the threat of bankruptcy or default is only strategic, thus if a penalty is harsh enough to prevent bankruptcy it is efficient. But in an economy with exogenous uncertainty the bankruptcy penalty is in essence a public good reflecting society's joint willingness to risk default.[12]

On the other hand, the gearing ratio k controls the money supply in the economy and determines an upper bound on prices. The imposition of a sufficiently harsh bankruptcy penalty and a high enough gearing ratio guarantees efficiency and no default in equilibrium. We also need the upper bound on prices determined by k to be greater than the lower bound on prices determined by μ; otherwise equilibrium price vectors without bankruptcies do *not* exist. Alternatively, we could in the one period model set exogenously the interest rate and allow the money supply to adjust endogenously by the optimizing behavior of the players.

As there is no exogenous uncertainty in the model above, our remarks can be confined to three levels of the penalty. They are when the penalty is so weak that default pays; when it is just strong enough to

---

[10] Except indirectly in terms of its eventual redemption.

[11] With the money distributed or loaned at zero interest.

[12] The nature of the sharing of damage caused by a default is critical (see Dubey et al., 1988).

discourage strategic default, and when it is more than strong enough to discourage default. These three levels of μ should be considered jointly with three levels of k; not enough to provide a sufficient money supply; just enough to provide a sufficient money supply, and more than enough to do so. There are nine cases analyzed below and summarized in Table 2.

Table 2

	$k < k^*$	$k = k^*$	$k > k^*$
$\mu < \mu^*$	(1) Strategic Default/ Deflation $\rho > 0$	(2) Strategic Default/ Price Shift $\rho > 0$	(3) Possibly Strategic Default/Inflation $\rho \geq 0$
$\mu = \mu^*$	(4) Strategic Default/ Deflation $\rho > 0$	(5) No Default $\rho = 0$	(6) No Default/ Inflation $\rho = 0$
$\mu > \mu^*$	(7) Possibly Strategic Default/Deflation $\rho \geq 0$	(8) No Default/ Potential Deflation $\rho = 0$	(9) No Default/ Deflation or Inflation $\rho = 0$

We assume at the minimal gold transaction W.E. (x, p):

$$\sum_{i=1}^{n}\sum_{j=1}^{m} p_j \max[0, (x_j^i - a_j^i)] \geq \sum_{i=1}^{n} a_{m+1}^i \,.$$

The gearing ratio k will be just sufficient if:

$$\sum_{i=1}^{n}\sum_{j=1}^{m} p_j \max[0, (x_j^i - a_j^i)] = k \sum_{i=1}^{n} a_{m+1}^i \,.$$

We call this k, k^*.

If the amount of money is $\bar{M} = k^* \sum_{i=1}^{n} a_{m+1}^i$ there will be a default penalty such that at $\mu = \mu^*$ no one defaults at the price level which uses all of M in transactions (see footnote 7). The possible cases are contrasted with this benchmark where $k = k^*$ and $\mu = \mu^*$ (Case 5).

Case 1: $\mu < \mu^*$ and $k < k^*$.
Since $k < k^*$, we get deflation (i.e., lower prices from the ones of the benchmark Case 5). The marginal utility of an extra unit of gold is

higher than the marginal cost incurred by the bankruptcy penalty, since $\mu < \mu^*$. This induces strategic default and overbidding for loans so that $\rho > 0$. The T.S.N.E. is characterized by default and the deadweight loss of the penalty.

Case 2: $\mu < \mu^*$ and $k = k^*$.
Since $\mu < \mu^*$, strategic default will be induced and overbidding for loans so that $\rho > 0$. We will have a shift in the price vector and no conclusion for deflated or inflated prices can be drawn. The T.S.N.E. is again characterized by default and the deadweight loss of the penalty.

Case 3: $\mu < \mu^*$ and $k > k^*$.
The fact that $\mu < \mu^*$ creates the possibility of strategic default but *not* certainty because $k > k^*$. If prices are sufficiently inflated traders do not overbid for loans and consequently $\rho = 0$ and no default is incurred. However, if prices are not sufficiently inflated then $\rho > 0$ and default results. The T.S.N.E. is possibly characterized by default and the deadweight loss of the penalty. Also, one or more W.E.'s as T.S.N.E.'s with a continuum of inflated prices.

Case 4: $\mu = \mu^*$ and $k < k^*$.
Since $k < k^*$, there is not enough money so traders overbid and consequently $\rho > 0$. So, we have strategic default and since $k < k^*$ we have deflation. The T.S.N.E. is transaction constrained and is characterized by strategic default and a unique price level.

Case 5: $\mu = \mu^*$ and $k = k^*$.
A unique price level and transactions summing to $\bar{M} = k \sum_{i=1}^{n} a_{m+1}^{i}$. Moreover, $\rho = 0$ and no strategic default as already has been argued.

Case 6: $\mu = \mu^*$ and $k > k^*$.
No strategic default but one or more W.E.'s as T.S.N.E.'s with a continuum of inflated prices because $k > k^*$. Prices are bounded above by transactions summing to $M = k \sum_{i=1}^{n} a_{m+1}^{i}$. Also, no default induces $\rho = 0$.

Case 7: $\mu > \mu^*$ and $k < k^*$.
As in Case 4, since $k < k^*$, there is not enough money so traders may possibly overbid for loans and strategically default. Only if $\mu \gg \mu^*$ there will be no default. The fact that $k < k^*$ causes deflation. Also, if $\rho > 0$ then we have default; otherwise if $\mu \gg \mu^*$ then we end up with $\rho = 0$. The T.S.N.E. is characterized by possible default and a continuum of deflated prices.

Case 8: $\mu > \mu^*$ and $k = k^*$.
No strategic default and therefore $\rho = 0$ but we have potential deflation

with boarding because the high bankruptcy penalty. It is hoarding that may potentially cause deflation. So, one or more W.E.'s can be attained as T.S.N.E.'s with upper bound price level as in Case 5, but a continuum of deflated prices.

Case 9: $\mu > \mu^*$ and $k > k^*$.
Again, we have no strategic default which produces $\rho = 0$. As in Case 3, one or more W.E.'s as T.S.N.E.'s and continua of deflated and inflated prices.

Appendix 1

Type I: *No* bankruptcy
- Case (a): Loans paid back by banknotes
- Case (b): Loans paid back by final gold holdings
- Case (c): Loans paid back by banknotes and final gold holdings

Type II: Bankruptcy
- Case (a): Loans paid back *partially* by banknotes
- Case (b): Loans paid back *partially* by final gold holdings
- Case (c): Loans paid back *partially* by banknotes and final gold holdings

Example: I (a)

$$G^i = \{a^i_{m+1} - u^i\} + \left\{\frac{v^i}{1+\rho} - v^i\right\} + \left\{\sum_{j=1}^{m+1} p_j q^i_j - \sum_{j=1}^{m+1} b^i_j w^i\right\} + \\ + \theta^i \left\{M - \frac{\sum_{i=1}^n v^i}{1+\rho} + \frac{\sum_{i=1}^n v^i(1+\rho)}{1+\rho}\right\} .$$

After arranging and substituting the last term,

$$\boxed{G^i = \{a^i_{m+1} - u^i\} + \left\{\frac{v^i}{1+\rho} - v^i\right\} + \\ + \left\{\sum_{j=1}^{m+1} p_j q^i_j - \sum_{j=1}^{m+1} b^i_j w^i\right\} + ku^i(1+\rho) .}$$

Note that since there does *not* exist bankruptcy, $\rho = \rho^*$, $\hat{v}^i = v^i$, $g^i_L = 0$.

Appendix 2

A.

B.

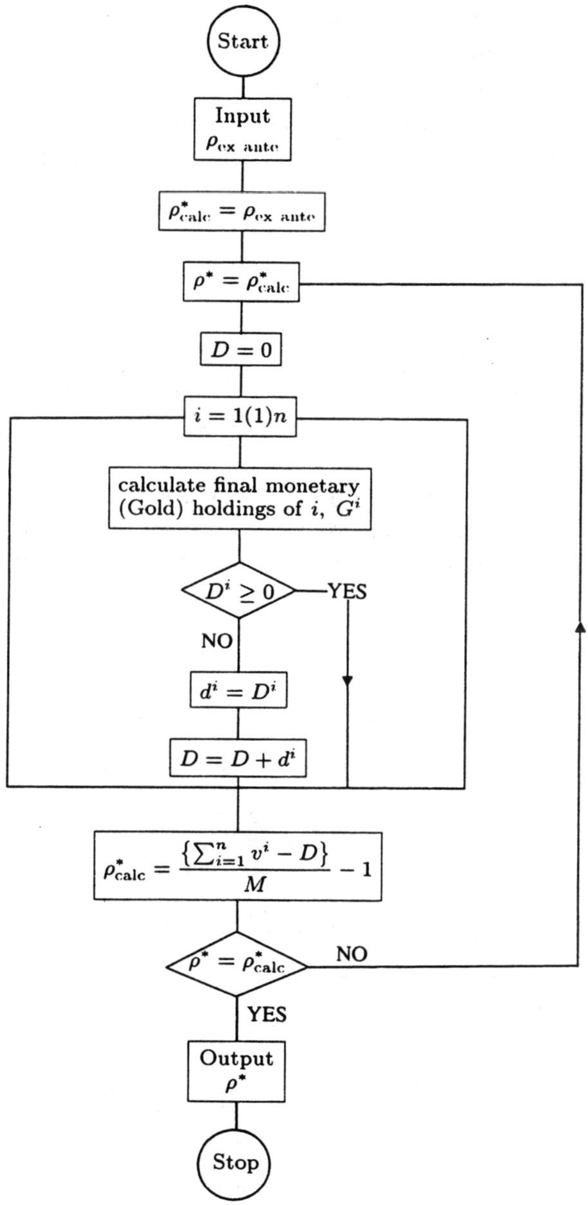

Calculation of the *ex post* interest rate [Discrete case]

Appendix 3

The inner loop A and the iterative formation of ρ^* have to be replaced by the following routine.

3.A

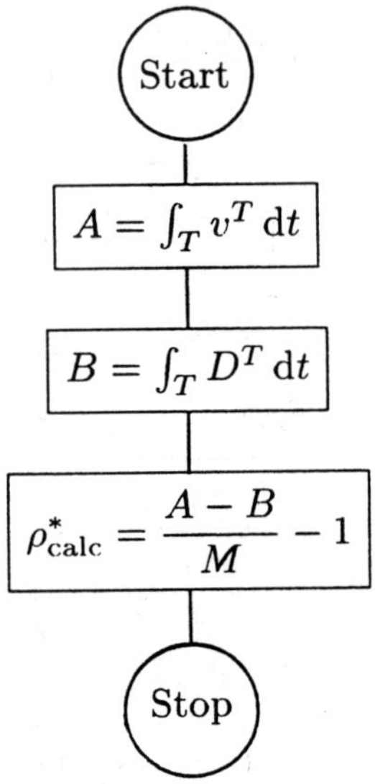

We now need a routine simple enough for the evaluation of the integral.
Let for example

$$I = \int_T f(t) \, dt$$

3.B

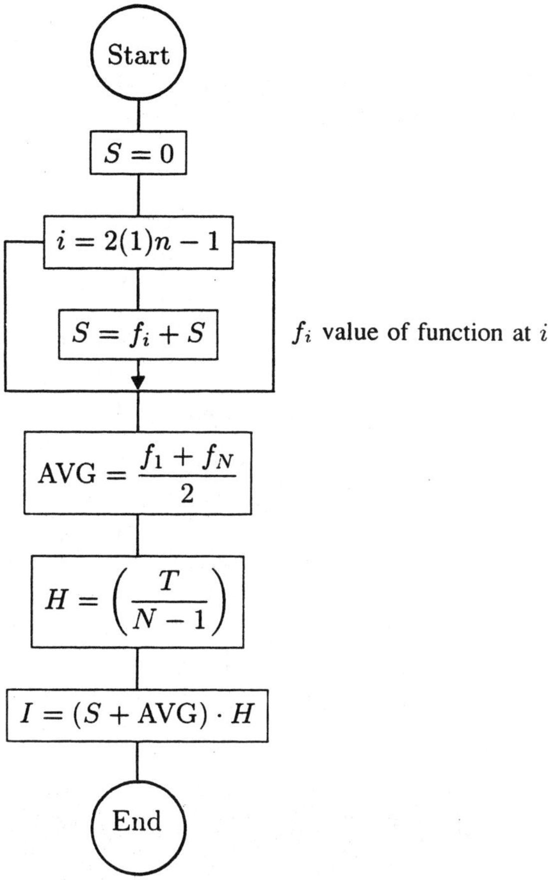

where $S = \sum_{i=1}^{N-1} f_i$, $\text{AVG} = (f_1 + f_N)/2$.

Calculation of the *ex post* interest rate [Continuum case]

Lemma:[13] $I = (S + \text{AVG}) \cdot H$

[13] More sophisticated algorithms for the evaluation of integrals can be constructed. One such algorithm is the Romberg Algorithm (see Burden and Douglas Faires, 1985, pp. 177–182).

Proof:

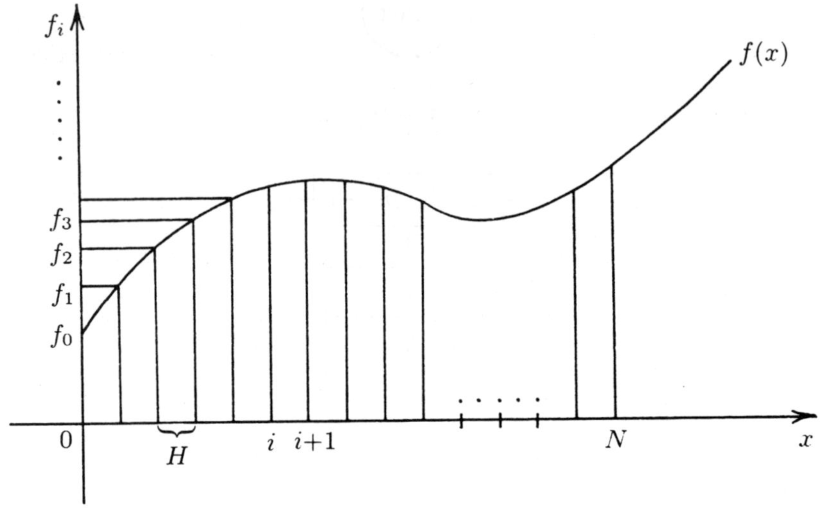

Therefore the area of the sum of the trapezoids gives us the integral

$$I = \sum_{i=0}^{N-1} \left(\frac{f_i + f_{i+1}}{2} \right) H$$

$$= \frac{H}{2} \left[f_1 + 2 \sum_{i=1}^{N-1} f_i + f_N \right]$$

$$= H \left[\underbrace{\frac{f_1 + f_N}{2}}_{\text{AVG}} + \underbrace{\sum_{i=1}^{N-1} g_i}_{S} \right] ,$$

therefore $I = (S + \text{AVG}) \cdot H$. Q.E.D.

References

Berge, C. (1963): *Topological Spaces*. New York: Macmillan.

Burden, R. L., and Faires, J. D. (1985): *Numerical Analysis*. Boston: P.W.S. Publishers.

Dubey, P., and Geanakoplos, J. (1989a): "Existence of Walras Equilibrium Without a Price Player or Generalized Game." CFDP # 912.

—— (1989b): "Liquidity and Bankruptcy with Incomplete Markets: Pure Exchange." CFDP # 900.

Dubey, P., and Shubik, M. (1978): "The Non-cooperative Equilibria of a Closed Trading Economy with Market Supply and Bidding Strategies." *Journal of Economic Theory* 17: 1–20.

—— (1988): "A Note on an Optimal Garnishing Rule." *Economics Letters* 27: 5–6.

Dubey, P., Geanakoplos, J., and Shubik, M. (1987): "The Revelation of Information in Strategic Market Games: A Critique of Rational Expectations Equilibrium." *Journal of Mathematical Economics* 16: 105–137.

—— (1988): "Bankruptcy and Efficiency in a General Equilibrium Model with Incomplete Markets." CFDP # 879.

Grandmont, J. M. (1983): *Money and Value*. Cambridge: Cambridge University Press.

Grandmont, J. M., and Younes, Y. (1972): "On the Role of Money and the Existence of a Monetary Equilibrium." *Review of Economic Studies* 39: 355–372.

Keynes, J. M. (1958): *A Treatise on Money, Vol. I*. New York: Macmillan (first ed. 1930).

Knapp, G. F. (1905): *Staatliche Theorie des Geldes*. Leipzig: Duncker & Humblot.

Ricardo, D. (1965): *The Principles of Political Economy and Taxation*. London: J. M. Dent & Sons (based on 3rd ed. 1821).

Shapley, L. S., and Shubik, M. (1977): "Trade Using One Commodity as a Means of Payment." *Journal of Political Economy* 85: 937–968.

Shubik, M. (1972): "Commodity Money, Oligopoly, Credit and Bankruptcy in a General Equilibrium Model." *Western Economic Journal* 10: 24–38.

—— (1985): "A Note on Enough Money in a Strategic Market Game with Complete or Fewer Markets." *Economics Letters* 19: 231–235.

—— (1986): "Enough Commodity Money and the Selection of a Unique Competitive Equilibrium." CFDP # 804.

—— (1987): "The Unique Minimal Cash Flow Competitive Equilibrium." *Economics Letters* 25: 303–306.

—— (1990): "The Transactions Trust Demand for Money." *Journal of Economics* 52: 211–232.

Shubik, M., and Wilson, C. (1977): "Optimal Bankruptcy Rule in a Trading Economy Using Fiat Money." *Journal of Economics* 47: 337–354.

Shubik, M., and Yao, S. (1989): "The Transactions Cost of Money (A Strategic Market Game Analysis)." *Mathematical Social Sciences* 20: 99–114.

Vicker, R. (1975): *The Realms of Gold*. New York: Scribner's.

Address of authors: Professor Martin Shubik and Professor Dimitris P. Tsomocos, Cowles Foundation for Research in Economics, Yale University, Box 2125, New Haven, CT 06520, USA.

Name index

Economists of the Twentieth Century

Monetarism and Macroeconomic Policy
Thomas Mayer

Studies in Fiscal Federalism
Wallace E. Oates

The World Economy in Perspective
Essays in International Trade and European Integration
Herbert Giersch

Towards a New Economics
Critical Essays on Ecology, Distribution and Other Themes
Kenneth E. Boulding

Studies in Positive and Normative Economics
Martin J. Bailey

The Collected Essays of Richard E. Quandt (2 volumes)
Richard E. Quandt

International Trade Theory and Policy
Selected Essays of W. Max Corden
W. Max Corden

Organization and Technology in Capitalist Development
William Lazonick

Studies in Human Capital
Collected Essays of Jacob Mincer, Volume 1
Jacob Mincer

Studies in Labor Supply
Collected Essays of Jacob Mincer, Volume 2
Jacob Mincer

Macroeconomics and Economic Policy
The Selected Essays of Assar Lindbeck
Volume I
Assar Lindbeck

The Welfare State
The Selected Essays of Assar Lindbeck
Volume II
Assar Lindbeck

Classical Economics, Public Expenditure and Growth
Walter Eltis

Money, Interest Rates and Inflation
Frederic S. Mishkin

The Public Choice Approach to Politics
Dennis C. Mueller

The Liberal Economic Order
Volume I Essays on International Economics
Volume II Money, Cycles and Related Themes
Gottfried Haberler
Edited by Anthony Y.C. Koo

Economic Growth and Business Cycles
Prices and the Process of Cyclical Development
Paolo Sylos Labini

International Adjustment, Money and Trade
Theory and Measurement for Economic Policy
Volume I
Herbert G. Grubel

International Capital and Service Flows
Theory and Measurement for Economic Policy
Volume II
Herbert G. Grubel

Unintended Effects of Government Policies
Theory and Measurement for Economic Policy
Volume III
Herbert G. Grubel

The Economics of Competitive Enterprise
Selected Essays of P.W.S. Andrews
Edited by Frederic S. Lee
and Peter E. Earl

The Repressed Economy
Causes, Consequences, Reform
Deepak Lal

Economic Theory and Market Socialism
Selected Essays of Oskar Lange
Edited by Tadeusz Kowalik

Trade, Development and Political Economy
Selected Essays of Ronald Findlay
Ronald Findlay

General Equilibrium Theory
The Collected Essays of Takashi Negishi
Volume I
Takashi Negishi

The History of Economics
The Collected Essays of Takashi Negishi
Volume II
Takashi Negishi

Studies in Econometric Theory
The Collected Essays of Takeshi Amemiya
Takeshi Amemiya

Exchange Rates and the Monetary System
Selected Essays of Peter B. Kenen
Peter B. Kenen

Econometric Methods and Applications
(2 volumes)
G.S. Maddala

National Accounting and Economic Theory
The Collected Papers of Dan Usher, Volume I
Dan Usher

Welfare Economics and Public Finance
The Collected Papers of Dan Usher, Volume II
Dan Usher

Economic Theory and Capitalist Society
The Selected Essays of Shigeto Tsuru, Volume I
Shigeto Tsuru

Methodology, Money and the Firm
The Collected Essays of D.P. O'Brien
(2 volumes)
D.P. O'Brien

Economic Theory and Financial Policy
The Selected Essays of Jacques J. Polak
(2 volumes)
Jacques J. Polak

Sturdy Econometrics
Edward E. Leamer

The Emergence of Economic Ideas
Essays in the History of Economics
Nathan Rosenberg

Productivity Change, Public Goods and Transaction Costs
Essays at the Boundaries of Microeconomics
Yoram Barzel

Reflections on Economic Development
The Selected Essays of Michael P. Todaro
Michael P. Todaro

The Economic Development of Modern Japan
The Selected Essays of Shigeto Tsuru
Volume II
Shigeto Tsuru

Money, Credit and Policy
Allan H. Meltzer

Macroeconomics and Monetary Theory
The Selected Essays of Meghnad Desai
Volume I
Meghnad Desai

Poverty, Famine and Economic Development
The Selected Essays of Meghnad Desai
Volume II
Meghnad Desai

Explaining the Economic Performance of Nations
Essays in Time and Space
Angus Maddison

Economic Doctrine and Method
Selected Papers of R.W. Clower
Robert W. Clower

Economic Theory and Reality
Selected Essays on their Disparities and Reconciliation
Tibor Scitovsky

Doing Economic Research
Essays on the Applied Methodology of Economics
Thomas Mayer

Institutions and Development Strategies
The Selected Essays of Irma Adelman
Volume I
Irma Adelman

Dynamics and Income Distribution
The Selected Essays of Irma Adelman
Volume II
Irma Adelman

The Economics of Growth and Development
Selected Essays of A.P. Thirlwall
A.P. Thirlwall

Theoretical and Applied Econometrics
The Selected Papers of Phoebus J. Dhrymes
Phoebus J. Dhrymes

Innovation, Technology and the Economy
The Selected Essays of Edwin Mansfield
(2 volumes)
Edwin Mansfield

Economic Theory and Policy in Context
The Selected Essays of R.D. Collison Black
R.D. Collison Black

Location Economics
Theoretical Underpinnings and Applications
Melvin L. Greenhut

Spatial Microeconomics
Theoretical Underpinnings and Applications
Melvin L. Greenhut

Capitalism, Socialism and Post-Keynesianism
Selected Essays of G.C. Harcourt
G.C. Harcourt

Time Series Analysis and Macroeconometric Modelling
The Collected Papers of Kenneth F. Wallis
Kenneth F. Wallis

Foundations of Modern Econometrics
The Selected Essays of Ragnar Frisch
(2 volumes)
Edited by Olav Bjerkholt

Growth, the Environment and the Distribution of Incomes
Essays by a Sceptical Optimist
Wilfred Beckerman

The Economics of Environmental Regulation
Wallace E. Oates

Econometrics, Macroeconomics and Economic Policy
Selected Papers of Carl F. Christ
Carl F. Christ

Strategic Approaches to the International Economy
Selected Essays of Koichi Hamada
Koichi Hamada

Economic Analysis and Political Ideology
The Selected Essays of Karl Brunner
Volume One
Edited by Thomas Lys

Growth Theory and Technical Change
The Selected Essays of Ryuzo Sato
Volume One
Ryuzo Sato

Industrialization, Inequality and Economic Growth
Jeffrey G. Williamson

Economic Theory and Public Decisions
Selected Essays of Robert Dorfman
Robert Dorfman

The Logic of Action One
Method, Money, and the Austrian School
Murray N. Rothbard

The Logic of Action Two
Applications and Criticism from the Austrian School
Murray N. Rothbard

Bayesian Analysis in Econometrics and Statistics
The Zellner View and Papers
Arnold Zellner

On the Foundations of Monopolistic Competition and Economic Geography
The Selected Essays of B. Curtis Eaton and Richard G. Lipsey
B. Curtis Eaton and Richard G. Lipsey

Microeconomics, Growth and Political Economy
The Selected Essays of Richard G. Lipsey
Volume One
Richard G. Lipsey

Macroeconomic Theory and Policy
The Selected Essays of Richard G. Lipsey
Volume Two
Richard G. Lipsey

Employment, Labor Unions and Wages
The Collected Essays of Orley Ashenfelter
Volume One
Edited by Kevin F. Hallock

Education, Training and Discrimination
The Collected Essays of Orley Ashenfelter
Volume Two
Edited by Kevin F. Hallock

Economic Institutions and the Demand and Supply of Labour
The Collected Essays of Orley Ashenfelter
Volume Three
Edited by Kevin F. Hallock

Monetary Theory and Monetary Policy
The Selected Essays of Karl Brunner
Volume Two
Edited by Thomas Lys

Macroeconomic Issues from a Keynesian Perspective
Selected Essays of A.P. Thirlwall
Volume Two
A.P. Thirlwall

Money and Macroeconomics
The Selected Essays of David Laidler
David Laidler

The Economics and Politics of Money
The Selected Essays of Alan Walters
Edited by Kent Matthews

Economics and Social Justice
Essays on Power, Labor and Institutional Change
David M. Gordon
Edited by Thomas E. Weisskopf and Samuel Bowles

Practicing Econometrics
Essays in Method and Application
Zvi Griliches

Economics Against the Grain
Volume One
Microeconomics, Industrial Organization and Related Themes
Julian L. Simon

Economics Against the Grain
Volume Two
Population Economics, Natural Resources and Related Themes
Julian L. Simon

Advances in Econometric Theory
The Selected Works of Halbert White
Halbert White

The Economics of Imperfect Knowledge
Collected Papers of G.B. Richardson
G.B. Richardson

Economic Performance and the Theory of the Firm
The Selected Papers of David J. Teece
Volume One
David J. Teece

Strategy, Technology and Public Policy
The Selected Papers of David J. Teece
Volume Two
David J. Teece

The Keynesian Revolution, Then and Now
The Selected Essays of Robert Eisner
Volume One
Robert Eisner

Investment, National Income and Economic Policy
The Selected Essays of Robert Eisner
Volume Two
Robert Eisner

International Trade Opening and the Formation of the Global Economy
Selected Essays of P. J. Lloyd
P. J. Lloyd

Production, Stability and Dynamic Symmetry
The Selected Essays of Ryuzo Sato
Volume Two
Ryuzo Sato

Variants in Economic Theory
Selected Works of Hal R. Varian
Hal R. Varian

Political Economy, Oligopoly and Experimental Game
The Selected Essays of Martin Shubik
Volume One
Martin Shubik

Money and Financial Institutions
A GAME THEORETIC APPROACH
The Selected Essays of Martin Shubik
Volume Two
Martin Shubik